Progress in Molecular and Subcellular Biology

Volume 57

This series gives an insight into the most current, cutting edge topics in molecular biology, including applications in biotechnology and molecular medicine. In the recent years, the progress of research in the frontier area of molecular and cell biology has resulted in an overwhelming amount of data on the structural components and molecular machineries of the cell and its organelles and the complexity of intra- and intercellular communication. The molecular basis of hereditary and acquired diseases is beginning to be unravelled, and profound new insights into development and evolutionary biology, as well as the genetically driven formation of 3D biological architectures, have been gained from molecular approaches. Topical volumes, written and edited by acknowledged experts in the field, present the most recent findings and their implications for future research. This series is indexed in PubMed.

More information about this series at http://www.springer.com/series/388

Christophe Lamaze · Ian Prior
Editors

Endocytosis and Signaling

Springer

Editors
Christophe Lamaze
Membrane Dynamics and Mechanics of Intracellular Signaling Laboratory, CNRS UMR3666, INSERM U1143, Institut Curie—Centre de Recherche
PSL Research University
Paris, France

Ian Prior
Department of Cellular and Molecular Physiology, Institute of Translational Medicine
University of Liverpool
Liverpool, UK

ISSN 0079-6484 ISSN 2197-8484 (electronic)
Progress in Molecular and Subcellular Biology
ISBN 978-3-030-07234-6 ISBN 978-3-319-96704-2 (eBook)
https://doi.org/10.1007/978-3-319-96704-2

This Springer imprint is published by the registered company Springer Nature Switzerland AG
The registered company address is: Gewerbestrasse 11, 6330 Cham, Switzerland

Preface

It is now widely appreciated that cell signalling is highly context dependent. The location and concentration of signalling nodes regulate their activation cycles and engagement with distinct effector pathways. Whilst many cell signalling pathways are initiated from the cell surface, endocytosis provides an opportunity for modulation of the output of signalling networks. In this book, we will highlight how the endosomal system helps to organise and regulate signalling pathways. In a series of reviews, we will firstly focus on the endocytic and endosomal system and describe how these subcellular platforms sort and regulate a wide range of signalling pathway components and phenotypic outputs. We will then review the latest scientific insights into how endocytic trafficking and subcellular location modulate a set of major pathways essential for normal cellular function and organism development.

Paris, France Christophe Lamaze
Liverpool, UK Ian Prior

Contents

Chapter 1
The Endosomal Network: Mediators and Regulators of Endosome Maturation

Maria Podinovskaia and Anne Spang

Contents

Abstract Endocytosis is a means for the cell to sample its environment for nutrients and regulate plasma membrane (PM) composition and area. Whereas the majority of internalized cargo is recycled back to the cell surface, select material is sent to the

M. Podinovskaia · A. Spang (✉)
Biozentrum, University of Basel, Klingelbergstrasse 70, 4056 Basel, Switzerland
e-mail: anne.spang@unibas.ch

M. Podinovskaia
e-mail: maria.podinovskaya@unibas.ch

C. Lamaze and I. Prior (eds.), *Endocytosis and Signaling*, Progress in Molecular and Subcellular Biology 57, https://doi.org/10.1007/978-3-319-96704-2_1

lysosome for degradation. Endosomes further play major roles in central cell activities as diverse as establishment of cell polarity and signaling, lysosomal storage and immunity. The complexity of endosomal functions is reflected by the extensive changes to endosome properties as they mature. The identity of individual endosomes is influenced by the presence of specific Rab GTPases and phosphoinositides (PIPs), which coordinate membrane traffic and facilitate endosomal functions. Motors and tethers direct the endosomes to the required locations and moderate fusion with other organelles. The maintenance of the elaborate endosomal network is supported by the ER and the *trans*-Golgi network (TGN), which promote the exchange of membrane components, provide enzymes, and assist with signaling. Additionally, V-ATPase is emerging as an underappreciated coordinator of endosome maturation and cell signaling. The inputs of the various mediators of endosome maturation are tightly regulated and coordinated to ensure appropriate maintenance and functioning of endosomes at each stage of the maturation process. Perturbations in endosome maturation are implicated in devastating diseases, such as neurodegeneration and cancer, and the endosome maturation processes are manipulated and exploited by intracellular pathogens to meet their own needs. A greater understanding of coordination and fine-tuning of endosome maturation will help us address various pathologies more effectively.

Abbreviations

EE	Early endosome
CI-M6PR	Cation-independent mannose-6-phosphate receptor
ILV	Intra luminal vesicle
LE	Late endosome
MVB	Multivesicular body
PIP	Phosphoinositide phosphate
PM	Plasma membrane
TfR	Transferrin receptor
TGN	*Trans*-Golgi network

1.1 Endosome Maturation Overview

Plasma membrane (PM) turnover and ingestion of extracellular material relies on a variety of endocytic pathways. The best studied pathway is clathrin-dependent endocytosis, which entails recognition of cargo via specific receptors, development of clathrin-coated pit, and formation of subsequent clathrin-coated vesicle (Doherty and McMahon 2009). Indiscriminate fluid sampling is concerted by macropinocytosis (Buckley and King 2017). A number of alternative endocytic pathways have been described and may be categorized by their PM-associated mediators, such as caveolin, IL2R, Arf6, flotillin, or CLIC/GEEC (Mayor and Pagano 2007; Doherty and McMahon 2009). These distinct pathways may be selective for specific cargoes and play unique roles in endocytosis directing the fate of cargoes (Doherty and McMahon 2009). Larger particles, such as debris and bacteria, are taken up via receptor-mediated phagocytosis, a process mainly reserved for professional phagocytes, such as amoebae and macrophages (McDermott and Kim 2015; Doherty and McMahon 2009). It is conceivable that the spatial organization required at the PM for the different endocytic pathways to proceed also initiates the early sorting steps, further dealt with in the early endosomes (EE).

The newly formed endocytic vesicles of all origins move away from the cell surface and fuse with EEs (Fig. 1.1). The heterogeneous endocytic vesicles do not fuse with each other and show different maturation kinetics; however, all converge at the EE (Danson et al. 2013; Mayor and Pagano 2007). At some point, the EE stops accepting endocytic vesicles and becomes a sorting endosome. The trigger for this transition is poorly characterized (Spang 2016). The sorting endosome consists of tubular and vacuolar regions. The large surface area of the tubules allows the majority of membrane to be recycled back to the cell surface. Recycling proceeds directly to the PM or indirectly via recycling endosomes or *trans*-Golgi network (TGN) retrograde pathways (Lakadamyali et al. 2006; Clague and Hammond 2006). Luminal cargo is concentrated in the vacuolar regions and is targeted for degradation. Any membrane cargo destined for degradation is ubiquitinated and internalized into intraluminal vesicles (ILVs), giving the endosomes their characteristic multivesicular body (MVB) morphology.

To enable cargo degradation, the EE first has to acquire late endosome (LE) properties to prepare itself for fusion with the highly acidic lysosomes. These maturation events include the Rab5–Rab7 conversion, which in turn governs specificity of fusion with other compartments, and are accompanied by acidification, PIP conversion, and gain of lysosomal hydrolases and the protective highly glycosylated proteins, such as LAMP1. LEs are trafficked toward the perinuclear space, where the lysosomes are situated. Coordination of these events is poorly understood.

The TGN is not considered as a part of the endosomal network, but participates in sorting, recycling, and delivery of lysosomal hydrolases. The TGN sorts not only the cargoes coming from endosomes, but also cargoes destined to endosomes and to the PM. Retrograde transport is important for retrieval of receptors for future reuse, whereas anterograde pathway is important for delivery of lysosomal hydrolases, e.g., via cation-independent mannose-6-phosphate receptors (CI-M6PR) (McDermott and Kim 2015), and also for cargo destined to the PM (Spang 2015). Interestingly, in plants, the TGN functions as an EE able to receive incoming cargo directly from the PM, highlighting the partially overlapping functions of the two organelles (Scheuring et al. 2011).

Once all sorting receptors are recycled and all ubiquitinated cargoes are internalized into ILVs, LE content is transferred to lysosomes. The LE may directly be able to fuse with the lysosome to form an endolysosome, or deliver the luminal content via a "kiss-and-run" mechanism. The hydrolytic environment of the lysosome breaks down the cargo, which is then transferred out of the lysosome via specific transporters

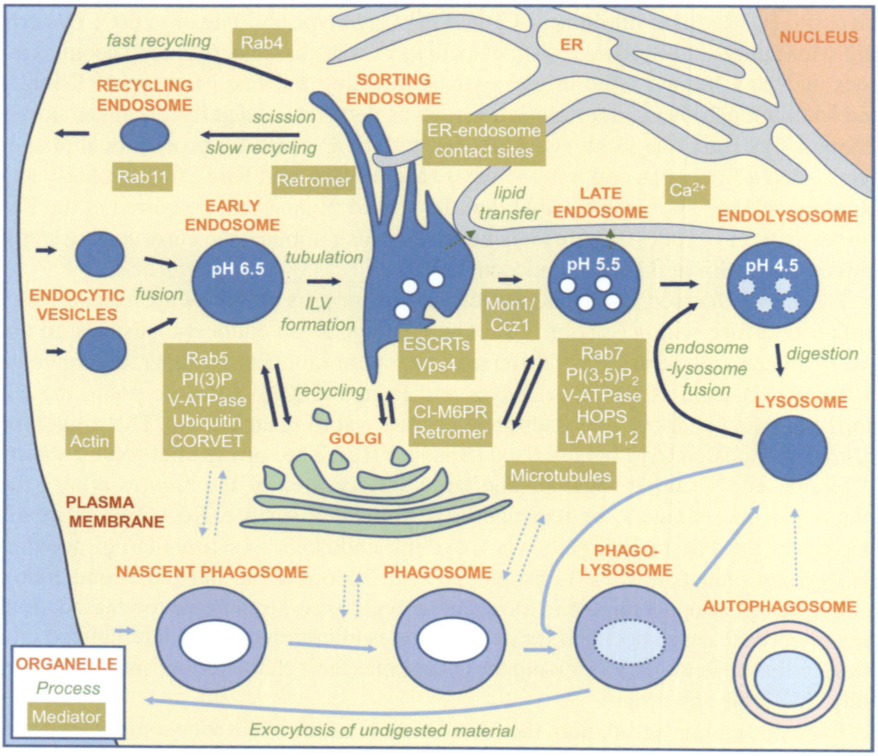

Fig. 1.1 Endosome maturation scheme. Endocytic vesicles containing endocytosed material fuse with early endosomes, which sort cargo for recycling and degradation. The cargo destined for recycling is sorted into tubular domains and sent directly to the plasma membrane or to recycling endosomes via fast and slow recycling pathways, respectively. Ubiquitinated membrane cargo destined for degradation is packaged into intraluminal vesicles (ILVs), which gives the late endosomes their characteristic multivesicular body appearance. The lipids are transferred to the ER via the ER–endosome contact sites. The continuous exchange of material with the TGN ensures supply of the required enzymes and membrane components to the maturing endosomes. Once the sorting steps are complete, the endosome fuses with the lysosome to form an endolysosome, where the cargo is digested and transported to the cytosol, to allow the lysosome to be regenerated and reused. Larger particles, such as bacteria, can be taken up by phagocytosis. As phagosomes mature, they interact with the endosomal pathway and the Golgi and, like endosomes, eventually fuse with lysosomes to facilitate cargo degradation. Undigested material is exocytosed. Lysosomes also receive material for degradation from the autophagy pathway. Endosome maturation is regulated by various Rabs and phosphoinositol phosphate (PIP) species, such as Rab5 and PI(3)P in early endosomes, Rab7 and PI(3,5)P_2 in late endosomes, and Rab4 and Rab11 in the recycling pathways. These regulators help recruit further essential mediators of endosome maturation, which include specific tethers (e.g., CORVET, HOPS), structural proteins (e.g., ESCRTs, SNX–BAR–retromer complexes), and cytoskeletal components (e.g., actin, dynein, kinesin), necessary for fusion, fission, and endosomal positioning, respectively. Gradual acidification accompanies endosome maturation, mediated by the proton pump V-ATPase, which is critical for endosomal function and maturation

and channels, allowing the lysosome to be regenerated and reused (Guerra and Bucci 2016). The lysosome also acts as a hub for integration of endocytic and autophagic pathways, thus being able to receive and digest cargo from both pathways. In yeast, lysosomes form one or more vacuoles.

Phagosomes display a lot of similarities with the endocytic pathway, their maturation consisting of early, late, and lysosome-interacting stages (Guerra and Bucci 2016). Unlike endosomes, the sorting stage is usually very short-lived and the majority of the phagosomal content gets targeted for degradation. The kinetics of phagosome maturation are strongly dependent on the phagocytic receptor, with the Fcγ receptor the one most extensively studied (Dill et al. 2015). Like endosomes, the phagosomes need PIP conversion to drive maturation and acquire LAMP1 and hydrolytic enzymes to form a phagolysosome (Kim et al. 2014). Indigestible material in phagosomes is exocytosed.

1.2 Endosome Functions

One of the primary functions of the endosomal network is nutrient acquisition and protein and lipid degradation. Sorting is an essential part of this process, allowing for extensive exchange of internalized cargo between compartments and ensuring only appropriate cargo reaches the lysosome (Villasenor et al. 2016; Stasyk and Huber 2016). Integrity of the endosomal system is required for lysosome biogenesis and its hydrolytic capacity to fulfill its functions to digest both extracellular material from the endocytic pathways and intracellular material from the autophagy pathways (Jacomin et al. 2016). Lysosomal degradation is also a central feature in immunity, allowing for inactivation of phagocytosed pathogens and partial digestion for antigen presentation via the MHC class II pathway (Boes et al. 2004).

Acidification is an important aspect of endosome maturation and function. Each stage of endosome maturation is characterized by a specific luminal pH, with EEs having pH of around 6.0–6.5, which drops to pH 5.0–5.5 in LEs and down to pH of below 4.5 in lysosomes, although the final pH in lysosomes may be highly heterogeneous (Wang et al. 2015; Johnson et al. 2016). Acidic pH is required for proper localization of many vacuolar proteins in yeast, including vacuolar hydrolases, proton pump subunits, enzymes for membrane biogenesis, and various transporters (Matsumoto et al. 2013). Acidification plays a role in uncoupling of internalized ligand–receptor complexes following receptor-mediated endocytosis, efficient sorting, inactivation of internalized pathogens, MVB formation, and activation of degradative enzymes (Matsuo et al. 2004; Fairn and Grinstein 2012; Kane 2006; De Luca and Bucci 2014; Kharitidi et al. 2015). Interestingly, pharmacological disruption of pH did not affect phagosome–lysosome fusion in mouse macrophages (Kissing et al. 2015). Low pH promotes acid sphingomyelinase (ASM)-mediated sphingomyelin conversion to ceramide, which regulates NPC2 participation in cholesterol transport and MVB maturation (Sandhoff 2016).

Additionally, endocytosis itself is a means to regulate PM composition, consequently contributing to fundamental cell properties such as cell polarity (Gao and

Kaestner 2010), cell adhesion (e.g., via integrin endocytosis) (Sandri et al. 2012), and cell motility (e.g., via endocytosis, recycling, and degradation of N-cadherin) (Tang 2016). Intercellular communication is mediated through exocytosis of MVBs containing DNA and RNA materials, which can function as messengers or participate in miRNA-mediated gene silencing (Soria et al. 2017; Gibbings et al. 2009). Endocrine and paracrine signaling, as well as cytotoxic immune functions, may also be mediated by exosomes generated in MVBs (Soria et al. 2017).

The endosomal system is a major regulator of cell signaling. Endocytosis modulates numbers of transmembrane proteins involved in signaling at PM, including growth factor receptors, such as EGFR. Internalization of activated signaling receptors is a mechanism to silence signaling. Once endocytosed, the signaling receptors may remain active, until inactivated through ubiquitination or ILV packaging. In neurons, the silencing of the Hedgehog receptor Patched by its targeting into ILVs was implicated in axon pruning (Issman-Zecharya and Schuldiner 2014). Surprisingly, accumulation of Notch and receptor tyrosine kinase (RTK) receptors at the enlarged EEs in Dmon1 mutant *Drosophila melanogaster* cells did not result in the expected overactive signaling, suggesting additional silencing mechanisms (Yousefian et al. 2013). Activity of signaling receptors at endosomes may also be spatiotemporally controlled, depending on access to their effectors. For example, hepatocyte growth factor receptor Met mediates cytoskeleton remodeling to control cell migration via two distinct pathways, which depends on endosome positioning (Stasyk and Huber 2016; Menard et al. 2014). Endosomal sensing and sorting are an important part of signaling regulation. Thus, the EGFR is recycled at low EGF concentration, but sent for lysosomal degradation when EGF levels are high (Sigismund et al. 2008). Additional Akt-dependent signaling functions have been demonstrated for a non-canonical subset of sorting endosomes marked with adaptor proteins APPL1 and APPL2 in zebrafish and mammalian cells (Urbanska et al. 2011; Kalaidzidis et al. 2015).

The lysosome acts as a major signaling hub to convey the nutrient status to the cell, which in turn influences the dynamics of endosome maturation, induction of autophagy, and cell growth and division, via the mTORC1 complex positioned at the lysosome. Lysosomes are able to store Ca^{2+} and participate in Ca^{2+}-dependent signaling. The transcription factor TFEB regulation is controlled through phosphorylation by mTORC1, which is sensitive to lysosomal amino acid levels, and dephosphorylation by the Ca^{2+}-dependent phosphatase calcineurin (Medina et al. 2015).

The endosome network also contributes to storage of nutrients, such as the reservoir of amino acids and carbohydrates in the yeast vacuole, or yolk granules in *Caenorhabditis elegans* and *D. melanogaster* oocytes (Teixeira et al. 2016; Armstrong 2010; Poteryaev et al. 2010). The integrity of the endosomal system also affects the dynamics of lipid droplets, an organelle used for lipid storage but also for signaling and hydrophobic protein metabolism (Bouchez et al. 2015). In case of PM damage, lysosomes are involved in injury-induced exocytosis and contribute to wound repair (Andrews et al. 2015). Harmful content, such as misfolded proteins, may be channeled through the endosomal system to be secreted in exosomes generated in MVBs (Soria et al. 2017). Secretory lysosomes are used by cells with antimicrobial, clotting, and other specialized functions to secrete proteins such as histamine, perforin, and von Willebrand factor (Griffiths 2016). Secretory lysosomes are also used during intercellular cell fusion to establish continuous tubular networks such as the *D. melanogaster* tracheal system (Caviglia et al. 2016).

1.3 Mediators of Endosome Maturation

1.3.1 Phosphoinositides

The endosomal network is maintained through continuous fusion and fission events, membrane exchange, and recruitment of various regulators to coordinate endosome maturation. PIPs play a central role in compartmental identity, including that of endosomes, serving as binding platforms for proteins, assisting with the properties and functions of each endosome stage and sustaining directionality of the maturation process (Santiago-Tirado and Bretscher 2011; Fili et al. 2006). The seven species of PIPs (PI(3)P, PI(4)P, PI(5)P, PI(3,4)P_2, PI(3,5)P_2, PI(4,5)P_2, PI(3,4,5)P_3) are tightly controlled spatially and temporally by specific phosphatases (myotubularins) and kinases. Disruption of these enzymes leads to aberrant endosome trafficking.

The role of PIPs in endosome formation and maturation starts at the PM. PI(4,5)P_2 is present at the PM and is required for maturation of late-stage clathrin-coated pits prior to fission (Antonescu et al. 2011). PIP4K, which contributes to the generation of PI(4,5)P_2, regulates EE dynamics during clathrin-mediated endocytosis (Kamalesh et al. 2017). PI(3,4)P_2 has also recently emerged as a major mediator in clathrin-mediated endocytosis (Posor et al. 2013). Subsequently, the early endocytic pathway is dominated by PI(3)P. It is unclear how this conversion to PI(3)P is mediated, but the steps are likely to involve the Sac2 PI(4)P phosphatase and class III PI3K (PI3KC3 or Vps34 in yeast) (Hsu et al. 2015). PI(3)P is recognized by proteins with FYVE or PX domains, such as the EE tether EEA1 and the mediator of ILV formation Hrs (Schink et al. 2013; Candiello et al. 2016; Katzmann et al. 2003). During recycling, hydrolysis of PI(3)P and generation of PI(4)P on secretory/recycling endosomes are required for exocyst recruitment to enable fusion with PM, as visualized by transferrin receptor (TfR) recycling in HeLa and COS-1 cells (Ketel et al. 2016). A further regulatory step is accomplished by PI(5)P, the enrichment of which can delay endosome maturation to ensure completion of EE events prior to LE transition (Boal et al. 2015).

During EE to LE transition, PIKfyve catalyzes phosphorylation of PI(3)P to PI(3,5)P_2 and leads to conversion of PI(3)P at EEs to PI(3,5)P_2 at LEs and lysosomes. PI(3,5)P_2 was found to be enriched in the vesicular domain of endosomes in HeLa cells and was required for delivery of endocytic cargo into MVB, potentially via its interaction with the ESCRT-III component Vps24 (Shaw et al. 2003; Takatori et al. 2016; Jin et al. 2016). One of the principal functions of PI(3,5)P_2 is to regulate the fission in LEs and lysosomes (Dove et al. 2009). PI(3,5)P_2 was proposed to be important for membrane retrieval from matured lysosomes in *C. elegans* (Nicot et al.

2006). Its role in retrograde trafficking is essential in yeast, plants, and mammalian cells (Jin et al. 2016). In plants, PI(3,5)P_2 has been shown to mediate maturation of LEs by promoting their organization along the cortical microtubules (Hirano et al. 2015). PIKfyve activity is likewise important for regulation of the reformation of terminal storage lysosomes from endolysosomes (Bissig et al. 2017). An additional PIP species PI(3,4)P_2 assists in fission events by localizing at tubular domains and aiding scission (Posor et al. 2013; Tan et al. 2015). In phagosomes, PIPs act as markers of phagosome maturity and are required for fusion of phagosomes with endosomes and lysosomes (Jeschke et al. 2015). Phagocytic receptors determine PIP dynamics and dictate phagosome fate (Sarantis and Grinstein 2012; Bohdanowicz and Grinstein 2013; Levin et al. 2015; Posor et al. 2015).

1.3.2 Rab GTPases

In parallel with PIPs, Rab GTPases have central roles in membrane organization and trafficking and likewise serve as platforms at membranes for effector protein recruitment (Jean and Kiger 2012). Inactive Rabs are recruited to membranes and are activated by specific Rab GEFs (GDP/GTP exchange factors). Active Rabs control specificity of endosome fusion by interacting with PIPs and tethering complexes, thereby determining functional identity of the endosome at which they are positioned (Balderhaar and Ungermann 2013). Rabs can be inactivated by inherent GTP hydrolysis, which is further enhanced by specific Rab GAPs (GTPase-activating proteins). Over 70 different Rabs have been identified in humans. Besides Rabs, other types of GTPases have been reported to mediate lysosomal interactions with LEs and phagosomes. Arf and Arf-like GTPases are recruited to endosomes and lysosomes and play a role in directing traffic to lysosomes and recycling endosomes (Garg et al. 2011; Marwaha et al. 2017). Septins, filamentous heteromeric GTPases, are also implicated in endocytic membrane traffic without affecting Rab5 and Rab7, by promoting LE–lysosome fusion (Dolat and Spiliotis 2016).

The main Rab in EEs, Rab5, is recruited to EEs and facilitates fusion with endocytic and Golgi-derived vesicles. Rab5 plays an essential role in the biogenesis of the endolysosomal system (Zeigerer et al. 2012). Rab5 has three isoforms in mammalian cells, with Rab5a being the major Rab5 species to facilitate EE progression to LEs (Chen et al. 2009). As endosomes mature, Rab5 is replaced by Rab7, which mediates fusion with LEs and lysosomes (Rink et al. 2005). Rab7 is directly or indirectly involved in every step between EEs and lysosomes, including EE transition to LEs, LE transport to lysosomes, lysosome biogenesis, and LE–lysosome fusion. Additional Rabs, such as Rab11, Rab4, and Rab22, are involved in recycling pathways at EEs and are displaced following, or concomitant with, Rab5–Rab7 conversion, reflecting restriction of sorting and recycling pathways to early stages of endosome maturation (Grant and Donaldson 2009; McDermott and Kim 2015; D'Souza et al. 2014; Szatmari et al. 2014). Other Rabs, such as Rab7b, Rab6, and Rab9, contribute to the transport between endosomes and Golgi (Progida and Bakke 2016; Gutierrez

2013; Progida et al. 2010). Phagosomes contain at least 20 different Rab species, but their contribution to phagosome maturation remains to be investigated (Fairn and Grinstein 2012). Rab dynamics in phagosomes differ from those in endosomes. For example, in phagosomes Rab10 might be acting upstream of Rab5 to facilitate recycling of GPI anchor (Cardoso et al. 2010), and phagosome–lysosome fusion might be enhanced by Rab34 independently of Rab7 (Kasmapour et al. 2012). In macropinosomes in Cos-7 cells, Rab7 started to accumulate around the timing of cup closure and continued to increase even after fusion with lysosomes (characterized by the decrease of CI-M6PR), with its activity highest in lysosomes (Yasuda et al. 2016).

1.3.3 Tethers

Rabs execute their functions in part through interaction with tethering complexes, such as CORVET and HOPS, which activate and proofread SNARE assembly to drive membrane fusion. CORVET, a Rab5 effector, is involved in homotypic fusion of EEs and is required for MVB formation, possibly by facilitating fusion to supply membrane for ILV generation (Balderhaar and Ungermann 2013). HOPS is a Rab7 effector and is involved in the trafficking and docking of LEs to lysosomes. CORVET–HOPS conversion may facilitate Rab conversion, as the Rab conversion mediator Mon1 also interacts with HOPS (Solinger and Spang 2013; Poteryaev et al. 2010). Replacement of CORVET with HOPS ensures fusion with LEs, autophagosomes, and lysosomes, but no longer with EEs. Accordingly, disruptions of CORVET and HOPS lead to premature and delayed endosome maturation, respectively (Solinger and Spang 2014). An additional role has been proposed for the Vps39 subunit of HOPS in assisting with the formation of contact sites between vacuoles and mitochondria to promote exchange of nutrients, lipids, and ions between the two organelles, although it is not clear whether this additional function is independent of its function in the tethering complex (Honscher et al. 2014; Elbaz-Alon et al. 2014). In yeast, CORVET and HOPS share four core subunits (Vps11, Vps16, Vps18, and Vps33) and differ in a further two subunits (Vps8/3 for CORVET and Vps41/39 for HOPS), which bind the corresponding Rabs and bring the vesicles together for fusion. The SNARE-interacting core subunit Vps33 facilitates the SNARE-mediated fusion between the two vesicles and provides an additional layer of specificity, with different SNAREs present on EEs and LEs (Spang 2016). Lower abundance hybrid tethering complexes also exist, but their role in tethering and endosome maturation remains to be explored. In metazoans, there are at least two isoforms of the core subunits Vps33 and Vps16, which provide additional diversity and roles to tethering complexes. For example, Vps33b and Vps16b (Spe39 or VIPAR) form a complex, either a dimer or part of a larger complex, CHEVI, which has been proposed to accept cargo from the TGN (Spang 2016). Vps16b also forms a complex with a Vps33 homolog, Vps45, and a Rab5 effector Rabenosyn-5, and may form part of a larger complex, FERARI, which is believed to be involved in recycling and retrograde traffic (Spang 2016; Tornieri et al. 2013; van der Kant et al. 2015). A further unrelated multi-subunit tethering complex called endosome-associated recycling protein (EARP) has been shown to play a role in recycling of endocytic receptors to the cell surface (Schindler et al. 2015). Additional prominent tethers, such as the EEA1 homodimeric protein at EEs, contribute to membrane exchange and endosome maturation in concert with specific Rabs and PIPs (Balderhaar and Ungermann 2013; McDermott and Kim 2015; Kummel and Ungermann 2014). EEA1 ubiquitination status governs its functionality, with perturbations leading to severe disruptions to endosome morphology and trafficking (Ramanathan et al. 2013).

1.3.4 ESCRTs

Membrane cargo destined for degradation is ubiquitinated and sorted into ILVs. Ubiquitinated cargo is recognized by the ESCRT-0 component Hrs (Vps27 in yeast), resulting in the recruitment of further ESCRT I–III complexes to assist with budding and fission during ILV formation. Cargo deubiquitination takes place prior to the ILV fission and internalization into the MVB (Huotari and Helenius 2011; Hubner and Peter 2012; Gschweitl et al. 2016; McDermott and Kim 2015; Katzmann et al. 2003; Santiago-Tirado and Bretscher 2011). The AAA-ATPase Vps4 cooperates with ESCRT-III to induce membrane scission and disassembles the ESCRT components from ILVs to recycle them back into the cytoplasm (Schmidt and Teis 2012; Adell et al. 2014).

1.3.5 V-ATPase

Critical to endosome maturation is the acidic pH of endosomal network. The pH drop and the maintenance of acidified compartments are achieved by the vacuolar proton pump V-ATPase. V-ATPases are multi-subunit complexes that consist of the cytosolic ATPase sector V_1 and a transmembrane proton-translocating sector V_0. The two sectors are assembled separately, V_1 in the cytosol and V_0 in the ER, and are brought together into a functional proton pump at the required organelles (Cotter et al. 2015). Nascent phagosomes may acquire V-ATPase directly from lysosomes (Sun-Wada et al. 2009). Blocking V-ATPase function with specific pharmacological inhibitors or by siRNA leads to accumulation of cargo in EEs and subsequent inhibition of endocytosis (Pena-Llopis et al. 2011; Smith et al. 2016). In AtT20-secreting cells, V-ATPase deficiency led to the formation of large hybrid organelles containing markers of immature granules, lysosomes, and autophagy, affecting both degradation and secretory pathways (Sobota et al. 2009) and reflecting broader functions of the V-ATPase besides the endosomal pathway. Although secretory functions and correct targeting of PM proteins in yeast depend on V-ATPase-mediated acidification (Huang and Chang 2011), the functionally overlapping recycling pathways appear not to require V-ATPase (Baravalle et al. 2005). Initially shown to be important for the delivery of ILVs to lysosomes (van Deurs et al. 1996), V-ATPase activity does not appear to be essential for lysosome ability to fuse with endosomes and phagosomes (Kissing et al. 2015; Mauvezin et al. 2015) or indeed may even negatively regulate vacuole fusion (Desfougeres et al. 2016). During exocytosis of indigestible material in the phagolysosome, V-ATPase is retrieved just prior to exocytosis allowing the phagosome lumen to neutralize. Any V-ATPase that ends up at PM gets rapidly endocytosed (Clarke et al. 2010).

Besides vacuolar acidification, V-ATPase is involved in extracellular acidification in specialized cells, such as renal intercalated cells and osteoclasts, to assist with pH homeostasis and bone resorption, respectively (Cotter et al. 2015). Apart from proton-pumping function, the V-ATPase appears to act as a sensor for the luminal pH and also helps to regulate cytosolic pH and proteasome behavior in the cytosol (Peters et al. 2013; Hurtado-Lorenzo et al. 2006). V-ATPase has been shown to modulate vesicular trafficking through recruitment of small GTPases such as Arf6 (Hurtado-Lorenzo et al. 2006). The V-ATPase can also sense lysosomal amino acid levels and control mTORC1 activity at lysosomes, although the mechanisms underlying this sensory function remain unclear (Maxson and Grinstein 2014; Zoncu et al. 2011). Additionally, the V_0 sector by itself is able to facilitate membrane fusion and plays a role in sorting and secretion independently of its acidification functions (Jefferies et al. 2008; Merz 2015; Sobota et al. 2009; Liegeois et al. 2006; Sreelatha et al. 2015). V-ATPase membrane fusion function appears to be important for the retrieval and recycling of the Rab7 effector HOPS from the lysosome, indirectly modulating downstream Rab7 functions. The V_1 sector subunit H might be involved in the phosphorylation of the HOPS subunit Vps33b to promote endosome maturation and endosome–lysosome fusion (Wong et al. 2011). It is feasible that other subunits may have further roles in endosome maturation and elsewhere in the cell. For example, the V_1 sector subunit B interaction with actin nucleation-promoting factor WASH was important in recycling functions and lysosome neutralization prior to exocytosis of undigested material in *D. melanogaster* cells, implicating V-ATPase in coordination with multiple aspects of endosome maturation (Nagel et al. 2017). Such an array of functions makes it challenging to dissect the contributions of V-ATPase-mediated acidification and alternative V-ATPase functions during endosome maturation.

1.3.6 Additional Mediators of Vacuolar pH

Proton pump activity is complemented and potentially regulated by proton channels, such as the Na^+/H^+ and Cl^-/H^+ exchangers (Xinhan et al. 2011; Satoh et al. 2016; Prasad and Rao 2015). These channels act as a proton leak, important for counter-balancing V-ATPase activity and fine-tuning pH at specific organelles. The chloride channel CLC-5 has been implicated in creating a proton flux necessary for V-ATPase activation, whereas the sodium channel NHE6 locates to EEs and affects TfR uptake and recycling. In yeast, Nhx1p sodium channel plays a role in luminal pH regulation, critical for recycling, retromer, and endocytic pathways (Kojima et al. 2012). Nhx1 activity was found to promote MVB fusogenicity, consistent with observations that MVB fusion with lysosomes is highly sensitive to manipulations of both luminal and cytosolic pH (Cao et al. 2015; Karim and Brett 2017). Besides V-ATPase, acidification in plants is mediated by an additional pump, the vacuolar pyrophosphatase (V-PPase), with a unique and essential role in endocytic and secretory trafficking. V-PPase is a homodimer of a single polypeptide, which uses energy from the biosynthetic by-product pyrophosphate (PPi) to drive proton transport. Combined activity of V-ATPase and V-PPase is required for vacuolar acidification (Kriegel et al. 2015; Zhou et al. 2016). Their functional coordination remains to be established.

1.3.7 Cytoskeleton

Actin organization plays an essential role in endocytosis and endosome maturation. Actin polymerization through WASH, a nucleation-promoting factor and an activator of Arp2/3 complex, powers assembly of endocytic vesicles and their movement toward EEs (Nannapaneni et al. 2010). Myosin VI interacts with the endocytic marker and Rab5 effector APPL1 through its adaptor GIPC and helps transport the nascent vesicles away from the actin-rich periphery of the cell toward the EEs (Granger et al. 2014; Nielsen et al. 1999). Moesin supports F-actin network formation on endosomes through its interaction with Rab7 and is required for recycling and endosome maturation, possibly through membrane remodeling and segregation of tubular and multivesicular structures (Duleh and Welch 2010; Muriel et al. 2016; Chirivino et al. 2011). Actin also interacts with HOPS, suggesting that actin cytoskeleton plays an important role in later maturation stages too, perhaps by helping lysosomes maintain their position within the cell (Solinger and Spang 2013), or for the delivery of protease- and lipase-containing vesicles to endosomes (King et al. 2013; Kirkbride et al. 2012). Actin cytoskeleton remodeling is likewise essential for phagocytosis and for phagosome maturation (Freeman and Grinstein 2014). While actin polymerization drives membrane expansion around the particle during its internalization, larger particles also require F-actin disassembly prior to phagosome cup closure, potentially to avoid substrate exhaustion for further polymerization to complete internalization (Schlam et al. 2015). Actin assembly at phagosomes promotes their fusion with LEs and lysosomes, thereby affecting phagosome pH, membrane recycling and delivery of hydrolytic enzymes, essential events during phagosome maturation (Marion et al. 2011; Dieckmann et al. 2012; Gopaldass et al. 2012).

Endosome spatial organization and exchange of internalized material between compartments are facilitated by endosome movement along the microtubule cytoskeleton. The molecular mechanisms underlying the transition from actin cytoskeleton at the cell periphery to the microtubule cytoskeleton during endosome maturation remain to be elucidated (McDermott and Kim 2015). Kinesins and dyneins move cargo, including endosomes, toward the peripherally anchored plus ends and centrally located minus ends of the microtubule cytoskeleton, respectively. Dyneins use adaptor molecules such as dynactin and the Rab7 effectors RILP or Hook1 to transport endosomes toward lysosomes, whereas kinesins may interact with Rab4 and Rab11 to recycle cargo back toward the cell periphery (Granger et al. 2014; Solinger and Spang 2013; McDermott and Kim 2015). Rab5 might regulate EE motility on microtubules (Nielsen et al. 1999), since Rab5 co-immunoprecipitates with dynein (Zajac et al. 2013); however, it is unclear how dynein binds EEs. Dynein activity was shown to be necessary for efficient endocytic uptake and endosome maturation in *D. melanogaster* oocytes (Liu et al. 2015) and likewise has been implicated in sorting TfR away from EGF-containing EEs in HeLa cells (Driskell et al. 2007). It is possible that dynein mediates sorting by facilitating elongation and scission at the sorting endosome and by maintaining its position while kinesins transport recycling endosomes to PM. Near the PM, actin takes over to assist with the final stages of exocytosis. In LEs, Rab7 can direct movement both ways, via RILP and dynein—dynactin interactions toward perinuclear location, or via the FYVE and coiled-coil (CC) domain-containing protein FYCO1 and kinesin toward the periphery (Guerra and Bucci 2016; Pankiv et al. 2010). In phagosomes, Rab7 interaction with RILP and dynein–dynactin promotes the formation of tubular extensions toward LEs and subsequent fusion (Harrison et al. 2003). Furthermore, in HeLa cells, Rab7 directly interacts with and phosphorylates vimentin, a type III protein of the intermediate filament cytoskeletal network, further reinforcing the role of Rab7 in endosome positioning within the cell (Cogli et al. 2013).

1.4 Coordination of Endosome Maturation

As endosomes mature, from early to sorting to late endosomes and ultimately to endolysosomes, their identities change to reflect their functions. For example, the peripheral location and the slightly acidic milieu of the EEs allow them to accept cargo from PM and sort it to the recycling pathways. In contrast, LEs translocate toward the perinuclear space, assemble a different set of tethers, and acquire components necessary for the lysosomal environment where the cargo is to be degraded. Rab5 and PI(3)P at early stages and Rab7 and PI(3,5)P_2 at later stages of endosome maturation seem to be major determinants of endosomal functions and are therefore tightly regulated. While Rab conversion and PIP conversion are recognized as major contributors to endosome maturation, V-ATPase activity is also participating in this transition. No single master coordinator of this transition has yet been identified. Rather, the coordination of these regulatory changes seems to exist at the level of significant cross talk between the key players in endosome maturation.

1.4.1 Sorting Compartments

During sorting, membrane domains destined for recycling organize into tubular domains, which then undergo fission to form recycling vesicles. Meanwhile, membrane domains destined for degradation bud in the opposite direction to form ILVs. During tubular-based membrane sorting, the tubular network formation requires various cargo adaptor sorting nexin (SNX) proteins, which step in at specific maturation stages to aid recycling to the PM and retrieval to Golgi (van Weering et al. 2012). Retrieval to the Golgi is mediated by the retromer complex, which interacts with PI(3)P, Rab5, and Rab7 for its recruitment to endosomes in mammalian cells (Seaman et al. 2009; Takatori et al. 2016; Bean et al. 2017). In yeast, retromer plays a dual role in retrograde cargo export and in control of LE–vacuole fusion dynamics (Liu et al. 2012). SNX–BAR recruitment to Rab7-positive membranes displaces Rab7 from the cargo-bound cargo selection complex (CSC), thus separating the retromer-positive tubule from the Rab7-positive vesicular domain (Purushothaman et al. 2017). In mammalian cells, the retromer has recently been shown to play a role in controlling Rab7 localization and activity via Rab7 GAP TBC1D15, with further research underway to dissect this unanticipated retromer function (Jimenez-Orgaz et al. 2017). As well as functioning as a structural component of the retromer-mediated tubule formation, SNX–BAR heterodimers independently can sense proteins for recycling or retrieving, for example, through association with the hydrophobic tripeptide WLM motif, as in the case of CI-M6PR (Kvainickas et al. 2017; Simonetti et al. 2017). In many instances, cargo retrieval proceeds independently of retromer and instead relies on the newly identified and functionally distinct retriever complex, such as in the case of $\alpha_5\beta_1$ integrin (McNally et al. 2017). Sorting of membrane cargo is facilitated by ubiquitination and deubiquitination steps. Ubiquitination at PM triggers internalization of the cargo and tags it for packaging into ILVs, whereas deubiquitinated cargo is targeted back to the cell surface via recycling pathways (MacDonald and Piper 2016). Catabolism of lipids and membrane-bound cargo in ILVs requires the help of lipid binding and transfer proteins that can distort membrane structures and bind membrane proteins and present them to the water-soluble lysosomal hydrolases. ILV maturation involves cholesterol removal by the glycoprotein NPC2, which transfers it to other vesicles or to NPC1 for efflux. This makes membrane cargo, such as sphingolipids, more accessible to lysosomal degradation (Sandhoff 2016). Rab7 effector and microtubule adaptor RILP interacts with ESCRT-II, thereby coordinating ILV formation with endosome motility (Solinger and Spang 2013). Interestingly, endosomes in *Aspergillus nidulans* were able to form MVBs in absence of Rab7 and were degradation competent but had a compromised vacuole in the form of multiple minivacuoles with EE features (Abenza et al. 2012).

1.4.2 Rab Conversion

Regulation of Rab5 and Rab7 on EEs and LEs, respectively, helps the two compartments maintain their identity and functions. The Rab5–Rab7 conversion is the hallmark of EE–LE transition, and its perturbations delay endosome–lysosome fusion, prolong homotypic fusion, and result in enlarged endosomes, such as Dmon1 mutation in *D. melanogaster* (Yousefian et al. 2013) or *SAND-1* mutation in *C. elegans* (both homologs of the mammalian Mon1) (Poteryaev et al. 2010). During Rab5–Rab7 conversion, Rab5 is displaced by Rab7, a process that is driven by a combination of repression of Rab5 activation, withdrawal of Rab5 recruitment factors, Rab7 recruitment and activation, and physical displacement of Rab5 by Rab7. The Mon1–Ccz-1 complex mediates all of these processes. In *C. elegans* coelomocytes, Mon1 homolog SAND-1 has been shown to be required for Rab5 repression and for Rab7 recruitment. SAND-1 interacts with the Rab5 GEF RABX-5, displacing it from the endosomes into the cytosol, hence leading to the repression of Rab5 recruitment and activation (Poteryaev et al. 2010). In absence of SAND-1, RABX-5 was trapped on EEs, leading to enlarged endosomes. Similarly, Mon1 siRNA in HeLa cells also led to enlarged EEs, suggesting a defect in endosome maturation progression. In parallel with Rab5 repression, the Mon1–Ccz1 complex acts as a Rab7 GEF and activates Rab7, which arrives to the endosomes concomitantly with the Mon1, leading to the displacement of Rab5 by the Rab7 (Nordmann et al. 2010). Mon1–Ccz1 also interacts with the HOPS complex, also thought to act as a Rab7 GEF alongside its tethering functions (Balderhaar and Ungermann 2013; Poteryaev et al. 2010). Additionally, Rab5 inactivation is coordinated with Rab7 activation in a cross talk involving Rab5 effector BLOC-1, which ensures that Rab5 is inactivated only when Rab7 and HOPS are present (John Peter et al. 2013; Rana et al. 2015). Such displacement of Rab5 by Rab7 has been described as a cutout switch, which ensures the mutually exclusive domains of the two Rabs and a unidirectional progression of endosome maturation. This model accommodates for the observations that Rab5 levels gradually increase and then drop suddenly concomitant with the Rab7 increase (Del Conte-Zerial et al. 2008). The onset and coordination of Rab5–Rab7 conversion are still poorly understood, but are likely influenced by endosome size and PI(3)P levels (Poteryaev et al. 2010). Indeed, PI(3)P levels, alongside Rabex-5, are important for Mon1 recruitment to EEs to initiate Rab5–Rab7 conversion (Vieira et al. 2003; Poteryaev et al. 2010; Lawrence et al. 2014). Interestingly, Mon1–Ccz1 is recruited to autophagosomes by PI(3)P independently of Rab5, where it functions to recruit Rab7 to enable autophagosome fusion with lysosomes. Rab5 is still required by those organelles for the acquisition of hydrolases (Hegedus et al. 2016).

1.4.3 Rab5 Regulation

Rab5 is recruited to EEs via endocytic vesicles from PM or via direct activation by the Rabaptin5–Rabex5 complex. The Rab5 GEF Rabex5 might already be present on endocytic vesicles prior to fusion with EEs (Balderhaar and Ungermann 2013) or be recruited to EEs by Rab22, Rab4, and ubiquitinated cargo (Kalin et al. 2015; Zhu et al. 2009). Stabilization and amplification of endosomal Rab5-GTP levels are achieved via the two positive feedback loops generated by the Rab5 interactions with its GEF Rabex5 as well as with the PI(3)P kinase Vps34, which is present at the EE (Zerial and McBride 2001). The PI(3)P allows for recruitment of the FYVE-domain-containing Rabenosyn-5, which recruits further Rabex5. The PM-located PI(4,5)P_2 was required for endocytosis and Rab5 recruitment in *Drosophila* oocytes and was subsequently released from endosomes in a Rab5-dependent manner, a step necessary for further endosome maturation (Compagnon et al. 2009). In interferon-γ-activated RAW264.7 macrophages, Rab20 prolonged Rab5 association with phagosomes, explaining the delay in phagosome maturation associated with macrophage activation (Pei et al. 2014). The Rab5–Rab7 conversion involves Rab5 inactivation, which is mediated by the displacement of Rabex5 by the Mon1–Ccz1 complex (Poteryaev et al. 2010). Inactivation of Rab5 can also be mediated by RabGAP5, which is in part coordinated by the GTP-binding protein DRG2 (Mani et al. 2016). Additionally, the decrease of ubiquitinated cargo following its internalization into ILVs, and inactivation of Rab4, may contribute to the loss of Rab5 activation (Kalin et al. 2015). Monoubiquitination of Rab5 itself has been proposed to prevent its interaction with downstream effectors such as EEA1 (Shin et al. 2017). The rising levels of Vps34 and its product PI(3)P in the maturing endosome lead to the recruitment of the Rab5 GAP TBC-2, facilitating the rapid drop of Rab5 levels at the endosome, as demonstrated in *C. elegans* and mammalian cells (Shin et al. 2005; Law et al. 2017). V-ATPase activity also contributes to Rab5 displacement, as clamping endosomal pH at 6.2, i.e., the typical pH of EEs, delayed Rab5 release, prolonged the EE state, and inhibited the EE–LE transition in HeLa cells (Wang et al. 2015). Thus, multiple parallel processes contribute to Rab5 regulation and turnover.

1.4.4 Rab7 Regulation

Rab7 recruitment is regulated by the Rab7 GEF, Mon1–Ccz1, and modulated by phosphorylation by the tumor suppressor PTEN, which promotes ILV formation and growth receptor inactivation (Shinde and Maddika 2016; Nordmann et al. 2010). Interestingly, the Mon1–Ccz1 complex promotes Rab7 activity on late endosomes only. In Cos-7 and HeLa cells, Mon1–Ccz1 has been shown to dissociate from lysosomes and not be required for Rab7 activity on those organelles. This may ensure unidirectionality in LE–lysosome transition (Yasuda et al. 2016). The protective lysosomal glycoprotein LAMP-2, which is recruited to LEs independently of Rab conversion or acidification, is required for the recruitment of Rab7 and fusion of LEs with lysosomes (Huynh et al. 2007; Endo et al. 2015). PI3KC3 was shown not to be necessary for Rab7 recruitment; however, it participates in subsequent Rab7 regulation through recruitment of Rab7 GAP, Armus (Vieira et al. 2003; Jaber et al. 2016). Furthermore, the PI3KC3 complex component and activator of PI(3)P production, UVRAG, interacts with HOPS and acts as a Rab7 GEF. Rab24 has been proposed to be in a complex with Rab7 and promote endosome maturation and degradation (Amaya et al. 2016). Various posttranslational modifications of Rab7, such as palmitoylation, phosphorylation, and ubiquitination, have been proposed to play a role in modifying and fine-tuning specific interactions with Rab7 effectors, facilitating coordination of different pathways at the LE (Modica and Lefrancois 2017; Modica et al. 2017; Lin et al. 2017).

1.4.5 PIP Conversion

Phosphorylation of PI(3)P to PI(3,5)P_2 during endosome maturation is catalyzed by the PIKfyve kinase (Fab1p in yeast), thought to be important for regulation of vacuole size (Dove et al. 2009). PI(3,5)P_2 promotes membrane fission during endosome maturation, whereas HOPS and the V-ATPase promote fusion. HOPS binding to PI(3,5)P_2 promotes coordination between the two opposing events (Solinger and Spang 2013). As expected, PIKfyve deficiency in mammalian cells depleted PI(3,5)P_2 levels and led to the formation of multiple enlarged endosomes, unable to fuse with lysosomes, which was rescued by addition of PI(3,5)P_2 (Compton et al. 2016). The PI(3,5)P_2 deficiency phenotype could also be countered by inhibition of Rab5, via suppression of homotypic fusion and hence prevention of vacuole enlargement (Compton et al. 2016).

1.4.6 PIP Regulation

PI(3)P, whose presence is necessary for the EE identity, is generated by Vps34 (PI3KC3), which is recruited to the membranes by Rab5-GTP (Shin et al. 2005; Law et al. 2017; Christoforidis et al. 1999). PI3KC3 localization, stability, and activity can be regulated through phosphorylation, ubiquitination, and SUMOylation (Schink et al. 2013). PI3KC3 activity is stimulated by Rabex-5, when it forms a complex with ArfGAP1, mediated by AP-1 clathrin-coated vesicle adaptor protein complexes (Candiello et al. 2016). The PI3KC3 activity levels are positively regulated by UVRAG, which also interacts with the C-VPS/HOPS complex, a Rab7 GEF. UVRAG in turn is regulated positively by Rab7, which competes with UVRAG for the binding partner Rubicon, another component of the endosomal PI3KC3 complex. The Rab7-sequestered Rubicon liberates UVRAG to bind HOPS in a feed-forward loop (Sun et al. 2010). mTORC1 provides an additional layer of regulation to endosome maturation by phosphorylating UVRAG and preventing its interaction with HOPS, leading to an inhibition of endosome maturation (Kim et al. 2015). In line with these observations, PI3KC3 deficiency did not compromise TfR recycling, but resulted in elevated levels of Rab5-GTP and Rab7-GTP at endosomes, enlarged LEs, and reduced EGFR degradation. Elevated Rab7-GTP led to a failure of ILV formation and lysosome maturation in mammalian cells (Jaber et al. 2016). PI(3)P serves as the main source for PI(3,5)P_2 production by phosphorylation by PIKfyve. PI(3,5)P_2 levels can be reversed by dephosphorylation by Fig4 (Sac3) phosphatase to yield PI(3)P or by myotubularin 3 to yield PI(5)P. Interestingly, Fig4 also acts as a PIK-fyve activator (Currinn and Wassmer 2016). Coordination of PI(3,5)P_2 production in relation to endosome maturation remains to be explored. A recent study identified a WD40-repeat protein WDR91 as a Rab7 effector that inhibits PI3KC3 activity, potentially coupling Rab conversion to PIP conversion in EE to LE transition (Liu et al. 2017). In phagosomes, association of PI3KC3 with the phagosomal membrane and the production of PI(3)P by PI3KC3 are pH-sensitive and stop when the phagosome is acidified, highlighting the role of acidification in coordinating phagosome maturation (Naufer et al. 2017).

1.4.7 V-ATPase Regulation

The V-ATPase complex consists of over a dozen different subunits, many of which have multiple isoforms and splice variants at least in higher eukaryotes, and is associated with several accessory proteins, which may regulate V-ATPase assembly and activity (Cotter et al. 2015; Maxson and Grinstein 2014). This degree of complexity allows control of V-ATPase activity and pH at multiple levels. The association–dissociation kinetics of the V_0 and V_1 sectors are determined by various stimuli, such as glucose availability and cytosolic pH, and provide immediate and direct control over V-ATPase activity. An intact microtubular network is required for dissociation, whereas the V-ATPase regulator protein RAVE is required for reassociation through interaction with the V_1C subunit (Jefferies et al. 2008; Xu and Forgac 2001). The expression of genes encoding V-ATPase subunits is at least partly regulated by the transcription factor TFEB. In response to nutrient deprivation or lysosomal dysfunction, inhibition of mTORC1 activity at lysosomes induces TFEB-mediated transcription of V-ATPase components (Ben-Sahra and Manning 2017). This regulation highlights the role of acidification in nutrient acquisition and ensures matching demand and supply. Phosphatidic acid phosphatase, whose main role is in lipid biosynthesis, is another recently discovered negative regulator of V-ATPase gene expression as shown for Pah1p in yeast (Sherr et al. 2017), reinforcing the connection between nutrient homeostasis and endosome acidification.

The different isoforms may be responsible for varying coupling efficiencies of the proton pump and are targeted to different compartments in the cell (Kane 2006; Sun-Wada et al. 2009). The spatial and functional separation of the different isoforms is evident with the V_0a subunit isoforms, such as in PC-3 cells, in which the V_0a1

subunit is responsible for TfR recycling, whereas V_0a3 is associated with the LEs and lysosomes (Smith et al. 2016). Isoform-specific assembly factors are involved in regulating V-ATPase in specific organelles. For example, RAVE assembly factor interacts only with the endosome-specific Vph1p but not with the Golgi-specific Stv1p isoform of the yeast V_0a subunit of the V-ATPase, hence allowing for regulation of V-ATPase activity specifically in endosomes (Smardon et al. 2014). Additionally, CORVET and HOPS are required for V-ATPase assembly at the lysosome, and Rab7 effector RILP controls the V_1G subunit stability and localization, potentially assisting with regulation of V-ATPase activity and lysosomal acidification (De Luca and Bucci 2014; De Luca et al. 2014; Bucci et al. 2000). When not bound to Rab7, free RILP is responsible for V_1G ubiquitination and proteasomal degradation. Thereby, functional Rab7 is required to sequester RILP and promote acidification (Yasuda et al. 2016).

PI3KC3 regulates V-ATPase assembly and trafficking in coordination with other V-ATPase regulators, such as mTORC1 and glucose (Stransky and Forgac 2015; Sautin et al. 2005). PI3KC3 has been shown to interact with the V_1B subunit in plants (Liu et al. 2016) and to regulate interaction between V-ATPase and F-actin; however, no known PIP binding sites exist in any of the V-ATPase subunits, so the mechanisms of V-ATPase regulation by PI3KC3 remain to be established (Chen et al. 2004). The downstream product of PI3KC3 and PIKfyve, $PI(3,5)P_2$, is required for the acidification activity of V-ATPase, at least in yeast, *D. melanogaster* and *C. elegans* (Dove et al. 2009). The mechanisms of this function are unclear, as $PI(3,5)P_2$ status of the cell does not appear to affect V-ATPase localization. However, it is likely to be involved in direct or indirect activation of the V-ATPase. Interestingly, phagosomes in mouse macrophages with defects in PIKfyve appeared to acidify normally but recruitment of lysosomal proteins LAMP1 and Cathepsin D was reduced and phagosome–lysosome fusion was compromised (Kim et al. 2014). This is supported by previous observations that fully functional PIKfyve does not appear to be essential for acidification in mice, probably because only small amount of $PI(3,5)P_2$ is required to support lysosomal acidification (Dove et al. 2009).

The downstream product of PI3KC3 and PIKfyve, $PI(3,5)P_2$, which is involved in endosome fission, does so in part by counteracting the membrane fusion function of the V-ATPase in LEs (Compton et al. 2016). Enlarged endosomes due to deficient $PI(3,5)P_2$ activity could be rescued by depletion of V-ATPase by RNAi, implicating V-ATPase in the cross talk with PIPs to regulate endosome size. Pharmacological alteration of endosomal pH did not rescue the $PI(3,5)P_2$ deficiency phenotype of aberrant vacuolation, suggesting that V-ATPase activity that is independent of its acidification function is responsible for PIP-mediated coordination of endosome maturation.

1.5 Contributions of Other Organelles to Endosome Maturation

1.5.1 ER

Besides the cytoskeleton, endosome positioning may also be affected by endosome interaction with the ER. The capture of endosomes by the ER might be mediated by various endosomal adaptors and scaffold proteins, which contain ubiquitin-binding domains and are ubiquitinated by the ER-located E3 ubiquitin ligase RNF26. This ensures that the endosomes are retained in the perinuclear region, where their functions can be efficiently supported by the proximal ER, TGN, and lysosomes (Jongsma et al. 2016). Such ubiquitination and the opposing deubiquitination regulate endosome motility and their position in the cell. These mechanisms allow for orchestration of endosomal maturation and cargo trafficking. Additionally, endosomes form contact sites with the ER, which increase in number as endosomes mature, with 50% EEs and 99% LEs associated with the ER, highlighting the importance of their interaction at later maturation stages (Friedman et al. 2013). The ER contact sites might mediate endosome association with the microtubule motors dynein and kinesin, thus influencing endosome traffic and maturation (Raiborg et al. 2015). Both endosome-localized PI(3)P and Rab7 are recognized by the ER protein protrudin at the ER–LE contact sites, which leads to the transfer of kinesin-1 from protrudin to FYCO1 on LEs, to regulate endosome traffic in mammalian cells (Raiborg et al. 2015). As endosomes traffic, their interaction with the ER causes a dramatic deformation in the ER tubules and formation of new ER tubules, which follow the endosome movement (Zajac et al. 2013). Such observations suggest that the ER–endosome interactions are long-lived and strong enough for mobile endosomes to restructure the ER. Both the endosomes and the ER connect to microtubules at or near ER–endosome contact sites, implying that at least some endosomes traffic while coupled to the ER rather than in isolation (Friedman et al. 2013).

The ER plays a major supportive and regulatory role in endosome maturation, with the endosome–ER contact sites permitting cholesterol transfer, inter-organellar communication through Ca^{2+} signaling, ILV biogenesis, and inactivation of receptors by ER-resident phosphatases, such as EGFR inactivation by PTP1B (van der Kant and Neefjes 2014). The ER also forms contact sites with other compartments, such as mitochondria, Golgi, lipid droplets, and PM, and allows lateral transfer of lipids from endosomes to other organelles. The ER contact sites have been proposed to play a role in endosome fission, with almost every fission event preceded by an ER contact site formation at the budding endosome (Friedman et al. 2013; Rowland et al. 2014; Zajac et al. 2013). ER contact sites might allow recruitment of necessary fission factors, such as the SNX–BAR complexes, or supply lipids to modify membrane curvature to promote fission, or increase local Ca^{2+} concentration to promote Ca^{2+}-dependent endosome fission. Whether all fission events are orchestrated by the ER or only a certain subset, such as recycling, remains to be investigated.

Ca^{2+} signaling provides an important layer of regulation to endosome maturation and is required for SNARE-mediated membrane docking and fusion (Hay 2007). Endosomal compartments are able to maintain Ca^{2+} at levels much higher than those in the cytosol, and its release activates secondary messengers such as calmodulin and mediators of endosome maturation such as the tethering factor Munc13-4 (He et al. 2016; Luzio et al. 2010). Pharmacological buffering of Ca^{2+} induces phenotypes associated with lysosomal storage disorders, highlighting the significance of endosomal Ca^{2+} in endosome function (Lloyd-Evans and Platt 2011). Endosomal and lysosomal Ca^{2+} channels include the two-pore channels TPCs and the mucolipins TRPML1, TRPML2, and TRPML3. Importantly, ER–LE contact sites might contribute to the maintenance of endosomal Ca^{2+} levels by providing Ca^{2+} from the ER storage. This is supported by observations of a secondary spike in Ca^{2+} from the ER following initial Ca^{2+} release from lysosomes (Garrity et al. 2016; van der Kant and Neefjes 2014; Kilpatrick et al. 2013). Given their role in membrane fission, it is tempting to speculate that PIKfyve and its product $PI(3,5)P_2$ might be involved in regulation of Ca^{2+} signaling. Indeed, Ca^{2+} channels including TPCs, TRPML1, and TRPML2 appear to be gated by $PI(3,5)P_2$, and their silencing prevents phagosome–lysosome fusion, which can be rescued by pharmacological Ca^{2+} release (Dong et al. 2010; van der Kant and Neefjes 2014; Dayam et al. 2015). PIKfyve might also directly regulate Ca^{2+} channels (Luzio et al. 2007). Thus, it seems that integration of Ca^{2+} signaling and PIPs is one of the key aspects of membrane traffic control (Marchant and Patel 2015). Ca^{2+} signaling seems to respond not only to inputs from PIKfyve activity but also to V-ATPase activity, as the endosomal calcium channel TRPML3 is inhibited by acidic pH (Kim et al. 2008). In contrast, Ca^{2+} refilling into Ca^{2+}-depleted lysosomes seems to proceed independently of V-ATPase activity (Garrity et al. 2016). In turn, luminal Ca^{2+} homeostasis is important for proper acidification and membrane fusion, with elevated Ca^{2+} levels inhibiting acidification and increasing membrane fusion (Lelouvier and Puertollano 2011).

1.5.2 Lipid Droplets

Lipid droplets may also form transient contact sites with endosomes, believed to supply membrane components required for endosome function and maturation (Liu et al. 2007). Lipid droplets recruit Rab5 and are able to interact with the Rab5 effector EEA1, and therefore with EEs, implicating this organelle in endosomal membrane traffic (Liu et al. 2007). Lipid droplets have also been reported to form close associations with endosomes in macrophages taking up high levels of high-density lipoprotein (Schmitz et al. 1985), supporting their role in lipid storage and metabolism. Lipid droplets are emerging as major regulatory and signaling hubs and may yet have unidentified roles in endosome maturation.

1.5.3 Cross talk with the Autophagy Pathway

While extracellular substrates are delivered to lysosomes by the endosomal network, intracellular substrates reach the lysosomes via the autophagic pathways. Although viewed as a separate degradation pathway, autophagy by no means acts independently, and there is a significant overlap and cross talk between the two pathways. As such, Rab5, Rab7, and Rab11 have been shown to act in both pathways (Ao et al. 2014). For example, Rab11 was observed to translocate from recycling endosomes to autophagosomes following autophagy induction in *D. melanogaster* cells (Szatmari et al. 2014). Likewise, PI(3)P endosomal distribution and levels may be maintained by the autophagy protein Beclin-1, as observed in mouse fibroblasts (McKnight et al. 2014). Furthermore, endosomes can take on autophagy functions, as damaged mitochondria, ubiquitinated by Parkin, can be taken up by Rab5-positive endosomes to form MVBs in ESCRT-dependent manner to be subsequently delivered to lysosomes (Hammerling et al. 2017).

1.6 Perturbations of Endosome Maturation in Disease

The endosomal system is emerging as a central player in a wide range of neurodegenerative diseases, such as Alzheimer's disease (AD), Parkinson's Disease (PD), Lewy body dementia, amyotrophic lateral sclerosis (ALS), and hereditary spastic paraplegia (HSP) (Luzio et al. 2010; Schreij et al. 2016). Neurodegeneration may be a result of impaired lysosomal degradation, such as the case of alpha-synuclein in PD (Miura et al. 2014), or a defect in broader endosomal functions, which may lead to protein mislocalization or aberrant secretion. Frequently, PIPs, Rabs and their effectors, proteins involved in V-ATPase assembly and function, and retromer components are involved in endosomal dysfunction-mediated neurodegeneration. There are many excellent reviews summarizing the effects of various aspects of the endosomal system on the vast number of neurodegenerative disorders (Schreij et al. 2016; Hu et al. 2015; Williamson and Hiesinger 2010; Borland and Vilhardt 2017; Progida and Bakke 2016; Colacurcio and Nixon 2016; Waugh 2015). Inherited neuropathies and myopathies also tend to be associated with mutations specific to the endosomal pathways, such as Rab7 in Charcot–Marie–Tooth disease, the PI(3)P and PI(3,5)P_2 3-phosphatase myotubularin in X-linked centronuclear myopathy, and PIKfyve in Francois–Neetens fleck corneal dystrophy (Nicot and Laporte 2008). The progressive neurological disorder Niemann–Pick type C (NPC) is caused by mutations to lysosomal cholesterol transporters NPC1 and NPC2 and indeed, like many other lysosomal storage disorders, is exacerbated by the resulting impairment to ER–endosome contact sites and lysosomal Ca^{2+} homeostasis (van der Kant and Neefjes 2014; Garrity et al. 2016; Jiang and Mizushima 2014; McDermott and Kim 2015).

Intracellular pathogens have long been known to subvert host defense pathways using a wide variety of mechanisms, the majority of which target the endosomal network that is designed to kill the pathogen. These include the *Mycobacterium* virulence factor PtpA, which targets the tethering factor subunit Vps33b for dephosphorylation, *Legionella* SidK, which inhibits the V-ATPase-mediated acidification, and *Salmonella* SopD, which interacts with Rab7, blocking its binding to RILP and FYCO1, hence blocking fusion with lysosomes (Wong et al. 2011; Bach et al. 2008; Akbar et al. 2016; Xu et al. 2010; D'Costa et al. 2015). Many intracellular pathogens also target PIPs to slow down phagosome maturation, either by directly phosphorylating or dephosphorylating specific PIP species, or by deregulating the host enzymes involved in PIP conversions (Pizarro-Cerda et al. 2015). *Mycobacterium* ManLAM and MptpB, *Legionella* SidP, and *Salmonella* SopB are the notable examples of pathogen virulence factors required for pathogen survival that act by disrupting PIP pathways. Another aspect of endosome maturation, actin cytoskeleton remodeling, is essential for phagocytosis and for phagosome maturation and may be used by pathogens, such as the protozoan parasite *Leishmania*, to stall maturation (Freeman and Grinstein 2014; Lodge and Descoteaux 2005). Viruses also exploit the endosome system at various stages of viral lifecycle, from gaining entry into cells and capsid uncoating to envelopment and transmission (Vale-Costa and Amorim 2016). For example, inhibition of endosome-related processes, such as actin polymerization or TRPML2 activity, led to a decrease in viral infectivity for rabies and Ebola viruses, respectively (Piccinotti et al. 2013; de Armas-Rillo et al. 2016; Grimm et al. 2017). ESCRT–Vps4 complexes are necessary for capsid uncoating during early infection by human papillomavirus (HPV) and are likewise hijacked by herpes simplex virus-1 (HSV-1) at later stages to orchestrate its envelopment and hence assembly of the infectious virion (Kharkwal et al. 2016; Broniarczyk et al. 2017).

Many immune functions, such as antigen presentation, are strongly dependent on a fully functional endosomal system. Silencing of Arl8b, which regulates LE–lysosome interaction via HOPS, results in impaired formation of CD1 antigen-presenting complex in lysosomes, its delivery to PM, and phagosome–lysosome fusion (Garg et al. 2011). The normal functioning of platelets, i.e., the activation-dependent release of effectors from storage granules, relies on normal PIKfyve function in endosome maturation (Min et al. 2014). The CHEVI tethering complex subunits Vps33b and VIPAS39 are implicated in the arthrogryposis–renal dysfunction–cholestasis (ARC) syndrome, which is a fatal recessive disorder characterized by trafficking defects in multiple organs, persistent infections, and sepsis. Specifically, the clearance of endosomes with internalized pathogen recognition receptors (PRRs) is affected, and hence, PRR inactivation and modulation of inflammatory responses are compromised (Akbar et al. 2016).

Cancer development is frequently assisted by endosomal upregulation of oncogenic signaling, through defective trafficking of growth factor receptors, increased recycling, or decreased degradation (Stasyk and Huber 2016). The components and regulators of the endosomal–lysosomal network, such as dynamin, Beclin-1, and mTORC1, which act as oncogenes and tumor suppressors, feature extensively in many cancers, whether stalling EE maturation to enhance growth factor receptor signaling or promoting further endosome maturation to boost recycling functions, efficient nutrient degradation, and subsequent cell growth and migration.

1.7 Conclusions

The endosomal system is a complex network of membrane-bound compartments that communicate and exchange components with each other. Its diverse functions span not only nutrient acquisition, but also regulation of membrane composition and signaling at many different levels, affecting fundamental cell functions such as growth and polarity. Endosome maturation is essential for its integrity and function and is regulated by a wide array of inputs, ranging from small GTPases and regulatory lipids to pH and Ca^{2+} signaling, from scaffolding proteins and tethers to the cytoskeletal and ER networks. Coordination of endosome maturation is mediated by the cross talk between Rab GTPases, PIPs, and V-ATPase, which ensure the identity of endosomes as well as unidirectionality and appropriate timing at each step of endosome maturation. A properly functioning endosomal network is essential to life, and even the slightest perturbations in the endocytic pathways lead to severe consequences in cellular functions, many of them associated with neurological diseases, infections, immune regulation, and cancers. Continuing discoveries in basic cell biology are acutely needed for the development of therapies to treat the neglected genetic disorders as well as the highly prevalent neurodegenerative diseases and cancer.

References

Abenza JF, Galindo A, Pinar M, Pantazopoulou A, de los Rios V, Penalva MA (2012) Endosomal maturation by Rab conversion in *Aspergillus nidulans* is coupled to dynein-mediated basipetal movement. Mol Biol Cell 23(10):1889–1901. https://doi.org/10.1091/mbc.E11-11-0925

Adell MA, Vogel GF, Pakdel M, Muller M, Lindner H, Hess MW, Teis D (2014) Coordinated binding of Vps4 to ESCRT-III drives membrane neck constriction during MVB vesicle formation. J Cell Biol 205(1):33–49. https://doi.org/10.1083/jcb.201310114

Akbar MA, Mandraju R, Tracy C, Hu W, Pasare C, Kramer H (2016) ARC syndrome-linked Vps33B protein is required for inflammatory endosomal maturation and signal termination. Immunity 45(2):267–279. https://doi.org/10.1016/j.immuni.2016.07.010

Amaya C, Militello RD, Calligaris SD, Colombo MI (2016) Rab24 interacts with the Rab7/Rab interacting lysosomal protein complex to regulate endosomal degradation. Traffic 17(11):1181–1196. https://doi.org/10.1111/tra.12431

Andrews NW, Corrotte M, Castro-Gomes T (2015) Above the fray: surface remodeling by secreted lysosomal enzymes leads to endocytosis-mediated plasma membrane repair. Semin Cell Dev Biol 45:10–17. https://doi.org/10.1016/j.semcdb.2015.09.022

Antonescu CN, Aguet F, Danuser G, Schmid SL (2011) Phosphatidylinositol-(4,5)-bisphosphate regulates clathrin-coated pit initiation, stabilization, and size. Mol Biol Cell 22(14):2588–2600. https://doi.org/10.1091/mbc.E11-04-0362

Ao X, Zou L, Wu Y (2014) Regulation of autophagy by the Rab GTPase network. Cell Death Differ 21(3):348–358. https://doi.org/10.1038/cdd.2013.187

Armstrong J (2010) Yeast vacuoles: more than a model lysosome. Trends Cell Biol 20(10):580–585. https://doi.org/10.1016/j.tcb.2010.06.010

Bach H, Papavinasasundaram KG, Wong D, Hmama Z, Av-Gay Y (2008) *Mycobacterium* tuberculosis virulence is mediated by PtpA dephosphorylation of human vacuolar protein sorting 33B. Cell Host Microbe 3(5):316–322. https://doi.org/10.1016/j.chom.2008.03.008

Balderhaar HJ, Ungermann C (2013) CORVET and HOPS tethering complexes—coordinators of endosome and lysosome fusion. J Cell Sci 126(Pt 6):1307–1316. https://doi.org/10.1242/jcs.107805

Baravalle G, Schober D, Huber M, Bayer N, Murphy RF, Fuchs R (2005) Transferrin recycling and dextran transport to lysosomes is differentially affected by bafilomycin, nocodazole, and low temperature. Cell Tissue Res 320(1):99–113. https://doi.org/10.1007/s00441-004-1060-x

Bean BD, Davey M, Conibear E (2017) Cargo selectivity of yeast sorting nexins. Traffic 18(2):110–122. https://doi.org/10.1111/tra.12459

Ben-Sahra I, Manning BD (2017) mTORC1 signaling and the metabolic control of cell growth. Curr Opin Cell Biol 45:72–82. https://doi.org/10.1016/j.ceb.2017.02.012

Bissig C, Hurbain I, Raposo G, van Niel G (2017) PIKfyve activity regulates reformation of terminal storage lysosomes from endolysosomes. Traffic 18(11):747–757. https://doi.org/10.1111/tra.12525

Boal F, Mansour R, Gayral M, Saland E, Chicanne G, Xuereb JM, Marcellin M, Burlet-Schiltz O, Sansonetti PJ, Payrastre B, Tronchere H (2015) TOM1 is a PI5P effector involved in the regulation of endosomal maturation. J Cell Sci 128(4):815–827. https://doi.org/10.1242/jcs.166314

Boes M, Cuvillier A, Ploegh H (2004) Membrane specializations and endosome maturation in dendritic cells and B cells. Trends Cell Biol 14(4):175–183. https://doi.org/10.1016/j.tcb.2004.02.004

Bohdanowicz M, Grinstein S (2013) Role of phospholipids in endocytosis, phagocytosis, and macropinocytosis. Physiol Rev 93(1):69–106. https://doi.org/10.1152/physrev.00002.2012

Borland H, Vilhardt F (2017) Prelysosomal compartments in the unconventional secretion of amyloidogenic seeds. Int J Mol Sci 18(1). https://doi.org/10.3390/ijms18010227

Bouchez I, Pouteaux M, Canonge M, Genet M, Chardot T, Guillot A, Froissard M (2015) Regulation of lipid droplet dynamics in *Saccharomyces cerevisiae* depends on the Rab7-like Ypt7p, HOPS complex and V1-ATPase. Biol Open 4(7):764–775. https://doi.org/10.1242/bio.20148615

Broniarczyk J, Pim D, Massimi P, Bergant M, Gozdzicka-Jozefiak A, Crump C, Banks L (2017) The VPS4 component of the ESCRT machinery plays an essential role in HPV infectious entry and capsid disassembly. Sci Rep 7:45159. https://doi.org/10.1038/srep45159

Bucci C, Thomsen P, Nicoziani P, McCarthy J, van Deurs B (2000) Rab7: a key to lysosome biogenesis. Mol Biol Cell 11(2):467–480

Buckley CM, King JS (2017) Drinking problems: mechanisms of macropinosome formation and maturation. FEBS J. https://doi.org/10.1111/febs.14115

Candiello E, Kratzke M, Wenzel D, Cassel D, Schu P (2016) AP-1/σ1A and AP-1/σ1B adaptor-proteins differentially regulate neuronal early endosome maturation via the Rab5/Vps34-pathway. Sci Rep 6:29950. https://doi.org/10.1038/srep29950

Cao Q, Zhong XZ, Zou Y, Murrell-Lagnado R, Zhu MX, Dong XP (2015) Calcium release through P2X4 activates calmodulin to promote endolysosomal membrane fusion. J Cell Biol 209(6):879–894. https://doi.org/10.1083/jcb.201409071

Cardoso CM, Jordao L, Vieira OV (2010) Rab10 regulates phagosome maturation and its overexpression rescues *Mycobacterium*-containing phagosomes maturation. Traffic 11(2):221–235. https://doi.org/10.1111/j.1600-0854.2009.01013.x

Caviglia S, Brankatschk M, Fischer EJ, Eaton S, Luschnig S (2016) Staccato/Unc-13-4 controls secretory lysosome-mediated lumen fusion during epithelial tube anastomosis. Nat Cell Biol 18(7):727–739. https://doi.org/10.1038/ncb3374

Chen PI, Kong C, Su X, Stahl PD (2009) Rab5 isoforms differentially regulate the trafficking and degradation of epidermal growth factor receptors. J Biol Chem 284(44):30328–30338. https://doi.org/10.1074/jbc.M109.034546

Chen SH, Bubb MR, Yarmola EG, Zuo J, Jiang J, Lee BS, Lu M, Gluck SL, Hurst IR, Holliday LS (2004) Vacuolar H+-ATPase binding to microfilaments: regulation in response to phosphatidylinositol 3-kinase activity and detailed characterization of the actin-binding site in subunit B. J Biol Chem 279(9):7988–7998. https://doi.org/10.1074/jbc.M305351200

Chirivino D, Del Maestro L, Formstecher E, Hupe P, Raposo G, Louvard D, Arpin M (2011) The ERM proteins interact with the HOPS complex to regulate the maturation of endosomes. Mol Biol Cell 22(3):375–385. https://doi.org/10.1091/mbc.E10-09-0796

Christoforidis S, Miaczynska M, Ashman K, Wilm M, Zhao L, Yip SC, Waterfield MD, Backer JM, Zerial M (1999) Phosphatidylinositol-3-OH kinases are Rab5 effectors. Nat Cell Biol 1(4):249–252. https://doi.org/10.1038/12075

Clague MJ, Hammond DE (2006) Membrane traffic: catching the lysosome express. Current Biol CB 16(11):R416–R418. https://doi.org/10.1016/j.cub.2006.05.009

Clarke M, Maddera L, Engel U, Gerisch G (2010) Retrieval of the vacuolar H-ATPase from phagosomes revealed by live cell imaging. PLoS ONE 5(1):e8585. https://doi.org/10.1371/journal.pone.0008585

Cogli L, Progida C, Bramato R, Bucci C (2013) Vimentin phosphorylation and assembly are regulated by the small GTPase Rab7a. Biochem Biophys Acta 1833(6):1283–1293. https://doi.org/10.1016/j.bbamcr.2013.02.024

Colacurcio DJ, Nixon RA (2016) Disorders of lysosomal acidification—the emerging role of v-ATPase in aging and neurodegenerative disease. Age Res Rev 32:75–88. https://doi.org/10.1016/j.arr.2016.05.004

Compagnon J, Gervais L, Roman MS, Chamot-Boeuf S, Guichet A (2009) Interplay between Rab5 and PtdIns(4,5)P2 controls early endocytosis in the *Drosophila* germline. J Cell Sci 122(Pt 1):25–35. https://doi.org/10.1242/jcs.033027

Compton LM, Ikonomov OC, Sbrissa D, Garg P, Shisheva A (2016) Active vacuolar H+ATPase and functional cycle of Rab5 are required for the vacuolation defect triggered by PtdIns(3,5)P2 loss under PIKfyve or Vps34 deficiency. Am J Physiol Cell Physiol 311(3):C366–C377. https://doi.org/10.1152/ajpcell.00104.2016

Cotter K, Stransky L, McGuire C, Forgac M (2015) Recent Insights into the structure, regulation, and function of the V-ATPases. Trends Biochem Sci 40(10):611–622. https://doi.org/10.1016/j.tibs.2015.08.005

Currinn H, Wassmer T (2016) The amyloid precursor protein (APP) binds the PIKfyve complex and modulates its function. Biochem Soc Trans 44(1):185–190. https://doi.org/10.1042/BST20150179

D'Costa VM, Braun V, Landekic M, Shi R, Proteau A, McDonald L, Cygler M, Grinstein S, Brumell JH (2015) *Salmonella* disrupts host endocytic trafficking by SopD2-mediated inhibition of Rab7. Cell Rep 12(9):1508–1518. https://doi.org/10.1016/j.celrep.2015.07.063

D'Souza RS, Semus R, Billings EA, Meyer CB, Conger K, Casanova JE (2014) Rab4 orchestrates a small GTPase cascade for recruitment of adaptor proteins to early endosomes. Current Biol CB 24(11):1187–1198. https://doi.org/10.1016/j.cub.2014.04.003

Danson C, Brown E, Hemmings OJ, McGough IJ, Yarwood S, Heesom KJ, Carlton JG, Martin-Serrano J, May MT, Verkade P, Cullen PJ (2013) SNX15 links clathrin endocytosis to the PtdIns3P early endosome independently of the APPL1 endosome. J Cell Sci 126(Pt 21):4885–4899. https://doi.org/10.1242/jcs.125732

Dayam RM, Saric A, Shilliday RE, Botelho RJ (2015) The Phosphoinositide-gated lysosomal Ca(2+) channel, TRPML1, is required for phagosome maturation. Traffic 16(9):1010–1026. https://doi.org/10.1111/tra.12303

de Armas-Rillo L, Valera MS, Marrero-Hernandez S, Valenzuela-Fernandez A (2016) Membrane dynamics associated with viral infection. Rev Med Virol 26(3):146–160. https://doi.org/10.1002/rmv.1872

De Luca M, Bucci C (2014) A new V-ATPase regulatory mechanism mediated by the Rab interacting lysosomal protein (RILP). Commun Integr Biol 7(5). https://doi.org/10.4161/cib.29616

De Luca M, Cogli L, Progida C, Nisi V, Pascolutti R, Sigismund S, Di Fiore PP, Bucci C (2014) RILP regulates vacuolar ATPase through interaction with the V1G1 subunit. J Cell Sci 127(Pt 12):2697–2708. https://doi.org/10.1242/jcs.142604

Del Conte-Zerial P, Brusch L, Rink JC, Collinet C, Kalaidzidis Y, Zerial M, Deutsch A (2008) Membrane identity and GTPase cascades regulated by toggle and cut-out switches. Mol Syst Biol 4:206. https://doi.org/10.1038/msb.2008.45

Desfougeres Y, Vavassori S, Rompf M, Gerasimaite R, Mayer A (2016) Organelle acidification negatively regulates vacuole membrane fusion in vivo. Sci Rep 6:29045. https://doi.org/10.1038/srep29045

Dieckmann R, Gueho A, Monroy R, Ruppert T, Bloomfield G, Soldati T (2012) The balance in the delivery of ER components and the vacuolar proton pump to the phagosome depends on myosin IK in *Dictyostelium*. Mol Cell Proteomics MCP 11(10):886–900. https://doi.org/10.1074/mcp.M112.017608

Dill BD, Gierlinski M, Hartlova A, Arandilla AG, Guo M, Clarke RG, Trost M (2015) Quantitative proteome analysis of temporally resolved phagosomes following uptake via key phagocytic receptors. Mol Cell Proteomics MCP 14(5):1334–1349. https://doi.org/10.1074/mcp.M114.044594

Doherty GJ, McMahon HT (2009) Mechanisms of endocytosis. Annu Rev Biochem 78:857–902. https://doi.org/10.1146/annurev.biochem.78.081307.110540

Dolat L, Spiliotis ET (2016) Septins promote macropinosome maturation and traffic to the lysosome by facilitating membrane fusion. J Cell Biol 214(5):517–527. https://doi.org/10.1083/jcb.201603030

Dong XP, Shen D, Wang X, Dawson T, Li X, Zhang Q, Cheng X, Zhang Y, Weisman LS, Delling M, Xu H (2010) PI(3,5)P(2) controls membrane trafficking by direct activation of mucolipin Ca(2+) release channels in the endolysosome. Nat Commun 1:38. https://doi.org/10.1038/ncomms1037

Dove SK, Dong K, Kobayashi T, Williams FK, Michell RH (2009) Phosphatidylinositol 3,5-bisphosphate and Fab1p/PIKfyve underPPIn endo-lysosome function. Biochem J 419(1):1–13. https://doi.org/10.1042/BJ20081950

Driskell OJ, Mironov A, Allan VJ, Woodman PG (2007) Dynein is required for receptor sorting and the morphogenesis of early endosomes. Nat Cell Biol 9(1):113–120. https://doi.org/10.1038/ncb1525

Duleh SN, Welch MD (2010) WASH and the Arp2/3 complex regulate endosome shape and trafficking. Cytoskeleton 67(3):193–206. https://doi.org/10.1002/cm.20437

Elbaz-Alon Y, Rosenfeld-Gur E, Shinder V, Futerman AH, Geiger T, Schuldiner M (2014) A dynamic interface between vacuoles and mitochondria in yeast. Dev Cell 30(1):95–102. https://doi.org/10.1016/j.devcel.2014.06.007

Endo Y, Furuta A, Nishino I (2015) Danon disease: a phenotypic expression of LAMP-2 deficiency. Acta Neuropathol 129(3):391–398. https://doi.org/10.1007/s00401-015-1385-4

Fairn GD, Grinstein S (2012) How nascent phagosomes mature to become phagolysosomes. Trends Immunol 33(8):397–405. https://doi.org/10.1016/j.it.2012.03.003

Fili N, Calleja V, Woscholski R, Parker PJ, Larijani B (2006) Compartmental signal modulation: endosomal phosphatidylinositol 3-phosphate controls endosome morphology and selective cargo sorting. Proc Natl Acad Sci USA 103(42):15473–15478. https://doi.org/10.1073/pnas.0607040103

Freeman SA, Grinstein S (2014) Phagocytosis: receptors, signal integration, and the cytoskeleton. Immunol Rev 262(1):193–215. https://doi.org/10.1111/imr.12212

Friedman JR, Dibenedetto JR, West M, Rowland AA, Voeltz GK (2013) Endoplasmic reticulum-endosome contact increases as endosomes traffic and mature. Mol Biol Cell 24(7):1030–1040. https://doi.org/10.1091/mbc.E12-10-0733

Gao N, Kaestner KH (2010) Cdx2 regulates endo-lysosomal function and epithelial cell polarity. Genes Dev 24(12):1295–1305. https://doi.org/10.1101/gad.1921510

Garg S, Sharma M, Ung C, Tuli A, Barral DC, Hava DL, Veerapen N, Besra GS, Hacohen N, Brenner MB (2011) Lysosomal trafficking, antigen presentation, and microbial killing are controlled by the Arf-like GTPase Arl8b. Immunity 35(2):182–193. https://doi.org/10.1016/j.immuni.2011.06.009

Garrity AG, Wang W, Collier CM, Levey SA, Gao Q, Xu H (2016) The endoplasmic reticulum, not the pH gradient, drives calcium refilling of lysosomes. eLife 5. https://doi.org/10.7554/elife.15887
Gibbings DJ, Ciaudo C, Erhardt M, Voinnet O (2009) Multivesicular bodies associate with components of miRNA effector complexes and modulate miRNA activity. Nat Cell Biol 11(9):1143–1149. https://doi.org/10.1038/ncb1929
Gopaldass N, Patel D, Kratzke R, Dieckmann R, Hausherr S, Hagedorn M, Monroy R, Kruger J, Neuhaus EM, Hoffmann E, Hille K, Kuznetsov SA, Soldati T (2012) Dynamin A, Myosin IB and Abp1 couple phagosome maturation to F-actin binding. Traffic 13(1):120–130. https://doi.org/10.1111/j.1600-0854.2011.01296.x
Granger E, McNee G, Allan V, Woodman P (2014) The role of the cytoskeleton and molecular motors in endosomal dynamics. Semin Cell Dev Biol 31:20–29. https://doi.org/10.1016/j.semcdb.2014.04.011
Grant BD, Donaldson JG (2009) Pathways and mechanisms of endocytic recycling. Nat Rev Mol Cell Biol 10(9):597–608. https://doi.org/10.1038/nrm2755
Griffiths GM (2016) Secretion from myeloid cells: secretory lysosomes. Microbiol Spectr 4(4). https://doi.org/10.1128/microbiolspec.mchd-0030-2016
Grimm C, Butz E, Chen CC, Wahl-Schott C, Biel M (2017) From mucolipidosis type IV to Ebola: TRPML and two-pore channels at the crossroads of endo-lysosomal trafficking and disease. Cell Calcium. https://doi.org/10.1016/j.ceca.2017.04.003
Gschweitl M, Ulbricht A, Barnes CA, Enchev RI, Stoffel-Studer I, Meyer-Schaller N, Huotari J, Yamauchi Y, Greber UF, Helenius A, Peter M (2016) A SPOPL/Cullin-3 ubiquitin ligase complex regulates endocytic trafficking by targeting EPS15 at endosomes. eLife 5:e13841. https://doi.org/10.7554/elife.13841
Guerra F, Bucci C (2016) Multiple roles of the small GTPase Rab7. Cells 5(3). https://doi.org/10.3390/cells5030034
Gutierrez MG (2013) Functional role(s) of phagosomal Rab GTPases. Small GTPases 4(3):148–158. https://doi.org/10.4161/sgtp.25604
Hammerling BC, Najor RH, Cortez MQ, Shires SE, Leon LJ, Gonzalez ER, Boassa D, Phan S, Thor A, Jimenez RE, Li H, Kitsis RN, Dorn Ii GW, Sadoshima J, Ellisman MH, Gustafsson AB (2017) A Rab5 endosomal pathway mediates Parkin-dependent mitochondrial clearance. Nat Commun 8:14050. https://doi.org/10.1038/ncomms14050
Harrison RE, Bucci C, Vieira OV, Schroer TA, Grinstein S (2003) Phagosomes fuse with late endosomes and/or lysosomes by extension of membrane protrusions along microtubules: role of Rab7 and RILP. Mol Cell Biol 23(18):6494–6506
Hay JC (2007) Calcium: a fundamental regulator of intracellular membrane fusion? EMBO Rep 8(3):236–240. https://doi.org/10.1038/sj.embor.7400921
He J, Johnson JL, Monfregola J, Ramadass M, Pestonjamasp K, Napolitano G, Zhang J, Catz SD (2016) Munc13-4 interacts with syntaxin 7 and regulates late endosomal maturation, endosomal signaling, and TLR9-initiated cellular responses. Mol Biol Cell 27(3):572–587. https://doi.org/10.1091/mbc.E15-05-0283
Hegedus K, Takats S, Boda A, Jipa A, Nagy P, Varga K, Kovacs AL, Juhasz G (2016) The Ccz1-Mon1-Rab7 module and Rab5 control distinct steps of autophagy. Mol Biol Cell 27(20):3132–3142. https://doi.org/10.1091/mbc.E16-03-0205
Hirano T, Munnik T, Sato MH (2015) Phosphatidylinositol 3-phosphate 5-kinase, FAB1/PIKfyve Kinase mediates endosome maturation to establish endosome-cortical microtubule interaction in *Arabidopsis*. Plant Physiol 169(3):1961–1974. https://doi.org/10.1104/pp.15.01368
Honscher C, Mari M, Auffarth K, Bohnert M, Griffith J, Geerts W, van der Laan M, Cabrera M, Reggiori F, Ungermann C (2014) Cellular metabolism regulates contact sites between vacuoles and mitochondria. Dev Cell 30(1):86–94. https://doi.org/10.1016/j.devcel.2014.06.006
Hsu F, Hu F, Mao Y (2015) Spatiotemporal control of phosphatidylinositol 4-phosphate by Sac2 regulates endocytic recycling. J Cell Biol 209(1):97–110. https://doi.org/10.1083/jcb.201408027

Hu YB, Dammer EB, Ren RJ, Wang G (2015) The endosomal-lysosomal system: from acidification and cargo sorting to neurodegeneration. Transl Neurodegeneration 4:18. https://doi.org/10.1186/s40035-015-0041-1

Huang C, Chang A (2011) pH-dependent cargo sorting from the Golgi. J Biol Chem 286(12):10058–10065. https://doi.org/10.1074/jbc.M110.197889

Hubner M, Peter M (2012) Cullin-3 and the endocytic system: new functions of ubiquitination for endosome maturation. Cell Logistics 2(3):166–168. https://doi.org/10.4161/cl.20372

Huotari J, Helenius A (2011) Endosome maturation. EMBO J 30(17):3481–3500. https://doi.org/10.1038/emboj.2011.286

Hurtado-Lorenzo A, Skinner M, El Annan J, Futai M, Sun-Wada GH, Bourgoin S, Casanova J, Wildeman A, Bechoua S, Ausiello DA, Brown D, Marshansky V (2006) V-ATPase interacts with ARNO and Arf6 in early endosomes and regulates the protein degradative pathway. Nat Cell Biol 8(2):124–136. https://doi.org/10.1038/ncb1348

Huynh KK, Eskelinen EL, Scott CC, Malevanets A, Saftig P, Grinstein S (2007) LAMP proteins are required for fusion of lysosomes with phagosomes. EMBO J 26(2):313–324. https://doi.org/10.1038/sj.emboj.7601511

Issman-Zecharya N, Schuldiner O (2014) The PI3K class III complex promotes axon pruning by downregulating a Ptc-derived signal via endosome-lysosomal degradation. Dev Cell 31(4):461–473. https://doi.org/10.1016/j.devcel.2014.10.013

Jaber N, Mohd-Naim N, Wang Z, DeLeon JL, Kim S, Zhong H, Sheshadri N, Dou Z, Edinger AL, Du G, Braga VM, Zong WX (2016) Vps34 regulates Rab7 and late endocytic trafficking through recruitment of the GTPase-activating protein Armus. J Cell Sci 129(23):4424–4435. https://doi.org/10.1242/jcs.192260

Jacomin AC, Fauvarque MO, Taillebourg E (2016) A functional endosomal pathway is necessary for lysosome biogenesis in *Drosophila*. BMC cell biology 17(1):36. https://doi.org/10.1186/s12860-016-0115-7

Jean S, Kiger AA (2012) Coordination between RAB GTPase and phosphoinositide regulation and functions. Nat Rev Mol Cell Biol 13(7):463–470. https://doi.org/10.1038/nrm3379

Jefferies KC, Cipriano DJ, Forgac M (2008) Function, structure and regulation of the vacuolar (H+)-ATPases. Arch Biochem Biophys 476(1):33–42. https://doi.org/10.1016/j.abb.2008.03.025

Jeschke A, Zehethofer N, Lindner B, Krupp J, Schwudke D, Haneburger I, Jovic M, Backer JM, Balla T, Hilbi H, Haas A (2015) Phosphatidylinositol 4-phosphate and phosphatidylinositol 3-phosphate regulate phagolysosome biogenesis. Proc Natl Acad Sci USA 112(15):4636–4641. https://doi.org/10.1073/pnas.1423456112

Jiang P, Mizushima N (2014) Autophagy and human diseases. Cell Res 24(1):69–79. https://doi.org/10.1038/cr.2013.161

Jimenez-Orgaz A, Kvainickas A, Nagele H, Denner J, Eimer S, Dengjel J, Steinberg F (2017) Control of RAB7 activity and localization through the retromer-TBC1D5 complex enables RAB7-dependent mitophagy. EMBO J. https://doi.org/10.15252/embj.201797128

Jin N, Lang MJ, Weisman LS (2016) Phosphatidylinositol 3,5-bisphosphate: regulation of cellular events in space and time. Biochem Soc Trans 44(1):177–184. https://doi.org/10.1042/BST20150174

John Peter AT, Lachmann J, Rana M, Bunge M, Cabrera M, Ungermann C (2013) The BLOC-1 complex promotes endosomal maturation by recruiting the Rab5 GTPase-activating protein Msb3. J Cell Biol 201(1):97–111. https://doi.org/10.1083/jcb.201210038

Johnson DE, Ostrowski P, Jaumouille V, Grinstein S (2016) The position of lysosomes within the cell determines their luminal pH. J Cell Biol 212(6):677–692. https://doi.org/10.1083/jcb.201507112

Jongsma ML, Berlin I, Wijdeven RH, Janssen L, Janssen GM, Garstka MA, Janssen H, Mensink M, van Veelen PA, Spaapen RM, Neefjes J (2016) An ER-associated pathway defines endosomal architecture for controlled cargo transport. Cell 166(1):152–166. https://doi.org/10.1016/j.cell.2016.05.078

Kalaidzidis I, Miaczynska M, Brewinska-Olchowik M, Hupalowska A, Ferguson C, Parton RG, Kalaidzidis Y, Zerial M (2015) APPL endosomes are not obligatory endocytic intermediates but act as stable cargo-sorting compartments. J Cell Biol 211(1):123–144. https://doi.org/10.1083/jcb.201311117

Kalin S, Hirschmann DT, Buser DP, Spiess M (2015) Rabaptin5 is recruited to endosomes by Rab4 and Rabex5 to regulate endosome maturation. J Cell Sci 128(22):4126–4137. https://doi.org/10.1242/jcs.174664

Kamalesh K, Trivedi D, Toscano S, Sharma S, Kolay S, Raghu P (2017) Phosphatidylinositol 5-phosphate 4-kinase regulates early endosomal dynamics during clathrin-mediated endocytosis. J Cell Sci 130(13):2119–2133. https://doi.org/10.1242/jcs.202259

Kane PM (2006) The where, when, and how of organelle acidification by the yeast vacuolar H+-ATPase. Microbiol Mol Biol Rev MMBR 70(1):177–191. https://doi.org/10.1128/MMBR.70.1.177-191.2006

Karim MA, Brett CL (2017) The Na(+)(K(+))/H(+) exchanger Nhx1 controls multivesicular body-vacuolar lysosome fusion. Mol Biol Cell. https://doi.org/10.1091/mbc.E17-08-0496

Kasmapour B, Gronow A, Bleck CK, Hong W, Gutierrez MG (2012) Size-dependent mechanism of cargo sorting during lysosome-phagosome fusion is controlled by Rab34. Proc Natl Acad Sci USA 109(50):20485–20490. https://doi.org/10.1073/pnas.1206811109

Katzmann DJ, Stefan CJ, Babst M, Emr SD (2003) Vps27 recruits ESCRT machinery to endosomes during MVB sorting. J Cell Biol 162(3):413–423. https://doi.org/10.1083/jcb.200302136

Ketel K, Krauss M, Nicot AS, Puchkov D, Wieffer M, Muller R, Subramanian D, Schultz C, Laporte J, Haucke V (2016) A phosphoinositide conversion mechanism for exit from endosomes. Nature 529(7586):408–412. https://doi.org/10.1038/nature16516

Kharitidi D, Apaja PM, Manteghi S, Suzuki K, Malitskaya E, Roldan A, Gingras MC, Takagi J, Lukacs GL, Pause A (2015) Interplay of endosomal pH and ligand occupancy in integrin α5β1 ubiquitination, endocytic sorting, and cell migration. Cell Rep 13(3):599–609. https://doi.org/10.1016/j.celrep.2015.09.024

Kharkwal H, Smith CG, Wilson DW (2016) Herpes simplex virus capsid localization to ESCRT-VPS4 complexes in the presence and absence of the large tegument protein UL36p. J Virol 90(16):7257–7267. https://doi.org/10.1128/JVI.00857-16

Kilpatrick BS, Eden ER, Schapira AH, Futter CE, Patel S (2013) Direct mobilisation of lysosomal Ca^{2+} triggers complex Ca^{2+} signals. J Cell Sci 126(Pt 1):60–66. https://doi.org/10.1242/jcs.118836

Kim GH, Dayam RM, Prashar A, Terebiznik M, Botelho RJ (2014) PIKfyve inhibition interferes with phagosome and endosome maturation in macrophages. Traffic 15(10):1143–1163. https://doi.org/10.1111/tra.12199

Kim HJ, Li Q, Tjon-Kon-Sang S, So I, Kiselyov K, Soyombo AA, Muallem S (2008) A novel mode of TRPML3 regulation by extracytosolic pH absent in the varitint-waddler phenotype. EMBO J 27(8):1197–1205. https://doi.org/10.1038/emboj.2008.56

Kim YM, Jung CH, Seo M, Kim EK, Park JM, Bae SS, Kim DH (2015) mTORC1 phosphorylates UVRAG to negatively regulate autophagosome and endosome maturation. Mol Cell 57(2):207–218. https://doi.org/10.1016/j.molcel.2014.11.013

King JS, Gueho A, Hagedorn M, Gopaldass N, Leuba F, Soldati T, Insall RH (2013) WASH is required for lysosomal recycling and efficient autophagic and phagocytic digestion. Mol Biol Cell 24(17):2714–2726. https://doi.org/10.1091/mbc.E13-02-0092

Kirkbride KC, Hong NH, French CL, Clark ES, Jerome WG, Weaver AM (2012) Regulation of late endosomal/lysosomal maturation and trafficking by cortactin affects Golgi morphology. Cytoskeleton 69(9):625–643. https://doi.org/10.1002/cm.21051

Kissing S, Hermsen C, Repnik U, Nesset CK, von Bargen K, Griffiths G, Ichihara A, Lee BS, Schwake M, De Brabander J, Haas A, Saftig P (2015) Vacuolar ATPase in phagosome-lysosome fusion. J Biol Chem 290(22):14166–14180. https://doi.org/10.1074/jbc.M114.628891

Kojima A, Toshima JY, Kanno C, Kawata C, Toshima J (2012) Localization and functional requirement of yeast Na+/H+ exchanger, Nhx1p, in the endocytic and protein recycling pathway. Biochem Biophys Acta 1823(2):534–543. https://doi.org/10.1016/j.bbamcr.2011.12.004

Kriegel A, Andres Z, Medzihradszky A, Kruger F, Scholl S, Delang S, Patir-Nebioglu MG, Gute G, Yang H, Murphy AS, Peer WA, Pfeiffer A, Krebs M, Lohmann JU, Schumacher K (2015) Job sharing in the endomembrane system: vacuolar acidification requires the combined activity of V-ATPase and V-PPase. Plant Cell 27(12):3383–3396. https://doi.org/10.1105/tpc.15.00733

Kummel D, Ungermann C (2014) Principles of membrane tethering and fusion in endosome and lysosome biogenesis. Curr Opin Cell Biol 29:61–66. https://doi.org/10.1016/j.ceb.2014.04.007

Kvainickas A, Jimenez-Orgaz A, Nagele H, Hu Z, Dengjel J, Steinberg F (2017) Cargo-selective SNX-BAR proteins mediate retromer trimer independent retrograde transport. J Cell Biol 216(11):3677–3693. https://doi.org/10.1083/jcb.201702137

Lakadamyali M, Rust MJ, Zhuang X (2006) Ligands for clathrin-mediated endocytosis are differentially sorted into distinct populations of early endosomes. Cell 124(5):997–1009. https://doi.org/10.1016/j.cell.2005.12.038

Law F, Seo JH, Wang Z, DeLeon JL, Bolis Y, Brown A, Zong WX, Du G, Rocheleau CE (2017) The VPS-34 PI3 kinase negatively regulates RAB-5 during endosome maturation. J Cell Sci. https://doi.org/10.1242/jcs.194746

Lawrence G, Brown CC, Flood BA, Karunakaran S, Cabrera M, Nordmann M, Ungermann C, Fratti RA (2014) Dynamic association of the PI3P-interacting Mon1-Ccz1 GEF with vacuoles is controlled through its phosphorylation by the type 1 casein kinase Yck3. Mol Biol Cell 25(10):1608–1619. https://doi.org/10.1091/mbc.E13-08-0460

Lelouvier B, Puertollano R (2011) Mucolipin-3 regulates luminal calcium, acidification, and membrane fusion in the endosomal pathway. J Biol Chem 286(11):9826–9832. https://doi.org/10.1074/jbc.M110.169185

Levin R, Grinstein S, Schlam D (2015) Phosphoinositides in phagocytosis and macropinocytosis. Biochem Biophys Acta 1851(6):805–823. https://doi.org/10.1016/j.bbalip.2014.09.005

Liegeois S, Benedetto A, Garnier JM, Schwab Y, Labouesse M (2006) The V0-ATPase mediates apical secretion of exosomes containing hedgehog-related proteins in *Caenorhabditis elegans*. Journal Cell Biol 173(6):949–961. https://doi.org/10.1083/jcb.200511072

Lin X, Zhang J, Chen L, Chen Y, Xu X, Hong W, Wang T (2017) Tyrosine phosphorylation of Rab7 by Src kinase. Cell Signal 35:84–94. https://doi.org/10.1016/j.cellsig.2017.03.006

Liu P, Bartz R, Zehmer JK, Ying YS, Zhu M, Serrero G, Anderson RG (2007) Rab-regulated interaction of early endosomes with lipid droplets. Biochem Biophys Acta 1773(6):784–793. https://doi.org/10.1016/j.bbamcr.2007.02.004

Liu TT, Gomez TS, Sackey BK, Billadeau DD, Burd CG (2012) Rab GTPase regulation of retromer-mediated cargo export during endosome maturation. Mol Biol Cell 23(13):2505–2515. https://doi.org/10.1091/mbc.E11-11-0915

Liu G, Sanghavi P, Bollinger KE, Perry L, Marshall B, Roon P, Tanaka T, Nakamura A, Gonsalvez GB (2015) Efficient endocytic uptake and maturation in *Drosophila* oocytes requires dynamitin/p50. Genetics 201(2):631–649. https://doi.org/10.1534/genetics.115.180018

Liu J, Ji Y, Zhou J, Xing D (2016) Phosphatidylinositol 3-Kinase promotes activation and vacuolar acidification and delays methyl jasmonate-induced leaf senescence. Plant Physiol 170(3):1714–1731. https://doi.org/10.1104/pp.15.00744

Liu K, Xing R, Jian Y, Gao Z, Ma X, Sun X, Li Y, Xu M, Wang X, Jing Y, Guo W, Yang C (2017) WDR91 is a Rab7 effector required for neuronal development. J Cell Biol 216(10):3307–3321. https://doi.org/10.1083/jcb.201705151

Lloyd-Evans E, Platt FM (2011) Lysosomal Ca(2+) homeostasis: role in pathogenesis of lysosomal storage diseases. Cell Calcium 50(2):200–205. https://doi.org/10.1016/j.ceca.2011.03.010

Lodge R, Descoteaux A (2005) Modulation of phagolysosome biogenesis by the lipophosphoglycan of *Leishmania*. Clin Immunol 114(3):256–265. https://doi.org/10.1016/j.clim.2004.07.018

Luzio JP, Bright NA, Pryor PR (2007) The role of calcium and other ions in sorting and delivery in the late endocytic pathway. Biochem Soc Trans 35(Pt 5):1088–1091. https://doi.org/10.1042/BST0351088

Luzio JP, Gray SR, Bright NA (2010) Endosome-lysosome fusion. Biochem Soc Trans 38(6):1413–1416. https://doi.org/10.1042/BST0381413

MacDonald C, Piper RC (2016) Cell surface recycling in yeast: mechanisms and machineries. Biochem Soc Trans 44(2):474–478. https://doi.org/10.1042/BST20150263

Mani M, Lee UH, Yoon NA, Kim HJ, Ko MS, Seol W, Joe Y, Chung HT, Lee BJ, Moon CH, Cho WJ, Park JW (2016) Developmentally regulated GTP-binding protein 2 coordinates Rab5 activity and transferrin recycling. Mol Biol Cell 27(2):334–348. https://doi.org/10.1091/mbc.E15-08-0558

Marchant JS, Patel S (2015) Two-pore channels at the intersection of endolysosomal membrane traffic. Biochem Soc Trans 43(3):434–441. https://doi.org/10.1042/BST20140303

Marion S, Hoffmann E, Holzer D, Le Clainche C, Martin M, Sachse M, Ganeva I, Mangeat P, Griffiths G (2011) Ezrin promotes actin assembly at the phagosome membrane and regulates phago-lysosomal fusion. Traffic 12(4):421–437. https://doi.org/10.1111/j.1600-0854.2011.01158.x

Marwaha R, Arya SB, Jagga D, Kaur H, Tuli A, Sharma M (2017) The Rab7 effector PLEKHM1 binds Arl8b to promote cargo traffic to lysosomes. J Cell Biol 216(4):1051–1070. https://doi.org/10.1083/jcb.201607085

Matsumoto R, Suzuki K, Ohya Y (2013) Organelle acidification is important for localisation of vacuolar proteins in *Saccharomyces cerevisiae*. Protoplasma 250(6):1283–1293. https://doi.org/10.1007/s00709-013-0510-2

Matsuo H, Chevallier J, Mayran N, Le Blanc I, Ferguson C, Faure J, Blanc NS, Matile S, Dubochet J, Sadoul R, Parton RG, Vilbois F, Gruenberg J (2004) Role of LBPA and Alix in multivesicular liposome formation and endosome organization. Science 303(5657):531–534. https://doi.org/10.1126/science.1092425

Mauvezin C, Nagy P, Juhasz G, Neufeld TP (2015) Autophagosome-lysosome fusion is independent of V-ATPase-mediated acidification. Nat Commun 6:7007. https://doi.org/10.1038/ncomms8007

Maxson ME, Grinstein S (2014) The vacuolar-type H(+)-ATPase at a glance—more than a proton pump. J Cell Sci 127(Pt 23):4987–4993. https://doi.org/10.1242/jcs.158550

Mayor S, Pagano RE (2007) Pathways of clathrin-independent endocytosis. Nat Rev Mol Cell Biol 8(8):603–612. https://doi.org/10.1038/nrm2216

McDermott H, Kim K (2015) Molecular dynamics at the endocytic portal and regulations of endocytic and recycling traffics. Eur J Cell Biol 94(6):235–248. https://doi.org/10.1016/j.ejcb.2015.04.003

McKnight NC, Zhong Y, Wold MS, Gong S, Phillips GR, Dou Z, Zhao Y, Heintz N, Zong WX, Yue Z (2014) Beclin 1 is required for neuron viability and regulates endosome pathways via the UVRAG-VPS34 complex. PLoS Genet 10(10):e1004626. https://doi.org/10.1371/journal.pgen.1004626

McNally KE, Faulkner R, Steinberg F, Gallon M, Ghai R, Pim D, Langton P, Pearson N, Danson CM, Nagele H, Morris LL, Singla A, Overlee BL, Heesom KJ, Sessions R, Banks L, Collins BM, Berger I, Billadeau DD, Burstein E, Cullen PJ (2017) Retriever is a multiprotein complex for retromer-independent endosomal cargo recycling. Nat Cell Biol 19(10):1214–1225. https://doi.org/10.1038/ncb3610

Medina DL, Di Paola S, Peluso I, Armani A, De Stefani D, Venditti R, Montefusco S, Scotto-Rosato A, Prezioso C, Forrester A, Settembre C, Wang W, Gao Q, Xu H, Sandri M, Rizzuto R, De Matteis MA, Ballabio A (2015) Lysosomal calcium signalling regulates autophagy through calcineurin and TFEB. Nat Cell Biol 17(3):288–299. https://doi.org/10.1038/ncb3114

Menard L, Parker PJ, Kermorgant S (2014) Receptor tyrosine kinase c-Met controls the cytoskeleton from different endosomes via different pathways. Nat Commun 5:3907. https://doi.org/10.1038/ncomms4907

Merz AJ (2015) What are the roles of V-ATPases in membrane fusion? Proc Natl Acad Sci USA 112(1):8–9. https://doi.org/10.1073/pnas.1422280112
Min SH, Suzuki A, Stalker TJ, Zhao L, Wang Y, McKennan C, Riese MJ, Guzman JF, Zhang S, Lian L, Joshi R, Meng R, Seeholzer SH, Choi JK, Koretzky G, Marks MS, Abrams CS (2014) Loss of PIKfyve in platelets causes a lysosomal disease leading to inflammation and thrombosis in mice. Nat Commun 5:4691. https://doi.org/10.1038/ncomms5691
Miura E, Hasegawa T, Konno M, Suzuki M, Sugeno N, Fujikake N, Geisler S, Tabuchi M, Oshima R, Kikuchi A, Baba T, Wada K, Nagai Y, Takeda A, Aoki M (2014) VPS35 dysfunction impairs lysosomal degradation of α-synuclein and exacerbates neurotoxicity in a *Drosophila* model of Parkinson's disease. Neurobiol Dis 71:1–13. https://doi.org/10.1016/j.nbd.2014.07.014
Modica G, Lefrancois S (2017) Post-translational modifications: how to modulate Rab7 functions. Small GTPases:0. https://doi.org/10.1080/21541248.2017.1387686
Modica G, Skorobogata O, Sauvageau E, Vissa A, Yip CM, Kim PK, Wurtele H, Lefrancois S (2017) Rab7 palmitoylation is required for efficient endosome-to-TGN trafficking. J Cell Sci 130(15):2579–2590. https://doi.org/10.1242/jcs.199729
Muriel O, Tomas A, Scott CC, Gruenberg J (2016) Moesin and cortactin control actin-dependent multivesicular endosome biogenesis. Mol Biol Cell 27(21):3305–3316. https://doi.org/10.1091/mbc.E15-12-0853
Nagel BM, Bechtold M, Rodriguez LG, Bogdan S (2017) *Drosophila* WASH is required for integrin-mediated cell adhesion, cell motility and lysosomal neutralization. J Cell Sci 130(2):344–359. https://doi.org/10.1242/jcs.193086
Nannapaneni S, Wang D, Jain S, Schroeder B, Highfill C, Reustle L, Pittsley D, Maysent A, Moulder S, McDowell R, Kim K (2010) The yeast dynamin-like protein Vps1: vps1 mutations perturb the internalization and the motility of endocytic vesicles and endosomes via disorganization of the actin cytoskeleton. Eur J Cell Biol 89(7):499–508. https://doi.org/10.1016/j.ejcb.2010.02.002
Naufer A, Hipolito VEB, Ganesan S, Prashar A, Zaremberg V, Botelho RJ, Terebiznik MR (2017) pH of endophagosomes controls association of their membranes with Vps34 and PtdIns(3)P levels. J Cell Biol. https://doi.org/10.1083/jcb.201702179
Nicot AS, Fares H, Payrastre B, Chisholm AD, Labouesse M, Laporte J (2006) The phosphoinositide kinase PIKfyve/Fab1p regulates terminal lysosome maturation in *Caenorhabditis elegans*. Mol Biol Cell 17(7):3062–3074. https://doi.org/10.1091/mbc.E06-12-1120
Nicot AS, Laporte J (2008) Endosomal phosphoinositides and human diseases. Traffic 9(8):1240–1249. https://doi.org/10.1111/j.1600-0854.2008.00754.x
Nielsen E, Severin F, Backer JM, Hyman AA, Zerial M (1999) Rab5 regulates motility of early endosomes on microtubules. Nat Cell Biol 1(6):376–382. https://doi.org/10.1038/14075
Nordmann M, Cabrera M, Perz A, Brocker C, Ostrowicz C, Engelbrecht-Vandre S, Ungermann C (2010) The Mon1-Ccz1 complex is the GEF of the late endosomal Rab7 homolog Ypt7. Curr Biol CB 20(18):1654–1659. https://doi.org/10.1016/j.cub.2010.08.002
Pankiv S, Alemu EA, Brech A, Bruun JA, Lamark T, Overvatn A, Bjorkoy G, Johansen T (2010) FYCO1 is a Rab7 effector that binds to LC3 and PI3P to mediate microtubule plus end-directed vesicle transport. J Cell Biol 188(2):253–269. https://doi.org/10.1083/jcb.200907015
Pei G, Repnik U, Griffiths G, Gutierrez MG (2014) Identification of an immune-regulated phagosomal Rab cascade in macrophages. J Cell Sci 127(Pt 9):2071–2082. https://doi.org/10.1242/jcs.144923
Pena-Llopis S, Vega-Rubin-de-Celis S, Schwartz JC, Wolff NC, Tran TA, Zou L, Xie XJ, Corey DR, Brugarolas J (2011) Regulation of TFEB and V-ATPases by mTORC1. EMBO J 30(16):3242–3258. https://doi.org/10.1038/emboj.2011.257
Peters LZ, Hazan R, Breker M, Schuldiner M, Ben-Aroya S (2013) Formation and dissociation of proteasome storage granules are regulated by cytosolic pH. J Cell Biol 201(5):663–671. https://doi.org/10.1083/jcb.201211146
Piccinotti S, Kirchhausen T, Whelan SP (2013) Uptake of rabies virus into epithelial cells by clathrin-mediated endocytosis depends upon actin. J Virol 87(21):11637–11647. https://doi.org/10.1128/JVI.01648-13

Pizarro-Cerda J, Kuhbacher A, Cossart P (2015) Phosphoinositides and host-pathogen interactions. Biochem Biophys Acta 1851(6):911–918. https://doi.org/10.1016/j.bbalip.2014.09.011
Posor Y, Eichhorn-Gruenig M, Puchkov D, Schoneberg J, Ullrich A, Lampe A, Muller R, Zarbakhsh S, Gulluni F, Hirsch E, Krauss M, Schultz C, Schmoranzer J, Noe F, Haucke V (2013) Spatiotemporal control of endocytosis by phosphatidylinositol-3,4-bisphosphate. Nature 499(7457):233–237. https://doi.org/10.1038/nature12360
Posor Y, Eichhorn-Grunig M, Haucke M (2015) Phosphoinositides in endocytosis. Biochem Biophys Acta 1851(6):794–804. https://doi.org/10.1016/j.bbalip.2014.09.014
Poteryaev D, Datta S, Ackema K, Zerial M, Spang A (2010) Identification of the switch in early-to-late endosome transition. Cell 141(3):497–508. https://doi.org/10.1016/j.cell.2010.03.011
Prasad H, Rao R (2015) The Na+/H+exchanger NHE6 modulates endosomal pH to control processing of amyloid precursor protein in a cell culture model of Alzheimer disease. J Biol Chem 290(9):5311–5327. https://doi.org/10.1074/jbc.M114.602219
Progida C, Bakke O (2016) Bidirectional traffic between the Golgi and the endosomes—machineries and regulation. J Cell Sci 129(21):3971–3982. https://doi.org/10.1242/jcs.185702
Progida C, Cogli L, Piro F, De Luca A, Bakke O, Bucci C (2010) Rab7b controls trafficking from endosomes to the TGN. J Cell Sci 123(Pt 9):1480–1491. https://doi.org/10.1242/jcs.051474
Purushothaman LK, Arlt H, Kuhlee A, Raunser S, Ungermann C (2017) Retromer-driven membrane tubulation separates endosomal recycling from Rab7/Ypt7-dependent fusion. Mol Biol Cell 28(6):783–791. https://doi.org/10.1091/mbc.E16-08-0582
Raiborg C, Wenzel EM, Pedersen NM, Olsvik H, Schink KO, Schultz SW, Vietri M, Nisi V, Bucci C, Brech A, Johansen T, Stenmark H (2015) Repeated ER-endosome contacts promote endosome translocation and neurite outgrowth. Nature 520(7546):234–238. https://doi.org/10.1038/nature14359
Ramanathan HN, Zhang G, Ye Y (2013) Monoubiquitination of EEA1 regulates endosome fusion and trafficking. Cell Biosci 3(1):24. https://doi.org/10.1186/2045-3701-3-24
Rana M, Lachmann J, Ungermann C (2015) Identification of a Rab GTPase-activating protein cascade that controls recycling of the Rab5 GTPase Vps21 from the vacuole. Mol Biol Cell 26(13):2535–2549. https://doi.org/10.1091/mbc.E15-02-0062
Rink J, Ghigo E, Kalaidzidis Y, Zerial M (2005) Rab conversion as a mechanism of progression from early to late endosomes. Cell 122(5):735–749. https://doi.org/10.1016/j.cell.2005.06.043
Rowland AA, Chitwood PJ, Phillips MJ, Voeltz GK (2014) ER contact sites define the position and timing of endosome fission. Cell 159(5):1027–1041. https://doi.org/10.1016/j.cell.2014.10.023
Sandhoff K (2016) Neuronal sphingolipidoses: membrane lipids and sphingolipid activator proteins regulate lysosomal sphingolipid catabolism. Biochimie 130:146–151. https://doi.org/10.1016/j.biochi.2016.05.004
Sandri C, Caccavari F, Valdembri D, Camillo C, Veltel S, Santambrogio M, Lanzetti L, Bussolino F, Ivaska J, Serini G (2012) The R-Ras/RIN2/Rab5 complex controls endothelial cell adhesion and morphogenesis via active integrin endocytosis and Rac signaling. Cell Res 22(10):1479–1501. https://doi.org/10.1038/cr.2012.110
Santiago-Tirado FH, Bretscher A (2011) Membrane-trafficking sorting hubs: cooperation between PI4P and small GTPases at the *trans*-Golgi network. Trends Cell Biol 21(9):515–525. https://doi.org/10.1016/j.tcb.2011.05.005
Sarantis H, Grinstein S (2012) Subversion of phagocytosis for pathogen survival. Cell Host Microbe 12(4):419–431. https://doi.org/10.1016/j.chom.2012.09.001
Satoh N, Yamada H, Yamazaki O, Suzuki M, Nakamura M, Suzuki A, Ashida A, Yamamoto D, Kaku Y, Sekine T, Seki G, Horita S (2016) A pure chloride channel mutant of CLC-5 causes Dent's disease via insufficient V-ATPase activation. Pflugers Arch 468(7):1183–1196. https://doi.org/10.1007/s00424-016-1808-7
Sautin YY, Lu M, Gaugler A, Zhang L, Gluck SL (2005) Phosphatidylinositol 3-kinase-mediated effects of glucose on vacuolar H+-ATPase assembly, translocation, and acidification of intracellular compartments in renal epithelial cells. Mol Cell Biol 25(2):575–589. https://doi.org/10.1128/MCB.25.2.575-589.2005

Scheuring D, Viotti C, Kruger F, Kunzl F, Sturm S, Bubeck J, Hillmer S, Frigerio L, Robinson DG, Pimpl P, Schumacher K (2011) Multivesicular bodies mature from the *trans*-Golgi network/early endosome in *Arabidopsis*. Plant Cell 23(9):3463–3481. https://doi.org/10.1105/tpc.111.086918

Schindler C, Chen Y, Pu J, Guo X, Bonifacino JS (2015) EARP is a multisubunit tethering complex involved in endocytic recycling. Nat Cell Biol 17(5):639–650. https://doi.org/10.1038/ncb3129

Schink KO, Raiborg C, Stenmark H (2013) Phosphatidylinositol 3-phosphate, a lipid that regulates membrane dynamics, protein sorting and cell signalling. BioEssays News Rev Mol Cell Dev Biol 35(10):900–912. https://doi.org/10.1002/bies.201300064

Schlam D, Bagshaw RD, Freeman SA, Collins RF, Pawson T, Fairn GD, Grinstein S (2015) Phosphoinositide 3-kinase enables phagocytosis of large particles by terminating actin assembly through Rac/Cdc42 GTPase-activating proteins. Nat Commun 6:8623. https://doi.org/10.1038/ncomms9623

Schmidt O, Teis D (2012) The ESCRT machinery. Current Biol CB 22(4):R116–R120. https://doi.org/10.1016/j.cub.2012.01.028

Schmitz G, Robenek H, Lohmann U, Assmann G (1985) Interaction of high density lipoproteins with cholesteryl ester-laden macrophages: biochemical and morphological characterization of cell surface receptor binding, endocytosis and resecretion of high density lipoproteins by macrophages. EMBO J 4(3):613–622

Schreij AM, Fon EA, McPherson PS (2016) Endocytic membrane trafficking and neurodegenerative disease. Cell Mol Life Sci CMLS 73(8):1529–1545. https://doi.org/10.1007/s00018-015-2105-x

Seaman MN, Harbour ME, Tattersall D, Read E, Bright N (2009) Membrane recruitment of the cargo-selective retromer subcomplex is catalysed by the small GTPase Rab7 and inhibited by the Rab-GAP TBC1D5. J Cell Sci 122(Pt 14):2371–2382. https://doi.org/10.1242/jcs.048686

Shaw JD, Hama H, Sohrabi F, DeWald DB, Wendland B (2003) PtdIns(3,5)P2 is required for delivery of endocytic cargo into the multivesicular body. Traffic 4(7):479–490

Sherr GL, LaMassa N, Li E, Phillips G, Shen CH (2017) Pah1p negatively regulates the expression of V-ATPase genes as well as vacuolar acidification. Biochem Biophys Res Commun 491(3):693–700. https://doi.org/10.1016/j.bbrc.2017.07.127

Shin HW, Hayashi M, Christoforidis S, Lacas-Gervais S, Hoepfner S, Wenk MR, Modregger J, Uttenweiler-Joseph S, Wilm M, Nystuen A, Frankel WN, Solimena M, De Camilli P, Zerial M (2005) An enzymatic cascade of Rab5 effectors regulates phosphoinositide turnover in the endocytic pathway. J Cell Biol 170(4):607–618. https://doi.org/10.1083/jcb.200505128

Shin D, Na W, Lee JH, Kim G, Baek J, Park SH, Choi CY, Lee S (2017) Site-specific monoubiquitination downregulates Rab5 by disrupting effector binding and guanine nucleotide conversion. eLife 6. https://doi.org/10.7554/elife.29154

Shinde SR, Maddika S (2016) PTEN modulates EGFR late endocytic trafficking and degradation by dephosphorylating Rab7. Nat Commun 7:10689. https://doi.org/10.1038/ncomms10689

Sigismund S, Argenzio E, Tosoni D, Cavallaro E, Polo S, Di Fiore PP (2008) Clathrin-mediated internalization is essential for sustained EGFR signaling but dispensable for degradation. Dev Cell 15(2):209–219. https://doi.org/10.1016/j.devcel.2008.06.012

Simonetti B, Danson CM, Heesom KJ, Cullen PJ (2017) Sequence-dependent cargo recognition by SNX-BARs mediates retromer-independent transport of CI-MPR. J Cell Biol 216(11):3695–3712. https://doi.org/10.1083/jcb.201703015

Smardon AM, Diab HI, Tarsio M, Diakov TT, Nasab ND, West RW, Kane PM (2014) The RAVE complex is an isoform-specific V-ATPase assembly factor in yeast. Mol Biol Cell 25(3):356–367. https://doi.org/10.1091/mbc.E13-05-0231

Smith GA, Howell GJ, Phillips C, Muench SP, Ponnambalam S, Harrison MA (2016) Extracellular and luminal pH regulation by vacuolar H+-ATPase isoform expression and targeting to the plasma membrane and endosomes. J Biol Chem 291(16):8500–8515. https://doi.org/10.1074/jbc.M116.723395

Sobota JA, Back N, Eipper BA, Mains RE (2009) Inhibitors of the V0 subunit of the vacuolar H+-ATPase prevent segregation of lysosomal- and secretory-pathway proteins. J Cell Sci 122(Pt 19):3542–3553. https://doi.org/10.1242/jcs.034298

Solinger JA, Spang A (2013) Tethering complexes in the endocytic pathway: CORVET and HOPS. FEBS J 280(12):2743–2757. https://doi.org/10.1111/febs.12151

Solinger JA, Spang A (2014) Loss of the Sec1/Munc18-family proteins VPS-33.2 and VPS-33.1 bypasses a block in endosome maturation in *Caenorhabditis elegans*. Mol Biol Cell 25(24):3909–3925. https://doi.org/10.1091/mbc.E13-12-0710

Soria FN, Pampliega O, Bourdenx M, Meissner WG, Bezard E, Dehay B (2017) Exosomes, an unmasked culprit in neurodegenerative diseases. Front Neurosci 11:26. https://doi.org/10.3389/fnins.2017.00026

Spang A (2015) The road not taken: less traveled roads from the TGN to the plasma membrane. Membranes 5(1):84–98. https://doi.org/10.3390/membranes5010084

Spang A (2016) Membrane tethering complexes in the endosomal system. Front Cell Dev Biol 4:35. https://doi.org/10.3389/fcell.2016.00035

Sreelatha A, Bennett TL, Carpinone EM, O'Brien KM, Jordan KD, Burdette DL, Orth K, Starai VJ (2015) Vibrio effector protein VopQ inhibits fusion of V-ATPase-containing membranes. Proc Natl Acad Sci USA 112(1):100–105. https://doi.org/10.1073/pnas.1413764111

Stasyk T, Huber LA (2016) Spatio-temporal parameters of endosomal signaling in cancer: implications for new treatment options. J Cell Biochem 117(4):836–843. https://doi.org/10.1002/jcb.25418

Stransky LA, Forgac M (2015) Amino acid availability modulates vacuolar H+-ATPase assembly. J Biol Chem 290(45):27360–27369. https://doi.org/10.1074/jbc.M115.659128

Sun-Wada GH, Tabata H, Kawamura N, Aoyama M, Wada Y (2009) Direct recruitment of H+-ATPase from lysosomes for phagosomal acidification. J Cell Sci 122(Pt 14):2504–2513. https://doi.org/10.1242/jcs.050443

Sun Q, Westphal W, Wong KN, Tan I, Zhong Q (2010) Rubicon controls endosome maturation as a Rab7 effector. Proc Natl Acad Sci USA 107(45):19338–19343. https://doi.org/10.1073/pnas.1010554107

Szatmari Z, Kis V, Lippai M, Hegedus K, Farago T, Lorincz P, Tanaka T, Juhasz G, Sass M (2014) Rab11 facilitates cross-talk between autophagy and endosomal pathway through regulation of Hook localization. Mol Biol Cell 25(4):522–531. https://doi.org/10.1091/mbc.E13-10-0574

Takatori S, Tatematsu T, Cheng J, Matsumoto J, Akano T, Fujimoto T (2016) Phosphatidylinositol 3,5-bisphosphate-rich membrane domains in endosomes and lysosomes. Traffic 17(2):154–167. https://doi.org/10.1111/tra.12346

Tan X, Thapa N, Choi S, Anderson RA (2015) Emerging roles of PtdIns(4,5)P2–beyond the plasma membrane. J Cell Sci 128(22):4047–4056. https://doi.org/10.1242/jcs.175208

Tang BL (2016) Rab, Arf, and Arl-regulated membrane traffic in cortical neuron migration. J Cell Physiol 231(7):1417–1423. https://doi.org/10.1002/jcp.25261

Teixeira V, Medeiros TC, Vilaca R, Ferreira J, Moradas-Ferreira P, Costa V (2016) Ceramide signaling targets the PP2A-like protein phosphatase Sit4p to impair vacuolar function, vesicular trafficking and autophagy in Isc1p deficient cells. Biochem Biophys Acta 1861(1):21–33. https://doi.org/10.1016/j.bbalip.2015.10.004

Tornieri K, Zlatic SA, Mullin AP, Werner E, Harrison R, L'Hernault SW, Faundez V (2013) Vps33b pathogenic mutations preferentially affect VIPAS39/SPE-39-positive endosomes. Hum Mol Genet 22(25):5215–5228. https://doi.org/10.1093/hmg/ddt378

Urbanska A, Sadowski L, Kalaidzidis Y, Miaczynska M (2011) Biochemical characterization of APPL endosomes: the role of annexin A2 in APPL membrane recruitment. Traffic 12(9):1227–1241. https://doi.org/10.1111/j.1600-0854.2011.01226.x

Vale-Costa S, Amorim MJ (2016) Recycling endosomes and viral infection. Viruses 8(3):64. https://doi.org/10.3390/v8030064

van der Kant R, Neefjes J (2014) Small regulators, major consequences—Ca(2)(+) and cholesterol at the endosome-ER interface. J Cell Sci 127(Pt 5):929–938. https://doi.org/10.1242/jcs.137539

van der Kant R, Jonker CT, Wijdeven RH, Bakker J, Janssen L, Klumperman J, Neefjes J (2015) Characterization of the mammalian CORVET and HOPS Complexes and their modular restruc-

turing for endosome specificity. J Biol Chem 290(51):30280–30290. https://doi.org/10.1074/jbc.M115.688440
van Deurs B, Holm PK, Sandvig K (1996) Inhibition of the vacuolar H(+)-ATPase with bafilomycin reduces delivery of internalized molecules from mature multivesicular endosomes to lysosomes in HEp-2 cells. Eur J Cell Biol 69(4):343–350
van Weering JR, Verkade P, Cullen PJ (2012) SNX-BAR-mediated endosome tubulation is co-ordinated with endosome maturation. Traffic 13(1):94–107. https://doi.org/10.1111/j.1600-0854.2011.01297.x
Vieira OV, Bucci C, Harrison RE, Trimble WS, Lanzetti L, Gruenberg J, Schreiber AD, Stahl PD, Grinstein S (2003) Modulation of Rab5 and Rab7 recruitment to phagosomes by phosphatidylinositol 3-kinase. Mol Cell Biol 23(7):2501–2514
Villasenor R, Kalaidzidis Y, Zerial M (2016) Signal processing by the endosomal system. Curr Opin Cell Biol 39:53–60. https://doi.org/10.1016/j.ceb.2016.02.002
Wang C, Wang Y, Li Y, Bodemann B, Zhao T, Ma X, Huang G, Hu Z, DeBerardinis RJ, White MA, Gao J (2015) A nanobuffer reporter library for fine-scale imaging and perturbation of endocytic organelles. Nat Commun 6:8524. https://doi.org/10.1038/ncomms9524
Waugh MG (2015) PIPs in neurological diseases. Biochem Biophys Acta 1851(8):1066–1082. https://doi.org/10.1016/j.bbalip.2015.02.002
Williamson WR, Hiesinger PR (2010) On the role of v-ATPase V0a1-dependent degradation in Alzheimer disease. Communicative Integr Biol 3(6):604–607. https://doi.org/10.4161/cib.3.6.13364
Wong D, Bach H, Sun J, Hmama Z, Av-Gay Y (2011) *Mycobacterium* tuberculosis protein tyrosine phosphatase (PtpA) excludes host vacuolar-H+-ATPase to inhibit phagosome acidification. Proc Natl Acad Sci USA 108(48):19371–19376. https://doi.org/10.1073/pnas.1109201108
Xinhan L, Matsushita M, Numaza M, Taguchi A, Mitsui K, Kanazawa H (2011) Na+/H+ exchanger isoform 6 (NHE6/SLC9A6) is involved in clathrin-dependent endocytosis of transferrin. Am J Physiol Cell Physiol 301(6):C1431–C1444. https://doi.org/10.1152/ajpcell.00154.2011
Xu T, Forgac M (2001) Microtubules are involved in glucose-dependent dissociation of the yeast vacuolar [H+]-ATPase in vivo. J Biol Chem 276(27):24855–24861. https://doi.org/10.1074/jbc.M100637200
Xu L, Shen X, Bryan A, Banga S, Swanson MS, Luo ZQ (2010) Inhibition of host vacuolar H+-ATPase activity by a *Legionella pneumophila* effector. PLoS Pathog 6(3):e1000822. https://doi.org/10.1371/journal.ppat.1000822
Yasuda S, Morishita S, Fujita A, Nanao T, Wada N, Waguri S, Schiavo G, Fukuda M, Nakamura T (2016) Mon1-Ccz1 activates Rab7 only on late endosomes and dissociates from the lysosome in mammalian cells. J Cell Sci 129(2):329–340. https://doi.org/10.1242/jcs.178095
Yousefian J, Troost T, Grawe F, Sasamura T, Fortini M, Klein T (2013) Dmon1 controls recruitment of Rab7 to maturing endosomes in Drosophila. J Cell Sci 126(Pt 7):1583–1594. https://doi.org/10.1242/jcs.114934
Zajac AL, Goldman YE, Holzbaur EL, Ostap EM (2013) Local cytoskeletal and organelle interactions impact molecular-motor- driven early endosomal trafficking. Current Biol CB 23(13):1173–1180. https://doi.org/10.1016/j.cub.2013.05.015
Zeigerer A, Gilleron J, Bogorad RL, Marsico G, Nonaka H, Seifert S, Epstein-Barash H, Kuchimanchi S, Peng CG, Ruda VM, Del Conte-Zerial P, Hengstler JG, Kalaidzidis Y, Koteliansky V, Zerial M (2012) Rab5 is necessary for the biogenesis of the endolysosomal system in vivo. Nature 485(7399):465–470. https://doi.org/10.1038/nature11133
Zerial M, McBride H (2001) Rab proteins as membrane organizers. Nat Rev Mol Cell Biol 2(2):107–117. https://doi.org/10.1038/35052055
Zhou A, Bu Y, Takano T, Zhang X, Liu S (2016) Conserved V-ATPase c subunit plays a role in plant growth by influencing V-ATPase-dependent endosomal trafficking. Plant Biotechnol J 14(1):271–283. https://doi.org/10.1111/pbi.12381

Zhu H, Liang Z, Li G (2009) Rabex-5 is a Rab22 effector and mediates a Rab22-Rab5 signaling cascade in endocytosis. Mol Biol Cell 20(22):4720–4729. https://doi.org/10.1091/mbc.E09-06-0453

Zoncu R, Bar-Peled L, Efeyan A, Wang S, Sancak Y, Sabatini DM (2011) mTORC1 senses lysosomal amino acids through an inside-out mechanism that requires the vacuolar H(+)-ATPase. Science 334(6056):678–683. https://doi.org/10.1126/science.1207056

Chapter 2
Integration of the Endocytic System into the Network of Cellular Functions

Noga Budick-Harmelin and Marta Miaczynska

Contents

Abstract Maintenance of physiologic cellular functions and homeostasis requires highly coordinated interactions between different cellular compartments. In this regard, the endocytic system, which plays a key role in cargo internalization and trafficking within the cell, participates in upkeep of intracellular dynamics, while communicating with multiple organelles. This chapter will discuss the function of endosomes from a standpoint of cellular integration. We will present examples of different types of interactions between endosomes and other cellular compartments, such as the endoplasmic reticulum (ER), mitochondria, the plasma membrane (PM), and the nuclear envelope. In addition, we will describe the incorporation of endocytic components, such as endosomal sorting complexes required for transport (ESCRT) proteins and Rab small GTPases, into cellular processes that operate outside of the

N. Budick-Harmelin · M. Miaczynska (✉)
Laboratory of Cell Biology, International Institute of Molecular and Cell Biology, Księcia Trojdena 4, 02-109 Warsaw, Poland
e-mail: miaczynska@iimcb.gov.pl

N. Budick-Harmelin (✉)
Cell Research and Immunology Department, George S. Wise Faculty of Life Sciences, Tel Aviv University, Tel Aviv, Israel
e-mail: nogabu@tauex.tau.ac.il

C. Lamaze and I. Prior (eds.), *Endocytosis and Signaling*, Progress in Molecular and Subcellular Biology 57, https://doi.org/10.1007/978-3-319-96704-2_2

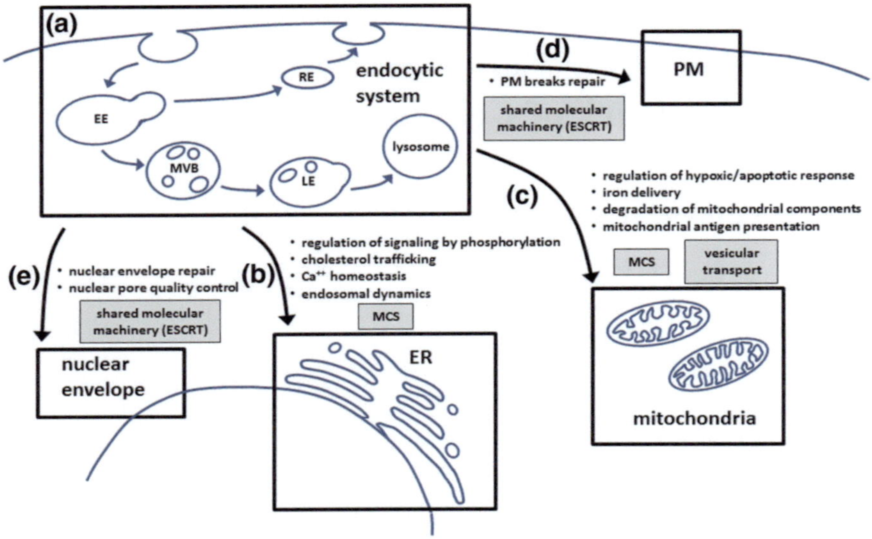

Fig. 2.1 Functional implications (bulleted lists) and the mechanisms involved (gray-filled boxes) of endosomal interactions with other cellular compartments. Selected examples for communication between organelles of the endocytic pathway (**a**) and the ER (**b**), mitochondria (**c**), PM (**d**), or the nuclear envelope (**e**)

endolysosomal pathway. The significance of endosomal interactions for processes such as signaling regulation, intracellular trafficking, organelle dynamics, metabolic control, and homeostatic responses will be reviewed. Accumulating data indicate that beyond its involvement in cargo transport, the endocytic pathway is comprehensively integrated into other systems of the cell and plays multiple roles in the complex net of cellular functions.

2.1 Introduction

The endolysosomal system is the cellular machinery specialized in uptake of (macro) molecules and sorting of the engulfed cargo either for degradation or for recycling back to the plasma membrane (PM). Hence, the two destinations of cargo internalized by membrane-bound vesicles within the endocytic compartments are well defined. Cargo designated for degradation travels within early endosomes (EE) during their maturation into late endosomes (LE) that fuse with lysosomes. Recycling back to the plasma membrane occurs by transport of cargo either directly from EE or via recycling endosomes (RE) (Fig. 2.1a).

However, when looking globally at cellular functions, accumulated data indicate that the endocytic system plays a broader role than simply serving as a trafficking and sorting platform. Endosomes are dynamic organelles, as they continuously move from budding sites at the PM toward the cell center, and constantly change their lipid and protein composition due to sorting, fusion, and fission events. Along this pathway, in the compact intracellular milieu, endosomes interact with different cellular systems and organelles at multiple levels and mediate various processes (Fig. 2.1b–e). In general, membrane contact sites (MCS) between the endomembrane system and other compartments serve as hubs for inter-organellar communication. Fission and fusion of endosomes occur in contact with other organelles. Moreover, endosomal proteins can also act in different cellular compartments and participate in processes that take place outside of the endocytic system, a phenomenon known as "moonlighting" (Huberts and van der Klei 2010; Jeffery 1999).

The cooperation between endosomes and other cellular compartments governs balanced membrane flow and endosome dynamics, metabolic control, cell signaling, transcription regulation, and other homeostasis-related functions. In this chapter, we will discuss primarily two topics demonstrating the function of endosomes as an integral part of the organellar system of a cell: (1) cross talk between the endosomal system and other organelles, and (2) the role of endosomes in regulation of metabolic pathways. While we are unable to comprehensively cover these topics within the scope of this chapter, we will provide a selection of examples highlighting the involvement of endosomes in homeostatic and adaptive functions of a cell.

2.2 Cross Talk Between the Endosomal System and Other Organelles

Various cellular processes are compartmentalized in different subcellular organelles, which ensure their efficient proceeding but at the same time poses the need for their proper coordination with each other. One mechanism of such inter-organellar communication involves MCS, which are microdomains of close apposition between two organelles. In these sites, close proximity of membranes facilitates communication by signaling and exchange of molecules between the two compartments (Prinz 2014).

By interactions with various cellular compartments, endosomes contribute to maintenance of inter-organellar cross talk, coordinated signaling and metabolic reactions, membrane dynamics and organelle identity [discussed in (Henne 2016)]. Indeed, the sites of interaction between endosomal membranes and other cellular membranes serve as "hot spots" for signal transduction and membrane remodeling events that affect structure and function of the involved organelles. To demonstrate how endosomes are integrated within the organellar system of a cell to mediate homeostatic and adaptive responses, we will discuss interactions of the endocytic system with the endoplasmic reticulum (ER) and with mitochondria and describe the role of endosomal sorting complexes required for transport (ESCRT) proteins in preserving membrane integrity.

2.2.1 *Endosomal-ER Contact Sites Regulate Multiple Cellular Processes*

The ER is a network of branched cisternae and tubules that extensively interact with other membranous organelles in the cell by various mechanisms. Communication of the endolysosomal system with the ER is facilitated by several types of MCS that differ in their molecular composition and functionality [reviewed in (Eden 2016; Raiborg et al. 2015b)]. Although the roles of endosome-ER MCS are not fully understood, they clearly influence a number of cellular processes to be mentioned below (Fig. 2.1b).

Regulation of receptor signaling. Interactions of endosomes with the ER are suggested to negatively regulate signaling from receptors internalized by the endosomal system. This was shown to be mediated by an ER-associated protein tyrosine phosphatase 1B (PTP1B) (Prinz 2014) that dephosphorylates two receptors located on endosomes: the cytokine receptor granulocyte colony-stimulating factor receptor (G-CSFR) on EE (Palande et al. 2011) and epidermal growth factor receptor (EGFR) on LE (Eden et al. 2010), a member of the receptor tyrosine kinases (RTK) family. The interaction of PTP1B with the RTK on the ER was found to be endocytosis-dependent (Haj et al. 2002) and seems to occur at the endosome-ER MCS, where the close apposition of two organelles allows such interaction. Since various RTK and cytokine receptors can still signal from endosomes after their internalization (Cendrowski et al. 2016; Platta and Stenmark 2011), this direct deactivation of receptors on endosomes may be important for controlling the magnitude and length of signaling transduced after the pathway stimulation. Furthermore, PTP1B decreased phosphorylation of G-CSFR even in the absence of its ligand (Palande et al. 2011), implying that interaction with the ER plays a role also in balancing basal signaling from non-activated receptors on endosomes.

In contrast, ER-associated PTP1B may also augment signaling of endosomal receptors in a non-receptor specific manner, by inhibiting activity of the ESCRT machinery. ESCRT components are recruited to the endosomal membrane for targeting internalized receptors toward lysosomal degradation. Two ESCRT components, hepatocyte growth factor-regulated tyrosine kinase substrate (HRS) and signal-transducing adaptor molecule 2 (STAM2), were identified as substrates of PTP1B (Eden et al. 2010; Stuible et al. 2010). Hence, decreased activity of the ESCRT machinery after dephosphorylation of its components can attenuate degradation and increase the signaling propagated from internalized receptors on endosomes.

Cholesterol transport. Another role of the endosome-ER communication is to coordinate transfer of metabolites, ions, and proteins between these compartments, which is of major importance for regulation of signaling and metabolic status at the whole-cell level. In this regard, the regulation of endosome-ER transfer of cholesterol will be discussed under the topic of "metabolic pathways regulation" below.

Ca^{2+} homeostasis. Cooperation between the endolysosomal system and the ER is important also for inter-organellar mobilization of Ca^{2+} ions and regulation of Ca^{2+} release to the cytoplasm. Ca^{2+} is a universal second messenger in many fundamental signaling cascades, and hence, its levels have to be tightly controlled in a spatiotemporal manner. Both the ER and lysosomes serve as cellular storage compartments of Ca^{2+}, and data from different reports imply that Ca^{2+} release from both these organelles is coordinated [discussed in (Eden 2016; Lam and Galione 2013)]. Experimental data support the "trigger hypothesis," arguing that a relatively small wave of Ca^{2+} that is released from lysosomes is required for activation of prominent Ca^{2+} efflux from the ER (Kilpatrick et al. 2013).

Interestingly, during Ca^{2+} signaling, Ca^{2+} is transferred also from the ER to lysosomes and this Ca^{2+} shuttling is assumed to be mediated via endosome-ER MCS (Morgan et al. 2013). Recently, it has even been proposed that the ER, rather than the pH gradient, is actually the primary source of Ca^{2+} for the lysosome (Garrity et al. 2016). This idea is supported by the fact that inhibition of Ca^{2+} uptake by the lysosome induced Ca^{2+} release from the ER (Lopez-Sanjurjo et al. 2013). The Ca^{2+} ions accumulated in lysosomes are rapidly returned to the ER (Lopez-Sanjurjo et al. 2014), indicating efficient bidirectional coordination between the compartments. Hence, cross talk between the ER and lysosomes seems to participate in maintenance of Ca^{2+} homeostasis of the cell.

Regulation of endosomal dynamics. In addition, the spatial relationship between endosomes and the ER modulates the dynamics of the endocytic system itself. Motile EE and LE keep close contacts with the ER during their movement (Friedman et al. 2013), and lysosomes move along ER tubules (Lopez-Sanjurjo et al. 2013). The contact of endosomes with the ER increases with their transition from EE into LE (Friedman et al. 2013), implying that maturation process of endosomes is regulated by the ER. Direct interaction of lysosomes and the ER in both juxtanuclear and peripheral cell regions was also recently demonstrated by systematic interactome analysis using multispectral fluorescence imaging (Valm et al. 2017).

Interestingly, several studies indicate that the ER is involved in defining the intracellular distribution of endosomes by modulating their motility. A key player in this modulation is the ER protein VAP [VAMP (vesicle-associated membrane protein)-associated ER protein] that interacts with several endosomal proteins. The binding of VAP to two LE-proteins, STARD3 [StAR (steroidogenic acute regulatory protein)-related lipid transfer (START) domain-3] and STARD3 N-terminal like (STARD3NL), was shown to alter endosomal dynamics (Alpy et al. 2013). In addition, a direct interaction of VAP with the endocytic protein ORPlL induced peripheral positioning of the endosomes (Rocha et al. 2009). This effect was a consequence of decreased association between endosomes and the minus-end-directed microtubule motor dynein. Conversely, interaction with the plus-end-directed microtubule motor kinesin 1 was increased by another type of MCS formed by binding of the ER protein protrudin to the small GTPase Rab7 and phosphatidylinositol 3-phosphate (PI3P) that reside in the endosomal membrane (Raiborg et al. 2015a). These two mechanisms of endosome-ER tethering are suggested to regulate the motility of endosomes from the perinuclear region toward the plasma membrane in a coordinated "gear shift" mechanism (Raiborg et al. 2016).

Finally, interaction with the ER also regulates constriction and fission of both EE and LE, which occur at the endosome-ER MCS (Rowland et al. 2014). Endosomal fission is an important mean for sorting of cargo and resident membrane proteins, and therefore, the involvement of the ER in this process represents another level of ER-mediated regulation over functionality of the endocytic pathway. Together, the changes in endosome motility, distribution, and fission indicate that interaction with the ER dramatically affects the intracellular architecture of the endosomal network. However, it is yet to be elucidated whether and how these effects influence the function of endosomes and other organelles.

2.2.2 Endolysosomal–Mitochondrial Interactions Are Mediated by Various Mechanisms

Mitochondria play an important homeostatic role by their crucial participation in various cellular processes, including energy metabolism, oxidative stress-related pathways, and Ca^{2+} signaling. Being integrative signaling hubs, mitochondria communicate with other organelles at physical and functional levels. The interaction of the endocytic system with mitochondria was shown to participate in different cellular processes. Most recent data indicate that the endosomal system is also directly involved in degradation of damaged mitochondria, after sequestrating them in EE by a Rab5 and ESCRT-dependent mechanism (Hammerling et al. 2017). Interestingly, communication of the endocytic system with mitochondria was found to be facilitated by both vesicular and non-vesicular mechanisms (Fig. 2.1c).

Endosomal–mitochondrial MCS. Although much less studied than the endosome-ER MCS, membranes of LE and lysosomes also form contacts with mitochondria. Multispectral imaging analysis indicates that lysosome-mitochondria contacts are stable over time and less abundant in the peripheral areas of the cell (Valm et al. 2017). In yeast, an MCS complex connecting mitochondria and vacuole (the yeast lysosomal compartment) was identified and named vacuole and mitochondria patch (vCLAMP) (Elbaz-Alon et al. 2014; Honscher et al. 2014). The exact role of these physical contacts remains to be clarified, but they are suggested to participate in lipid exchange and metabolic regulation in the cell (Daniele and Schiaffino 2016). Interestingly, vCLAMP was found to be reciprocally co-regulated and (at least partially) functionally redundant with the mitochondrial-ER MCS complex ERMES (Elbaz-Alon et al. 2014; Honscher et al. 2014).

In mammalian cells, until now no equivalent endosomal–mitochondrial MCS have been described to exist under physiological conditions, although similar MCS may occur between mitochondria and melanosomes, which are endosome-related organelles found in specific cell types [discussed in (Daniele and Schiaffino 2016)]. Nevertheless, in a hypoxic environment, local contacts and microfusion between mitochondria and LE/lysosome membranes have been reported (Brahimi-Horn et al. 2015). This fusional contact is suggested to mediate cleavage of the mitochondrial outer membrane protein voltage-dependent anion channel 1 (VDAC1) by endosomal asparaginyl endopeptidase (AEP). Another example of a close association between mitochondria and endosomes was detected in the course of apoptotic response to *Helicobacter pylori* infection (Calore et al. 2010). This MCS is induced in infected cells by vacuolating cytotoxin (VacA), a bacterial virulence factor, and is required for recruitment of the apoptotic mediator BAX to mitochondria.

Endosomal-derived vesicles. Recent studies indicate that inter-organellar communication between endosomes and mitochondria can also occur via direct vesicle trafficking between these compartments. Endosomal-derived vesicles targeted to mitochondria were described in both erythrocytes and non-erythroid cells as involved in intracellular delivery of transferrin-bound iron (Das et al. 2016; Hamdi et al. 2016). These studies revealed a direct transient interaction between transferrin-loaded endosomes and mitochondria, supporting the existence of a "kiss and run" mechanism for efficient delivery of iron to mitochondria, rather than uptake of iron by mitochondria from the cytosol after its release from endosomes. This is consistent with previous findings indicating that endocytosed iron can bypass the cytosol on its way to mitochondria and that this delivery requires vesicular motility (Sheftel et al. 2007; Zhang et al. 2005). Interestingly, the dissociation of endosomes from mitochondria was found to depend on release of iron from the endosomes, implying that intra-endosomal iron levels regulate this vesicular contact (Das et al. 2016; Hamdi et al. 2016).

Mitochondrial-derived vesicles (MDV). Intriguingly, vesicular transport in the opposite direction (from the mitochondrial reticulum to the endosomal system) was described recently, indicating that the endosomal–mitochondrial trafficking route is actually bidirectional. Reports available so far suggest the existence of a few different types of MDV. One kind of MDV can be stress-induced by parkin/PINK1-dependent machinery (McLelland et al. 2014) to selectively deliver oxidized or damaged components of the mitochondrial matrix and inner membrane to LE/lysosomes (Soubannier et al. 2012). In this route, fusion of MDV with endo-lysosomes is mediated by the homotypic fusion and vacuole protein sorting (HOPS) tethering complex and by the soluble NSF attachment protein receptor (SNARE) pairing machinery (McLelland et al. 2016). Another type of MDV targeted to the endolysosomal system delivers the large GTPase DLP1 (that controls mitochondrial fission) to lysosomal degradation, a process regulated by the retromer component, vacuolar protein sorting 35 (VPS35) (Wang et al. 2016). Together, these data indicate a role of the endolysosomal–mitochondrial interaction in mitochondrial quality control and regulation of mitochondrial division. In turn, biogenesis of another type of endolysosomal-directed MDV, that is induced by heat stress or bacterial lipopolysaccharide (LPS) and inhibited by parkin/PINK1, was reported to mediate mitochondrial antigen presentation, implying an important role for endolysosomal–mitochondrial interaction in immune response (Matheoud et al. 2016).

2.2.3 Endosomal ESCRT Machinery Maintains Integrity of the PM and Nuclear Envelope

Accumulating evidence documents that some endocytic proteins play additional roles, not related to the endolysosomal pathway. A prominent example for this type of "moonlighting" is various functions of the ESCRT protein machinery. Apart from their role in sorting of endocytosed cargo into intraluminal vesicles (ILV) in endosomes, ESCRT proteins act at the mitotic spindle during cellular abscission, at the PM during viral budding, and in the cell nucleus where they regulate transcription (Alonso et al. 2016; Christ et al. 2016; Pyrzynska et al. 2009).

Moreover, as it might be expected, the abilities of ESCRT proteins to remodel membranes are used for repair of membranes at additional cellular sites, apart from the endocytic system (Fig. 2.1d, e). Hence, ESCRT is unambiguously established as a general machinery for maintaining intactness of diverse membranes within the cell.

Repair of PM breaks. Breaks in the PM appear during various export and import functions, but also due to mechanical stress or exposure to toxins. Because of its dynamic nature and ability to exchange and modify membranes, the endocytic system is of major importance in PM damage repair. The PM breaks can be filled by fusion with endo-membranes or removed via internalization or extracellular budding. In addition, it was found that components of the endocytic ESCRT machinery are directly active in resealing PM breaks (Jimenez et al. 2014). ESCRT proteins are recruited to the site of PM wound to participate in its efficient repair by pinching out the broken membrane piece. Preserving intactness of the PM is essential for maintenance of cell integrity and survival, indicating an important homeostatic role played by the ESCRT machinery.

Nuclear envelope repair and quality control. Moreover, ESCRT proteins were demonstrated to participate in sealing of the nuclear membrane. The nuclear envelope has to be broken during cell division to allow the formation of two nuclei in daughter cells. ESCRT components were found to act at the post-mitotic stage, sealing the nuclear membranes (Olmos et al. 2015; Vietri et al. 2015). This indicates the essential involvement of endocytic proteins in ensuring proper re-establishment of nucleocytoplasmic compartmentalization during mitosis.

In the same fashion, ESCRT proteins act in resealing discontinuities that are formed in the nuclear envelope during cell migration (Denais et al. 2016; Raab et al. 2016). Interestingly, an inhibition of DNA-repair machinery in migrating cells had no effect on cell viability, whereas it led to cell death when combined with depletion of ESCRT components (Raab et al. 2016). This observation implies a possible cooperation between these two machineries and brings up a speculation that sensing DNA damage is the trigger for recruitment of ESCRT to the site of rupture (Ventimiglia and Martin-Serrano 2016).

In addition to closing membrane breaks, ESCRT components were found to participate in quality control during the assembly of nuclear pores in yeast (Webster et al. 2014). This study revealed another level of ESCRT-mediated regulation of nuclear membrane function. In this respect, ESCRT components take part in the cellular surveillance over formation of nuclear pores: When a defective pore is assembled, ESCRT proteins are recruited and exert its removal. In general, by preserving integrity and functionality of the nuclear envelope, the ESCRT machinery contributes to maintaining genomic stability and cell viability.

2.3 The Role of Endosomes in Regulation of Metabolic Pathways

The endocytic system regulates cellular metabolism by controlling uptake and trafficking of nutrient carriers, transporters, and signaling receptors (Antonescu et al. 2014). Since the metabolic regulation is highly dynamic, a well-coordinated interpretation of external cues is critical for inducing the proper cellular response, and endosomes serve as efficient mediators in this process. The best recognized endosomal organelles in metabolic regulation are the lysosomes, with much data emerging in recent years acknowledging their role as hubs for integration of metabolic sensing and signaling activation. The topic of lysosomes is covered by another chapter of this book; hence, it will not be discussed herein. We will describe the involvement of several other endocytic organelles and components in spatial and temporal regulation of metabolic pathways.

Endosomal machineries in metabolic signaling. The endosomal system regulates appropriate distribution of metabolite transporters between the PM and intracellular compartments. Intriguingly, the same extracellular cue can mobilize partly different endocytic machineries acting on individual transporters that lead to diverse metabolic consequences. This type of regulation was demonstrated for two different transporters, facilitating uptake of glucose and long-chain fatty acid (LCFA), in isolated cardiomyocytes (Steinbusch et al. 2010). Both transporters are recruited to the PM in response to insulin/oligomycin stimulation, yet distinct vesicular trafficking machineries are involved in their translocation. Specifically, actin filaments and endosomal acidification are required for the stimulated uptake of glucose, but not of LCFA. By this mechanism, the endocytic system is involved in selective regulation of nutrient uptake to directly control the metabolite status.

Cross talk between endosomal trafficking and metabolic signaling is also demonstrated by the widespread involvement of Rab proteins in both processes. Rab GTPases play a crucial role as molecular switches that direct intracellular vesicular transport, also through the endocytic system. Moreover, different Rab proteins were found to participate in signal transduction of multiple metabolic-related pathways [reviewed in (Chua and Tang 2015)]. A recent study identified a role for endosomal Rab5 in regulation of hepatic gluconeogenesis (Zeigerer et al. 2015). In this study, depletion of Rab5 resulted in loss of endosomes in liver cells and induction of severe metabolic abnormalities in vivo in mice. The findings indicated the existence of an endosomally controlled regulation of transcription factors responsible for expression

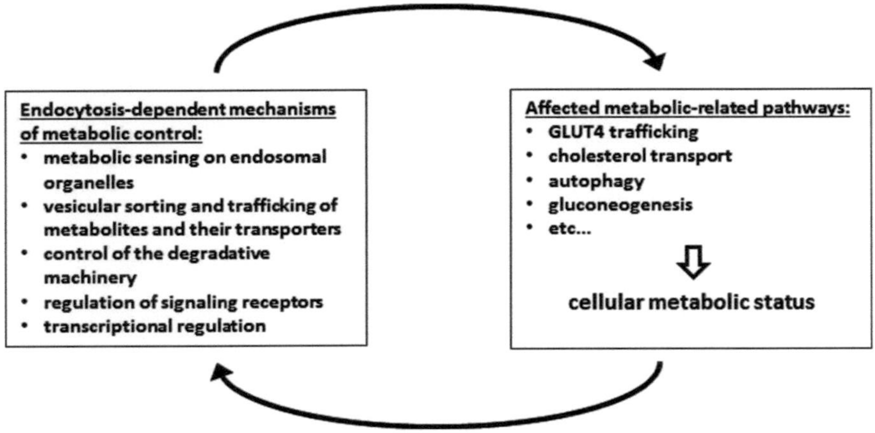

Fig. 2.2 Overview of a bidirectional interplay between endocytosis and cellular metabolism

of gluconeogenic genes. This implies that functional coordination between endocytosis and metabolism goes beyond a traditional concept of simple regulation exerted by the endocytic system controlling the abundance of receptors and transporters at the PM.

Trafficking regulation by the metabolic status. In a reciprocal manner, the metabolic state of a cell directly affects endolysosomal trafficking [discussed in (Antonescu et al. 2014)]. This was demonstrated by systematic quantitative assessment of changes observed in endocytic trafficking in a genome-wide siRNA screen (Collinet et al. 2010). Depletion of metabolism-related proteins induced endocytic phenotypes, most prominently upregulated endocytosis, as measured by internalization of transferrin receptor (a prototype marker for the endosomal recycling pathway). The authors suggested that this may reflect a cellular adaptive response, aiming to increase uptake of nutrients under conditions of metabolic deficiency. This view is supported by another study, demonstrating a metabolic adaptive response mediated by the endocytic system that is observed during glucose starvation in yeast (Lang et al. 2014). In this setting, modulation of the endocytic pathway is essential for cellular survival, facilitated by inhibition of endosomal recycling to the PM to allow vacuolar hydrolysis of cell components for energy production. Together, the data imply that the endolysosomal system and metabolic pathways are interdependent on one another. In the following subchapters, we will present how they intersect at various compartments and multiple functional levels and can bidirectionally affect cellular signaling and vesicular transport (summarized in Fig. 2.2). Specifically, we will discuss the cross talk between the endocytic system and three pathways of cellular metabolism: (i) trafficking of glucose transporter type 4 (GLUT4), (ii) intracellular cholesterol transfer, and (iii) autophagy, all of paramount importance for maintenance of metabolic homeostasis.

2.3.1 Involvement of the Endocytic System in Vesicular Trafficking of GLUT4

Glucose is taken up from an extracellular environment by a family of transmembrane transporters, named GLUT, which are expressed in different cell types. By controlling the trafficking of these transporters, the endocytic system determines their distribution on endo-membranes versus the PM and hence regulates the cellular dynamics of glucose uptake. A well-studied member of this family is GLUT4, which is primarily expressed in adipose and striated muscle tissues and plays a key role in the development of type 2 diabetes. Translocation of GLUT4 to the PM is induced by insulin to facilitate glucose uptake and control postprandial blood glucose levels. Trafficking of GLUT4 is a very dynamic process, and when blood insulin levels decrease it is directed back from the PM to unique endo-vesicles, named GLUT4 storage vesicles (GSV) (Bogan 2012).

Several different internalization routes of GLUT4 and mechanisms regulating its translocation have been described in various cell types under basal and insulin-stimulated conditions (Antonescu et al. 2014). Two mechanisms, termed "dynamic exchange" and "static retention," control the intracellular trafficking of GLUT4 and both require endosome-mediated transport (Muretta et al. 2008). Regulation of GLUT4 trafficking is of main importance for upkeep of normal glucose metabolism in the cell; therefore, this process remains a subject of intense research.

Intracellular routes of GLUT4. It is believed that after internalization into EE, GLUT4 is directed to an intermediate compartment, which could be RE and/or the trans-Golgi network (TGN), from where it is transported into GSV (Kandror and Pilch 2011). Some findings indicate that the TGN is the main site of GSV biogenesis [discussed in (Kandror and Pilch 2011)]. This view is in agreement with a unique role suggested for clathrin heavy chain 22 (CHC22) and syntaxin 10 (STX10) in retrograde sorting of internalized GLUT4 from EE to the TGN, a delivery step found to be crucial for intracellular GLUT4 storage (Esk et al. 2010). However, a partial role of RE in GSV assembly cannot be excluded and is even supported by the dual-mode working model of insulin action (Xu et al. 2011). According to this model, GLUT4 is delivered to the PM by two circuits: The first is activated after short exposure to insulin and represents TGN-originated GSV, and the second appears after prolonged insulin stimulation, involving exocytosis of larger vesicles which probably bud from RE. Hence, the endosomal recycling compartment plays a specialized role in insulin-regulated GLUT4 trafficking, coordinating GLUT4 translocation to the PM.

Involvement of Rab proteins. Several Rab proteins were found to regulate various steps of the GLUT4 trafficking itinerary. Translocation of GSV to the PM is facilitated by Rab10, which can be used as a marker to distinguish GSV from GLUT4-containing endosomes in adipocytes (Chen et al. 2012). This specificity of Rab10 is additionally confirmed by the fact that Rab10 is not essential for exocytosis of GLUT4 in fibroblasts which have no GSV and therefore deliver GLUT4 to the PM only via the constitutive endosomal recycling pathway (Brewer et al. 2016b). Using Rab10 as a GSV marker allowed demonstrating that GSV which fuse with the PM upon insulin stimulation do not merge with endosomes, but rather constitute the main source that directly conveys GLUT4 to the cell surface (Chen et al. 2012). Yet, fusion of GLUT4-containing endosomes with the PM was found to independently contribute another portion of GLUT4 redistributed to the PM in response to insulin, in a Rab14-dependent manner (Chen et al. 2012). More recent studies indicated that Rab14 actually functions in preceding GLUT4 trafficking steps, from EE to later compartments (RE and/or Golgi), rather than in fusion of GLUT4-endosomal vesicles with the PM (Brewer et al. 2016a; Reed et al. 2013). A similar role, in sorting of GLUT4 from RE to GSV, was suggested also for Rab11 (Zhang et al. 2005), a general regulator of endocytic recycling. In agreement with the involvement of Rab proteins in GLUT4 trafficking, the insulin-sensitive Rab GTPase-activating protein AS160 (Akt substrate of 160 kDa) was introduced as the main switch of GLUT4 redistribution, regulating both sorting of GLUT4 from EE into GSV (through Rab14) and GLUT4 exocytosis from GSV (through Rab10) (Brewer et al. 2016a).

In muscle cells, Rab10 appears to be less critical for GLUT4 exocytosis after insulin stimulation, while Rab8A, Rab13, and Rab14 are required for the translocation of GLUT4 to the PM (Ishikura and Klip 2008; Sun et al. 2010). Here, Rab8A functions to mobilize GLUT4 vesicles through association with myosin motors (Ishikura and Klip 2008; Sun et al. 2014), and Rab13 assembles the molecular complex necessary for GLUT4 exocytosis (Sun et al. 2016). AS160 is suggested to act upstream of Rab proteins to mediate GLUT4 translocation also in muscle cells, since over-expression or silencing of Rab8A reverses the effects observed after AS160 activation or depletion, respectively (Ishikura and Klip 2008; Sun et al. 2010).

The reason for some redundancy of Rab proteins and discrepancies in their action between cell types is unknown, but it seems that GLUT4 trafficking is regulated by different endocytic components in different tissues. The involvement of multiple Rab proteins in various steps of the GLUT4 routing emphasizes that, in addition to their general roles in the endosomal pathway, these GTPases have specific functions in trafficking of GLUT4. However, while aberrant function of Rab proteins was suggested to contribute to diabetes (Bogan 2012), the global alterations of the endolysosomal system under pathological conditions still remain to be studied.

2.3.2 The Endolysosomal Pathway Mediates Intracellular Cholesterol Transfer

Exogenous cholesterol is internalized into the cell via endocytosis of circulating low-density lipoproteins (LDL) which bind to the LDL receptor (LDLR). Subsequently, LDLR is recycled back to the PM and LDL-bound cholesterol is delivered to late endosomal compartments for hydrolysis of lipoprotein particles, after which free cholesterol is redistributed from LE to several other cellular locations [reviewed in (Ikonen 2008)]. Indeed, LE contains many different cholesterol-binding proteins, such as NPC1 and NPC2 (Niemann–Pick type C1 and C2), indicating the existence of various pathways for cholesterol handling in this organelle. Although it is of main importance for metabolic homeostasis, the regulatory mechanisms of intra-endosomal cholesterol sensing are poorly understood. Interestingly, apoptosis-linked gene 2-interacting protein X (Alix), a regulator of endocytic trafficking, was found to control endosomal cholesterol levels by an interaction with an unconventional phospholipid lysobisphosphatidic acid (LBPA) (Chevallier et al. 2008). According to the suggested model, the amount of cholesterol stored in LE is determined by the buffering capacity of the LBPA-rich internal membranes of the multi-vesicular endosomes (MVE).

In addition, only partly known are the mechanisms of cholesterol redistribution from endosomes to other cellular destinations which includes the ER, the PM, and mitochondria (Luo et al. 2017; Pfisterer et al. 2016). The ER is the main site for cholesterol sensing and de novo synthesis, and for regulation of cellular sterol homeostasis. Cholesterol from the ER can be delivered to the PM to fulfill several essential functions, including determination of membrane rigidity, assembly of subdomains, and regulation of signal transduction. The PM can also function as an intermediate location for cholesterol redistribution to other organelles. In the mitochondria, cholesterol ensures maintenance of mitochondrial membranes and is in some cases needed for production of steroids and oxysterols.

Cholesterol transfer between endosomes and the ER. Cholesterol transfer between the endocytic system and the ER is an emerging subject of research, since some of the MCS between these compartments are mediated by cholesterol-binding proteins. These include the endocytic proteins STARD3 and ORPlL, which both bind to the ER protein VAP (Alpy et al. 2013; Rocha et al. 2009). Moreover, the ORPlL–VAP interaction was shown to be induced by low levels of cholesterol in endosomes. This effect is mediated by a conformational change of ORPlL which induces enhancement of the LE-ER tethering (Rocha et al. 2009). Since STARD3-VAP associations were found to be assembled and to function independently of the lipid transfer ability of STARD3 (Alpy et al. 2013), it was proposed that their primary role is sensing cholesterol levels. A more recent study, however, indicates that STARD3-VAP complexes actually transport cholesterol from the ER to endosomes (Wilhelm et al. 2017). Cholesterol transfer from the ER to endosomes was also shown to be mediated by another subpopulation of MCS formed between ER and EGFR-positive endosomes (Eden et al. 2016). These MCS are tethered by annexin A1 and its ligand S100A11, while sterol trafficking via these sites depends upon ORP1L–VAP interaction. The findings indicate that ER-to-endosome cholesterol delivery occurs when endosomal cholesterol levels are low, to support the formation of EGF-stimulated MVE. Yet, it is unknown whether any of these interactions are bidirectional and also mediate cholesterol transfer in the opposite direction, from LE to the ER.

Interestingly, formation of endosomal-ER MCS also affects the spatial distribution of LE, directing them toward the cell periphery (Rocha et al. 2009). Since these MCS are induced when cholesterol levels in LE are low, possible roles of this positioning are to facilitate uptake of exogenous cholesterol and/or egress of cholesterol from endosomes to the PM.

Endosome-to-PM cholesterol delivery. Regardless of the role played by STARD3 in the formation of endosomal-ER MCS (Alpy et al. 2013), this endosomal protein was proposed to mediate cholesterol transfer between LE and the PM (van der Kant et al. 2013). It was suggested that STARD3-containing endosomes differ from ORP1L-containing endosomes, representing a distinct "earlier" subpopulation of LE (van der Kant et al. 2013). Accordingly, internalized cholesterol is transferred from EE to the STARD3 "early" LE, from where it can be delivered to the PM or transported to the ORP1L "late" LE for further distribution to the ER. This possible existence of specialized LE subpopulations, used for differential sorting of cholesterol into various destinations, highlights the importance of endosomes in ensuring cellular homeostasis of cholesterol trafficking.

Furthermore, the results of siRNA-based screening imply that appropriate supply of cholesterol from endosomes to the PM is essential also for maintenance of functional clathrin-mediated endocytosis (Kozik et al. 2013). In addition, inhibition of the endosomal V-ATPase in HeLa cells was found to increase biogenesis of cholesterol-rich extracellular vesicles (exosomes) which are intraluminal vesicles generated inside endosomes and released from cells by fusion of MVE with the PM (Edgar et al. 2016). This finding indicates that by influencing endosome-PM fusion, the endocytic system also regulates cholesterol export from the cells. It was further discovered that exosomes induced by manipulating the endolysosomal function are attached to the PM by a protein named tetherin (Edgar et al. 2016). The authors suggested that the tetherin content in the exosome membrane is increased in cholesterol-enriched vesicles, representing a mechanism for selective release of certain vesicles while tethering others to the PM. This implies that the endosomal system controls exosome release via a mechanism regulated by cholesterol.

Cholesterol transfer via peroxisomes and via TGN. Interestingly, indirect routes, passing via other organelles, are also suggested to facilitate cholesterol delivery from the endolysosomal system to the ER and the PM. A recent study revealed direct transfer of cholesterol from lysosomes to peroxisomes via transient MCS formed by binding of a lysosomal protein synaptotagmin VII to PI(4,5)P_2 on the peroxisomal membrane (Chu et al. 2015). The role of this trafficking route is still ill-defined, but it is suggested to participate in distribution of cholesterol to the ER and the PM. Additionally, cholesterol can be transported from NPC1-positive endosomes to the TGN through SNARE-mediated vesicular trafficking (Urano et al. 2008).

Endosomal–mitochondrial cholesterol transfer. Finally, the endocytic system participates also in cholesterol transport to mitochondria. Interestingly, delivery of cholesterol from LE/lysosomes to mitochondria can be mediated by the cholesterol-binding protein STARD3, which, as described above, was suggested to be involved in cholesterol transfer between endosomes, the ER and the PM (Charman et al. 2010; Zhang et al. 2002).

To summarize, the endocytic system facilitates different aspects of the complex network of cellular cholesterol transport, storage, and distribution. Hence, endocytosis has multiple direct and indirect effects on cellular cholesterol homeostasis.

2.3.3 Interplay Between the Endocytic System and Autophagy

Macroautophagy (henceforth referred to as autophagy) is the main cellular pathway mediating lysosomal degradation of cell components, to eliminate damaged organelles and invading pathogens. In addition, as an important pathway for cellular housekeeping of nutrient supply, autophagy is activated by metabolic stress and starvation. During autophagy, the contents designated to be degraded are surrounded by a small portion of cup-shaped isolation membrane, termed phagophore, which later elongates to form a double-membrane vacuole known as autophagosome. Autophagosomes then fuse with lysosomes, creating autolysosomes, where lysosomal enzymes catalyze degradation [described in (Glick et al. 2010)]. Hence, autophagy directly depends on a stable cohort of lysosomes.

The endocytic and autophagic systems serve as cellular degradative pathways and share the same endpoint. As it may be expected, both routes are interconnected by several overlapping molecular mechanisms and functional interactions. Discussing all aspects of this multifaceted interplay is beyond the scope of the current chapter [for review, see (Barth and Kohler 2014; Tooze et al. 2014)]. Herein we will present some key findings indicating the complex communication between the endolysosomal and autophagy systems.

Diverging routes of autophagy. The autophagy pathway in yeast was described as a sequential maturation process of a phagophore, converging with the endolysosomal pathway at the point of autophagosome fusion with lysosome. However, the existence of another bypass route was discovered in mammalian cells, where autophagosomes do not fuse directly with lysosomes, but rather with LE, generating intermediate organelles named amphisomes (Fader and Colombo 2009). Amphisomes were detected also in flies and nematode (Djeddi et al. 2012; Rusten et al. 2007) and identified as prelysosomal compartments, containing both endocytic and autophagic cargo, which continue to fuse with lysosomes to form autolysosomes. Interestingly, amphisomes were found to contain not only LE markers, but also proteins typically located on EE (Berg et al. 1998), raising the possibility that fusion of autophagosome with EE may also occur. Hence, several intertwined routes may connect different endosomal organelles with the autophagy pathway. Understanding the regulation and importance of these endocytic-related routes in the multistep process of autophagy is still a challenge.

Autophagosome formation. The cellular sources of membrane material for formation of autophagosome as well as their assembly mechanisms are being intensely studied. Among contributors, such as the ER, mitochondria, the Golgi apparatus, and the PM, endosomes are also suggested to take part in incorporation of membrane fragments during formation and elongation of the isolation membrane [discussed in (Chan and Tang 2013)]. A recent report, using ultrastructural investigation, identified MCS formed between the phagophore membrane and other organelles, including endosomes, implying that these contacts may participate in cross talk or lipid transport, contributing to autophagosome formation (Biazik et al. 2015).

Indeed, fusion of vesicular and multi-vesicular endocytic organelles with nascent autophagosomes has already been shown earlier (Liou et al. 1997). In agreement, RE marked with Rab11 were found to participate in formation of autophagosomes by delivery of membrane to the expanding phagophore (Longatti et al. 2012). It was suggested that vesicular transfer from RE is activated by amino acid deficiency and repressed by the Rab11 effector TBC1D14, which shuttles between RE and the Golgi apparatus depending on the nutrient availability (RE under fed conditions, the Golgi during starvation) (Longatti et al. 2012).

Interestingly, a recent study revealed a role for vesicular transport from the endosomal system in facilitating the delivery of autophagy-related 9A (Atg9A), an integral membrane protein which is required for the formation of autophagosome (Imai et al. 2016). The sorting of Atg9A from RE and the Golgi is regulated via interaction with the adaptor protein AP-2. Interrupting this interaction resulted in accumulation of Atg9A in RE and dysregulated autophagy. These findings indicate that proper trafficking of Atg9A in the endolysosomal pathway is essential for functional activation of autophagy. This assumption is in agreement with an earlier study implying that Atg9A transport mediated by the endosomal retromer–WASH complex is required for autophagy (Zavodszky et al. 2014).

Endolysosomal regulation of autophagy. The central regulator of autophagy is mammalian target of rapamycin (mTOR), a component of the mTOR complex 1 (mTORC1) activated according to nutrient availability at the LE/lysosomal surface (Lim and Zoncu 2016). Hence, the endolysosomal system is directly interconnected to the autophagy pathway not only via endocytic organelles facilitating degradation, but also at a regulatory level controlling autophagy induction.

Actually, dynamics of the endolysosomal pathway was shown to directly influence mTOR signaling and therefore activation of autophagy (Korolchuk et al. 2011). In this study, the intracellular positioning of endosomes was found to modulate mTOR activation and hence the autophagic flux. Moreover, it was suggested by the authors that nutrient levels dictate lysosomal distribution and that peripheral localization of lysosomes favors mTOR activation due to a closer proximity to upstream activators near the PM. This demonstrates a spatial mechanism of metabolic regulation coordinated by the endocytic system.

Another level of autophagy regulation by the endocytic system was revealed by a recent work using proteomics analysis to study how endolysosomal proteolysis is coordinated with activation of autophagy-mediated degradation in response to starvation in yeast (Muller et al. 2015). The findings confirmed that autophagy was induced immediately upon introduction of starvation. However, during the first hours of nutrient depletion the destruction of membrane proteins by the endocytic system was the main source for amino acid supply to maintain cellular functions and activate an adaptive response. This adaptation included the de novo synthesis of vacuolar hydrolases that allowed potentiation of autophagy, which was suggested to be essential for restoration of amino acid levels during extended periods of starvation. Interestingly, this also demonstrates that selective (ubiquitin-dependent) degradation of membrane proteins via the endocytic system activates a catabolic cascade during starvation, by induction of the non-selective autophagic pathway. This timely regulated response was found to be required for cell survival. These findings uncover another homeostatic role played by the endocytic system, which requires intimate cooperation with the autophagic pathway and evokes a survival response at the whole-cell level.

Shared molecular machineries. Similarly to the endosomal pathway, autophagy requires membrane remodeling events such as elongation, closure of gaps, and fusion. Indeed, many molecular mechanisms are shared between the endocytic and autophagic pathways, adding another complexity level to the interactions between these two systems [discussed in (Fader and Colombo 2009)]. As one may expect, an example of effector molecules common for endocytosis and autophagy are the components of ESCRT.

Several alternative scenarios have been proposed to explain the involvement of ESCRT proteins in autophagy [presented in (Rusten and Stenmark 2009)]. The possible mechanisms include the involvement of ESCRT in phagophore closure, lysosome biogenesis, lysosomal fusion or in preventing proautophagic signaling by maintaining proper endosomal flux. None of these options can be ruled out, and it is possible that the mechanisms differ between distinct biological systems and/or that several mechanisms coexist.

Indeed, depletion of ESCRT components was found to inhibit autolysosome formation and impair autophagy (Filimonenko et al. 2007). The authors of this study proposed that autophagy was diminished due to dysfunction of endosomal transport caused by loss of ESCRT, indicating the dependence of autophagic degradation on a functional endocytic pathway. These findings are in agreement with accumulation of autophagosomes observed upon ESCRT-III disruption in cultured cells and in flies in vivo (Lee et al. 2007; Oshima et al. 2016; Rusten et al. 2007), all supportive of a role for ESCRT in fusion of lysosomes with both endosomal and autophagic organelles. The ESCRT-I subunit TSG101 was also implicated in vesicular fusion of lysosomes with LE and with amphisomes (Majumder and Chakrabarti 2015).

Similarly, autophagosome accumulation was observed also in ESCRT-depleted nematode (Djeddi et al. 2012). However, in this model autophagosomes were claimed to accumulate due to increased autophagic flux, rather than inhibition of lysosome fusion. In this setting, increased levels of aberrant ESCRT-depleted endosomes induced autophagy that played a protective, pro-survival role. These findings support the existence of cross talk between the endosomal and autophagy pathways, responsible for preserving cellular and organismal homeostasis.

Of note is that although the ESCRT machinery is highly conserved from yeast to humans, it was not found to be involved in the autophagic function in yeast, with no accumulation of autophagosomes observed in ESCRT mutants (Muller et al. 2015).

In addition to ESCRT, many other molecular players were suggested to be shared between the endosomal and autophagy systems and possibly to mediate the interplay between these pathways, which include ESCRT-associated protein Alix (Murrow et al. 2015), Rab machinery (Chen et al. 2014; Ganley et al. 2011; Longatti et al. 2012; Popovic et al. 2012; Szatmari et al. 2014), HOPS tethering complex (Jiang et al. 2014; Liang et al. 2008; Lindmo et al. 2006; Wartosch et al. 2015), and others [also discussed in (Fader and Colombo 2009)].

2.4 Conclusions

The cumulative evidence discussed above indicates that the endolysosomal pathway acts in coordination with other cellular systems, to support essential life processes and cell homeostasis. Hence, the endocytic system should not be referred to as a stand-alone machinery for uptake and degradation of vesicle-engulfed cargo, but rather as an integrated part of the cellular network of organelles and processes that are interconnected.

As demonstrated by several examples throughout this chapter, while playing their established roles in vesicle trafficking, endosomes keep close relations with other organelles and are dynamically involved in functions that have global cellular implications such as membrane flow, signaling control, and metabolic regulation. Yet, in many of these processes, the molecular mechanisms underlying the functional involvement of the endocytic system are not fully understood and remain an important challenge for future research.

Moreover, some functions of endosomes are universal, while others are specialized and vary between different cell types. Thus, it is highly possible that the interaction mechanisms of endosomes with other organelles and pathways are in part tissue-specific, which adds another level of complexity to their mutual co-regulation. Finally, the roles played by endosomal trafficking in various cellular processes are affected in pathological conditions involving dysregulated nutrient metabolism, improper immunological responses, developmental disorders or tumorigenesis. Therefore, understanding the participation of the endocytic system in the cellular network of organelles and pathways has important implications for human health and disease.

References

Alonso YAM, Migliano SM, Teis D (2016) ESCRT-III and Vps4: a dynamic multipurpose tool for membrane budding and scission. FEBS J 283(18):3288–3302. https://doi.org/10.1111/febs.13688

Alpy F, Rousseau A, Schwab Y, Legueux F, Stoll I, Wendling C, Spiegelhalter C, Kessler P, Mathelin C, Rio MC, Levine TP, Tomasetto C (2013) STARD3 or STARD3NL and VAP form a novel molecular tether between late endosomes and the ER. J Cell Sci 126(Pt 23):5500–5512. https://doi.org/10.1242/jcs.139295 jcs.139295 [pii]

Antonescu CN, McGraw TE, Klip A (2014) Reciprocal regulation of endocytosis and metabolism. Cold Spring Harb Perspect Biol 6(7):a016964. https://doi.org/10.1101/cshperspect.a016964 a016964, /7/a016964 [pii]

Barth JM, Kohler K (2014) How to take autophagy and endocytosis up a notch. Biomed Res Int 2014:960803. https://doi.org/10.1155/2014/960803

Berg TO, Fengsrud M, Stromhaug PE, Berg T, Seglen PO (1998) Isolation and characterization of rat liver amphisomes. Evidence for fusion of autophagosomes with both early and late endosomes. J Biol Chem 273(34):21883–21892

Biazik J, Yla-Anttila P, Vihinen H, Jokitalo E, Eskelinen EL (2015) Ultrastructural relationship of the phagophore with surrounding organelles. Autophagy 11(3):439–451. https://doi.org/10.1080/15548627.2015.1017178

Bogan JS (2012) Regulation of glucose transporter translocation in health and diabetes. Annu Rev Biochem 81:507–532. https://doi.org/10.1146/annurev-biochem-060109-094246

Brahimi-Horn MC, Lacas-Gervais S, Adaixo R, Ilc K, Rouleau M, Notte A, Dieu M, Michiels C, Voeltzel T, Maguer-Satta V, Pelletier J, Ilie M, Hofman P, Manoury B, Schmidt A, Hiller S, Pouyssegur J, Mazure NM (2015) Local mitochondrial-endolysosomal microfusion cleaves voltage-dependent anion channel 1 to promote survival in hypoxia. Mol Cell Biol 35(9):1491–1505. https://doi.org/10.1128/MCB.01402-14 MCB.01402-14 [pii]

Brewer PD, Habtemichael EN, Romenskaia I, Coster AC, Mastick CC (2016a) Rab14 limits the sorting of Glut4 from endosomes into insulin-sensitive regulated secretory compartments in adipocytes. Biochem J 473(10):1315–1327. https://doi.org/10.1042/BCJ20160020 BCJ20160020 [pii]

Brewer PD, Habtemichael EN, Romenskaia I, Mastick CC, Coster AC (2016b) Glut4 is sorted from a Rab10 GTPase-independent constitutive recycling pathway into a highly insulin-responsive Rab10 GTPase-dependent sequestration pathway after adipocyte differentiation. J Biol Chem 291(2):773–789. https://doi.org/10.1074/jbc.M115.694919 M115.694919 [pii]

Calore F, Genisset C, Casellato A, Rossato M, Codolo G, Esposti MD, Scorrano L, de Bernard M (2010) Endosome-mitochondria juxtaposition during apoptosis induced by H. pylori VacA. Cell Death Differ 17(11):1707–1716. https://doi.org/10.1038/cdd.2010.42 cdd201042 [pii]

Cendrowski J, Maminska A, Miaczynska M (2016) Endocytic regulation of cytokine receptor signaling. Cytokine Growth Factor Rev. https://doi.org/10.1016/j.cytogfr.2016.07.002 S1359-6101(16)30086-7 [pii]

Chan SN, Tang BL (2013) Location and membrane sources for autophagosome formation—from ER-mitochondria contact sites to Golgi-endosome-derived carriers. Mol Membr Biol 30(8):394–402. https://doi.org/10.3109/09687688.2013.850178

Charman M, Kennedy BE, Osborne N, Karten B (2010) MLN64 mediates egress of cholesterol from endosomes to mitochondria in the absence of functional Niemann-Pick Type C1 protein. J Lipid Res 51(5):1023–1034. https://doi.org/10.1194/jlr.M002345 jlr.M002345 [pii]

Chen Y, Wang Y, Zhang J, Deng Y, Jiang L, Song E, Wu XS, Hammer JA, Xu T, Lippincott-Schwartz J (2012) Rab10 and myosin-Va mediate insulin-stimulated GLUT4 storage vesicle translocation in adipocytes. J Cell Biol 198(4):545–560. https://doi.org/10.1083/jcb.201111091 jcb.201111091 [pii]

Chen Y, Zhou F, Zou S, Yu S, Li S, Li D, Song J, Li H, He Z, Hu B, Bjorn LO, Lipatova Z, Liang Y, Xie Z, Segev N (2014) A Vps21 endocytic module regulates autophagy. Mol Biol Cell 25(20):3166–3177. https://doi.org/10.1091/mbc.E14-04-0917 mbc.E14-04-0917 [pii]

Chevallier J, Chamoun Z, Jiang G, Prestwich G, Sakai N, Matile S, Parton RG, Gruenberg J (2008) Lysobisphosphatidic acid controls endosomal cholesterol levels. J Biol Chem 283(41):27871–27880. https://doi.org/10.1074/jbc.M801463200 M801463200 [pii]

Christ L, Raiborg C, Wenzel EM, Campsteijn C, Stenmark H (2016) Cellular functions and molecular mechanisms of the ESCRT membrane-scission machinery. Trends Biochem Sci. https://doi.org/10.1016/j.tibs.2016.08.016 S0968-0004(16)30146-3 [pii]

Chu BB, Liao YC, Qi W, Xie C, Du X, Wang J, Yang H, Miao HH, Li BL, Song BL (2015) Cholesterol transport through lysosome-peroxisome membrane contacts. Cell 161(2):291–306. https://doi.org/10.1016/j.cell.2015.02.019 10.1016/j.cell.2015.02.019

Chua CE, Tang BL (2015) Role of Rab GTPases and their interacting proteins in mediating metabolic signalling and regulation. Cell Mol Life Sci 72(12):2289–2304. https://doi.org/10.1007/s00018-015-1862-x

Collinet C, Stoter M, Bradshaw CR, Samusik N, Rink JC, Kenski D, Habermann B, Buchholz F, Henschel R, Mueller MS, Nagel WE, Fava E, Kalaidzidis Y, Zerial M (2010) Systems survey of endocytosis by multiparametric image analysis. Nature 464(7286):243–249. https://doi.org/10.1038/nature08779 nature08779 [pii]

Daniele T, Schiaffino MV (2016) Lipid transfer and metabolism across the endolysosomal-mitochondrial boundary. Biochim Biophys Acta 1861(8 Pt B):880–894. https://doi.org/10.1016/j.bbalip.2016.02.001 S1388-1981(16)30020-8 [pii]

Das A, Nag S, Mason AB, Barroso MM (2016) Endosome-mitochondria interactions are modulated by iron release from transferrin. J Cell Biol 214(7):831–845. https://doi.org/10.1083/jcb.201602069 jcb.201602069 [pii]

Denais CM, Gilbert RM, Isermann P, McGregor AL, te Lindert M, Weigelin B, Davidson PM, Friedl P, Wolf K, Lammerding J (2016) Nuclear envelope rupture and repair during cancer cell migration. Science 352(6283):353–358. https://doi.org/10.1126/science.aad7297 science.aad7297 [pii]

Djeddi A, Michelet X, Culetto E, Alberti A, Barois N, Legouis R (2012) Induction of autophagy in ESCRT mutants is an adaptive response for cell survival in *C. elegans*. J Cell Sci 125(Pt 3):685–694. https://doi.org/10.1242/jcs.091702 125/3/685 [pii]

Eden ER (2016) The formation and function of ER-endosome membrane contact sites. Biochim Biophys Acta 1861(8 Pt B):874–879. https://doi.org/10.1016/j.bbalip.2016.01.020 S1388-1981(16)30019-1 [pii]

Eden ER, White IJ, Tsapara A, Futter CE (2010) Membrane contacts between endosomes and ER provide sites for PTP1B-epidermal growth factor receptor interaction. Nat Cell Biol 12(3):267–272. https://doi.org/10.1038/ncb2026 ncb2026 [pii]

Eden ER, Sanchez-Heras E, Tsapara A, Sobota A, Levine TP, Futter CE (2016) Annexin A1 tethers membrane contact sites that mediate ER to endosome cholesterol transport. Dev Cell 37(5):473–483. https://doi.org/10.1016/j.devcel.2016.05.005 S1534-5807(16)30282-9 [pii]

Edgar JR, Manna PT, Nishimura S, Banting G, Robinson MS (2016) Tetherin is an exosomal tether. Elife 5. https://doi.org/10.7554/elife.17180 e17180 [pii]

Elbaz-Alon Y, Rosenfeld-Gur E, Shinder V, Futerman AH, Geiger T, Schuldiner M (2014) A dynamic interface between vacuoles and mitochondria in yeast. Dev Cell 30(1):95–102. https://doi.org/10.1016/j.devcel.2014.06.007 S1534-5807(14)00378-5 [pii]

Esk C, Chen CY, Johannes L, Brodsky FM (2010) The clathrin heavy chain isoform CHC22 functions in a novel endosomal sorting step. J Cell Biol 188(1):131–144. https://doi.org/10.1083/jcb.200908057 jcb.200908057 [pii]

Fader CM, Colombo MI (2009) Autophagy and multivesicular bodies: two closely related partners. Cell Death Differ 16(1):70–78. https://doi.org/10.1038/cdd.2008.168 cdd2008168 [pii]

Filimonenko M, Stuffers S, Raiborg C, Yamamoto A, Malerod L, Fisher EM, Isaacs A, Brech A, Stenmark H, Simonsen A (2007) Functional multivesicular bodies are required for autophagic clearance of protein aggregates associated with neurodegenerative disease. J Cell Biol 179(3):485–500. https://doi.org/10.1083/jcb.200702115 jcb.200702115 [pii]

Friedman JR, Dibenedetto JR, West M, Rowland AA, Voeltz GK (2013) Endoplasmic reticulum-endosome contact increases as endosomes traffic and mature. Mol Biol Cell 24(7):1030–1040. https://doi.org/10.1091/mbc.E12-10-0733 mbc.E12-10-0733 [pii]

Ganley IG, Wong PM, Gammoh N, Jiang X (2011) Distinct autophagosomal-lysosomal fusion mechanism revealed by thapsigargin-induced autophagy arrest. Mol Cell 42(6):731–743. https://doi.org/10.1016/j.molcel.2011.04.024 S1097-2765(11)00370-4 [pii]

Garrity AG, Wang W, Collier CM, Levey SA, Gao Q, Xu H (2016) The endoplasmic reticulum, not the pH gradient, drives calcium refilling of lysosomes. Elife 5. https://doi.org/10.7554/elife.15887 e15887 [pii]

Glick D, Barth S, Macleod KF (2010) Autophagy: cellular and molecular mechanisms. J Pathol 221(1):3–12. https://doi.org/10.1002/path.2697

Haj FG, Verveer PJ, Squire A, Neel BG, Bastiaens PI (2002) Imaging sites of receptor dephosphorylation by PTP1B on the surface of the endoplasmic reticulum. Science 295(5560):1708–1711. https://doi.org/10.1126/science.1067566

Hamdi A, Roshan TM, Kahawita TM, Mason AB, Sheftel AD, Ponka P (2016) Erythroid cell mitochondria receive endosomal iron by a “kiss-and-run” mechanism. Biochim Biophys Acta 1863(12):2859–2867. https://doi.org/10.1016/j.bbamcr.2016.09.008 S0167-4889(16)30228-2 [pii]

Hammerling BC, Najor RH, Cortez MQ, Shires SE, Leon LJ, Gonzalez ER, Boassa D, Phan S, Thor A, Jimenez RE, Li H, Kitsis RN, Dorn Ii GW, Sadoshima J, Ellisman MH, Gustafsson AB (2017) A Rab5 endosomal pathway mediates Parkin-dependent mitochondrial clearance. Nat Commun 8:14050. https://doi.org/10.1038/ncomms14050 ncomms14050 [pii]
Henne WM (2016) Organelle remodeling at membrane contact sites. J Struct Biol 196(1):15–19. https://doi.org/10.1016/j.jsb.2016.05.003 S1047-8477(16)30088-0 [pii]
Honscher C, Mari M, Auffarth K, Bohnert M, Griffith J, Geerts W, van der Laan M, Cabrera M, Reggiori F, Ungermann C (2014) Cellular metabolism regulates contact sites between vacuoles and mitochondria. Dev Cell 30(1):86–94. https://doi.org/10.1016/j.devcel.2014.06.006 S1534-5807(14)00377-3 [pii]
Huberts DH, van der Klei IJ (2010) Moonlighting proteins: an intriguing mode of multitasking. Biochim Biophys Acta 1803(4):520–525. https://doi.org/10.1016/j.bbamcr.2010.01.022
Ikonen E (2008) Cellular cholesterol trafficking and compartmentalization. Nat Rev Mol Cell Biol 9(2):125–138. https://doi.org/10.1038/nrm2336 nrm2336 [pii]
Imai K, Hao F, Fujita N, Tsuji Y, Oe Y, Araki Y, Hamasaki M, Noda T, Yoshimori T (2016) Atg9A trafficking through the recycling endosomes is required for autophagosome formation. J Cell Sci 129(20):3781–3791. https://doi.org/10.1242/jcs.196196 jcs.196196 [pii]
Ishikura S, Klip A (2008) Muscle cells engage Rab8A and myosin Vb in insulin-dependent GLUT4 translocation. Am J Physiol Cell Physiol 295(4):C1016–C1025. https://doi.org/10.1152/ajpcell.00277.2008 00277.2008 [pii]
Jeffery CJ (1999) Moonlighting proteins. Trends Biochem Sci 24(1):8–11
Jiang P, Nishimura T, Sakamaki Y, Itakura E, Hatta T, Natsume T, Mizushima N (2014) The HOPS complex mediates autophagosome-lysosome fusion through interaction with syntaxin 17. Mol Biol Cell 25(8):1327–1337. https://doi.org/10.1091/mbc.E13-08-0447 mbc.E13-08-0447 [pii]
Jimenez AJ, Maiuri P, Lafaurie-Janvore J, Divoux S, Piel M, Perez F (2014) ESCRT machinery is required for plasma membrane repair. Science 343(6174):1247136. https://doi.org/10.1126/science.1247136 science.1247136 [pii]
Kandror KV, Pilch PF (2011) The sugar is sIRVed: sorting Glut4 and its fellow travelers. Traffic 12(6):665–671. https://doi.org/10.1111/j.1600-0854.2011.01175.x
Kilpatrick BS, Eden ER, Schapira AH, Futter CE, Patel S (2013) Direct mobilisation of lysosomal Ca^{2+} triggers complex Ca^{2+} signals. J Cell Sci 126(Pt 1):60–66. https://doi.org/10.1242/jcs.118836 jcs.118836 [pii]
Korolchuk VI, Saiki S, Lichtenberg M, Siddiqi FH, Roberts EA, Imarisio S, Jahreiss L, Sarkar S, Futter M, Menzies FM, O'Kane CJ, Deretic V, Rubinsztein DC (2011) Lysosomal positioning coordinates cellular nutrient responses. Nat Cell Biol 13(4):453–460. https://doi.org/10.1038/ncb2204 ncb2204 [pii]
Kozik P, Hodson NA, Sahlender DA, Simecek N, Soromani C, Wu J, Collinson LM, Robinson MS (2013) A human genome-wide screen for regulators of clathrin-coated vesicle formation reveals an unexpected role for the V-ATPase. Nat Cell Biol 15(1):50–60. https://doi.org/10.1038/ncb2652 ncb2652 [pii]
Lam AK, Galione A (2013) The endoplasmic reticulum and junctional membrane communication during calcium signaling. Biochim Biophys Acta 1833(11):2542–2559. https://doi.org/10.1016/j.bbamcr.2013.06.004 S0167-4889(13)00227-9 [pii]
Lang MJ, Martinez-Marquez JY, Prosser DC, Ganser LR, Buelto D, Wendland B, Duncan MC (2014) Glucose starvation inhibits autophagy via vacuolar hydrolysis and induces plasma membrane internalization by down-regulating recycling. J Biol Chem 289(24):16736–16747. https://doi.org/10.1074/jbc.M113.525782 M113.525782 [pii]
Lee JA, Beigneux A, Ahmad ST, Young SG, Gao FB (2007) ESCRT-III dysfunction causes autophagosome accumulation and neurodegeneration. Curr Biol 17(18):1561–1567. https://doi.org/10.1016/j.cub.2007.07.029 S0960-9822(07)01707-1 [pii]

Liang C, Lee JS, Inn KS, Gack MU, Li Q, Roberts EA, Vergne I, Deretic V, Feng P, Akazawa C, Jung JU (2008) Beclin1-binding UVRAG targets the class C Vps complex to coordinate autophagosome maturation and endocytic trafficking. Nat Cell Biol 10(7):776–787. https://doi.org/10.1038/ncb1740 ncb1740 [pii]

Lim CY, Zoncu R (2016) The lysosome as a command-and-control center for cellular metabolism. J Cell Biol 214(6):653–664. https://doi.org/10.1083/jcb.201607005

Lindmo K, Simonsen A, Brech A, Finley K, Rusten TE, Stenmark H (2006) A dual function for deep orange in programmed autophagy in the Drosophila melanogaster fat body. Exp Cell Res 312(11):2018–2027. https://doi.org/10.1016/j.yexcr.2006.03.002 S0014-4827(06)00088-7 [pii]

Liou W, Geuze HJ, Geelen MJ, Slot JW (1997) The autophagic and endocytic pathways converge at the nascent autophagic vacuoles. J Cell Biol 136(1):61–70

Longatti A, Lamb CA, Razi M, Yoshimura S, Barr FA, Tooze SA (2012) TBC1D14 regulates autophagosome formation via Rab11- and ULK1-positive recycling endosomes. J Cell Biol 197(5):659–675. https://doi.org/10.1083/jcb.201111079 jcb.201111079 [pii]

Lopez Sanjurjo CI, Tovey SC, Taylor CW (2014) Rapid recycling of Ca^{2+} between IP_3-sensitive stores and lysosomes. PLoS ONE 9(10):e111275. https://doi.org/10.1371/journal.pone.0111275 PONE-D-14-31104 [pii]

Lopez-Sanjurjo CI, Tovey SC, Prole DL, Taylor CW (2013) Lysosomes shape Ins(1,4,5) P3-evoked Ca^{2+} signals by selectively sequestering Ca^{2+} released from the endoplasmic reticulum. J Cell Sci 126(Pt 1):289–300. https://doi.org/10.1242/jcs.116103 jcs.116103 [pii]

Luo J, Jiang L, Yang H, Song BL (2017) Routes and mechanisms of post-endosomal cholesterol trafficking: a story that never ends. Traffic. https://doi.org/10.1111/tra.12471

Majumder P, Chakrabarti O (2015) Mahogunin regulates fusion between amphisomes/MVBs and lysosomes via ubiquitination of TSG101. Cell Death Dis 6:e1970. https://doi.org/10.1038/cddis.2015.257 cddis2015257 [pii]

Matheoud D, Sugiura A, Bellemare-Pelletier A, Laplante A, Rondeau C, Chemali M, Fazel A, Bergeron JJ, Trudeau LE, Burelle Y, Gagnon E, McBride HM, Desjardins M (2016) Parkinson's disease-related proteins PINK1 and Parkin repress mitochondrial antigen presentation. Cell 166(2):314–327. https://doi.org/10.1016/j.cell.2016.05.039 S0092-8674(16)30590-6 [pii]

McLelland GL, Soubannier V, Chen CX, McBride HM, Fon EA (2014) Parkin and PINK1 function in a vesicular trafficking pathway regulating mitochondrial quality control. EMBO J 33(4):282–295. https://doi.org/10.1002/embj.201385902 embj.201385902 [pii]

McLelland GL, Lee SA, McBride HM, Fon EA (2016) Syntaxin-17 delivers PINK1/parkin-dependent mitochondrial vesicles to the endolysosomal system. J Cell Biol 214(3):275–291. https://doi.org/10.1083/jcb.201603105 jcb.201603105 [pii]

Morgan AJ, Davis LC, Wagner SK, Lewis AM, Parrington J, Churchill GC, Galione A (2013) Bidirectional Ca^{2+} signaling occurs between the endoplasmic reticulum and acidic organelles. J Cell Biol 200(6):789–805. https://doi.org/10.1083/jcb.201204078 jcb.201204078 [pii]

Muller M, Schmidt O, Angelova M, Faserl K, Weys S, Kremser L, Pfaffenwimmer T, Dalik T, Kraft C, Trajanoski Z, Lindner H, Teis D (2015) The coordinated action of the MVB pathway and autophagy ensures cell survival during starvation. Elife 4:e07736. https://doi.org/10.7554/eLife.07736

Muretta JM, Romenskaia I, Mastick CC (2008) Insulin releases Glut4 from static storage compartments into cycling endosomes and increases the rate constant for Glut4 exocytosis. J Biol Chem 283(1):311–323. https://doi.org/10.1074/jbc.M705756200 M705756200 [pii]

Murrow L, Malhotra R, Debnath J (2015) ATG12-ATG3 interacts with Alix to promote basal autophagic flux and late endosome function. Nat Cell Biol 17(3):300–310. https://doi.org/10.1038/ncb3112 ncb3112 [pii]

Olmos Y, Hodgson L, Mantell J, Verkade P, Carlton JG (2015) ESCRT-III controls nuclear envelope reformation. Nature 522(7555):236–239. https://doi.org/10.1038/nature14503 nature14503 [pii]

Oshima R, Hasegawa T, Tamai K, Sugeno N, Yoshida S, Kobayashi J, Kikuchi A, Baba T, Futatsugi A, Sato I, Satoh K, Takeda A, Aoki M, Tanaka N (2016) ESCRT-0 dysfunction compromises autophagic degradation of protein aggregates and facilitates ER stress-mediated neurodegeneration via apoptotic and necroptotic pathways. Sci Rep 6:24997. https://doi.org/10.1038/srep24997 srep24997 [pii]

Palande K, Roovers O, Gits J, Verwijmeren C, Iuchi Y, Fujii J, Neel BG, Karisch R, Tavernier J, Touw IP (2011) Peroxiredoxin-controlled G-CSF signalling at the endoplasmic reticulum-early endosome interface. J Cell Sci 124(Pt 21):3695–3705. https://doi.org/10.1242/jcs.089656 jcs.089656 [pii]

Pfisterer SG, Peranen J, Ikonen E (2016) LDL-cholesterol transport to the endoplasmic reticulum: current concepts. Curr Opin Lipidol 27(3):282–287. https://doi.org/10.1097/MOL.0000000000000292

Platta HW, Stenmark H (2011) Endocytosis and signaling. Curr Opin Cell Biol 23(4):393–403. https://doi.org/10.1016/j.ceb.2011.03.008 S0955-0674(11)00025-1 [pii]

Popovic D, Akutsu M, Novak I, Harper JW, Behrends C, Dikic I (2012) Rab GTPase-activating proteins in autophagy: regulation of endocytic and autophagy pathways by direct binding to human ATG8 modifiers. Mol Cell Biol 32(9):1733–1744. https://doi.org/10.1128/MCB.06717-11 MCB.06717-11 [pii]

Prinz WA (2014) Bridging the gap: membrane contact sites in signaling, metabolism, and organelle dynamics. J Cell Biol 205(6):759–769. https://doi.org/10.1083/jcb.201401126 jcb.201401126 [pii]

Pyrzynska B, Pilecka I, Miaczynska M (2009) Endocytic proteins in the regulation of nuclear signaling, transcription and tumorigenesis. Mol Oncol 3(4):321–338. https://doi.org/10.1016/j.molonc.2009.06.001 S1574-7891(09)00081-7 [pii]

Raab M, Gentili M, de Belly H, Thiam HR, Vargas P, Jimenez AJ, Lautenschlaeger F, Voituriez R, Lennon-Dumenil AM, Manel N, Piel M (2016) ESCRT III repairs nuclear envelope ruptures during cell migration to limit DNA damage and cell death. Science 352(6283):359–362. https://doi.org/10.1126/science.aad7611 science.aad7611 [pii]

Raiborg C, Wenzel EM, Pedersen NM, Olsvik H, Schink KO, Schultz SW, Vietri M, Nisi V, Bucci C, Brech A, Johansen T, Stenmark H (2015a) Repeated ER-endosome contacts promote endosome translocation and neurite outgrowth. Nature 520(7546):234–238. https://doi.org/10.1038/nature14359 nature14359 [pii]

Raiborg C, Wenzel EM, Stenmark H (2015b) ER-endosome contact sites: molecular compositions and functions. EMBO J 34(14):1848–1858. https://doi.org/10.15252/embj.201591481 embj.201591481 [pii]

Raiborg C, Wenzel EM, Pedersen NM, Stenmark H (2016) ER-endosome contact sites in endosome positioning and protrusion outgrowth. Biochem Soc Trans 44(2):441–446. https://doi.org/10.1042/BST20150246 BST20150246 [pii]

Reed SE, Hodgson LR, Song S, May MT, Kelly EE, McCaffrey MW, Mastick CC, Verkade P, Tavare JM (2013) A role for Rab14 in the endocytic trafficking of GLUT4 in 3T3-L1 adipocytes. J Cell Sci 126(Pt 9):1931–1941. https://doi.org/10.1242/jcs.104307 jcs.104307 [pii]

Rocha N, Kuijl C, van der Kant R, Janssen L, Houben D, Janssen H, Zwart W, Neefjes J (2009) Cholesterol sensor ORP1L contacts the ER protein VAP to control Rab7-RILP-p150 Glued and late endosome positioning. J Cell Biol 185(7):1209–1225. https://doi.org/10.1083/jcb.200811005 jcb.200811005 [pii]

Rowland AA, Chitwood PJ, Phillips MJ, Voeltz GK (2014) ER contact sites define the position and timing of endosome fission. Cell 159(5):1027–1041. https://doi.org/10.1016/j.cell.2014.10.023 S0092-8674(14)01311-7 [pii]

Rusten TE, Stenmark H (2009) How do ESCRT proteins control autophagy? J Cell Sci 122(Pt 13):2179–2183. https://doi.org/10.1242/jcs.050021 122/13/2179 [pii]

Rusten TE, Vaccari T, Lindmo K, Rodahl LM, Nezis IP, Sem-Jacobsen C, Wendler F, Vincent JP, Brech A, Bilder D, Stenmark H (2007) ESCRTs and Fab1 regulate distinct steps of autophagy. Curr Biol 17(20):1817–1825. https://doi.org/10.1016/j.cub.2007.09.032 S0960-9822(07)01991-4 [pii]

Sheftel AD, Zhang AS, Brown C, Shirihai OS, Ponka P (2007) Direct interorganellar transfer of iron from endosome to mitochondrion. Blood 110(1):125–132. https://doi.org/10.1182/blood-2007-01-068148 blood-2007-01-068148 [pii]

Soubannier V, Rippstein P, Kaufman BA, Shoubridge EA, McBride HM (2012) Reconstitution of mitochondria derived vesicle formation demonstrates selective enrichment of oxidized cargo. PLoS ONE 7(12):e52830. https://doi.org/10.1371/journal.pone.0052830 PONE-D-12-27184 [pii]

Steinbusch LK, Wijnen W, Schwenk RW, Coumans WA, Hoebers NT, Ouwens DM, Diamant M, Bonen A, Glatz JF, Luiken JJ (2010) Differential regulation of cardiac glucose and fatty acid uptake by endosomal pH and actin filaments. Am J Physiol Cell Physiol 298(6):C1549–C1559. https://doi.org/10.1152/ajpcell.00334.2009 ajpcell.00334.2009 [pii]

Stuible M, Abella JV, Feldhammer M, Nossov M, Sangwan V, Blagoev B, Park M, Tremblay ML (2010) PTP1B targets the endosomal sorting machinery: dephosphorylation of regulatory sites on the endosomal sorting complex required for transport component STAM2. J Biol Chem 285(31):23899–23907. https://doi.org/10.1074/jbc.M110.115295 M110.115295 [pii]

Sun Y, Bilan PJ, Liu Z, Klip A (2010) Rab8A and Rab13 are activated by insulin and regulate GLUT4 translocation in muscle cells. Proc Natl Acad Sci U S A 107(46):19909–19914. https://doi.org/10.1073/pnas.1009523107 1009523107 [pii]

Sun Y, Chiu TT, Foley KP, Bilan PJ, Klip A (2014) Myosin Va mediates Rab8A-regulated GLUT4 vesicle exocytosis in insulin-stimulated muscle cells. Mol Biol Cell 25(7):1159–1170. https://doi.org/10.1091/mbc.E13-08-0493 mbc.E13-08-0493 [pii]

Sun Y, Jaldin-Fincati J, Liu Z, Bilan PJ, Klip A (2016) A complex of Rab13 with MICAL-L2 and alpha-actinin-4 is essential for insulin-dependent GLUT4 exocytosis. Mol Biol Cell 27(1):75–89. https://doi.org/10.1091/mbc.E15-05-0319 mbc.E15-05-0319 [pii]

Szatmari Z, Kis V, Lippai M, Hegedus K, Farago T, Lorincz P, Tanaka T, Juhasz G, Sass M (2014) Rab11 facilitates cross-talk between autophagy and endosomal pathway through regulation of Hook localization. Mol Biol Cell 25(4):522–531. https://doi.org/10.1091/mbc.E13-10-0574 mbc.E13-10-0574 [pii]

Tooze SA, Abada A, Elazar Z (2014) Endocytosis and autophagy: exploitation or cooperation? Cold Spring Harb Perspect Biol 6(5):a018358. https://doi.org/10.1101/cshperspect.a018358 a018358, /5/a018358 [pii]

Urano Y, Watanabe H, Murphy SR, Shibuya Y, Geng Y, Peden AA, Chang CC, Chang TY (2008) Transport of LDL-derived cholesterol from the NPC1 compartment to the ER involves the trans-Golgi network and the SNARE protein complex. Proc Natl Acad Sci U S A 105(43):16513–16518. https://doi.org/10.1073/pnas.0807450105 0807450105 [pii]

Valm AM, Cohen S, Legant WR, Melunis J, Hershberg U, Wait E, Cohen AR, Davidson MW, Betzig E, Lippincott-Schwartz J (2017) Applying systems-level spectral imaging and analysis to reveal the organelle interactome. Nature 546(7656):162–167. https://doi.org/10.1038/nature22369

van der Kant R, Zondervan I, Janssen L, Neefjes J (2013) Cholesterol-binding molecules MLN64 and ORP1L mark distinct late endosomes with transporters ABCA3 and NPC1. J Lipid Res 54(8):2153–2165. https://doi.org/10.1194/jlr.M037325 jlr.M037325 [pii]

Ventimiglia LN, Martin-Serrano J (2016) ESCRT machinery: damage control at the nuclear membrane. Cell Res 26(6):641–642. https://doi.org/10.1038/cr.2016.52 cr201652 [pii]

Vietri M, Schink KO, Campsteijn C, Wegner CS, Schultz SW, Christ L, Thoresen SB, Brech A, Raiborg C, Stenmark H (2015) Spastin and ESCRT-III coordinate mitotic spindle disassembly and nuclear envelope sealing. Nature 522(7555):231–235. https://doi.org/10.1038/nature14408 nature14408 [pii]

Wang W, Wang X, Fujioka H, Hoppel C, Whone AL, Caldwell MA, Cullen PJ, Liu J, Zhu X (2016) Parkinson's disease-associated mutant VPS35 causes mitochondrial dysfunction by recycling DLP1 complexes. Nat Med 22(1):54–63. https://doi.org/10.1038/nm.3983 nm.3983 [pii]

Wartosch L, Gunesdogan U, Graham SC, Luzio JP (2015) Recruitment of VPS33A to HOPS by VPS16 is required for lysosome fusion with endosomes and autophagosomes. Traffic 16(7):727–742. https://doi.org/10.1111/tra.12283

Webster BM, Colombi P, Jager J, Lusk CP (2014) Surveillance of nuclear pore complex assembly by ESCRT-III/Vps4. Cell 159(2):388–401. https://doi.org/10.1016/j.cell.2014.09.012 S0092-8674(14)01161-1 [pii]

Wilhelm LP, Wendling C, Vedie B, Kobayashi T, Chenard MP, Tomasetto C, Drin G, Alpy F (2017) STARD3 mediates endoplasmic reticulum-to-endosome cholesterol transport at membrane contact sites. EMBO J 36(10):1412–1433. https://doi.org/10.15252/embj.201695917

Xu Y, Rubin BR, Orme CM, Karpikov A, Yu C, Bogan JS, Toomre DK (2011) Dual-mode of insulin action controls GLUT4 vesicle exocytosis. J Cell Biol 193(4):643–653. https://doi.org/10.1083/jcb.201008135 jcb.201008135 [pii]

Zavodszky E, Seaman MN, Moreau K, Jimenez-Sanchez M, Breusegem SY, Harbour ME, Rubinsztein DC (2014) Mutation in VPS35 associated with Parkinson's disease impairs WASH complex association and inhibits autophagy. Nat Commun 5:3828. https://doi.org/10.1038/ncomms4828 ncomms4828 [pii]

Zeigerer A, Bogorad RL, Sharma K, Gilleron J, Seifert S, Sales S, Berndt N, Bulik S, Marsico G, D'Souza RC, Lakshmanaperumal N, Meganathan K, Natarajan K, Sachinidis A, Dahl A, Holzhutter HG, Shevchenko A, Mann M, Koteliansky V, Zerial M (2015) Regulation of liver metabolism by the endosomal GTPase Rab5. Cell Rep 11(6):884–892. https://doi.org/10.1016/j.celrep.2015.04.018 S2211-1247(15)00405-2 [pii]

Zhang M, Liu P, Dwyer NK, Christenson LK, Fujimoto T, Martinez F, Comly M, Hanover JA, Blanchette-Mackie EJ, Strauss JF 3rd (2002) MLN64 mediates mobilization of lysosomal cholesterol to steroidogenic mitochondria. J Biol Chem 277(36):33300–33310. https://doi.org/10.1074/jbc.M200003200 M200003200 [pii]

Zhang AS, Sheftel AD, Ponka P (2005) Intracellular kinetics of iron in reticulocytes: evidence for endosome involvement in iron targeting to mitochondria. Blood 105(1):368–375. https://doi.org/10.1182/blood-2004-06-2226 2004-06-2226 [pii]

Chapter 3
GTPases Rac1 and Ras Signaling from Endosomes

Francesc Tebar, Carlos Enrich, Carles Rentero and Thomas Grewal

Contents

Abstract The endocytic compartment is not only the functional continuity of the plasma membrane but consists of a diverse collection of intracellular heterogeneous complex structures that transport, amplify, sustain, and/or sort signaling molecules. Over the years, it has become evident that early, late, and recycling endosomes represent an interconnected vesicular-tubular network able to form signaling platforms that dynamically and efficiently translate extracellular signals into biological outcome. Cell activation, differentiation, migration, death, and survival are some of the endpoints of endosomal signaling. Hence, to understand the role of the endosomal system in signal transduction in space and time, it is therefore necessary to dissect and identify the plethora of decoders that are operational in the different steps along the endocytic pathway. In this chapter, we focus on the regulation of spatiotemporal signaling in cells, considering endosomes as central platforms, in which several small GTPases proteins of the Ras superfamily, in particular Ras and Rac1, actively participate to control cellular processes like proliferation and cell mobility.

F. Tebar (✉) · C. Enrich · C. Rentero
Departament de Biomedicina, Unitat de Biologia Cel·lular, Facultat de Medicina i Ciències de la Salut, Centre de Recerca Biomèdica CELLEX, Institut d'Investigacions Biomèdiques August Pi i Sunyer (IDIBAPS), Universitat de Barcelona, Casanova 143, 08036 Barcelona, Spain
e-mail: tebar@ub.edu

T. Grewal
School of Pharmacy, Faculty of Medicine and Health, University of Sydney, Sydney, NSW 2006, Australia

C. Lamaze and I. Prior (eds.), *Endocytosis and Signaling*, Progress in Molecular and Subcellular Biology 57, https://doi.org/10.1007/978-3-319-96704-2_3

3.1 Introduction

3.1.1 An Overview of the Endosomal System

Endocytosis describes the internalization of nutrients, receptor–ligand complexes, fluids, lipids, extracellular proteins and viruses, and many other biomolecules. Once inside cells, endosomes are responsible for the fine-tuning of multiple pathways that enable the degradation, recycling, storage, activation, or signaling of incoming molecules. Since their discovery in the late 70s, using biochemical analysis and cellular fractionation techniques on extracts from rat liver (Bergeron et al. 1978; Posner et al. 1980; Marsh et al. 1983; Debanne et al. 1982), the knowledge on the complexity of the endosomal compartment has vastly increased. Over the years, a huge number of publications revealed an astonishing complex system of vacuolar-tubular endosomal structures with distinct characteristics that are dynamically interconnected, communicating with the plasma membrane and other cellular compartments like the trans-Golgi network (TGN). The proper functioning of this endocytic system guarantees balanced cellular homeostasis and essential cellular processes like proliferation, migration, invasion, division, polarity among many others (Scita and Di Fiore 2010; Huotari and Helenius 2011; Gould and Lippincott-Schwartz 2009).

The endosomal membrane system consists of several different compartments. Early endosomes (EEs) are the first and main sorting station in the endocytic pathway where endocytosed molecules that enter the cell via clathrin-dependent (CDE) or clathrin-independent (CIE) pathways are routed to different cellular destinations. From EEs, most internalized molecules are recycled back to the plasma membrane directly by transport vesicles or indirectly through passage via recycling endosomes (REs). Molecules in EE that are neither segregated for recycling nor directed to the TGN, where it can be delivered to the cell surface via the secretory pathway, are destined for degradation along the lysosomal pathway. For this to occur, EEs mature to late endosomes (LEs)/multivesicular bodies (MVBs) and thereafter fusion with lysosomes (Scita and Di Fiore 2010; Huotari and Helenius 2011; Mayor and Pagano 2007). There is also a continuous exchange between the TGN and the EEs in order to direct acid hydrolases from the Golgi apparatus to lysosomes. The structure, identity, and functionality of each endosomal compartment is characterized by a specific luminal pH regulated by the vacuolar proton pump V-ATPase and most relevant to this review, a specific combination of proteins, lipids, and signaling complexes attached to its cytosolic membrane surface. In particular, the family of Rab GTPases is critical for the organization of microenvironments that determine endosomal functions. Rab proteins define the identity of endosomal subdomains by recruiting effectors and facilitate membrane flux along the endocytic pathway (Jovic et al. 2010; Zerial and McBride 2001). In the following, we will briefly outline the characteristics of the various endosomal compartments, their respective Rab proteins and the Rab effectors that define their identity. Intruigingly, some of these proteins and lipids listed below and required to determine uniqueness and guiding of vesicle transport within the endosomal compartment are intimately linked to the trafficking and signaling of rat sarcoma (Ras) and Ras-related C3 botulinum toxin substrate 1 (Rac1) proteins from EEs, REs, and LEs (see Sect. 3.1.2 onwards).

EEs represent a weakly acidic (pH 6.8–6.1) compartment that is mainly located in the periphery of the cell. This compartment is principally characterized by a tubulo-vesicular morphology and the presence of Rab5 and its effector Vps34, a phosphatidylinositol 3-kinase type III that generates the phosphoinositide (PI) PtdIns(3)P. The presence of active Rab5 (Rab5-GTP), which is generated in a guanine exchange factor (GEF) [rabaptin 5-associated exchange factor for RAB5 (Rabex5)] and PtdIns(3)P-dependent manner, allows the recruitment of the effector early endosome antigen 1 (EEA1), which together with SNAREs (syntaxin 6 and 13) enables endosome fusion. Rab5-GTP and PtdIns(3)P also recruit rabenosin-5, which interacts with EH domain-containing protein 1 (EHD1) and regulates recycling from EEs. In addition, EEs contain distinctive membrane microdomains enriched in specific proteins [including Rab4, Rab11, ADP ribosylation factor 1 (Arf1), complex protein I (COPI), retromer, Rab9, or Rab7] that together with the cytoskeleton, actin, and microtubules, regulate the subcellular trafficking of certain cargoes through slow and fast recycling routes to the plasma membrane, the degradative pathway to lysosomes or retrograde transport to the TGN (Huotari and Helenius 2011; Vonderheit and Helenius 2005; Rojas et al. 2008; Bonifacino and Rojas 2006; Hayer et al. 2010; Bonifacino and Hurley 2008; Pfeffer 2009; Zerial and McBride 2001; Sigismund et al. 2012; Johannes and Popoff 2008). There is also a subpopulation of EEs that instead of EEA1 contain the Rab5 effector Adaptor protein containing PH domain, PTB domain, and Leucine zipper motif 1/2 (APPL1/2). As outlined below, the presence of APPL1/2 on EE might be responsible to create specific endosomal subpopulations that can trigger signaling events related to cell growth (see Sect. 3.1.2) (Miaczynska et al. 2004; Schenck et al. 2008; Zoncu et al. 2009).

Rab5 also promotes the transformation of EEs to LEs by ensuing conversion from a Rab5-positive to a Rab7-positive (LE) compartment. This Rab conversion is achieved when the Mon1-complex binds to PtdIns(3)P and facilitates the exchange of the Rab5-GEF Rabex5 for the Rab7-GEF that is associated with the homotypic fusion and protein sorting (HOPS) complex (Poteryaev et al. 2010; Rink et al. 2005). In this LE maturation, intraluminal vesicles (ILVs) are acquired through the participation of the endosomal sorting complex required for Transport (ESCRT) complexes by a regulated posttranslational ubiquitination and de-ubiquitination modification. Although a simple maturation that confers the conversion of EE to LE compartment is possible, it seems that a fission of domains within EE that acquired LE features participates in the formation of the LE compartment. The resulting endosomal carrier vesicles (ECVs) from this fission process move to the center of the cell via microtubules and fuse with the LE compartment. The vacuolar LEs, arranged in a perinuclear location, are more acidic (pH: 6.0–5.0), contain Rab7 and the PtdIns(3,5)P_2, which is synthesized by the PtdIns(3)P 5-kinase FAB1/PIKfyve. Finally, LEs will mature or fuse with lysosomes containing acidic hydrolases (proteases and lipases among others). This compartment has a pH around 5.0–4.5, and its membrane is protected by the presence of lysosome-associated membrane proteins (LAMP) proteins (Huotari and Helenius 2011; Poteryaev et al. 2010; Jovic et al. 2010; Scott et al. 2014; Platta and Stenmark 2011).

3.1.2 Endocytosis and Signaling from Endosomal Compartments

Over the last decades, it has become apparent that signaling within the endosomal system contributes to an enormous variety of events that participate in a range of cellular processes. This has led to the widely accepted concept of the "signaling endosome."

This concept recognizes endocytosis to play a key role attenuating signals generated by activated receptors at the plasma membrane, directing them into the degradative-lysosomal pathway. On the other hand, on route to lysosomes, these active receptors can continue to signal in EEs and LEs/MVBs. Therefore, endosomes act as signaling platforms to maintain or prolong signals generated at the plasma membrane. Sustained stimulation can be also enhanced by increased recycling of receptors to the plasma membrane, which is often observed in tumor cells in order to intensify proliferative signals. In addition, endosomes also seem to promote localized and selective recruitment of scaffold and effectors proteins, thereby assembling specific modules. These signaling building blocks can also be transported and directed, through endocytosis-mediated recycling, to specific places in the cell or to specific domains at the plasma membrane to achieve functionality (Sigismund et al. 2012; Lobert and Stenmark 2011; Jones et al. 2006; Frittoli et al. 2011; Taub et al. 2007; Teis and Huber 2003; Palamidessi et al. 2008; White et al. 2006; Sorkin and von Zastrow 2009; Platta and Stenmark 2011; Ohashi et al. 2011; Puthenveedu et al. 2010; Dobrowolski and De Robertis 2011; Schiefermeier et al. 2014; Villasenor et al. 2016).

A substantial number of publications, using biochemical cellular fractionation techniques as well as new live cell imaging techniques such as fluorescence resonance energy transfer (FRET), fluorescence-lifetime imaging microscopy (FLIM), photoactivatable fluorescent proteins (PAFPs), or Ras and interacting protein chimeric unit (Raichu) biosensors (Gonnord et al. 2012; Miaczynska and Bar-Sagi 2010; Murphy et al. 2009), have highlighted the role of the endocytic system to attenuate or sustain signaling, to participate in specific outcomes and to direct modules or complexes to specific subcellular sites or microdomains. The potential relevance of the endocytic system is highlighted in cancer, where altered dynamics of the endocytic pathway are often associated with an inability to properly internalize, recycle, or degrade key cancer drivers such as receptor tyrosine kinases, leading to aberrant proliferation and metastasis of tumor cells (Stasyk and Huber 2016; Lanzetti and Di Fiore 2008; Porther and Barbieri 2015).

Before embarking on the focus of this chapter, the signaling of the small GTPases Ras and Rac on endosomes (Sect. 3.2), it is essential to outline the overarching role of cell surface receptors in signal transduction. Hence, in the following, we will first illustrate in more detail the endosome signaling outputs for some receptors.

Specific endosomal signaling has been demonstrated for several families of cell surface receptors, in particular, receptor tyrosine kinases (RTKs) such as the epidermal growth factor receptor (EGFR), platelet-derived growth factor receptor (PDGFR), hepatocyte growth factor receptor (HGF-R or Met), tropomyosin receptor kinase A (TrkA), and insulin receptor (Ins-R). In addition, serine/threonine kinases such as transforming growth factor beta (TGFβ) receptors, G-protein coupled receptors (GPCRs), or Wnt and Notch receptors have also been reported to signal from endosomes (Stasyk and Huber 2016; Joffre et al. 2011; Wang et al. 2002; Sorkin and von Zastrow 2009; Dobrowolski and De Robertis 2011; Kermorgant and Parker 2008; Murphy et al. 2009; Flinn et al. 2010; Gould and Lippincott-Schwartz 2009; Lanzetti and Di Fiore 2008; Le Roy and Wrana 2005; Tomas et al. 2014; Mellman and Yarden 2013; Barrow-McGee and Kermorgant 2014).

To reach the endosomal compartment, the above-mentioned receptors appear to employ different internalization routes, either via CDE or CIE. Interestingly, depending on ligand concentration, some receptors can be internalized by both entry routes, which then differentially impacts on their signaling output. For instance, while CIE for TGFβ-R and EGFR directs ligand–receptor complexes to lysosomes for degradation, the CDE routes these receptors to signaling competent endosomal vesicles distinct from the lysosomal pathway (Sigismund et al. 2008, 2012). Moreover, signaling of these TGFβ- and EGF receptors, that remain ligand-bound and active after internalization, are principal examples for the creation of a very localized and specific signal transduction elicited from EE microdomains exclusively located within a population of Smad anchor for receptor activation (SARA)- and APPL-positive endosomes, respectively. SARA, which binds PtdIns(3)P on EEs, is a scaffold protein that interacts with TGFβ-R and Mothers against decapentaplegic (SMAD) family member 2 enabling the phosphorylation of the former by TGFβ and its translocation to the nucleus to regulate gene transcription (Hayes et al. 2002; Tsukazaki et al. 1998; Di Guglielmo et al. 2003). On the other hand, APPL, which recruits protein kinase B (PKB or AKT) and its substrate glycogen synthase kinase 3 (GSK3), is activated and translocated to the nucleus by the endocytosed EGFR (Miaczynska et al. 2004).

Other examples for the relevance of endosomal signaling establishing discrete cellular functions include the sustained signaling of GPCRs on endosomes, which is important to ensure continuous cyclic AMP production and mitogen-activated protein kinase (MEK)—extracellular signal-regulated kinases (ERK) signaling beyond its initial activation at the cell surface (Vilardaga et al. 2014; Shenoy and Lefkowitz 2011; Sorkin and von Zastrow 2002). Internalization of nerve growth factor (NGF) bound to its TrkA receptor has also been proven necessary to prolong activation of Rap1 GTPase and to promote neuronal survival via ERK5 activation and cAMP response element binding protein (CREB)-dependent transcription (Wu et al. 2001;

Watson et al. 2001). In addition, the NGF-TrkA receptor–ligand complex can also activate the Ras/mitogen-activated protein kinase (MAPK) pathway from endosomes in neuronal cells (Howe et al. 2001).

EGFR exemplifies a classical example of specific and sustained signaling on endosomes. EGF induces the accumulation of EGFR and downstream signaling molecules such as SH2-containing collagen related (Shc), growth factor receptor-bound protein 2 (Grb2), son of sevenless (Sos), Ras, as well as the serine/threonine kinases rapidly growing fibrosarcoma protein 1 (Raf-1), Mek, and Erk1/2 in endosome compartments (Balbis et al. 2007; Lu et al. 2009; Moreto et al. 2008; Pol et al. 1998; Sorkin and von Zastrow 2009; Teis et al. 2006; Di Guglielmo et al. 1994; Nada et al. 2009; Teis et al. 2002; Wang et al. 2002). In fact, early pioneering work from Vieira et al. elegantly demonstrated the importance of CDE for the control of the EGFR signaling cascade (Vieira et al. 1996). In these studies, inhibition of EGFR internalization, through overexpression of a dominant-negative mutant of the vesicular fission protein dynamin (Dynk44A), drastically interfered with activation of Raf-1 kinase downstream Ras and consequently MAPK signal output, indicating a crucial role of endosomes in the EGFR/Ras/Raf-1/MAPK signaling cascade (Vieira et al. 1996; Moreto et al. 2008). Furthermore, in elegant experiments from Wang et al., the EGFR tyrosine kinase inhibitor AG1478 was used to inhibit EGFR signaling at the cell surface, leading to the internalization of nonactive EGF-EGFR complexes into endosomes. Subsequent removal of AG1478 then enabled activation of endosome-associated EGFR followed by Ras, ERK1/2, and Akt signaling to promote cell proliferation and survival (Wang et al. 2002).

Under certain conditions, EGFR signaling likely occurs in the LE compartment, as the late endosomal adaptors p14 and Mek partner-1 (MP1) can recruit Mek1 to participate in EGFR-induced MAPK activation (Teis et al. 2002, 2006). Indeed, in primary hepatocytes late endosomal EGFR signaling has recently been demonstrated to participate in cell cycle progression (Luo et al. 2011). The p14/MP1 complex also serves as a scaffold to recruit Rag GTPases, which sense amino acid levels and together with Ras homolog enriched in brain (RheB) and hVps34, activate mammalian target of rapamycin complex 1 (mTORC1) in the endo-lysosomal (LE/Lys) compartment to promote cell growth via protein synthesis (Sancak et al. 2010; Zoncu et al. 2011; Duan et al. 2015).

In addition, p14/MP1 in the LE/Lys compartment is also critical for cell migration, as the adaptor complex is directed to the cell periphery, in a Rab7-regulated manner, in order to promote focal adhesion (FA) turnover for cell motility and tumor invasion (Schiefermeier et al. 2014). Furthermore, in the context of cell migration, EEs, and LE/Lys have also been identified to facilitate receptor-mediated signaling events that activate and transport Rac1 GTPase to the leading edge (Palamidessi et al. 2008; Menard et al. 2014; Joffre et al. 2011). Specific guanine exchange factors (GEFs) of the small GTPase Rac1 are known to activate Rac1 in EEs and LEs. This is followed by vesicular transport to the plasma membrane, where active Rac1 then regulates the actin cytoskeleton to promote the formation of lamellipodia in the leading edge, altogether creating forward movement. In the rear of migratory cells, EE and LE control the delivery of the pro-migratory Endo180 receptor to promote adhesion disassembly by the Rho kinase-derived contractile signals (Gould and Lippincott-Schwartz 2009; Sturge et al. 2006).

Further to the trafficking routes of Rac1 and Endo180 receptor to the front and rear of moving cells, respectively, several recent publications have highlighted the importance of recycling for the delivery of integrins from endosomes to specific plasma membrane domains, which in combination with the export of membrane type-1 matrix metalloproteinase 1 (MT1-MMP) also regulates motility and invasiveness (De Franceschi et al. 2015; Alanko and Ivaska 2016). Along these lines, impaired cholesterol export from LEs in Niemann–Pick type C1 (NPC1) mutant cells or upon overexpression of annexin A6 interfered with the task of the RE compartment to deliver cargo to the cell surface. This imbalanced distribution of intracellular cholesterol strongly reduced integrin recycling from RE to plasma membrane, and consequently, inhibited cell migration (Garcia-Melero et al. 2016; Reverter et al. 2014). Integrins are transmembrane adhesion proteins that by forming FAs connect the F-actin cytoskeleton to the extracellular matrix (ECM), thus attaching cells to their surroundings. In migrating cells, integrin recycling via a fast and Rab4-dependent, as well as a slow and Rab11-dependent recycling route, is the driving force to continuously assemble and disassemble FAs in the leading edge enabling forward movement (De Franceschi et al. 2015; Shafaq-Zadah et al. 2016). Additionally, integrin-dependent adhesion to collagen type-1 matrix protein stimulates MT1-MMP translocation, from the intracellular biosynthetic-storage compartment and by a Rab8-dependent exocytosis, to surface structures that promote invasion of MDA-MB-231 tumor cells (Bravo-Cordero et al. 2007). However, additional Rab GTPases (Rab2, Rab5, Rab7, Rab14) and the internalization process could also regulate MT1-MMP activity at the plasma membrane in other settings (Frittoli et al. 2011; Wiesner et al. 2013; Williams and Coppolino 2011; Castro-Castro et al. 2016; Kajiho et al. 2016). Interestingly, from the point of view of endosomal signaling, integrins bind distinct sets of proteins when located at the plasma membrane or in endosomes. Consequently, this elicits different signal output in endosomes: activation of focal adhesion kinase (FAK), Akt, ERK, and suppression of anoikis (Alanko et al. 2015; Alanko and Ivaska 2016).

In summary, in this section we have given an overview how the endo/lysosomal system provides membranous platforms to regulate spatiotemporal signaling in cells (Flinn et al. 2010; Gould and Lippincott-Schwartz 2009; Kermorgant and Parker 2008; Lobert and Stenmark 2011; Ohashi et al. 2011; Palamidessi et al. 2008; Platta and Stenmark 2011; Sorkin and von Zastrow 2009; Taub et al. 2007). In the following, we will discuss in more detail how the endosomal localization of several small GTPases proteins of the Ras superfamily actively participates to control cellular processes like proliferation and cell mobility.

3.2 Signaling of the Small GTPases Ras and Rac1 from Endosomes

The Ras superfamily of GTPases comprising a total of 150 members is also known as small GTPases due to their small molecular weight (20–40 kDa). They are classified into six subfamilies: Ras, Rho, Rab, Arf, Ran, and Galpha subunits (Flinn et al. 2010; Rojas et al. 2012). The Rho family members control cytoskeleton dynamics and cellular mobility, while Rab and Arf proteins modulate formation and transport of intracellular vesicles, including exo- and endocytosis. The Ran family coordinates nuclear transport processes, and Galpha subunits manage GPCR signaling. Finally, Ras GTPases are located upstream signaling cascades such as the Raf-1/MAPK pathway that is known to regulate transcription relevant for proliferation and differentiation processes among others.

Rac1, which will be discussed in more detail in Sect. 3.2.2, belongs to the Rho subfamily and within the Ras subfamily there are four isoforms (HRAS, NRAS, KRAS4A, and KRas4B). All of them have in common a module called the G-domain, which adopts an α/β topology and contains 60–180 residues, responsible for nucleotide (GTP)-dependent conformational changes of two internal regions named switch region I (residues 30–38) and switch region II (residues 59–67) (Wittinghofer and Vetter 2011). The ability to bind GTP and the concomitant conformational changes associated with GTP or GDP binding enables Ras proteins to act as binary molecular switches, active when GTP is bound to the G-domain and inactive when GDP is associated with the G-domain. In the active state (GTP-bound), GTPases interact and activate a plethora of different effectors with different functions, including Raf-1, phosphoinositide 3–kinase (PI3K), RAL guanine nucleotide dissociation stimulator (RalGDS), T lymphoma invasion and metastasis-inducing 1 (TIAM1), WASP-family verprolin-homologous protein (WAVE,) and others. The active and inactive state cycle is controlled by guanine exchange factors (GEFs) which facilitate the exchange of GDP for GTP. The ability of GEFs to remove GDP enables more efficient GTP binding which is found at approximately tenfold higher concentrations compared to GDP in the cytosol. On the other hand, GTPase activating proteins (GAPs) increase the intrinsic GTP hydrolyzing capacities of GTPases to ensure rapid inactivation (Rajalingam et al. 2007; Downward 1996; Marshall 1996). Both, GEFs and GAPs, are spatially and temporally modulated by external stimuli and signaling molecules (Bos et al. 2007).

Given their prominent localization at the plasma membrane and proximity to RTKs and GPCRs, Ras GTPases are key players in the initial steps of signaling cascades generated at the cell surface and therefore control important processes such as proliferation, differentiation, apoptosis, cytoskeleton dynamics or cell motility (Malumbres and Barbacid 2003). However, critical for the "signaling endosome" concept, the presence of Ras proteins in the endosomal compartment makes them the key transducers of the signaling events generated by active receptors along endocytic pathways. As outlined in the following sections in more detail, Ras and Rac1 signaling from endomembranes, in particular from endosomes, have been associated with proliferation, apoptosis, and/or cellular migration. Nevertheless, it should be noted that some caution should be taken when referring to the conclusions based on Ras and Rac1 signaling from endosomes as the majority of published data has been obtained from model systems that often express unphysiologically high levels of receptors and signaling components.

3.2.1 Ras Signaling from Endosomes

3.2.1.1 The Ras Family

In human cells, Ras isoforms are encoded by three genes: *KRAS* (Kirsten rat sarcoma viral oncogene homolog, *NRAS* (Neuroblastoma RAS viral (v-ras) oncogene homolog), and *HRAS* (Harvey rat sarcoma viral oncogene homolog). This gives rise to 4 different Ras isoforms (~21 kDa): HRas, NRas, KRas4A, and KRas4B (referred to as KRas), the latter two being derived from alternative RNA splicing. Since Ras proteins control proliferation, survival and migration, missense mutations at position G12, G13, and Q61 result in gain-of-function Ras mutants that confer oncogenic activity. All these mutations impair Ras-GTP hydrolysis by inhibiting its intrinsic GTPase activity. Consequently, this interferes with the action of GAPs and Ras remains in a constitutively active state. Hence, oncogenic Ras mutant hyperactivity contributes to the initiation and progression of a large variety of human cancers (~25% of human cancers, with KRas being the most frequently mutated isoform) (Hobbs et al. 2016; Barbacid 1987; Bos 1989; Malumbres and Barbacid 2003; Rajalingam et al. 2007; Newlaczyl et al. 2014).

Given their prominent contribution to the development of tumorigenic events, much effort over the years aimed to unravel the structure, localization and differential function of the Ras isoforms. Structural analysis first revealed that all Ras isoforms shared a highly conserved and nearly identical globular N-terminal domain (residues 1–165), which binds nucleotides and the majority of Ras effectors (see Sects. 3.2.1.3 and 3.2.1.4). However, as described in more detail below, important differences were found in the C-terminal domain (last 24–25 residues), named the hypervariable region (HVR), which is posttranslational modified and responsible for differential subcellular localization and as a result differential signaling among the Ras isoforms (Barbacid 1987; Hancock 2003; Mor and Philips 2006; Fotiadou et al. 2007; Calvo et al. 2010; Eisenberg and Henis 2008; Prior and Hancock 2012). For instance, in cell culture, KRas activates the small GTPase Rac1 more efficiently than HRas because of their differently membrane anchoring and localization, which is consequently translated in more efficient KRas induction of membrane ruffling, pinocytosis, cell motility, and cell survival than HRas (Walsh and Bar-Sagi 2001).

3.2.1.2 Synthesis, Processing, and Trafficking of Ras Isoforms to the Plasma Membrane

Ras proteins are synthesized in the cytosol as globular hydrophilic proteins (188 amino acids in the case of KRas and 189 for all other Ras isoforms) containing a C-terminal CAAX motif (C, cysteine; A, aliphatic residue; X, any residue); being CVLS, CVVM and CVIM for H-, N-, and KRas, respectively. This CAAX sequence is essential for subsequent posttranslational modifications of Ras and its successive targeting to different cellular membranes. Because Ras proteins contain a methionine or a serine as last amino acid (X residue), this CAAX sequence is specifically recognized by a farnesyl transferase in the cytosol, which irreversibly incorporates a farnesyl group (15-carbons isoprenyl) to the cysteine C186, and C185 in KRas. Farnesylation allows Ras proteins to insert and localize on the endoplasmic reticulum membrane for subsequent CAAX modifications which includes AAX hydrolysis by Ras-converting enzyme 1 (Rce1) and the methylation of the remaining farnesylcysteine residue by Isoprenylcysteine carboxyl methyltransferase (Icmt) (Fehrenbacher et al. 2009) (see [1] in Fig. 3.1). Strikingly, despite all these posttranslational modifications occurring for each Ras isoform, H-, N-, and KRas then follow different routes from the ER to the plasma membrane (Mor and Philips 2006).

Although farnesylcysteine methylation is an essential prerequisite, it is not sufficient for cell surface delivery, and Ras needs a second signal to finally reach the plasma membrane or other organelles like endosomes. In the case of KRas, this second signal is a polybasic region (PBR) composed of six basic lysine residues near the farnesylcysteine in the HVR. This PBR motif electrostatically interacts with anionic phospholipids in membranes, such as phosphatidylinositol-4,5-bisphosphate PtdIns $(4,5)P_2$, the phosphatidylinositol-3,4,5-trisphosphate PtdIns $(3,4,5)P_3$ or phophatidylserine (PS) (Hancock et al. 1990; Cho et al. 2012; Apolloni et al. 2000; Heo et al. 2006; Yeung et al. 2008; Gelabert-Baldrich et al. 2014). Hence, through a yet still not well-defined Golgi exocytic-independent route that likely involves the above-mentioned electrostatic interactions both signals (farnesyl lipid group and PBR) contribute to direct KRas from the ER to the plasma membrane (Apolloni et al. 2000; Magee and Marshall 1999) (see [2] in Fig. 3.1). The molecular machinery that regulates this trafficking route is also not well known and includes the possibility of a passive electrostatic switch. Alternatively, recent results from the Bastiaens group suggest that the delta-subunit of phosphodiesterase 6 (PDE6-δ), or maybe other chaperone-like proteins, act as a cytosolic solubilization factor and its binding to the farnesyl moiety of KRas could then facilitate KRas trafficking to the plasma membrane. Once at the plasma membrane, the highest electrostatic interaction exerted by acid phospholipids could compete and displace PDE6-δ from KRas, favoring its incorporation or insertion into the inner leaflet of the cell surface phospholipid bilayer (Schmick et al. 2015).

Much different to the complex regulation of KRas translocation to the cell surface, the second signal for targeting H-, N-, and spliced KRas4A proteins to the plasma membrane is represented by the reversible addition of one or two palmitoyl groups at additional cysteine residues adjacent to the CAAX motif. A palmitoyl-transferase

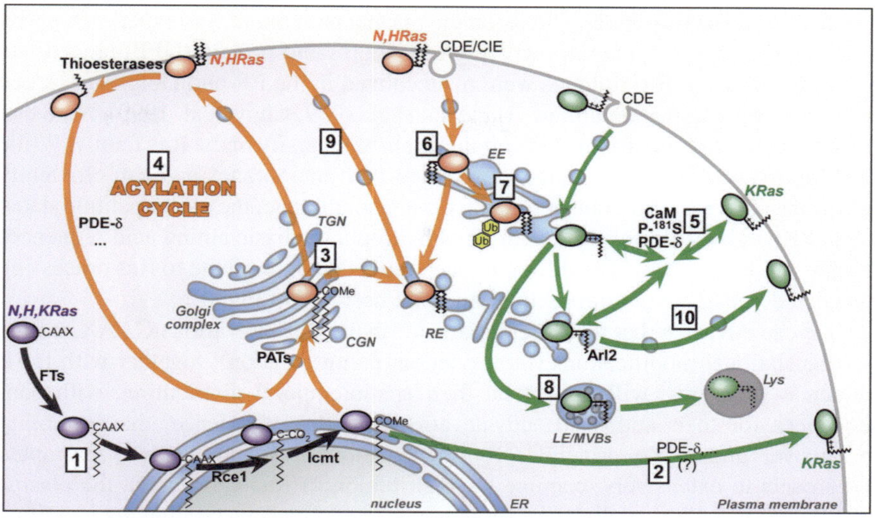

Fig. 3.1 Different routes to the plasma membrane and through the endocytic compartment followed by the Ras isoforms. Once synthesized in the cytosol, Ras isoforms are cysteine farnesylated in their terminal CAAX motif by farnesyl transferases (FTs). This is followed by further posttranslational modifications mediated by Ras-converting enzyme 1 (Rce1) and Isoprenylcysteine carboxyl methyltransferase (Icmt) in the endoplasmic reticulum (ER) [1]. This scheme recapitulates the following traffic routes for the different Ras isoforms: H- or NRas trafficking is shown in orange and KRas trafficking is depicted in green. KRas reaches the PM by still not well-defined non-vesicular routes that likely involve chaperone proteins like PDE-δ [2]. In contrast, H- and NRas use the vesicular exocytic pathway via Golgi and/or Golgi and recycling endosomes (RE) to the plasma membrane (PM) [3]. H- and NRas switch between PM and the Golgi by the action of thioesterases at the PM and palmitoyl-transferases (PATs) at the Golgi, which is known as the acylation cycle [4]. Moreover, from the PM, H-, and NRas can be internalized via CDE or CIE pathways to reach early endosomes (EEs) [6], where they can either be recycled back to PM trough RE [9] or remain retained when ubiquitinated [7]. On the other hand, KRas can also use the vesicular CDE pathway but the bidirectional movement of KRas between PM and endosomal compartments is mainly accomplished through an electrostatic switch modulated by serine phosphorylation (P-^{181}S) or binding proteins, like CaM or PDE-δ, to its hypervariable region [5]. KRas can also return back from RE to the PM via a Arl2-dependent vesicle transport route [10]. In contrast to H- and NRas, KRas moves to late endosome/multivesicular bodies (LE/MVB$_S$) on route for degradation in lysosomes (Lys) [8]

enzyme localized in the Golgi/ER is responsible for palmitoylation of HRas at C181 and C184, NRas at C181and KRas4A at C180 (Hancock et al. 1989; Swarthout et al. 2005). This modification stabilizes the membrane interaction of these Ras isoforms and favors its transport to the plasma membrane through vesicles carriers following the exocytic pathway from the TGN (Choy et al. 1999). In some cases this includes trafficking through REs, which act as a way-station for palmitoylated H- and NRas proteins as they move along the post-Golgi exocytic pathway to the plasma membrane (Misaki et al. 2010) (see [3] in Fig. 3.1). Palmitoylation being decisive for these Ras proteins to reach the plasma membrane was experimentantlly evidenced in live cell imaging, where green fluorescent protein (GFP)-tagged and palmitoyl-deficient Ras mutants were mislocalized in the ER and Golgi and lacked delivery to the plasma membrane (Hancock et al. 1989; Choy et al. 1999). Note that palmitoylation also contributes to certain levels of diversity in the Ras family. While di-palmitoylated HRas is strongly associated with membranes and can efficiently follow the slow exocytic transport to the plasma membrane, the mono-palmitoylated N- and KRas4A isoforms need additional hydrophobic/basic amino acid sequences (Laude and Prior 2008). The different mechanisms that contribute to Ras processing associated with different trafficking routes are depicted in Fig. 3.1.

One can envisage that the differential contribution of the various CAAX motifs, posttranslational modifications (farnesylation, palmitoylation), together with HVR sequence variations, will impact on their spatiotemporal distribution, with consequences for their ability to activate effector pathways. Hence, the possibility to uncover underlying principles for the creation of signal diversity prompted researchers to extensively examine the distribution of Ras isoforms at the plasma membrane. Indeed, subcellular fractionation as well as advanced imaging identified that once at the plasma membrane, Ras isoforms display distinct localizations in specific membrane subdomains. Moreover, it is now believed that active and inactive Ras proteins are organized in nanoclusters, containing 6–7 Ras proteins per nanocluster. Over the years, it has become clear that these clusters highly depend on the distribution of lipids within the membrane. HRas is the best-characterized member of the Ras family in this context, with HRas-GTP nanoclusters being found in disordered (fluid, cholesterol-poor) domains (Prior et al. 2003; Zhou and Hancock 2015). In contrast, HRas GDP clusters are mainly localized in cholesterol- and sphingolipid-rich domains (lipid rafts). While NRas seems to distribute similar to HRas, KRas is predominantly localized outside lipid rafts. In addition, a distinct cohort of phospholipids, in particular phosphatidylserine, but also phosphatidic acid and phosphatidylinositides contribute to the spatial segregation of Ras isoforms. The different distribution of active and inactive Ras proteins implicates lateral movement of Ras isoforms in the plasma membrane. As Ras activity is determined by GTP/GDP exchange, the localized recruitment and activity of GEFs and GAPs appears critical to ensure the transient nature of Ras nanoclusters (Grewal and Enrich 2006). Interestingly, a specialized form of lipid raft, caveolae, seems to critically translate environmental cues, such as mechanical stress, into the remodeling of lipids at the plasma membrane that in turn extensively modulate the organization of Ras nanoclusters (Ariotti et al. 2014). As such, caveolin-1 deficiency or downregulation of cavin-1, another structural component of caveolae, alters phosphatidylserine distribution at the plasma membrane, which correlates with enhanced KRasG12V nanoclustering and MAPK signaling. Yet lateral segregation of HRas was abolished, thereby compromising signal output from HRasG12V nanoclusters (Ariotti et al. 2014).

In addition to their differential distribution at the cell surface outline above, the different posttranslational modifications of each Ras isoforms affect also the association dynamics of cytosolic Ras protein pools with the plasma membrane. It was originally postulated that all Ras isoforms display stable membrane association, moving by lateral diffusion as described above (Niv et al. 1999, 2002). However, under certain conditions or upon specific posttranslational modifications, Ras localization at the plasma membrane switches from lateral diffusion to exchange dynamics with cytoplasmic Ras protein pools (Vartak and Bastiaens 2010). As described above, activation of H- and NRas, via GTP-induced conformational changes, leads to their exchange between lipid rafts and non-rafts (Gorfe et al. 2007). Yet, H- and NRas can also dissociate from the plasma membrane in a process known as the acylation cycle, which involves the depalmitoylation of cysteine residues in their HVR domain that is independent of the activation state of Ras (Rocks et al. 2005, 2010). This non-vesicular pathway is mediated by acyl protein thioesterases in the plasma membrane rapidly returning Ras to the diffusing low-membrane affinity farnesylated Ras pool in the cytoplasm. It is yet unclear how the cell translates the intricate network of cellular and environmental signals to provide the balance between lateral diffusion at the plasma membrane or diffusion into the cytosol, but one modulating factor could be the interaction of the Ras-farnesyl group with PDE-δ when Ras proteins are depalmitoylated (Vartak and Bastiaens 2010; Chandra et al. 2011; Goodwin et al. 2005). This could enable protein acetyltransferase (PAT) enzymes to trap depalmitoylated Ras specifically in Golgi/ER membranes, where another round of palmitoylation and trafficking through the exocytic pathway may then be required to ensure the return to the plasma membrane (Rocks et al. 2005; Schmick et al. 2015) (see [4] in Fig. 3.1).

While a general concept of diffusion dynamics for H- and NRas seems to be emerging, the data on trafficking dynamics of KRas to and from the plasma membrane is more complex and in part controversial. Initially, it was proposed that the half-time for KRas membrane desorption was in the order of minutes or even longer, which was in consonance with a postulated stable plasma membrane association regulated by lateral diffusion (Silvius et al. 2006; Niv et al. 1999, 2002). Lately, Silvius et al. (2006), using an inducible heterodimerization technique between ectopically expressed KRas and a complementary binding partner confirmed predominant KRas association with the plasma membrane under basal conditions. Yet, this included a continuous cycling of KRas on and off the membrane in an interval of minutes. Further results from Yokoe and Meyer indicate that the rate of KRas exchange between plasma membrane and cytoplasm may be even more rapid, within seconds (Yokoe and Meyer 1996). In line with these findings, in vitro studies demonstrated that the PBR of KRas associates with lipid bilayers in a rapidly reversible manner with a half-time of seconds or less (Leventis and Silvius 1998). On the other hand, depending on the experimental settings and stimuli, desorption of KRas from the plasma membrane is also modulated by several KRas-interacting proteins, as such as PDE-δ, prenylated Rab protein acceptor protein 1, calmodulin (CaM), and galectin-3 (Chandra et al. 2011; Bhagatji et al. 2010; Elad-Sfadia et al. 2004; Figueroa et al. 2001; Fivaz and Meyer 2005; Lopez-Alcala et al. 2008; Nancy et al. 2002; Villalonga et al. 2001; Philips 2012). In addition, the protein kinase C (PKC)-mediated phosphorylation of active KRas at serine 181 (Ballester et al. 1987) affects the PBR net charge of KRas, thereby inducing an electrostatic switch that displaces KRas from the plasma membrane to endomembranes, including endosomes (Bivona et al. 2006;

Chandra et al. 2011). KRas phosphorylation at serine 181 inhibits CaM binding and accordingly, CaM efficiently interacts with non-phosphorylated KRas-GTP (Lopez-Alcala et al. 2008), an interaction that also favors KRas desorption from the plasma membrane (Bhagatji et al. 2010; Fivaz and Meyer 2005). In this context, elegant studies from Fivaz and Meyer revealed the physiological significance of these interactions in neuronal cells, as cell activation caused the translocation of KRas from the plasma membrane to endosomes through sequestration of the PBR-farnesyl motif in a Ca^{2+}/CaM-dependent manner (Fivaz and Meyer 2005) (see [5] in Fig. 3.1).

3.2.1.3 Ras Trafficking Along Endocytic Routes

Although Ras proteins are predominantly found at the plasma membrane to transduce signals received from surface receptors, substantial amounts of Ras proteins have also been identified in subcellular organelles where they can elicit several functions. Over the last decades, the combination of cellular fractionation techniques with newly developed advanced microscopy revealed the presence of Ras proteins in endosomal fractions, visualized their trafficking to endosomal structures in live cells, and provided extensive colocalization data with established markers of the different endocytic compartments (Lu et al. 2009; Moreto et al. 2008, 2009; Pol et al. 1998; Howe et al. 2001; Fivaz and Meyer 2005; Jiang and Sorkin 2002; Gomez and Daniotti 2007; Roy et al. 2002; Yeung et al. 2008; Hancock 2003; Gelabert-Baldrich et al. 2014; Choy et al. 1999; Zheng et al. 2012b; Prior and Hancock 2012). Most of the results have been obtained with ectopic expression of fluorescently tagged wildtype, active and inactive Ras mutant proteins in different cell types. Using advanced imaging like FRET, FRAP technology or in the case of KRas, RAICHU probes, these fusion constructs were critical tools to develop experimental approaches that were able to monitor the location and activity of Ras proteins over time in live and fixed cells. As the availability of sensitive antibodies specific for Ras isoforms and their activity remains an issue in the field, these fluorescently labeled Ras proteins have been proven very valuable. However, despite the great insights obtained with these methodologies, it should be noted that a lot of the conclusions drawn from these overexpression studies still lack validation for their endogenous counterparts.

In general, Ras proteins can be translocated from the plasma membrane to endosomes via two different routes. The pathway mainly used by H-, N-, and KRas4A is through endocytic vesicles derived from CDE and CIE pathways (Porat-Shliom et al. 2008; Howe et al. 2001; Gomez and Daniotti 2005; Jiang and Sorkin 2002; Roy et al. 2002) (see [6] in Fig. 3.1). The other route predominantly used by KRas involves desorption from the plasma membrane into the cytosol, followed by non-vesicular diffusion shuttling mechanism to endosomes (Gelabert-Baldrich et al. 2014; Fivaz and Meyer 2005; Yeung et al. 2008; Schmick et al. 2015) (see [5] in Fig. 3.1).

Endocytosed palmitoylated Ras proteins, via CDE or Arf6-dependent CIE encounter the EE or RE compartment in a process regulated by Rab5 or Rab11 GTPases (Gomez and Daniotti 2005; Porat-Shliom et al. 2008; Howe et al. 2001) (see [6] in Fig. 3.1). In EEs, H- and NRas can be mono- and di-ubiquitinated via a lysine 63-linked chains on lysine 117, lysine 147, and lysine 170 independently of their activation state (Jura et al. 2006). This is not a degradative posttranslational modification but rather stabilizes these Ras isoforms in endosomes and inhibits their recycling to plasma membrane (Jura et al. 2006) (see [7] in Fig. 3.1). The HVR of these Ras proteins is not an obligatory acceptor for ubiquitin but participates in ubiquitination by the E3 ubiquitin ligase Rabex5 (Xu et al. 2010). Interestingly, activated Ras induces the recruitment of Rabex-5 to endosomes through a RAS and RAB interactor 1 (RIN1)-containing machinery. Consequently, this leads to Rab5 activation. Hence, through this complex interplay of Ras with ubiquitination and Rab5 activity, Ras modulates its own stability in endosomes (Zheng et al. 2012b; Ahearn et al. 2011). In addition, retention of Ras proteins in endosomes is also regulated by several members of the sorting nexin (SNX) family such as SNX17, SNX27, and SNX31, which bind PtdIns(3)P as well as active Ras on endosomes (Ghai et al. 2011).

In contrast to H- and NRas, KRas is mono-ubiquitinated via lysine 45-linked chain on lysine 104 and lysine 147 and this modification does not affect its endosomal trafficking but enhances interactions with several effectors (Sasaki et al. 2011). Overall, the endosomal localization of KRas is less prominent compared to H- and NRas isoforms, probably because KRas is less retained on endosomes and as a result faster recycling to the plasma membrane may occur (Vigil et al. 2010; Jiang and Sorkin 2002; Roy et al. 2002). Yet, despite the small amount of KRas in EE, KRas can also continue its journey to LE/MVBs and eventually into lysosomes following the degradative endocytic pathway (Lu et al. 2009) (see [8] in Fig. 3.1). This feature is strikingly different to H- and NRas, which are essentially found in EEs and REs following the recycling route to the plasma membrane (see [9] in Fig. 3.1). The trafficking of GFP-tagged KRas through these different compartments was confirmed by colocalization with established endocytic markers for EE, LE and LE/lysosomes (Lu et al. 2009). Moreover, subcellular fractionation identified approximately 10% of ectopically expressed GFP-KRas, but also endogenous KRas independent of its activation state, in purified EE and LE fractions (Lu et al. 2009; Gelabert-Baldrich et al. 2014). The trafficking routes that deliver KRas to LE and lysosomes are not fully resolved. As GFP-KRas is observed in clathrin-coated pits and vesicles, KRas may reach EE/LE/MVB/lysosomes via CDE and endosomal transport vesicles (Lu et al. 2009). However, FRAP microscopy implicated that delivery through transport vesicles accounted for only a minor proportion of GFP-KRas on endosomal membranes (Gelabert-Baldrich et al. 2014).

FRAP analysis identified a fast replenishment (half-time 1.3 s) of approximately 80% of the bleached GFP-KRas pool on endosomes. This indicates that KRas is highly dynamic, which is in agreement with a model of rapid diffusional incorporation from the cytoplasm. Strikingly different from KRas, the fluorescence recovery was negligible for GFP-H- and GFP-KRas4A, strongly supporting their dependence in slow vesicular transport (Gelabert-Baldrich et al. 2014).

Based on these findings, one can assume that cellular stimuli that lead to KRas serine 181 phosphorylation or promote association of known PBR/farnesyl-binding proteins to the active KRas (CaM, PRA1, PDE-δ) could regulate KRas interaction dynamics with endosomes by inducing electrostatic switch or membrane dissociation, respectively (Alvarez-Moya et al. 2011; Bhagatji et al. 2010; Fivaz and Meyer 2005; Lopez-Alcala et al. 2008; Chandra et al. 2011; Bivona et al. 2006). Indeed, FRAP microscopy in COS1 cells identified that the constitutively inactive KRas mutant KRasS17N was more immobile than the active mutant KRasG12V as a consequence that active KRas, but not the inactive, can bind CaM or can be posttranslationally modified via serine 181 phosphorylation (Gelabert-Baldrich et al. 2014) (see [5] Fig. 3.1).

As previously mentioned, the PDE-δ protein, through its interaction with the farnesyl group of KRas, regulates KRas localization and dynamics on endosomal membranes and plasma membrane. This interaction solubilizes KRas from endomembranes with low negatively charged surfaces, including EE membranes, leading to a redistribution of KRas to perinuclear RE membranes. The underlying mechanism for this redistribution was based on an activity in RE that displaced KRas from PDE-δ. Further studies then identified the Arf-like GTPase Arl2, which in its active form binds to an allosteric site on PDE-δ, thereby inducing a conformational change that unloads farnesylated cargo (Ismail et al. 2011). Finally, from RE, KRas follows the recycling pathway to the plasma membrane through vesicular transport (Schmick et al. 2015) (see [10] in Fig. 3.1). Trafficking of the different isoforms is summarized in Fig. 3.1.

3.2.1.4 Ras Signaling Pathways from Endosomes

Several studies identified upstream components of the Ras activation pathway on endosomes, including activated EGFR and adaptor proteins Shc/Grb2. This complex is able to recruit Sos1, facilitating GDP/GTP exchange for increasing Ras activity in the endosomal compartment (von Zastrow and Sorkin 2007; Vieira et al. 1996; Herbst et al. 1994; Jiang and Sorkin 2002; Wang et al. 2002). Ras activation on endosomes in live cells has been elegantly demonstrated using FRET microscopy, which allows the spatiotemporal analysis of interaction between molecules inside cells (Jiang and Sorkin 2002; Moreto et al. 2008; Miaczynska and Bar-Sagi 2010). This methodology is based on the energy transfer between two spectrally overlapping GFP variants in cells, for instance, Cyan/cerulean-FP acting as a donor and Yellow/Venus-FP as an acceptor of energy. In order to determine Ras activation utilizing FRET, one of the GFP variants were fused with Ras, while the other GFP variants was fused with the Ras-binding domain of the Raf1 effector. This approach appeared appropriate to detect Ras activation on endosomes (Lu et al. 2009; Jiang and Sorkin 2002; Gomez and Daniotti 2005; Misaki et al. 2010), in particular, as phosphorylated and activated Raf1 was present in purified endosomes from rat liver (Di Guglielmo et al. 1994; Pol et al. 1998). In addition, bimolecular fluorescent complementation (BiFC) microscopy has been also very useful to characterize several other Ras effectors that interact directly or in a protein complex with active Ras on endosomes, such as PI3K and Cdc42 (Tsutsumi et al. 2009; Chang and Philips 2006; Cheng et al. 2011). The BiFC technique relies on the fusion of the N- and C-terminal Venus fluorescent protein with Ras and the RBD of the selected effector, respectively. Once both proteins interact, the Venus fluorophore rebuilds by complementation, generating a fluorescence signal that can be monitored using microscopy (Zheng and Chang 2014).

The FRET technology described above is based on the overexpression of two fluorescent proteins, often creating false FRET signals that did not correlate with endogenous Ras activation. Unimolecular FRET technology using Raichu probes provided a great advancement for the field. In this approach, YFP-KRas is fused to the CFP-tagged RBD of Raf-1 to provide a single ('biosensor') construct that upon KRas activation, allows intramolecular binding to Raf-RBD. This ultimately brought CFP and YFP in close proximity to create a detectable FRET signal. Using this Raichu probe, it has been demonstrated that EGF-induced activation of KRas in live cells (Lu et al. 2009; Kiyokawa et al. 2006; Mochizuki et al. 2001; Miaczynska and Bar-Sagi 2010).

Despite these findings, the general perception in the field still considers Ras activation at the plasma membrane rather than endosomal signaling as the driver of oncogenic events. However, in NIH3T3 cells, inhibition of Ras internalization to exclusively examine Ras signaling emanating from the cell surface was inefficient to effectively induce cell transformation (Cheng et al. 2011). In addition, a focus formation assay identified expression of endosomal GFP-Ras to produce more foci than the GFP transfected control (Aran and Prior 2013). Hence, endosomal Ras signaling seems to substantially contribute to oncogenic events. In the following, we will dissect some of the Ras isoforms and their effector pathways and endosomal locations that possibly contribute to cell transformation.

The identification of signal specificity within the Ras family has captivated the field for a long time, as the highly conserved effector binding G-domain in all active Ras (GTP-bound) isoforms is capable to interact with the same set of effectors (Wittinghofer and Herrmann 1995) to potentially elicit the same signaling output. Out of more than 20 known Ras effectors, Raf1 and PI3K have been the most extensively studied, driving MEK/MAPK and Akt signaling cascades that control proliferation and cell survival, respectively. Based on their similar binding behavior, it was initially proposed that Ras isoforms would have no preference to couple with either Raf1/MAPK or PI3K/Akt (Omerovic et al. 2008). However, more recent work revealed that depending on the activated Ras isoform, Raf1/MAPK and PI3K/Akt effector pathways appear to be differentially regulated on endosomes. This observation seems to be a consequence of the Ras isoform-specific posttranslational ubiquitin modifications discussed previously. The ubiquitination of H- and NRas on endosomes, which stabilizes their association with EE and simultaneously, reduces recycling to the plasma membrane (see Sect. 3.2.1.3), impairs interaction of H- and NRas with Raf1. Consequently, MAPK activation is reduced, yet PI3K/Akt activation is not affected (Xu et al. 2010; Yan et al. 2010; Jura et al. 2006). Hence, inhibition or overexpression of the ubiquitin ligase Rabex5, which is responsible for H- and NRas ubiquitination on endosomes, resulted in increased or attenuated Raf1/MAPK activation, respectively (Xu et al. 2010; Yan et al. 2010). In a much more complex scenario, Rin1, the GEF for Rab5, can bind HRas on endosomes and simultaneously stimulates Rab5-dependent endocytosis (Cheng et al. 2011; Tall et al. 2001). If the latter then triggers Rabex5 recruitment, this could also lead to HRas ubiquitination, thereby generating a negative feedback mechanism for HRas/MAPK activation.

Despite the substantial amount of data generated from multiple research groups in recent years on Ras signaling from endosomes, including FRET microscopy demonstrating interaction of ectopically expressed HRas with Raf1 on endosomes (Jiang and Sorkin 2002), Sorkin and coworkers recently reported that endogenous HRas signals from receptors activated at the plasma membrane and not from internal membranes (Pinilla-Macua et al. 2016). These findings clearly challenge the now widely accepted concept of HRas-mediated Raf1/MAPK activation on endosomes highlighting the need for cautious interpretation of data based on ectopic Ras overexpression, which generate highly elevated Ras levels compared to their endogenous counterpart. Hence, future research with advanced technology that would clarify the signaling activities of endogenous HRas on endosomes is still required.

In contrast to the inhibitory impact of ubiquitination on H- and NRas activity, the same posttranslational modification enhances KRas activity. This increases interaction with Raf1 and PI3K, thus elevating signal output of MAPK and Akt pathways (Sasaki et al. 2011). Interestingly, HRas- and KRas-mediated activation of the Raf1/MAPK cascade varies in its dependence on endocytosis, CaM and PI3K activity (Roy et al. 2002; Moreto et al. 2008, 2009). Actually, inhibition of CaM and/or PI3K impairs recycling from EEs, which seems to be linked to HRas/MAPK inhibition (Roy et al. 2002). In fact, CaM inhibition generates enlarged endosomes by preventing the exit of endocytosed molecules, such as the EGFR, from EEs (Tebar et al. 2002) by a molecular mechanism that involves PKC-δ activity (Llado et al. 2004) and actin polymerization (Llado et al. 2008).

In COS1 cells, inhibition of endocytosis via overexpression of the dominant-negative dynamin mutant dynK44A, negatively affects HRas-dependent activation of Raf1 (Moreto et al. 2008). Although dynK44A may also affect other cellular processes, Omevoric et al. also demonstrated that inhibition of receptor internalization reduced H- and N-, but not KRas-mediated Raf1 activation (Omerovic et al. 2008). Together with reports demonstrating HRas and Raf1 interaction in Rab11-positive recycling endosomes after serum stimulation (Gomez and Daniotti 2005), this may suggest that Raf1 activation by HRas indeed occurs in endosomes. Taken together, current views favor divergent roles of endosomes in H/NRas-mediated Raf1/MAPK activation. Signal outcome, either abrogation or stimulation, could be explained by the existence of two different pools of H/NRas on endosomes: ubiquitinated Ras (Raf1/MAPK incompetent) and non-ubiquitinated Ras (Raf1/MAPK competent). The final signal output from H- and NRas on endosomes would then depend on the balance between these two Ras pools, which could be modulated differently by environmental signals, experimental settings, and the cell type.

Besides the differential regulation of Ras isoforms in EE described above, an additional layer of signal specificity is achieved through the localization of KRas, but not H- and NRas in LE/MVBs/lysosomes. In COS1 cells, Raichu probes revealed that EGF stimulation increased KRas activity in LE/MVBs, which in combination with Raf1 and the p14/MP1 scaffold for MEK and ERK (Teis et al. 2002), activates MAPK in this compartment (Lu et al. 2009).

Most interestingly, although the pool of ubiquitinated H- and NRas displays impaired MAPK activation in EEs, these isoforms are perfectly competent to activate the Rho-GTPase Cdc42, thus regulating Cdc42-dependent cytoskeletal rearrangements and cell transformation when they are endocytosed (Cheng et al. 2011; Cheng and Chang 2011). These findings complement Ras function in the fission Yeast pombe, which only express one Ras protein, Ras1. At the plasma membrane, Ras1 activates Byr2 (a MEKK homolog)/MAPK but on endomembranes, Ras1 activates Sed1, a GEF for Cdc42 but not MAPK. Thus, in endosomes Ras also activates the cytoskeleton to maintain an elongated morphology, cell polarity, and mitosis (Chang and Philips 2006; Onken et al. 2006). Strikingly, HRas-mediated Cdc42 activation also seems relevant in vivo, as expression of a constitutively active HRas mutant restricted to endomembranes induced tumors in nude mice by a mechanism comprising HRas/Cdc42 complex formation (Cheng et al. 2011). This interaction may be mediated by Dbl, one of several GEFs for Cdc42 (Cheng et al. 2011; Cerione 2004). Moreover, the presence of NRas-, and to a minor extent, KRas-containing Cdc42 complexes in these studies could indicate their contribution to tumor initiation (Cheng et al. 2011).

That Ras isoform signaling from endosomal compartments is intimately linked to the endocytic machinery is further highlighted by a screen for HRas effectors in endosomes using BiFC methodology. This study confirmed interaction of ubiquitinated HRas with Cdc42 and PI3K in this compartment (Zheng and Chang 2014; Tsutsumi et al. 2009), but also identified CHMP6/VPS20 and VPS4A as HRas interaction partners, all proteins of the ESCRT-III complex that controls recycling from endosomes to the plasma membrane (Zheng et al. 2012a). One could speculate that the latter interaction could increase recycling of HRas, but also other components of endosomes, such as growth factor receptors, to enhance and sustain growth factor signaling.

Finally, among the different Ras effectors, it is noteworthy to mention T lymphoma invasion and metastasis-inducing 1 (Tiam1) (Lambert et al. 2002), which is a specific GEF for the small GTPase Rac1 (Habets et al. 1994). Interestingly, endosomal Tiam1 activates Rac1 in this compartment, a prerequisite for the subsequent transport of active Rac1 to specific plasma membrane domains and the generation of lamellipodia in migrating cells (Palamidessi et al. 2008). Together with Tiam1 being required for the development of Ras-induced skin tumors (Malliri et al. 2002), this points at Tiam1 as a critical link between Ras and Rac1 in metastasis. The role of Rac1 on endosomes will be explained in more detail in the following section. Figure 3.2 summarizes endosome signaling outcome of different Ras isoforms.

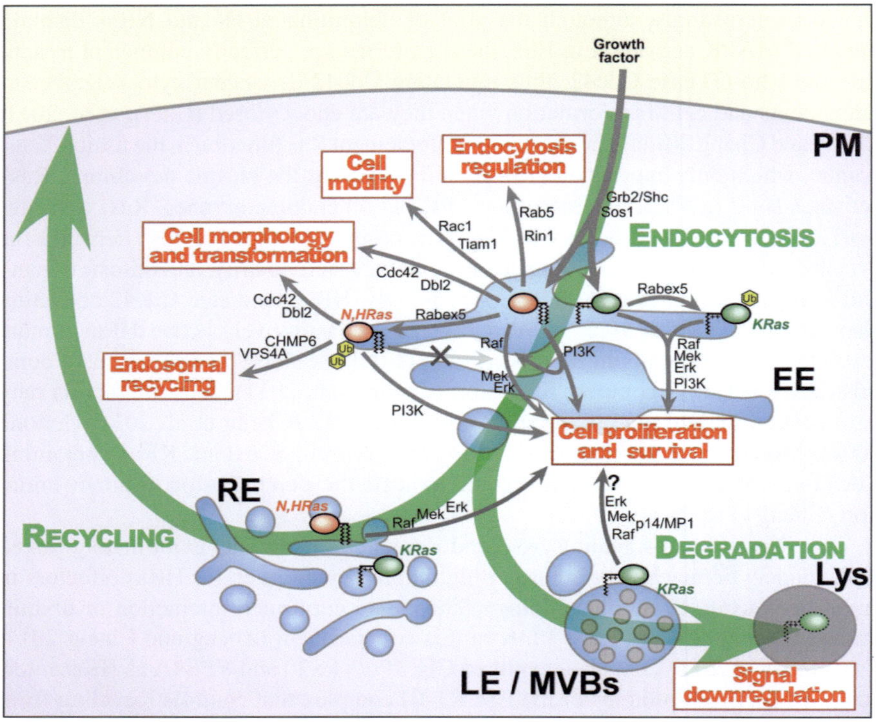

Fig. 3.2 Signaling pathways and outputs from endosomes generated by Ras GTPases. Growth factors induce Ras activation in early endosomes (EE) by recruiting the molecular machinery required for Ras-GTP loading, including adaptors Grb2, Shc, and the guanine exchange factor Sos1. In EE, activated H- or NRas (orange) and KRas (green) then signal through Raf1/Mek/Erk and PI3K/Akt pathways to regulate cell proliferation and survival, respectively. In the case of KRas, Erk activation is also elicited in late endosome/multivesicular bodies (LE/MVBs) through the Raf1/p14/MP1/Mek signaling cascade, which may also support cell proliferation. Furthermore, trafficking of KRas along the degradative route via LE/MVBs and lysosomes (Lys) finally down-regulates KRas signaling. In addition, once in EEs, Ras isoforms then also follow the recycling pathway back to the plasma membrane (PM). Along this route, activation of the Raf1/Mek/Erk cascade can occur in the recycling endosomal compartment (RE), which consequently impacts on cell proliferation. In the EE compartment, Ras regulates several cellular processes by activating GEFs of other signaling proteins. In particular, Ras stimulates Tiam1-mediated Rac1 activation for cell motility and Rin1-mediated Rab5 activation to increase endocytosis. In EEs, Ras isoforms could also become substrates for the ubiquitin ligase Rabex5 for ubiquitination. This posttranslational modification enables all Ras isoforms to enhance PI3K activation but differentially affects the ability of H-, N-, or KRas to regulate Raf1 activity. On the other hand, ubiquitinated H- and NRas, through CHMP6 and VPS4A, increase overall transport through the recycling pathway. Additionally, both ubiquitinated as well as non-ubiquitinated H- and NRas activate Cdc42, through the GEF Dbl2, which may control cell morphology and transformation

3.2.2 Rac1 Signaling from Endosomes

Rac proteins belong to the Rho subfamily within the Ras superfamily of small GTPases (Bishop and Hall 2000; Didsbury et al. 1989). Mammalian organisms express three Rac isoforms: Rac1, the focus of this chapter, is ubiquitously expressed, while Rac2 is mostly expressed in the hematopoietic lineage and Rac3 is only found in the central nervous system (Didsbury et al. 1989; Bolis et al. 2003). Together with Rho and Cdc42, Rac1 represents one of the three most extensively studied Rho family members, all well known to coordinate cytoskeleton dynamics that control the formation of stress fibers, filopodia, and lamellipodia, respectively, for cell mobility (Bishop and Hall 2000; Hall 1998; Ridley 2001).

By controlling the dynamics of the actin cytoskeleton, Rac1 regulates many cellular processes linked to cell motility, including cell spreading, adhesion, migration, and axonal growth, but also phagocytosis, macropinocytosis, pinocytosis or vesicular transport (Bosco et al. 2009; Ridley 2006; Heasman and Ridley 2008; Bustelo et al. 2007). Moreover, a wide range of cellular functions, not strictly related with the control of the cytoskeleton, have also been described for Rac1, such as cell survival, the control of gene expression, cell cycle, cell differentiation as well as proliferation (Bosco et al. 2009; Sahai and Marshall 2002; Bishop and Hall 2000). In this context, some mitogenic Rac1 activities can be attributed to signaling events downstream of oncogenic Ras (Samuel et al. 2011; Joneson et al. 1996; Qiu et al. 1995).

Hence, given the multiple functions of Rac1 in fundamental cellular processes, deregulation of Rac1 has been identified to significantly contribute to pathogenic events in several human diseases, including cardiovascular diseases, and metastatic dissemination during cancer progression (Marei and Malliri 2016; Sahai and Marshall 2002). Indeed, Rac1 is a potent regulator of epithelial-mesenchymal transition (EMT) and its reverse process, mesenchymal-epithelial transition (MET), both considered critical to guide cell migration and metastasis of epithelial tumors (Marei and Malliri 2016).

Although activating mutations of Rac1 have only been found with a very low frequency in lung, skin or breast cancer (Schnelzer et al. 2000; Davis et al. 2013), its significant contribution to cell motility, metastasis and cancer progression has been thoroughly demonstrated (Parri and Chiarugi 2010; Bosco et al. 2009; Marei and Malliri 2016). However, despite these pro-oncogenic activities, Rac1 and its GEF Tiam1 also protect against invasion by stabilizing cadherin-mediated cell—cell contacts (Marei and Malliri 2016), and restoring epithelial morphology in Ras-transformed Madin Darby canine kidney cells (Hordijk et al. 1997). It is believed that these pro- and anti-invasive Rac1 activities could be triggered by different stimuli, or response to changes in the surrounding microenvironment, such as cell interaction with the ECM or reflect differential effects of Rac1 activity in relation to the stage of tumor progression (Sander et al. 1998; Sahai and Marshall 2002; De Franceschi et al. 2015; Marei and Malliri 2016; Bosco et al. 2009).

3.2.2.1 Synthesis, Processing and Trafficking of Rac1

In the cytosol, Rac1 is synthesized as a hydrophilic protein and like the Ras isoforms, contains a C-terminal CAAX motif (CLLL). The final leucine residue allows its recognition by a geranylgeranyltransferase type I that covalently incorporates a geranylgeranyl group (20-carbons isoprenyl) to the cysteine residue of the CLLL recognition sequence (Reid et al. 2004). This first cytosolic posttranslational modification enables incorporation of Rac1 into ER membranes and the subsequent AAX hydrolysis, followed by methylation of the geranylgeranylated cysteine residue mediated by Rce1 and Icmt enzymes, respectively, similar to the processing of Ras isoforms described above (Sect. 3.2.1.2). Adding additional complexity in the regulation of Rac1 processing, localization and activity, interaction with Rho-GDP dissociation inhibitor (RhoGDI) proteins then facilitates solubilization of ER-associated Rac1 and consequently, Rac1 release into the cytosol (Bustelo et al. 2007; Hoffman et al. 2000; Marei and Malliri 2016).

Like all other Ras GTPases, Rac1 switches between the active GTP-bound and inactive GDP-bound form, a cycle controlled by several Rho-GEFs and Rho-GAPs that ensure the spatiotemporal regulation of Rac1 activity (Sahai and Marshall 2002). However, in striking contrast to the Ras isoforms, the interaction of RhoGDIs with Rac1, as well as the other members of the Rho family, provides an additional layer to control Rac1 activity. RhoGDIs mainly interact with the inactive Rac1 (GDP-bound) via the G-domain and the geranylgeranyl group, which weakens the membrane anchoring provided by the geranylgeranyl group, thereby solubilizing and sequestering the inactive Rac1 GTPase into the cytosol (Olofsson 1999; DerMardirossian and Bokoch 2005; Grizot et al. 2001). Extensive research over the years identified multiple regulatory circuits driven by growth factors and other external stimuli that can modify this interaction. For instance, RhoGDI can be phosphorylated by PKC or p21-activated kinase (PAK), decreasing its affinity for Rac1-GDP and allowing the insertion of the Rac1 prenyl group into the plasma membrane, followed by GEF-mediated activation of Rac1 (DerMardirossian et al. 2004; Price et al. 2003). Also, integrins can displace RhoGDIs by favoring Rac1 insertion into lipid rafts, specialized plasma membrane domains associated with Rac1 activation (del Pozo et al. 2000, 2002). This complex regulatory circuit is initiated by ECM proteins, fibronectins, first activating integrins in lipid rafts. This stimulates phospholipase D (PLD) and leads to the localized generation of PA in this membrane domain, which competes with RhoGDI for the binding to Rac1. Ultimately, this ensures the Rac1 prenyl group insertion into this membrane domain, which favors Rac1 activation and consequently cell spreading, lamellipodia and migration (Chae et al. 2008).

Within the hypervariable region and adjacent to the CAAX motif, Rac1 also harbors a polybasic region. Several reports demonstrate that this PBR electrostatically interacts with anionic membrane phospholipids, like PS, PtdIns(3,4,5)P_3, PtdIns(3,5)P_2, or PA, determining its localization in specific membrane domains under different physiological conditions (Michaelson et al. 2001; ten Klooster and Hordijk 2007; Finkielstein et al. 2006; Yeung et al. 2006; Chae et al. 2008). In addition, a proline-rich domain neighboring the PBR has been demonstrated to contribute to the targeting of Rac1 to cellular focal adhesions (ten Klooster et al. 2006).

Finally, Rac1 is also palmitoylated at the cysteine 178 amino acid, which greatly enhances Rac1 membrane stability and promotes its localization and functionality in cholesterol-rich plasma membrane domains (lipid rafts) (Navarro-Lerida et al. 2012). Similar to the Ras isoforms, at the plasma membrane Rac1 can be internalized through CIE transport vesicles reaching EEs and then LEs. The role of Rac1 endocytosis in signaling and its implication in the regulation of different cellular processes is detailed in the next section.

3.2.2.2 Signaling Pathways Regulated by Rac1 from Endosomes

As the genetic ablation of Rac1 in mice results in embryonic lethality (Sugihara et al. 1998) limited information of Rac1 function in vivo is available up to date, mostly relying on studies modulating Rac1 effector gene expression. Therefore, the current knowledge in the field is still based on the vast majority of Rac1 signaling studies performed in cell culture experiments using different cell types.

Signaling elicited by Rac1 effectors commonly are related to actin cytoskeleton rearrangements. One of the most extensively studied Rac1 effectors is PAK which phosphorylates LIM kinase and cortactin, among others, to coordinate actin polymerization (F-actin) at the plasma membrane in a multifactorial process that includes the actin-related protein-2/3 (Arp2/3) complex, Neural Wiskott–Aldrich syndrome protein (N-WASP)/*WASP-family verprolin-homologous* protein (WAVE), cofilin or dynamin proteins (Frost et al. 1998; Yang et al. 1998; Vidal et al. 2002; Webb et al. 2006; Sauvonnet et al. 2005; Schafer et al. 2002; Grassart et al. 2010). In addition, another Rac1 effector determining actin dynamics is phosphatidylinositol-4-phosphate 5-kinase (Tolias and Carpenter 2000; Tolias et al. 2000; Vidal-Quadras et al. 2011; Weernink et al. 2004; Chao et al. 2010), which upon activation leads to increased production of PtdIns(4,5)P_2 at the plasma membrane (Doughman et al. 2003; Tolias et al. 2000; van den Bout and Divecha 2009; Shibasaki et al. 1997). On the other hand, Rac1 interacts and activates phospholipase C enzymes (Illenberger et al. 1998; Jezyk et al. 2006; Li et al. 2009), which hydrolyzes PtdIns(4,5)P_2 to generate diacylglycerol and PtdIns(3)P both well established second messengers. This rapid turnover of PtdIns(4,5)P_2 has been demonstrated to promote F-actin polymerization and cell migration (Li et al. 2009). The diversity and complexity of PIs modulating Rac1 signaling are further underscored with PI3K generating PtdIns(3,4,5)P_3 from PtdIns(4,5)P_2, which affects cell migration and cell survival controlled by Rac1 (Yang et al. 2011; Murga et al. 2002). Moreover, PtdIns(3,4,5)P_3 is able to recruit several Rac1-GEFs to the plasma membrane, providing multiple opportunities for positive feedback mechanisms between PI3K and Rac1 (Zhu et al. 2015; Ebi et al. 2013).

All of these Rac1 effectors are critical for actin organization at the plasma membrane, which in addition to controlling formation of membrane protrusions such as lamellipodia in migrating cells, also influences endocytosis (Lamaze et al. 1996;

Soriano-Castell et al. 2017). For instance, Rac1 impacts on CIE of receptors, such as IL-2R (Grassart et al. 2008; Lamaze et al. 2001), fluid phase ingestion, TrkA receptor internalization via macropinocytosis (Valdez et al. 2007), as well as phagocytosis of pathogens (Etienne-Manneville and Hall 2002; Criss et al. 2001). Interestingly, the recently identified novel Rac1 effector amyotrophic lateral sclerosis 2 (ALS2) gene, which is vital for motor neurons, is a Rab5-GEF driving Rac1 activation for macropinocytosis and the subsequent fusion of macropinosomes with EEs (Kunita et al. 2007).

As outlined above, membrane recruitment of Rac1 effectors is intimately linked to cholesterol- and sphingolipid-enriched domains, but is also strongly influenced by the distribution of phosphorylated derivatives of PI. All of these membrane lipids localize to distinct membrane domains at the plasma membrane and in endosomal compartments, contributing to the recruitment of distinct effectors that not only establish signaling platforms, but also control membrane dynamics. Thus, Rac1 does not only control endocytic transport (see above), but vice versa endocytic trafficking is also instrumental in modulating Rac1 activity. This intermingled connection between endocytic trafficking and signal output provides opportunity for localized Rac1 signaling within the endocytic compartment and has been thoroughly demonstrated for ectopically expressed and fluorescently tagged Rac1, Raichu sensors, or photoactivatable GFP-Rac1 by means of biochemical and microscopy techniques. For instance, endocytosis of activated growth factor receptor enabled Rac1 activation in EEs and LEs, while RE ensured the translocation of active Rac1 to specific plasma membrane domains to control decisive events enabling forward movement, such as cell–cell contact, focal adhesion dynamics, assembly and disassembly of invadopodia or lamellipodia membrane protrusion at the leading edge (Menard et al. 2014; Revach et al. 2016; Garcia-Weber and Millan 2016; Stasyk and Huber 2016; Miaczynska and Bar-Sagi 2010; De Franceschi et al. 2015; Zhou et al. 2007; Harrington et al. 2011). Indeed, Kermorgant and coworkers demonstrated that the signaling output from endosomes of activated HGF-R (Met) not only enables Rac1 activation in this compartment, but subsequently ensures redirection of active (GTP-bound) Rac1 to the plasma membrane to control cell migration (Barrow-McGee and Kermorgant 2014). This implicates Rac1 signaling events from endosomes as well as from the plasma membrane to cooperatively determine the migratory behavior of cells. Indeed, PI3K activity and the Rac1-GEF Vav2 in perinuclear endosomes are required to sustain Rac1 signaling output to efficiently activate cell migration and invasion (Menard et al. 2014; Joffre et al. 2011). Most strikingly, constitutively active and oncogenic Met mutants (M1268T and D1246N), which accumulate in endosomal compartments, are characterized by enhanced endosomal Rac1 activity, reduced actin stress fibers, and increased cell migration, highlighting the significant contribution of endocytosis, and endosomal Rac1 signaling to tumor progression and metastatic events (Joffre et al. 2011; Barrow-McGee and Kermorgant 2014).

Further underscoring endocytic trafficking substantially contributing to Rac1 activation, after growth factor-induced activation of motogenic receptors such as HGF-R, Rab5 activity, and CDE is required for Rac1 activation on EEs through the Rho-GEF Tiam1. Moreover, the recycling of active endosomal Rac1 back to the plasma membrane, via the small GTPase Arf6, triggered the formation of actin-based migratory protrusions. This endocytic trafficking route of active Rac1 through Rab5- and Arf6-positive compartments seems to contribute to cell motility in a variety of tumor cells (Palamidessi et al. 2008). In support of the latter, it has been described that active Arf6 induces Rac1 activation through endosomal trafficking (Donaldson et al. 2009).

In other cell types and settings, the initiation of Rac1 signaling from endosomes does not always require growth factor activation. For example, in endothelial cells ECM-bound β1-integrins recruit another member of the Ras superfamily, R-Ras, to nascent adhesions in lamellipodia, which promotes β1-integrin and R-Ras internalization by a Rab5-dependent pathway. Once endocytosed, R-Ras then activates Rac1 through Tiam1, followed by active Rac1 redirection to the plasma membrane to control cell adhesion and morphogenesis (Sandri et al. 2012). Similarly, TrkB receptor-mediated activation of PI3K and Tiam1/Rac in endosomes is essential to mediate the chemotactic response of the brain-derived neurotrophic factor (Zhou et al. 2007).

While the majority of Rac1 signaling events described above seem to occur predominantly in EE and RE compartments, Rac1 activity has also been associated with proteins and events located in LE. Along this line, the small GTPase Rab7, an established LE marker that controls late endocytic trafficking, directly interacts with Rac1 and both proteins colocalize in endosomes at the perinuclear region and on vesicles near the plasma membrane. It has been recently demonstrated that Rab7 enables Rac1 activation and promotes Rac1 delivery to the plasma membrane to stimulate cell migration (Margiotta et al. 2017). Moreover, Rab7 and Rac1 association also facilitates endosomal transport, through microtubules and actin filaments, in the context of ruffled border formation in osteoclasts, E-cadherin turnover, and stability of cell-cell contacts (Sun et al. 2005; Frasa et al. 2010).

In the LE compartment, the small GTPase RhoB acts as a negative regulator of Rac1 activity. Hence, inhibition of RhoB induced Rac1 activity and consequently lamellipodia protrusion (Garcia-Weber and Millan 2016). In contrast, active RhoB retained Rac1 in intracellular endosomal localizations and prevented Rac1 activation and its recycling to the cell border, blocking Rac1-dependent endothelial barrier reformation and stabilization of cell-cell junctions (Marcos-Ramiro et al. 2016). These findings are in consonance with the protective effect of HGF-induced and Tiam1-dependent Rac1 activation on endothelial cell barrier function, also requiring the Rac1 effector cortactin and a formation of a cortical actin ring (Birukova et al. 2007). Other negative Rac1 regulators include Rab11-family interacting protein 3 (FIP3). In T-cells, this interaction diverts Rac1 to a Rab11-positive recycling perinuclear endosomes, restricting access of Rac1 to the plasma membrane. In fact, FIP3 silencing induced T-cell spreading, a process that is controlled by Rac1, suggesting endosomal trafficking of Rac1 to regulate T-cell spreading in the immunological synapse (Bouchet et al. 2016).

Finally, it should be noted that the localization of intracellular Rac1 regulators may include players not restricted to the EE, RE, and LE compartments. This includes Rab8, which has been implicated in exocytic/recycling membrane trafficking, but also actin and microtubule cytoskeletal rearrangements (Hattula et al. 2006; Huber

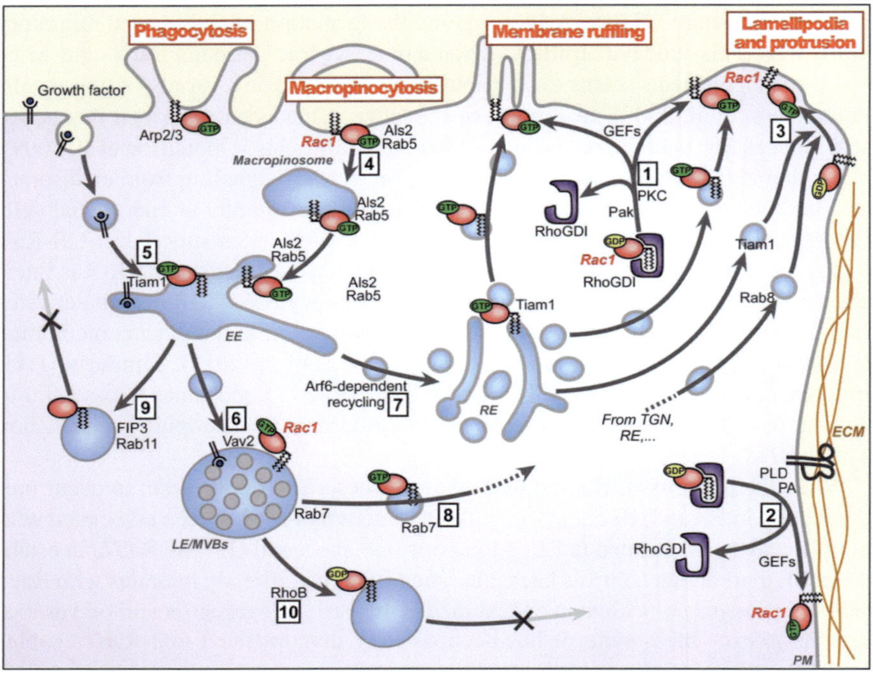

Fig. 3.3 Involvement of the endosomal compartment in Rac1 activation and trafficking to plasma membrane. This scheme illustrates the current knowledge on Rac1 activation on endosomes and Rac1 transport to the plasma membrane (PM) to regulate actin dynamics. In the cytosol, inactive Rac1 (GDP-bound, yellow) is sequestered and maintained soluble by RhoGDI. RhoGDI could be released by phosphorylation through activated p21-activated kinase (PAK)/protein kinase C (PKC) [1]. Alternatively, RhoGDI can be released by activated integrins bound to the extracellular matrix (ECM) upon phospholipase D (PLD) activation and generation of phosphatidic acid (PA), which competes with RhoGDI for Rac1 binding [2]. This allows the insertion of Rac1 at the PM and its interaction with GEFs, which are recruited to the PM by vesicular transport [3] or by interaction with elevated phosphoinositides at the PM. This interaction consequently leads to Rac1 activation (GTP-bound, green). At the PM, active Rac1 regulates cytoskeleton dynamics which controls phagocytosis, macropinocytosis, membrane ruffling, protrusion and lamellipodia formation. Exemplifying the multiple links between Rac1 activity and endocytosis, the Rac1-effector Als2 is a Rab5-GEF, which together with Rac1-induced actin rearrangements, activates macropinocytosis and the subsequent fusion of macropinosomes with early endosomes (EE) [4]. Endocytic membrane compartments and vesicular transport are also critical for Rac1 activation and its redirection to specific PM domains to exert the above-mentioned functions. Growth factor stimulation induces Rac1 activation in EEs and late endosomes (LEs/MVBs) by the GEFs Tiam1 [5] and Vav2 [6], respectively. Thereafter, recycling from these compartments by Arf6 [7], via the recycling compartment (RE), and Rab7 [8] redirects active Rac1 to specific domains at the PM. In addition, the Rab11 effector FIP3 in EEs [9] and the GTPase RhoB in LE/MVBs [10] have also been reported to deliver Rac1 to associated sequestering endocytic vesicles, which inactivate Rac1 by impairing its recycling to the PM

et al. 1993; Peranen et al. 1996; Roland et al. 2007). Rab8 localizes to vesicular structures, including RE and LE, but also the Golgi region, and peripheral membrane ruffles, and increases Rac1 activity and Tiam1/Rac1 mobilization from intracellular compartments to cortical locations to maintain directionality of migrating cells by enabling focal adhesion turnover and actin polymerization (Bravo-Cordero et al. 2016).

In summary, the aforementioned results exemplify and highlight the role of the endocytic system to regulate spatiotemporal Rac1 functionality. Several of these results have been illustrated in Fig. 3.3.

References

Ahearn IM, Haigis K, Bar-Sagi D, Philips MR (2011) Regulating the regulator: post-translational modification of RAS. Nat Rev Mol Cell Biol 13(1):39–51. https://doi.org/10.1038/nrm3255

Alanko J, Ivaska J (2016) Endosomes: emerging platforms for integrin-mediated FAK signalling. Trends Cell Biol 26(6):391–398. https://doi.org/10.1016/j.tcb.2016.02.001

Alanko J, Mai A, Jacquemet G, Schauer K, Kaukonen R, Saari M, Goud B, Ivaska J (2015) Integrin endosomal signalling suppresses anoikis. Nat Cell Biol 17(11):1412–1421. https://doi.org/10.1038/ncb3250

Alvarez-Moya B, Barcelo C, Tebar F, Jaumot M, Agell N (2011) CaM interaction and Ser181 phosphorylation as new K-Ras signaling modulators. Small GTPases 2(2):99–103. https://doi.org/10.4161/sgtp.2.2.15555

Apolloni A, Prior IA, Lindsay M, Parton RG, Hancock JF (2000) H-ras but not K-ras traffics to the plasma membrane through the exocytic pathway. Mol Cell Biol 20(7):2475–2487

Aran V, Prior IA (2013) Compartmentalized Ras signaling differentially contributes to phenotypic outputs. Cell Signal 25(9):1748–1753. https://doi.org/10.1016/j.cellsig.2013.05.004

Ariotti N, Fernandez-Rojo MA, Zhou Y, Hill MM, Rodkey TL, Inder KL, Tanner LB, Wenk MR, Hancock JF, Parton RG (2014) Caveolae regulate the nanoscale organization of the plasma membrane to remotely control Ras signaling. J Cell Biol 204(5):777–792. https://doi.org/10.1083/jcb.201307055

Balbis A, Parmar A, Wang Y, Baquiran G, Posner BI (2007) Compartmentalization of signaling-competent epidermal growth factor receptors in endosomes. Endocrinology 148(6):2944–2954. https://doi.org/10.1210/en.2006-1674

Ballester R, Furth ME, Rosen OM (1987) Phorbol ester- and protein kinase C-mediated phosphorylation of the cellular Kirsten ras gene product. J Biol Chem 262(6):2688–2695

Barbacid M (1987) ras genes. Annu Rev Biochem 56:779–827. https://doi.org/10.1146/annurev.bi.56.070187.004023

Barrow-McGee R, Kermorgant S (2014) Met endosomal signalling: in the right place, at the right time. Int J Biochem Cell Biol 49:69–74. https://doi.org/10.1016/j.biocel.2014.01.009

Bergeron JJ, Posner BI, Josefsberg Z, Sikstrom R (1978) Intracellular polypeptide hormone receptors. The demonstration of specific binding sites for insulin and human growth hormone in Golgi fractions isolated from the liver of female rats. J Biol Chem 253(11):4058–4066

Bhagatji P, Leventis R, Rich R, Lin CJ, Silvius JR (2010) Multiple cellular proteins modulate the dynamics of K-ras association with the plasma membrane. Biophys J 99(10):3327–3335. https://doi.org/10.1016/j.bpj.2010.10.001

Birukova AA, Alekseeva E, Mikaelyan A, Birukov KG (2007) HGF attenuates thrombin-induced endothelial permeability by Tiam1-mediated activation of the Rac pathway and by Tiam1/Rac-dependent inhibition of the Rho pathway. Faseb J 21(11):2776–2786. https://doi.org/10.1096/fj.06-7660com

Bishop AL, Hall A (2000) Rho GTPases and their effector proteins. Biochem J 348(Pt 2):241–255
Bivona TG, Quatela SE, Bodemann BO, Ahearn IM, Soskis MJ, Mor A, Miura J, Wiener HH, Wright L, Saba SG, Yim D, Fein A, Perez de Castro I, Li C, Thompson CB, Cox AD, Philips MR (2006) PKC regulates a farnesyl-electrostatic switch on K-Ras that promotes its association with Bcl-XL on mitochondria and induces apoptosis. Mol Cell 21(4):481–493. https://doi.org/10.1016/j.molcel.2006.01.012
Bolis A, Corbetta S, Cioce A, de Curtis I (2003) Differential distribution of Rac1 and Rac3 GTPases in the developing mouse brain: implications for a role of Rac3 in Purkinje cell differentiation. Eur J Neurosci 18(9):2417–2424
Bonifacino JS, Hurley JH (2008) Retromer. Curr Opin Cell Biol 20(4):427–436. https://doi.org/10.1016/j.ceb.2008.03.009
Bonifacino JS, Rojas R (2006) Retrograde transport from endosomes to the trans-Golgi network. Nat Rev Mol Cell Biol 7(8):568–579. https://doi.org/10.1038/nrm1985
Bos JL (1989) ras oncogenes in human cancer: a review. Cancer Res 49(17):4682–4689
Bos JL, Rehmann H, Wittinghofer A (2007) GEFs and GAPs: critical elements in the control of small G proteins. Cell 129(5):865–877. https://doi.org/10.1016/j.cell.2007.05.018
Bosco EE, Mulloy JC, Zheng Y (2009) Rac1 GTPase: a "Rac" of all trades. Cell Mol Life Sci 66(3):370–374. https://doi.org/10.1007/s00018-008-8552-x
Bouchet J, Del Rio-Iniguez I, Lasserre R, Aguera-Gonzalez S, Cuche C, Danckaert A, McCaffrey MW, Di Bartolo V, Alcover A (2016) Rac1-Rab11-FIP3 regulatory hub coordinates vesicle traffic with actin remodeling and T-cell activation. EMBO J 35(11):1160–1174. https://doi.org/10.15252/embj.201593274
Bravo-Cordero JJ, Marrero-Diaz R, Megias D, Genis L, Garcia-Grande A, Garcia MA, Arroyo AG, Montoya MC (2007) MT1-MMP proinvasive activity is regulated by a novel Rab8-dependent exocytic pathway. EMBO J 26(6):1499–1510. https://doi.org/10.1038/sj.emboj.7601606
Bravo-Cordero JJ, Cordani M, Soriano SF, Diez B, Munoz-Agudo C, Casanova-Acebes M, Boullosa C, Guadamillas MC, Ezkurdia I, Gonzalez-Pisano D, Del Pozo MA, Montoya MC (2016) A novel high-content analysis tool reveals Rab8-driven cytoskeletal reorganization through Rho GTPases, calpain and MT1-MMP. J Cell Sci 129(8):1734–1749. https://doi.org/10.1242/jcs.174920
Bustelo XR, Sauzeau V, Berenjeno IM (2007) GTP-binding proteins of the Rho/Rac family: regulation, effectors and functions in vivo. BioEssays 29(4):356–370. https://doi.org/10.1002/bies.20558
Calvo F, Agudo-Ibanez L, Crespo P (2010) The Ras-ERK pathway: understanding site-specific signaling provides hope of new anti-tumor therapies. BioEssays 32(5):412–421. https://doi.org/10.1002/bies.200900155
Castro-Castro A, Marchesin V, Monteiro P, Lodillinsky C, Rosse C, Chavrier P (2016) Cellular and molecular mechanisms of MT1-MMP-dependent cancer cell invasion. Annu Rev Cell Dev Biol 32:555–576. https://doi.org/10.1146/annurev-cellbio-111315-125227
Cerione RA (2004) Cdc42: new roads to travel. Trends Cell Biol 14(3):127–132. https://doi.org/10.1016/j.tcb.2004.01.008
Chae YC, Kim JH, Kim KL, Kim HW, Lee HY, Heo WD, Meyer T, Suh PG, Ryu SH (2008) Phospholipase D activity regulates integrin-mediated cell spreading and migration by inducing GTP-Rac translocation to the plasma membrane. Mol Biol Cell 19(7):3111–3123. https://doi.org/10.1091/mbc.E07-04-0337
Chandra A, Grecco HE, Pisupati V, Perera D, Cassidy L, Skoulidis F, Ismail SA, Hedberg C, Hanzal-Bayer M, Venkitaraman AR, Wittinghofer A, Bastiaens PI (2011) The GDI-like solubilizing factor PDEdelta sustains the spatial organization and signalling of Ras family proteins. Nat Cell Biol 14(2):148–158. https://doi.org/10.1038/ncb2394
Chang EC, Philips MR (2006) Spatial segregation of Ras signaling: new evidence from fission yeast. Cell Cycle 5(17):1936–1939. https://doi.org/10.4161/cc.5.17.3187
Chao WT, Daquinag AC, Ashcroft F, Kunz J (2010) Type I PIPK-alpha regulates directed cell migration by modulating Rac1 plasma membrane targeting and activation. J Cell Biol 190(2):247–262. https://doi.org/10.1083/jcb.200911110

Cheng CM, Chang EC (2011) Busy traveling Ras. Cell Cycle 10(8):1180–1181. https://doi.org/10.4161/cc.10.8.15259
Cheng CM, Li H, Gasman S, Huang J, Schiff R, Chang EC (2011) Compartmentalized Ras proteins transform NIH 3T3 cells with different efficiencies. Mol Cell Biol 31(5):983–997. https://doi.org/10.1128/MCB.00137-10
Cho KJ, Park JH, Piggott AM, Salim AA, Gorfe AA, Parton RG, Capon RJ, Lacey E, Hancock JF (2012) Staurosporines disrupt phosphatidylserine trafficking and mislocalize Ras proteins. J Biol Chem 287(52):43573–43584. https://doi.org/10.1074/jbc.M112.424457
Choy E, Chiu VK, Silletti J, Feoktistov M, Morimoto T, Michaelson D, Ivanov IE, Philips MR (1999) Endomembrane trafficking of ras: the CAAX motif targets proteins to the ER and Golgi. Cell 98(1):69–80. https://doi.org/10.1016/S0092-8674(00)80607-8
Criss AK, Ahlgren DM, Jou TS, McCormick BA, Casanova JE (2001) The GTPase Rac1 selectively regulates Salmonella invasion at the apical plasma membrane of polarized epithelial cells. J Cell Sci 114(Pt 7):1331–1341
Davis MJ, Ha BH, Holman EC, Halaban R, Schlessinger J, Boggon TJ (2013) RAC1P29S is a spontaneously activating cancer-associated GTPase. Proc Natl Acad Sci U S A 110(3):912–917. https://doi.org/10.1073/pnas.1220895110
De Franceschi N, Hamidi H, Alanko J, Sahgal P, Ivaska J (2015) Integrin traffic—the update. J Cell Sci 128(5):839–852. https://doi.org/10.1242/jcs.161653
Debanne MT, Evans WH, Flint N, Regoeczi E (1982) Receptor-rich intracellular membrane vesicles transporting asialotransferrin and insulin in liver. Nature 298(5872):398–400
del Pozo MA, Price LS, Alderson NB, Ren XD, Schwartz MA (2000) Adhesion to the extracellular matrix regulates the coupling of the small GTPase Rac to its effector PAK. EMBO J 19(9):2008–2014. https://doi.org/10.1093/emboj/19.9.2008
del Pozo MA, Kiosses WB, Alderson NB, Meller N, Hahn KM, Schwartz MA (2002) Integrins regulate GTP-Rac localized effector interactions through dissociation of Rho-GDI. Nat Cell Biol 4(3):232–239. https://doi.org/10.1038/ncb759
DerMardirossian C, Bokoch GM (2005) GDIs: central regulatory molecules in Rho GTPase activation. Trends Cell Biol 15(7):356–363. https://doi.org/10.1016/j.tcb.2005.05.001
DerMardirossian C, Schnelzer A, Bokoch GM (2004) Phosphorylation of RhoGDI by Pak1 mediates dissociation of Rac GTPase. Mol Cell 15(1):117–127. https://doi.org/10.1016/j.molcel.2004.05.019
Di Guglielmo GM, Baass PC, Ou WJ, Posner BI, Bergeron JJ (1994) Compartmentalization of SHC, GRB2 and mSOS, and hyperphosphorylation of Raf-1 by EGF but not insulin in liver parenchyma. EMBO J 13(18):4269–4277
Di Guglielmo GM, Le Roy C, Goodfellow AF, Wrana JL (2003) Distinct endocytic pathways regulate TGF-beta receptor signalling and turnover. Nat Cell Biol 5(5):410–421. https://doi.org/10.1038/ncb975
Didsbury J, Weber RF, Bokoch GM, Evans T, Snyderman R (1989) rac, a novel ras-related family of proteins that are botulinum toxin substrates. J Biol Chem 264(28):16378–16382
Dobrowolski R, De Robertis EM (2011) Endocytic control of growth factor signalling: multivesicular bodies as signalling organelles. Nat Rev Mol Cell Biol 13(1):53–60. https://doi.org/10.1038/nrm3244
Donaldson JG, Porat-Shliom N, Cohen LA (2009) Clathrin-independent endocytosis: a unique platform for cell signaling and PM remodeling. Cell Signal 21(1):1–6. https://doi.org/10.1016/j.cellsig.2008.06.020
Doughman RL, Firestone AJ, Wojtasiak ML, Bunce MW, Anderson RA (2003) Membrane ruffling requires coordination between type Ialpha phosphatidylinositol phosphate kinase and Rac signaling. J Biol Chem 278(25):23036–23045. https://doi.org/10.1074/jbc.M211397200
Downward J (1996) Control of ras activation. Cancer Surv 27:87–100
Duan Y, Li F, Tan K, Liu H, Li Y, Liu Y, Kong X, Tang Y, Wu G, Yin Y (2015) Key mediators of intracellular amino acids signaling to mTORC1 activation. Amino Acids 47(5):857–867. https://doi.org/10.1007/s00726-015-1937-x

Ebi H, Costa C, Faber AC, Nishtala M, Kotani H, Juric D, Della Pelle P, Song Y, Yano S, Mino-Kenudson M, Benes CH, Engelman JA (2013) PI3K regulates MEK/ERK signaling in breast cancer via the Rac-GEF, P-Rex1. Proc Natl Acad Sci U S A 110(52):21124–21129. https://doi.org/10.1073/pnas.1314124110

Eisenberg S, Henis YI (2008) Interactions of Ras proteins with the plasma membrane and their roles in signaling. Cell Signal 20(1):31–39. https://doi.org/10.1016/j.cellsig.2007.07.012

Elad-Sfadia G, Haklai R, Balan E, Kloog Y (2004) Galectin-3 augments K-Ras activation and triggers a Ras signal that attenuates ERK but not phosphoinositide 3-kinase activity. J Biol Chem 279(33):34922–34930. https://doi.org/10.1074/jbc.M312697200

Etienne-Manneville S, Hall A (2002) Rho GTPases in cell biology. Nature 420(6916):629–635. https://doi.org/10.1038/nature01148

Fehrenbacher N, Bar-Sagi D, Philips M (2009) Ras/MAPK signaling from endomembranes. Mol Oncol 3(4):297–307. https://doi.org/10.1016/j.molonc.2009.06.004

Figueroa C, Taylor J, Vojtek AB (2001) Prenylated Rab acceptor protein is a receptor for prenylated small GTPases. J Biol Chem 276(30):28219–28225. https://doi.org/10.1074/jbc.M101763200

Finkielstein CV, Overduin M, Capelluto DG (2006) Cell migration and signaling specificity is determined by the phosphatidylserine recognition motif of Rac1. J Biol Chem 281(37):27317–27326. https://doi.org/10.1074/jbc.M605560200

Fivaz M, Meyer T (2005) Reversible intracellular translocation of KRas but not HRas in hippocampal neurons regulated by Ca^{2+}/calmodulin. J Cell Biol 170(3):429–441. https://doi.org/10.1083/jcb.200409157

Flinn RJ, Yan Y, Goswami S, Parker PJ, Backer JM (2010) The late endosome is essential for mTORC1 signaling. Mol Biol Cell 21(5):833–841. https://doi.org/10.1091/mbc.E09-09-0756

Fotiadou PP, Takahashi C, Rajabi HN, Ewen ME (2007) Wild-type NRas and KRas perform distinct functions during transformation. Mol Cell Biol 27(19):6742–6755. https://doi.org/10.1128/MCB.00234-07

Frasa MA, Maximiano FC, Smolarczyk K, Francis RE, Betson ME, Lozano E, Goldenring J, Seabra MC, Rak A, Ahmadian MR, Braga VM (2010) Armus is a Rac1 effector that inactivates Rab7 and regulates E-cadherin degradation. Curr Biol 20(3):198–208. https://doi.org/10.1016/j.cub.2009.12.053

Frittoli E, Palamidessi A, Disanza A, Scita G (2011) Secretory and endo/exocytic trafficking in invadopodia formation: the MT1-MMP paradigm. Eur J Cell Biol 90(2–3):108–114. https://doi.org/10.1016/j.ejcb.2010.04.007

Frost JA, Khokhlatchev A, Stippec S, White MA, Cobb MH (1998) Differential effects of PAK1-activating mutations reveal activity-dependent and -independent effects on cytoskeletal regulation. J Biol Chem 273(43):28191–28198

Garcia-Melero A, Reverter M, Hoque M, Meneses-Salas E, Koese M, Conway JR, Johnsen CH, Alvarez-Guaita A, Morales-Paytuvi F, Elmaghrabi YA, Pol A, Tebar F, Murray RZ, Timpson P, Enrich C, Grewal T, Rentero C (2016) Annexin A6 and late endosomal cholesterol modulate integrin recycling and cell migration. J Biol Chem 291(3):1320–1335. https://doi.org/10.1074/jbc.M115.683557

Garcia-Weber D, Millan J (2016) Parallels between single cell migration and barrier formation: the case of RhoB and Rac1 trafficking. Small GTPases 1–7. https://doi.org/10.1080/21541248.2016.1231655

Gelabert-Baldrich M, Soriano-Castell D, Calvo M, Lu A, Vina-Vilaseca A, Rentero C, Pol A, Grinstein S, Enrich C, Tebar F (2014) Dynamics of KRas on endosomes: involvement of acidic phospholipids in its association. Faseb J 28(7):3023–3037. https://doi.org/10.1096/fj.13-241158

Ghai R, Mobli M, Norwood SJ, Bugarcic A, Teasdale RD, King GF, Collins BM (2011) Phox homology band 4.1/ezrin/radixin/moesin-like proteins function as molecular scaffolds that interact with cargo receptors and Ras GTPases. Proc Natl Acad Sci U S A 108(19):7763–7768. https://doi.org/10.1073/pnas.1017110108

Gomez GA, Daniotti JL (2005) H-Ras dynamically interacts with recycling endosomes in CHO-K1 cells: involvement of Rab5 and Rab11 in the trafficking of H-Ras to this pericentriolar endocytic compartment. J Biol Chem 280(41):34997–35010. https://doi.org/10.1074/jbc.M506256200
Gomez GA, Daniotti JL (2007) Electrical properties of plasma membrane modulate subcellular distribution of K-Ras. FEBS J 274(9):2210–2228. https://doi.org/10.1111/j.1742-4658.2007.05758.x
Gonnord P, Blouin CM, Lamaze C (2012) Membrane trafficking and signaling: two sides of the same coin. Semin Cell Dev Biol 23(2):154–164. https://doi.org/10.1016/j.semcdb.2011.11.002
Goodwin JS, Drake KR, Rogers C, Wright L, Lippincott-Schwartz J, Philips MR, Kenworthy AK (2005) Depalmitoylated Ras traffics to and from the Golgi complex via a nonvesicular pathway. J Cell Biol 170(2):261–272. https://doi.org/10.1083/jcb.200502063
Gorfe AA, Hanzal-Bayer M, Abankwa D, Hancock JF, McCammon JA (2007) Structure and dynamics of the full-length lipid-modified H-Ras protein in a 1,2-dimyristoylglycero-3-phosphocholine bilayer. J Med Chem 50(4):674–684. https://doi.org/10.1021/jm061053f
Gould GW, Lippincott-Schwartz J (2009) New roles for endosomes: from vesicular carriers to multi-purpose platforms. Nat Rev Mol Cell Biol 10(4):287–292. https://doi.org/10.1038/nrm2652
Grassart A, Dujeancourt A, Lazarow PB, Dautry-Varsat A, Sauvonnet N (2008) Clathrin-independent endocytosis used by the IL-2 receptor is regulated by Rac1, Pak1 and Pak2. EMBO Rep 9(4):356–362. https://doi.org/10.1038/embor.2008.28
Grassart A, Meas-Yedid V, Dufour A, Olivo-Marin JC, Dautry-Varsat A, Sauvonnet N (2010) Pak1 phosphorylation enhances cortactin-N-WASP interaction in clathrin-caveolin-independent endocytosis. Traffic 11(8):1079–1091. https://doi.org/10.1111/j.1600-0854.2010.01075.x
Grewal T, Enrich C (2006) Molecular mechanisms involved in Ras inactivation: the annexin A6-p120GAP complex. BioEssays 28(12):1211–1220. https://doi.org/10.1002/bies.20503
Grizot S, Faure J, Fieschi F, Vignais PV, Dagher MC, Pebay-Peyroula E (2001) Crystal structure of the Rac1-RhoGDI complex involved in nadph oxidase activation. Biochemistry 40(34):10007–10013
Habets GG, Scholtes EH, Zuydgeest D, van der Kammen RA, Stam JC, Berns A, Collard JG (1994) Identification of an invasion-inducing gene, Tiam-1, that encodes a protein with homology to GDP-GTP exchangers for Rho-like proteins. Cell 77(4):537–549
Hall A (1998) Rho GTPases and the actin cytoskeleton. Science 279(5350):509–514
Hancock JF (2003) Ras proteins: different signals from different locations. Nat Rev Mol Cell Biol 4(5):373–384. https://doi.org/10.1038/nrm1105
Hancock JF, Magee AI, Childs JE, Marshall CJ (1989) All ras proteins are polyisoprenylated but only some are palmitoylated. Cell 57(7):1167–1177
Hancock JF, Paterson H, Marshall CJ (1990) A polybasic domain or palmitoylation is required in addition to the CAAX motif to localize p21ras to the plasma membrane. Cell 63(1):133–139
Harrington AW, St Hillaire C, Zweifel LS, Glebova NO, Philippidou P, Halegoua S, Ginty DD (2011) Recruitment of actin modifiers to TrkA endosomes governs retrograde NGF signaling and survival. Cell 146(3):421–434. https://doi.org/10.1016/j.cell.2011.07.008
Hattula K, Furuhjelm J, Tikkanen J, Tanhuanpaa K, Laakkonen P, Peranen J (2006) Characterization of the Rab8-specific membrane traffic route linked to protrusion formation. J Cell Sci 119(Pt 23):4866–4877. https://doi.org/10.1242/jcs.03275
Hayer A, Stoeber M, Ritz D, Engel S, Meyer HH, Helenius A (2010) Caveolin-1 is ubiquitinated and targeted to intralumenal vesicles in endolysosomes for degradation. J Cell Biol 191(3):615–629. https://doi.org/10.1083/jcb.201003086
Hayes S, Chawla A, Corvera S (2002) TGF beta receptor internalization into EEA1-enriched early endosomes: role in signaling to Smad2. J Cell Biol 158(7):1239–1249
Heasman SJ, Ridley AJ (2008) Mammalian Rho GTPases: new insights into their functions from in vivo studies. Nat Rev Mol Cell Biol 9(9):690–701. https://doi.org/10.1038/nrm2476

Heo WD, Inoue T, Park WS, Kim ML, Park BO, Wandless TJ, Meyer T (2006) PI(3,4,5)P_3 and PI(4,5)P_2 lipids target proteins with polybasic clusters to the plasma membrane. Science 314(5804):1458–1461. https://doi.org/10.1126/science.1134389

Herbst JJ, Opresko LK, Walsh BJ, Lauffenburger DA, Wiley HS (1994) Regulation of postendocytic trafficking of the epidermal growth factor receptor through endosomal retention. J Biol Chem 269(17):12865–12873

Hobbs GA, Der CJ, Rossman KL (2016) RAS isoforms and mutations in cancer at a glance. J Cell Sci 129(7):1287–1292. https://doi.org/10.1242/jcs.182873

Hoffman GR, Nassar N, Cerione RA (2000) Structure of the Rho family GTP-binding protein Cdc42 in complex with the multifunctional regulator RhoGDI. Cell 100(3):345–356

Hordijk PL, ten Klooster JP, van der Kammen RA, Michiels F, Oomen LC, Collard JG (1997) Inhibition of invasion of epithelial cells by Tiam1-Rac signaling. Science 278(5342):1464–1466

Howe CL, Valletta JS, Rusnak AS, Mobley WC (2001) NGF signaling from clathrin-coated vesicles: evidence that signaling endosomes serve as a platform for the Ras-MAPK pathway. Neuron 32(5):801–814

Huber LA, Pimplikar S, Parton RG, Virta H, Zerial M, Simons K (1993) Rab8, a small GTPase involved in vesicular traffic between the TGN and the basolateral plasma membrane. J Cell Biol 123(1):35–45

Huotari J, Helenius A (2011) Endosome maturation. EMBO J 30(17):3481–3500. https://doi.org/10.1038/emboj.2011.286

Illenberger D, Schwald F, Pimmer D, Binder W, Maier G, Dietrich A, Gierschik P (1998) Stimulation of phospholipase C-beta2 by the Rho GTPases Cdc42Hs and Rac1. EMBO J 17(21):6241–6249. https://doi.org/10.1093/emboj/17.21.6241

Ismail SA, Chen YX, Rusinova A, Chandra A, Bierbaum M, Gremer L, Triola G, Waldmann H, Bastiaens PI, Wittinghofer A (2011) Arl2-GTP and Arl3-GTP regulate a GDI-like transport system for farnesylated cargo. Nat Chem Biol 7(12):942–949. https://doi.org/10.1038/nchembio.686

Jezyk MR, Snyder JT, Gershberg S, Worthylake DK, Harden TK, Sondek J (2006) Crystal structure of Rac1 bound to its effector phospholipase C-beta2. Nat Struct Mol Biol 13(12):1135–1140. https://doi.org/10.1038/nsmb1175

Jiang X, Sorkin A (2002) Coordinated traffic of Grb2 and Ras during epidermal growth factor receptor endocytosis visualized in living cells. Mol Biol Cell 13(5):1522–1535. https://doi.org/10.1091/mbc.01-11-0552

Joffre C, Barrow R, Menard L, Calleja V, Hart IR, Kermorgant S (2011) A direct role for Met endocytosis in tumorigenesis. Nat Cell Biol 13(7):827–837. https://doi.org/10.1038/ncb2257

Johannes L, Popoff V (2008) Tracing the retrograde route in protein trafficking. Cell 135(7):1175–1187. https://doi.org/10.1016/j.cell.2008.12.009

Jones MC, Caswell PT, Norman JC (2006) Endocytic recycling pathways: emerging regulators of cell migration. Curr Opin Cell Biol 18(5):549–557. https://doi.org/10.1016/j.ceb.2006.08.003

Joneson T, White MA, Wigler MH, Bar-Sagi D (1996) Stimulation of membrane ruffling and MAP kinase activation by distinct effectors of RAS. Science 271(5250):810–812

Jovic M, Sharma M, Rahajeng J, Caplan S (2010) The early endosome: a busy sorting station for proteins at the crossroads. Histol Histopathol 25(1):99–112. https://doi.org/10.14670/HH-25.99

Jura N, Scotto-Lavino E, Sobczyk A, Bar-Sagi D (2006) Differential modification of Ras proteins by ubiquitination. Mol Cell 21(5):679–687. https://doi.org/10.1016/j.molcel.2006.02.011

Kajiho H, Kajiho Y, Frittoli E, Confalonieri S, Bertalot G, Viale G, Di Fiore PP, Oldani A, Garre M, Beznoussenko GV, Palamidessi A, Vecchi M, Chavrier P, Perez F, Scita G (2016) RAB2A controls MT1-MMP endocytic and E-cadherin polarized Golgi trafficking to promote invasive breast cancer programs. EMBO Rep 17(7):1061–1080. https://doi.org/10.15252/embr.201642032

Kermorgant S, Parker PJ (2008) Receptor trafficking controls weak signal delivery: a strategy used by c-Met for STAT3 nuclear accumulation. J Cell Biol 182(5):855–863. https://doi.org/10.1083/jcb.200806076

Kiyokawa E, Hara S, Nakamura T, Matsuda M (2006) Fluorescence (Forster) resonance energy transfer imaging of oncogene activity in living cells. Cancer Sci 97(1):8–15. https://doi.org/10.1111/j.1349-7006.2006.00141.x

Kunita R, Otomo A, Mizumura H, Suzuki-Utsunomiya K, Hadano S, Ikeda JE (2007) The Rab5 activator ALS2/alsin acts as a novel Rac1 effector through Rac1-activated endocytosis. J Biol Chem 282(22):16599–16611. https://doi.org/10.1074/jbc.M610682200

Lamaze C, Chuang TH, Terlecky LJ, Bokoch GM, Schmid SL (1996) Regulation of receptor-mediated endocytosis by Rho and Rac. Nature 382(6587):177–179. https://doi.org/10.1038/382177a0

Lamaze C, Dujeancourt A, Baba T, Lo CG, Benmerah A, Dautry-Varsat A (2001) Interleukin 2 receptors and detergent-resistant membrane domains define a clathrin-independent endocytic pathway. Mol Cell 7(3):661–671

Lambert JM, Lambert QT, Reuther GW, Malliri A, Siderovski DP, Sondek J, Collard JG, Der CJ (2002) Tiam1 mediates Ras activation of Rac by a PI(3)K-independent mechanism. Nat Cell Biol 4(8):621–625. https://doi.org/10.1038/ncb833

Lanzetti L, Di Fiore PP (2008) Endocytosis and cancer: an 'insider' network with dangerous liaisons. Traffic 9(12):2011–2021. https://doi.org/10.1111/j.1600-0854.2008.00816.x

Laude AJ, Prior IA (2008) Palmitoylation and localisation of RAS isoforms are modulated by the hypervariable linker domain. J Cell Sci 121(Pt 4):421–427. https://doi.org/10.1242/jcs.020107

Le Roy C, Wrana JL (2005) Clathrin- and non-clathrin-mediated endocytic regulation of cell signalling. Nat Rev Mol Cell Biol 6(2):112–126. https://doi.org/10.1038/nrm1571

Leventis R, Silvius JR (1998) Lipid-binding characteristics of the polybasic carboxy-terminal sequence of K-ras4B. Biochemistry 37(20):7640–7648. https://doi.org/10.1021/bi973077h

Li S, Wang Q, Wang Y, Chen X, Wang Z (2009) PLC-gamma1 and Rac1 coregulate EGF-induced cytoskeleton remodeling and cell migration. Mol Endocrinol 23(6):901–913. https://doi.org/10.1210/me.2008-0368

Llado A, Tebar F, Calvo M, Moreto J, Sorkin A, Enrich C (2004) Protein kinaseCdelta-calmodulin crosstalk regulates epidermal growth factor receptor exit from early endosomes. Mol Biol Cell 15(11):4877–4891. https://doi.org/10.1091/mbc.E04-02-0127

Llado A, Timpson P, Vila de Muga S, Moreto J, Pol A, Grewal T, Daly RJ, Enrich C, Tebar F (2008) Protein kinase Cdelta and calmodulin regulate epidermal growth factor receptor recycling from early endosomes through Arp2/3 complex and cortactin. Mol Biol Cell 19(1):17–29. https://doi.org/10.1091/mbc.E07-05-0411

Lobert VH, Stenmark H (2011) Cell polarity and migration: emerging role for the endosomal sorting machinery. Physiology (Bethesda) 26(3):171–180. https://doi.org/10.1152/physiol.00054.2010

Lopez-Alcala C, Alvarez-Moya B, Villalonga P, Calvo M, Bachs O, Agell N (2008) Identification of essential interacting elements in K-Ras/calmodulin binding and its role in K-Ras localization. J Biol Chem 283(16):10621–10631. https://doi.org/10.1074/jbc.M706238200

Lu A, Tebar F, Alvarez-Moya B, Lopez-Alcala C, Calvo M, Enrich C, Agell N, Nakamura T, Matsuda M, Bachs O (2009) A clathrin-dependent pathway leads to KRas signaling on late endosomes en route to lysosomes. J Cell Biol 184(6):863–879. https://doi.org/10.1083/jcb.200807186

Luo Y, Cheng Z, Dixon CJ, Hall JF, Taylor E, Boarder MR (2011) Endosomal signalling of epidermal growth factor receptors contributes to EGF-stimulated cell cycle progression in primary hepatocytes. Eur J Pharmacol 654(2):173–180. https://doi.org/10.1016/j.ejphar.2010.11.038

Magee T, Marshall C (1999) New insights into the interaction of Ras with the plasma membrane. Cell 98(1):9–12. https://doi.org/10.1016/S0092-8674(00)80601-7

Malliri A, van der Kammen RA, Clark K, van der Valk M, Michiels F, Collard JG (2002) Mice deficient in the Rac activator Tiam1 are resistant to Ras-induced skin tumours. Nature 417(6891):867–871. https://doi.org/10.1038/nature00848

Malumbres M, Barbacid M (2003) RAS oncogenes: the first 30 years. Nat Rev Cancer 3(6):459–465. https://doi.org/10.1038/nrc1097

Marcos-Ramiro B, Garcia-Weber D, Barroso S, Feito J, Ortega MC, Cernuda-Morollon E, Reglero-Real N, Fernandez-Martin L, Duran MC, Alonso MA, Correas I, Cox S, Ridley AJ, Millan

J (2016) RhoB controls endothelial barrier recovery by inhibiting Rac1 trafficking to the cell border. J Cell Biol 213(3):385–402. https://doi.org/10.1083/jcb.201504038
Marei H, Malliri A (2016) Rac1 in human diseases: the therapeutic potential of targeting Rac1 signaling regulatory mechanisms. Small GTPases 1–25. https://doi.org/10.1080/21541248.2016.1211398
Margiotta A, Progida C, Bakke O, Bucci C (2017) Rab7a regulates cell migration through Rac1 and vimentin. Biochim Biophys Acta 2(1864):367–381. https://doi.org/10.1016/j.bbamcr.2016.11.020
Marsh M, Bolzau E, Helenius A (1983) Penetration of Semliki Forest virus from acidic prelysosomal vacuoles. Cell 32(3):931–940
Marshall CJ (1996) Ras effectors. Curr Opin Cell Biol 8(2):197–204
Mayor S, Pagano RE (2007) Pathways of clathrin-independent endocytosis. Nat Rev Mol Cell Biol 8(8):603–612. https://doi.org/10.1038/nrm2216
Mellman I, Yarden Y (2013) Endocytosis and cancer. Cold Spring Harb Perspect Biol 5(12):a016949. https://doi.org/10.1101/cshperspect.a016949
Menard L, Parker PJ, Kermorgant S (2014) Receptor tyrosine kinase c-Met controls the cytoskeleton from different endosomes via different pathways. Nat Commun 5:3907. https://doi.org/10.1038/ncomms4907
Miaczynska M, Bar-Sagi D (2010) Signaling endosomes: seeing is believing. Curr Opin Cell Biol 22(4):535–540. https://doi.org/10.1016/j.ceb.2010.05.007
Miaczynska M, Christoforidis S, Giner A, Shevchenko A, Uttenweiler-Joseph S, Habermann B, Wilm M, Parton RG, Zerial M (2004) APPL proteins link Rab5 to nuclear signal transduction via an endosomal compartment. Cell 116(3):445–456
Michaelson D, Silletti J, Murphy G, D'Eustachio P, Rush M, Philips MR (2001) Differential localization of Rho GTPases in live cells: regulation by hypervariable regions and RhoGDI binding. J Cell Biol 152(1):111–126
Misaki R, Morimatsu M, Uemura T, Waguri S, Miyoshi E, Taniguchi N, Matsuda M, Taguchi T (2010) Palmitoylated Ras proteins traffic through recycling endosomes to the plasma membrane during exocytosis. J Cell Biol 191(1):23–29. https://doi.org/10.1083/jcb.200911143
Mochizuki N, Yamashita S, Kurokawa K, Ohba Y, Nagai T, Miyawaki A, Matsuda M (2001) Spatio-temporal images of growth-factor-induced activation of Ras and Rap1. Nature 411(6841):1065–1068. https://doi.org/10.1038/35082594
Mor A, Philips MR (2006) Compartmentalized Ras/MAPK signaling. Annu Rev Immunol 24:771–800. https://doi.org/10.1146/annurev.immunol.24.021605.090723
Moreto J, Llado A, Vidal-Quadras M, Calvo M, Pol A, Enrich C, Tebar F (2008) Calmodulin modulates H-Ras mediated Raf-1 activation. Cell Signal 20(6):1092–1103. https://doi.org/10.1016/j.cellsig.2008.01.022
Moreto J, Vidal-Quadras M, Pol A, Santos E, Grewal T, Enrich C, Tebar F (2009) Differential involvement of H- and K-Ras in Raf-1 activation determines the role of calmodulin in MAPK signaling. Cell Signal 21(12):1827–1836. https://doi.org/10.1016/j.cellsig.2009.07.018
Murga C, Zohar M, Teramoto H, Gutkind JS (2002) Rac1 and RhoG promote cell survival by the activation of PI3K and Akt, independently of their ability to stimulate JNK and NF-kappaB. Oncogene 21(2):207–216. https://doi.org/10.1038/sj.onc.1205036
Murphy JE, Padilla BE, Hasdemir B, Cottrell GS, Bunnett NW (2009) Endosomes: a legitimate platform for the signaling train. Proc Natl Acad Sci U S A 106(42):17615–17622. https://doi.org/10.1073/pnas.0906541106
Nada S, Hondo A, Kasai A, Koike M, Saito K, Uchiyama Y, Okada M (2009) The novel lipid raft adaptor p18 controls endosome dynamics by anchoring the MEK-ERK pathway to late endosomes. EMBO J 28(5):477–489. https://doi.org/10.1038/emboj.2008.308
Nancy V, Callebaut I, El Marjou A, de Gunzburg J (2002) The delta subunit of retinal rod cGMP phosphodiesterase regulates the membrane association of Ras and Rap GTPases. J Biol Chem 277(17):15076–15084. https://doi.org/10.1074/jbc.M109983200

Navarro-Lerida I, Sanchez-Perales S, Calvo M, Rentero C, Zheng Y, Enrich C, Del Pozo MA (2012) A palmitoylation switch mechanism regulates Rac1 function and membrane organization. EMBO J 31(3):534–551. https://doi.org/10.1038/emboj.2011.446

Newlaczyl AU, Hood FE, Coulson JM, Prior IA (2014) Decoding RAS isoform and codon-specific signalling. Biochem Soc Trans 42(4):742–746. https://doi.org/10.1042/BST20140057

Niv H, Gutman O, Henis YI, Kloog Y (1999) Membrane interactions of a constitutively active GFP-Ki-Ras 4B and their role in signaling. Evidence from lateral mobility studies. J Biol Chem 274(3):1606–1613

Niv H, Gutman O, Kloog Y, Henis YI (2002) Activated K-Ras and H-Ras display different interactions with saturable nonraft sites at the surface of live cells. J Cell Biol 157(5):865–872. https://doi.org/10.1083/jcb.200202009

Ohashi E, Tanabe K, Henmi Y, Mesaki K, Kobayashi Y, Takei K (2011) Receptor sorting within endosomal trafficking pathway is facilitated by dynamic actin filaments. PLoS ONE 6(5):e19942. https://doi.org/10.1371/journal.pone.0019942

Olofsson B (1999) Rho guanine dissociation inhibitors: pivotal molecules in cellular signalling. Cell Signal 11(8):545–554

Omerovic J, Hammond DE, Clague MJ, Prior IA (2008) Ras isoform abundance and signalling in human cancer cell lines. Oncogene 27(19):2754–2762. https://doi.org/10.1038/sj.onc.1210925

Onken B, Wiener H, Philips MR, Chang EC (2006) Compartmentalized signaling of Ras in fission yeast. Proc Natl Acad Sci U S A 103(24):9045–9050. https://doi.org/10.1073/pnas.0603318103

Palamidessi A, Frittoli E, Garre M, Faretta M, Mione M, Testa I, Diaspro A, Lanzetti L, Scita G, Di Fiore PP (2008) Endocytic trafficking of Rac is required for the spatial restriction of signaling in cell migration. Cell 134(1):135–147. https://doi.org/10.1016/j.cell.2008.05.034

Parri M, Chiarugi P (2010) Rac and Rho GTPases in cancer cell motility control. Cell Commun Signal 8:23. https://doi.org/10.1186/1478-811X-8-23

Peranen J, Auvinen P, Virta H, Wepf R, Simons K (1996) Rab8 promotes polarized membrane transport through reorganization of actin and microtubules in fibroblasts. J Cell Biol 135(1):153–167

Pfeffer SR (2009) Multiple routes of protein transport from endosomes to the trans Golgi network. FEBS Lett 583(23):3811–3816. https://doi.org/10.1016/j.febslet.2009.10.075

Philips MR (2012) Ras hitchhikes on PDE6delta. Nat Cell Biol 14(2):128–129. https://doi.org/10.1038/ncb2429

Pinilla-Macua I, Watkins SC, Sorkin A (2016) Endocytosis separates EGF receptors from endogenous fluorescently labeled HRas and diminishes receptor signaling to MAP kinases in endosomes. Proc Natl Acad Sci U S A 113(8):2122–2127. https://doi.org/10.1073/pnas.1520301113

Platta HW, Stenmark H (2011) Endocytosis and signaling. Curr Opin Cell Biol 23(4):393–403. https://doi.org/10.1016/j.ceb.2011.03.008

Pol A, Calvo M, Enrich C (1998) Isolated endosomes from quiescent rat liver contain the signal transduction machinery. Differential distribution of activated Raf-1 and Mek in the endocytic compartment. FEBS Lett 441(1):34–38

Porat-Shliom N, Kloog Y, Donaldson JG (2008) A unique platform for H-Ras signaling involving clathrin-independent endocytosis. Mol Biol Cell 19(3):765–775. https://doi.org/10.1091/mbc.E07-08-0841

Porther N, Barbieri MA (2015) The role of endocytic Rab GTPases in regulation of growth factor signaling and the migration and invasion of tumor cells. Small GTPases 6(3):135–144. https://doi.org/10.1080/21541248.2015.1050152

Posner BI, Patel B, Verma AK, Bergeron JJ (1980) Uptake of insulin by plasmalemma and Golgi subcellular fractions of rat liver. J Biol Chem 255(2):735–741

Poteryaev D, Datta S, Ackema K, Zerial M, Spang A (2010) Identification of the switch in early-to-late endosome transition. Cell 141(3):497–508. https://doi.org/10.1016/j.cell.2010.03.011

Price LS, Langeslag M, ten Klooster JP, Hordijk PL, Jalink K, Collard JG (2003) Calcium signaling regulates translocation and activation of Rac. J Biol Chem 278(41):39413–39421. https://doi.org/10.1074/jbc.M302083200

Prior IA, Hancock JF (2012) Ras trafficking, localization and compartmentalized signalling. Semin Cell Dev Biol 23(2):145–153. https://doi.org/10.1016/j.semcdb.2011.09.002
Prior IA, Muncke C, Parton RG, Hancock JF (2003) Direct visualization of Ras proteins in spatially distinct cell surface microdomains. J Cell Biol 160(2):165–170. https://doi.org/10.1083/jcb.200209091
Puthenveedu MA, Lauffer B, Temkin P, Vistein R, Carlton P, Thorn K, Taunton J, Weiner OD, Parton RG, von Zastrow M (2010) Sequence-dependent sorting of recycling proteins by actin-stabilized endosomal microdomains. Cell 143(5):761–773. https://doi.org/10.1016/j.cell.2010.10.003
Qiu RG, Chen J, McCormick F, Symons M (1995) A role for Rho in Ras transformation. Proc Natl Acad Sci U S A 92(25):11781–11785
Rajalingam K, Schreck R, Rapp UR, Albert S (2007) Ras oncogenes and their downstream targets. Biochim Biophys Acta 1773(8):1177–1195. https://doi.org/10.1016/j.bbamcr.2007.01.012
Reid TS, Terry KL, Casey PJ, Beese LS (2004) Crystallographic analysis of CaaX prenyltransferases complexed with substrates defines rules of protein substrate selectivity. J Mol Biol 343(2):417–433. https://doi.org/10.1016/j.jmb.2004.08.056
Revach OY, Winograd-Katz SE, Samuels Y, Geiger B (2016) The involvement of mutant Rac1 in the formation of invadopodia in cultured melanoma cells. Exp Cell Res 343(1):82–88. https://doi.org/10.1016/j.yexcr.2016.02.003
Reverter M, Rentero C, Garcia-Melero A, Hoque M, Vila de Muga S, Alvarez-Guaita A, Conway JR, Wood P, Cairns R, Lykopoulou L, Grinberg D, Vilageliu L, Bosch M, Heeren J, Blasi J, Timpson P, Pol A, Tebar F, Murray RZ, Grewal T, Enrich C (2014) Cholesterol regulates Syntaxin 6 trafficking at trans-Golgi network endosomal boundaries. Cell Rep 7(3):883–897. https://doi.org/10.1016/j.celrep.2014.03.043
Ridley AJ (2001) Rho GTPases and cell migration. J Cell Sci 114(Pt 15):2713–2722
Ridley AJ (2006) Rho GTPases and actin dynamics in membrane protrusions and vesicle trafficking. Trends Cell Biol 16(10):522–529. https://doi.org/10.1016/j.tcb.2006.08.006
Rink J, Ghigo E, Kalaidzidis Y, Zerial M (2005) Rab conversion as a mechanism of progression from early to late endosomes. Cell 122(5):735–749. https://doi.org/10.1016/j.cell.2005.06.043
Rocks O, Peyker A, Kahms M, Verveer PJ, Koerner C, Lumbierres M, Kuhlmann J, Waldmann H, Wittinghofer A, Bastiaens PI (2005) An acylation cycle regulates localization and activity of palmitoylated Ras isoforms. Science 307(5716):1746–1752. https://doi.org/10.1126/science.1105654
Rocks O, Gerauer M, Vartak N, Koch S, Huang ZP, Pechlivanis M, Kuhlmann J, Brunsveld L, Chandra A, Ellinger B, Waldmann H, Bastiaens PI (2010) The palmitoylation machinery is a spatially organizing system for peripheral membrane proteins. Cell 141(3):458–471. https://doi.org/10.1016/j.cell.2010.04.007
Rojas R, van Vlijmen T, Mardones GA, Prabhu Y, Rojas AL, Mohammed S, Heck AJ, Raposo G, van der Sluijs P, Bonifacino JS (2008) Regulation of retromer recruitment to endosomes by sequential action of Rab5 and Rab7. J Cell Biol 183(3):513–526. https://doi.org/10.1083/jcb.200804048
Rojas AM, Fuentes G, Rausell A, Valencia A (2012) The Ras protein superfamily: evolutionary tree and role of conserved amino acids. J Cell Biol 196(2):189–201. https://doi.org/10.1083/jcb.201103008
Roland JT, Kenworthy AK, Peranen J, Caplan S, Goldenring JR (2007) Myosin Vb interacts with Rab8a on a tubular network containing EHD1 and EHD3. Mol Biol Cell 18(8):2828–2837. https://doi.org/10.1091/mbc.E07-02-0169
Roy S, Wyse B, Hancock JF (2002) H-Ras signaling and K-Ras signaling are differentially dependent on endocytosis. Mol Cell Biol 22(14):5128–5140
Sahai E, Marshall CJ (2002) RHO-GTPases and cancer. Nat Rev Cancer 2(2):133–142. https://doi.org/10.1038/nrc725
Samuel MS, Lourenco FC, Olson MF (2011) K-Ras mediated murine epidermal tumorigenesis is dependent upon and associated with elevated Rac1 activity. PLoS ONE 6(2):e17143. https://doi.org/10.1371/journal.pone.0017143

Sancak Y, Bar-Peled L, Zoncu R, Markhard AL, Nada S, Sabatini DM (2010) Ragulator-Rag complex targets mTORC1 to the lysosomal surface and is necessary for its activation by amino acids. Cell 141(2):290–303. https://doi.org/10.1016/j.cell.2010.02.024

Sander EE, van Delft S, ten Klooster JP, Reid T, van der Kammen RA, Michiels F, Collard JG (1998) Matrix-dependent Tiam1/Rac signaling in epithelial cells promotes either cell-cell adhesion or cell migration and is regulated by phosphatidylinositol 3-kinase. J Cell Biol 143(5):1385–1398

Sandri C, Caccavari F, Valdembri D, Camillo C, Veltel S, Santambrogio M, Lanzetti L, Bussolino F, Ivaska J, Serini G (2012) The R-Ras/RIN2/Rab5 complex controls endothelial cell adhesion and morphogenesis via active integrin endocytosis and Rac signaling. Cell Res 22(10):1479–1501. https://doi.org/10.1038/cr.2012.110

Sasaki AT, Carracedo A, Locasale JW, Anastasiou D, Takeuchi K, Kahoud ER, Haviv S, Asara JM, Pandolfi PP, Cantley LC (2011) Ubiquitination of K-Ras enhances activation and facilitates binding to select downstream effectors. Sci Signal 4(163):ra13. https://doi.org/10.1126/scisignal.2001518

Sauvonnet N, Dujeancourt A, Dautry-Varsat A (2005) Cortactin and dynamin are required for the clathrin-independent endocytosis of gammac cytokine receptor. J Cell Biol 168(1):155–163. https://doi.org/10.1083/jcb.200406174

Schafer DA, Weed SA, Binns D, Karginov AV, Parsons JT, Cooper JA (2002) Dynamin2 and cortactin regulate actin assembly and filament organization. Curr Biol 12(21):1852–1857

Schenck A, Goto-Silva L, Collinet C, Rhinn M, Giner A, Habermann B, Brand M, Zerial M (2008) The endosomal protein Appl1 mediates Akt substrate specificity and cell survival in vertebrate development. Cell 133(3):486–497. https://doi.org/10.1016/j.cell.2008.02.044

Schiefermeier N, Scheffler JM, de Araujo ME, Stasyk T, Yordanov T, Ebner HL, Offterdinger M, Munck S, Hess MW, Wickstrom SA, Lange A, Wunderlich W, Fassler R, Teis D, Huber LA (2014) The late endosomal p14-MP1 (LAMTOR2/3) complex regulates focal adhesion dynamics during cell migration. J Cell Biol 205(4):525–540. https://doi.org/10.1083/jcb.201310043

Schmick M, Kraemer A, Bastiaens PI (2015) Ras moves to stay in place. Trends Cell Biol 25(4):190–197. https://doi.org/10.1016/j.tcb.2015.02.004

Schnelzer A, Prechtel D, Knaus U, Dehne K, Gerhard M, Graeff H, Harbeck N, Schmitt M, Lengyel E (2000) Rac1 in human breast cancer: overexpression, mutation analysis, and characterization of a new isoform, Rac1b. Oncogene 19(26):3013–3020. https://doi.org/10.1038/sj.onc.1203621

Scita G, Di Fiore PP (2010) The endocytic matrix. Nature 463(7280):464–473. https://doi.org/10.1038/nature08910

Scott CC, Vacca F, Gruenberg J (2014) Endosome maturation, transport and functions. Semin Cell Dev Biol 31:2–10. https://doi.org/10.1016/j.semcdb.2014.03.034

Shafaq-Zadah M, Gomes-Santos CS, Bardin S, Maiuri P, Maurin M, Iranzo J, Gautreau A, Lamaze C, Caswell P, Goud B, Johannes L (2016) Persistent cell migration and adhesion rely on retrograde transport of beta(1) integrin. Nat Cell Biol 18(1):54–64. https://doi.org/10.1038/ncb3287

Shenoy SK, Lefkowitz RJ (2011) Beta-Arrestin-mediated receptor trafficking and signal transduction. Trends Pharmacol Sci 32(9):521–533. https://doi.org/10.1016/j.tips.2011.05.002

Shibasaki Y, Ishihara H, Kizuki N, Asano T, Oka Y, Yazaki Y (1997) Massive actin polymerization induced by phosphatidylinositol-4-phosphate 5-kinase in vivo. J Biol Chem 272(12):7578–7581

Sigismund S, Argenzio E, Tosoni D, Cavallaro E, Polo S, Di Fiore PP (2008) Clathrin-mediated internalization is essential for sustained EGFR signaling but dispensable for degradation. Dev Cell 15(2):209–219. https://doi.org/10.1016/j.devcel.2008.06.012

Sigismund S, Confalonieri S, Ciliberto A, Polo S, Scita G, Di Fiore PP (2012) Endocytosis and signaling: cell logistics shape the eukaryotic cell plan. Physiol Rev 92(1):273–366. https://doi.org/10.1152/physrev.00005.2011

Silvius JR, Bhagatji P, Leventis R, Terrone D (2006) K-ras4B and prenylated proteins lacking "second signals" associate dynamically with cellular membranes. Mol Biol Cell 17(1):192–202. https://doi.org/10.1091/mbc.E05-05-0408

Soriano-Castell D, Chavero A, Rentero C, Bosch M, Vidal-Quadras M, Pol A, Enrich C, Tebar F (2017) ROCK1 is a novel Rac1 effector to regulate tubular endocytic membrane formation during clathrinindependent endocytosis. Sci Rep 7(1):6866. https://doi.org/10.1038/s41598-017-07130-x

Sorkin A, von Zastrow M (2002) Signal transduction and endocytosis: close encounters of many kinds. Nat Rev Mol Cell Biol 3(8):600–614. https://doi.org/10.1038/nrm883

Sorkin A, von Zastrow M (2009) Endocytosis and signalling: intertwining molecular networks. Nat Rev Mol Cell Biol 10(9):609–622. https://doi.org/10.1038/nrm2748

Stasyk T, Huber LA (2016) Spatio-temporal parameters of endosomal signaling in cancer: implications for new treatment options. J Cell Biochem 117(4):836–843. https://doi.org/10.1002/jcb.25418

Sturge J, Wienke D, Isacke CM (2006) Endosomes generate localized Rho-ROCK-MLC2-based contractile signals via Endo180 to promote adhesion disassembly. J Cell Biol 175(2):337–347. https://doi.org/10.1083/jcb.200602125

Sugihara K, Nakatsuji N, Nakamura K, Nakao K, Hashimoto R, Otani H, Sakagami H, Kondo H, Nozawa S, Aiba A, Katsuki M (1998) Rac1 is required for the formation of three germ layers during gastrulation. Oncogene 17(26):3427–3433. https://doi.org/10.1038/sj.onc.1202595

Sun Y, Buki KG, Ettala O, Vaaraniemi JP, Vaananen HK (2005) Possible role of direct Rac1-Rab7 interaction in ruffled border formation of osteoclasts. J Biol Chem 280(37):32356–32361. https://doi.org/10.1074/jbc.M414213200

Swarthout JT, Lobo S, Farh L, Croke MR, Greentree WK, Deschenes RJ, Linder ME (2005) DHHC9 and GCP16 constitute a human protein fatty acyltransferase with specificity for H- and N-Ras. J Biol Chem 280(35):31141–31148. https://doi.org/10.1074/jbc.M504113200

Tall GG, Barbieri MA, Stahl PD, Horazdovsky BF (2001) Ras-activated endocytosis is mediated by the Rab5 guanine nucleotide exchange activity of RIN1. Dev Cell 1(1):73–82

Taub N, Teis D, Ebner HL, Hess MW, Huber LA (2007) Late endosomal traffic of the epidermal growth factor receptor ensures spatial and temporal fidelity of mitogen-activated protein kinase signaling. Mol Biol Cell 18(12):4698–4710. https://doi.org/10.1091/mbc.E07-02-0098

Tebar F, Villalonga P, Sorkina T, Agell N, Sorkin A, Enrich C (2002) Calmodulin regulates intracellular trafficking of epidermal growth factor receptor and the MAPK signaling pathway. Mol Biol Cell 13(6):2057–2068. https://doi.org/10.1091/mbc.01-12-0571

Teis D, Huber LA (2003) The odd couple: signal transduction and endocytosis. Cell Mol Life Sci 60(10):2020–2033. https://doi.org/10.1007/s00018-003-3010-2

Teis D, Wunderlich W, Huber LA (2002) Localization of the MP1-MAPK scaffold complex to endosomes is mediated by p14 and required for signal transduction. Dev Cell 3(6):803–814

Teis D, Taub N, Kurzbauer R, Hilber D, de Araujo ME, Erlacher M, Offterdinger M, Villunger A, Geley S, Bohn G, Klein C, Hess MW, Huber LA (2006) p14-MP1-MEK1 signaling regulates endosomal traffic and cellular proliferation during tissue homeostasis. J Cell Biol 175(6):861–868. https://doi.org/10.1083/jcb.200607025

ten Klooster JP, Hordijk PL (2007) Targeting and localized signalling by small GTPases. Biol Cell 99(1):1–12. https://doi.org/10.1042/BC20060071

ten Klooster JP, Jaffer ZM, Chernoff J, Hordijk PL (2006) Targeting and activation of Rac1 are mediated by the exchange factor beta-Pix. J Cell Biol 172(5):759–769. https://doi.org/10.1083/jcb.200509096

Tolias K, Carpenter CL (2000) In vitro interaction of phosphoinositide-4-phosphate 5-kinases with Rac. Methods Enzymol 325:190–200

Tolias KF, Hartwig JH, Ishihara H, Shibasaki Y, Cantley LC, Carpenter CL (2000) Type Ialpha phosphatidylinositol-4-phosphate 5-kinase mediates Rac-dependent actin assembly. Curr Biol 10(3):153–156

Tomas A, Futter CE, Eden ER (2014) EGF receptor trafficking: consequences for signaling and cancer. Trends Cell Biol 24(1):26–34. https://doi.org/10.1016/j.tcb.2013.11.002

Tsukazaki T, Chiang TA, Davison AF, Attisano L, Wrana JL (1998) SARA, a FYVE domain protein that recruits Smad2 to the TGFbeta receptor. Cell 95(6):779–791

Tsutsumi K, Fujioka Y, Tsuda M, Kawaguchi H, Ohba Y (2009) Visualization of Ras-PI3K interaction in the endosome using BiFC. Cell Signal 21(11):1672–1679. https://doi.org/10.1016/j.cellsig.2009.07.004

Valdez G, Philippidou P, Rosenbaum J, Akmentin W, Shao Y, Halegoua S (2007) Trk-signaling endosomes are generated by Rac-dependent macroendocytosis. Proc Natl Acad Sci U S A 104(30):12270–12275. https://doi.org/10.1073/pnas.0702819104

van den Bout I, Divecha N (2009) PIP5K-driven PtdIns(4,5)P_2 synthesis: regulation and cellular functions. J Cell Sci 122(Pt 21):3837–3850. https://doi.org/10.1242/jcs.056127

Vartak N, Bastiaens P (2010) Spatial cycles in G-protein crowd control. EMBO J 29(16):2689–2699. https://doi.org/10.1038/emboj.2010.184

Vidal C, Geny B, Melle J, Jandrot-Perrus M, Fontenay-Roupie M (2002) Cdc42/Rac1-dependent activation of the p21-activated kinase (PAK) regulates human platelet lamellipodia spreading: implication of the cortical-actin binding protein cortactin. Blood 100(13):4462–4469. https://doi.org/10.1182/blood.V100.13.4462

Vidal-Quadras M, Gelabert-Baldrich M, Soriano-Castell D, Llado A, Rentero C, Calvo M, Pol A, Enrich C, Tebar F (2011) Rac1 and calmodulin interactions modulate dynamics of ARF6-dependent endocytosis. Traffic 12(12):1879–1896. https://doi.org/10.1111/j.1600-0854.2011.01274.x

Vieira AV, Lamaze C, Schmid SL (1996) Control of EGF receptor signaling by clathrin-mediated endocytosis. Science 274(5295):2086–2089

Vigil D, Cherfils J, Rossman KL, Der CJ (2010) Ras superfamily GEFs and GAPs: validated and tractable targets for cancer therapy? Nat Rev Cancer 10(12):842–857. https://doi.org/10.1038/nrc2960

Vilardaga JP, Jean-Alphonse FG, Gardella TJ (2014) Endosomal generation of cAMP in GPCR signaling. Nat Chem Biol 10(9):700–706. https://doi.org/10.1038/nchembio.1611

Villalonga P, Lopez-Alcala C, Bosch M, Chiloeches A, Rocamora N, Gil J, Marais R, Marshall CJ, Bachs O, Agell N (2001) Calmodulin binds to K-Ras, but not to H- or N-Ras, and modulates its downstream signaling. Mol Cell Biol 21(21):7345–7354. https://doi.org/10.1128/MCB.21.21.7345-7354.2001

Villasenor R, Kalaidzidis Y, Zerial M (2016) Signal processing by the endosomal system. Curr Opin Cell Biol 39:53–60. https://doi.org/10.1016/j.ceb.2016.02.002

von Zastrow M, Sorkin A (2007) Signaling on the endocytic pathway. Curr Opin Cell Biol 19(4):436–445. https://doi.org/10.1016/j.ceb.2007.04.021

Vonderheit A, Helenius A (2005) Rab7 associates with early endosomes to mediate sorting and transport of Semliki forest virus to late endosomes. PLoS Biol 3(7):e233. https://doi.org/10.1371/journal.pbio.0030233

Walsh AB, Bar-Sagi D (2001) Differential activation of the Rac pathway by Ha-Ras and K-Ras. J Biol Chem 276(19):15609–15615. https://doi.org/10.1074/jbc.M010573200

Wang Y, Pennock S, Chen X, Wang Z (2002) Endosomal signaling of epidermal growth factor receptor stimulates signal transduction pathways leading to cell survival. Mol Cell Biol 22(20):7279–7290

Watson FL, Heerssen HM, Bhattacharyya A, Klesse L, Lin MZ, Segal RA (2001) Neurotrophins use the Erk5 pathway to mediate a retrograde survival response. Nat Neurosci 4(10):981–988. https://doi.org/10.1038/nn720

Webb BA, Zhou S, Eves R, Shen L, Jia L, Mak AS (2006) Phosphorylation of cortactin by p21-activated kinase. Arch Biochem Biophys 456(2):183–193. https://doi.org/10.1016/j.abb.2006.06.011

Weernink PA, Meletiadis K, Hommeltenberg S, Hinz M, Ishihara H, Schmidt M, Jakobs KH (2004) Activation of type I phosphatidylinositol 4-phosphate 5-kinase isoforms by the Rho GTPases, RhoA, Rac1, and Cdc42. J Biol Chem 279(9):7840–7849. https://doi.org/10.1074/jbc.M312737200

White IJ, Bailey LM, Aghakhani MR, Moss SE, Futter CE (2006) EGF stimulates annexin 1-dependent inward vesiculation in a multivesicular endosome subpopulation. EMBO J 25(1):1–12. https://doi.org/10.1038/sj.emboj.7600759

Wiesner C, El Azzouzi K, Linder S (2013) A specific subset of RabGTPases controls cell surface exposure of MT1-MMP, extracellular matrix degradation and three-dimensional invasion of macrophages. J Cell Sci 126(Pt 13):2820–2833. https://doi.org/10.1242/jcs.122358

Williams KC, Coppolino MG (2011) Phosphorylation of membrane type 1-matrix metalloproteinase (MT1-MMP) and its vesicle-associated membrane protein 7 (VAMP7)-dependent trafficking facilitate cell invasion and migration. J Biol Chem 286(50):43405–43416. https://doi.org/10.1074/jbc.M111.297069

Wittinghofer A, Herrmann C (1995) Ras-effector interactions, the problem of specificity. FEBS Lett 369(1):52–56

Wittinghofer A, Vetter IR (2011) Structure-function relationships of the G domain, a canonical switch motif. Annu Rev Biochem 80:943–971. https://doi.org/10.1146/annurev-biochem-062708-134043

Wu C, Lai CF, Mobley WC (2001) Nerve growth factor activates persistent Rap1 signaling in endosomes. J Neurosci 21(15):5406–5416

Xu L, Lubkov V, Taylor LJ, Bar-Sagi D (2010) Feedback regulation of Ras signaling by Rabex-5-mediated ubiquitination. Curr Biol 20(15):1372–1377. https://doi.org/10.1016/j.cub.2010.06.051

Yan H, Jahanshahi M, Horvath EA, Liu HY, Pfleger CM (2010) Rabex-5 ubiquitin ligase activity restricts Ras signaling to establish pathway homeostasis in Drosophila. Curr Biol 20(15):1378–1382. https://doi.org/10.1016/j.cub.2010.06.058

Yang N, Higuchi O, Ohashi K, Nagata K, Wada A, Kangawa K, Nishida E, Mizuno K (1998) Cofilin phosphorylation by LIM-kinase 1 and its role in Rac-mediated actin reorganization. Nature 393(6687):809–812. https://doi.org/10.1038/31735

Yang Y, Du J, Hu Z, Liu J, Tian Y, Zhu Y, Wang L, Gu L (2011) Activation of Rac1-PI3K/Akt is required for epidermal growth factor-induced PAK1 activation and cell migration in MDA-MB-231 breast cancer cells. J Biomed Res 25(4):237–245. https://doi.org/10.1016/S1674-8301(11)60032-8

Yeung T, Terebiznik M, Yu L, Silvius J, Abidi WM, Philips M, Levine T, Kapus A, Grinstein S (2006) Receptor activation alters inner surface potential during phagocytosis. Science 313(5785):347–351. https://doi.org/10.1126/science.1129551

Yeung T, Gilbert GE, Shi J, Silvius J, Kapus A, Grinstein S (2008) Membrane phosphatidylserine regulates surface charge and protein localization. Science 319(5860):210–213. https://doi.org/10.1126/science.1152066

Yokoe H, Meyer T (1996) Spatial dynamics of GFP-tagged proteins investigated by local fluorescence enhancement. Nat Biotechnol 14(10):1252–1256. https://doi.org/10.1038/nbt1096-1252

Zerial M, McBride H (2001) Rab proteins as membrane organizers. Nat Rev Mol Cell Biol 2(2):107–117. https://doi.org/10.1038/35052055

Zheng ZY, Chang EC (2014) A bimolecular fluorescent complementation screen reveals complex roles of endosomes in Ras-mediated signaling. Methods Enzymol 535:25–38. https://doi.org/10.1016/B978-0-12-397925-4.00002-X

Zheng ZY, Cheng CM, Fu XR, Chen LY, Xu L, Terrillon S, Wong ST, Bar-Sagi D, Songyang Z, Chang EC (2012a) CHMP6 and VPS4A mediate the recycling of Ras to the plasma membrane to promote growth factor signaling. Oncogene 31(43):4630–4638. https://doi.org/10.1038/onc.2011.607

Zheng ZY, Xu L, Bar-Sagi D, Chang EC (2012b) Escorting Ras. Small GTPases 3(4):236–239. https://doi.org/10.4161/sgtp.20460

Zhou Y, Hancock JF (2015) Ras nanoclusters: Versatile lipid-based signaling platforms. Biochim Biophys Acta 1853(4):841–849. https://doi.org/10.1016/j.bbamcr.2014.09.008

Zhou P, Porcionatto M, Pilapil M, Chen Y, Choi Y, Tolias KF, Bikoff JB, Hong EJ, Greenberg ME, Segal RA (2007) Polarized signaling endosomes coordinate BDNF-induced chemotaxis of cerebellar precursors. Neuron 55(1):53–68. https://doi.org/10.1016/j.neuron.2007.05.030

Zhu G, Fan Z, Ding M, Zhang H, Mu L, Ding Y, Zhang Y, Jia B, Chen L, Chang Z, Wu W (2015) An EGFR/PI3K/AKT axis promotes accumulation of the Rac1-GEF Tiam1 that is critical in EGFR-driven tumorigenesis. Oncogene 34(49):5971–5982. https://doi.org/10.1038/onc.2015.45
Zoncu R, Perera RM, Balkin DM, Pirruccello M, Toomre D, De Camilli P (2009) A phosphoinositide switch controls the maturation and signaling properties of APPL endosomes. Cell 136(6):1110–1121. https://doi.org/10.1016/j.cell.2009.01.032
Zoncu R, Efeyan A, Sabatini DM (2011) mTOR: from growth signal integration to cancer, diabetes and ageing. Nat Rev Mol Cell Biol 12(1):21–35. https://doi.org/10.1038/nrm3025

Chapter 4
ESCRT and Membrane Protein Ubiquitination

Simona M. Migliano and David Teis

Contents

Abstract The ubiquitin-dependent degradation of membrane proteins via the multivesicular body (MVB) pathway requires the Endosomal Sorting Complexes Required for Transport (ESCRT). This molecular machinery is composed of five distinct multisubunit complexes. On the surface of endosomes, ESCRT-0, -I and -II bind to ubiquitinated membrane proteins, while ESCRT-III and Vps4 bud intraluminal vesicles (ILVs) into the lumen of the endosomes. By working together, ESCRTs package membrane proteins into ILVs and thereby generate MVBs. The fusion of mature MVBs with lysosomes delivers ILVs into the lysosomal lumen where the membrane proteins are degraded. Besides generating ILVs, the ESCRT machinery mediates for topologically related membrane budding processes at the plasma membrane and the nuclear envelop. In this chapter, we briefly discuss membrane protein ubiquitination, endocytosis, and summarize current knowledge on the ESCRT machinery in the MVB pathway.

S. M. Migliano · D. Teis (✉)
Division of Cell Biology, Biocenter, Medical University of Innsbruck,
Innrain 80/82, CCB Building, 6020 Innsbruck, Austria
e-mail: david.teis@i-med.ac.at

C. Lamaze and I. Prior (eds.), *Endocytosis and Signaling*, Progress in Molecular and Subcellular Biology 57, https://doi.org/10.1007/978-3-319-96704-2_4

4.1 Ubiquitination

Ubiquitination is a modular, reversible and transferable protein modification (Piper and Lehner 2011). Ubiquitin itself is a highly conserved, cytoplasmic, 76-amino acid polypeptide. It can be covalently attached (conjugated) to lysine residues in target proteins by ubiquitin ligases and removed by deubiquitinating enzymes.

Ubiquitin conjugation is a three-step process. A ubiquitin-activating enzyme (E1) takes up the ubiquitin and transfers it to a ubiquitin-conjugating enzyme (E2). Ubiquitin-ligating (E3) enzymes conduct the final step of ubiquitination and conjugate the carboxy-terminal glycine of ubiquitin to the amino group of a lysine in the target protein (Fig. 4.1a).

E2 and E3 enzymes are directly involved in transferring ubiquitin to the target protein and therefore contribute to the specificity of the ligation process (Hicke et al. 2005; Hofmann and Pickart 2001; Komander 2009). Eukaryotic genomes encode only few E1 enzymes (1 in yeast, approx. 10 in human), while E2 enzymes are more abundant (11 in yeast, approx. 40 in human). E3 enzymes are even more diverse (54 in yeast, approx. 600 in human (Hershko and Ciechanover 1998; Hicke et al. 2005; Metzger et al. 2012). About 100 deubiquitinating, enzymes (DUBs) can cleave the isopeptide ubiquitin and the target protein and thereby make ubiquitination reversible and help to recycle ubiquitin (Komander 2009) (Figs. 4.1a, 4.2a, b).

E3 enzymes can be grouped in two families: HECT ligases and RING ligases. The HECT E3 ligase (homologous to the E6AP carboxyl terminus) family members pick up the ubiquitin from the E2 enzyme before they conjugate it to the target protein. Therefore, a prototypical HECT domain interacts with the E2 enzyme, as well as an active-site cysteine, that forms the thioester with ubiquitin (Metzger et al. 2012; Huang et al. 1999) and then transfers the ubiquitin to the target protein (Fig. 4.1a).

In contrast, the really interesting new gene (RING) E3 ligases mediate a direct transfer from the E2 carrier to the target protein. RING ligases function as a scaffold, that brings the E2 enzyme and the target protein in close proximity for the ubiquitin transfer process (Lorick et al. 1999; Metzger et al. 2012) (Fig. 4.1a).

From yeast to human ubiquitin has remained evolutionary almost invariable except for three conservative changes. It has an overall molecular mass of 8.5 kDA and contains a compact β-grasp fold with a flexible C-terminal tail. Ubiquitin contains seven lysine residues and they are ubiquitinated as well (Ravid and Hochstrasser 2007; Wang and Pickart 2005). This implies that target proteins can be mono- (one ubiquitin on one lysine), multi- (single ubiquitin molecules on several lysines within one protein), or poly-ubiquitinated (>10 ubiquitin moieties on one lysine) (Fig. 4.1b). Monoubiquitin can be extended into eight different homotypic polyubiquitin chains that always consist of the same linkage: K6, K11, K27, K29, K33, K48, K63 and Met1 (Dikic and Robertson 2012; Hofmann and Pickart 2001; Kerscher et al. 2006; Piper and Lehner 2011). Heterotypic polyubiquitin chains have mixed linkages. Moreover, conjugated ubiquitin molecules can be subject to posttranslational modifications, such as acetylation on 6 of 7 Lys and phosphorylation on Ser, Thr and Tyr residues (Swatek and Komander 2016; Mevissen and Komander 2016).

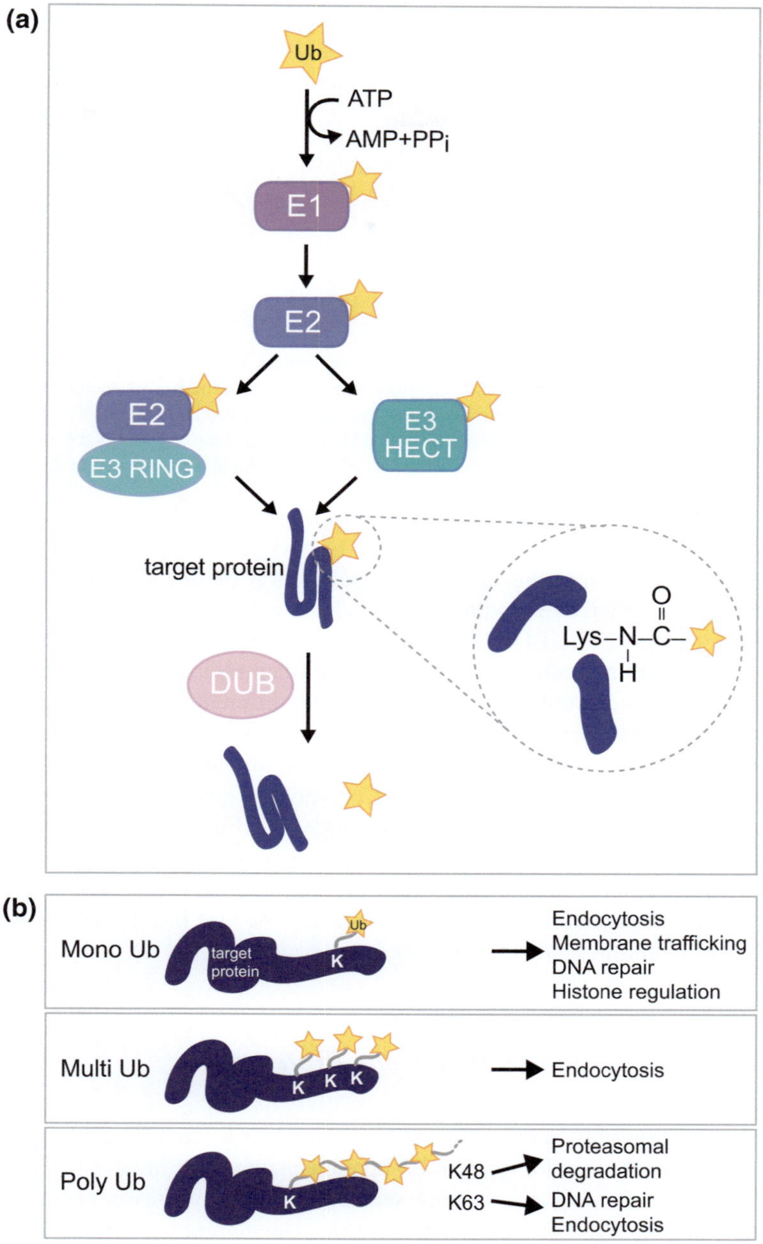

Fig. 4.1 Ubiquitination of proteins a Ubiquitination involves ubiquitin-activating enzymes (E1), ubiquitin-conjugating enzymes (E2) and ubiquitin ligases (E3). Ubiquitination can be reversed by deubiquitinating enzymes (DUBs). **b** Mono-, multi- and poly-ubiquitination of target proteins. Ubiquitin is covalently attached to lysine residues of the target protein via an isopeptide bond. The form of ubiquitination determines the destination of the protein

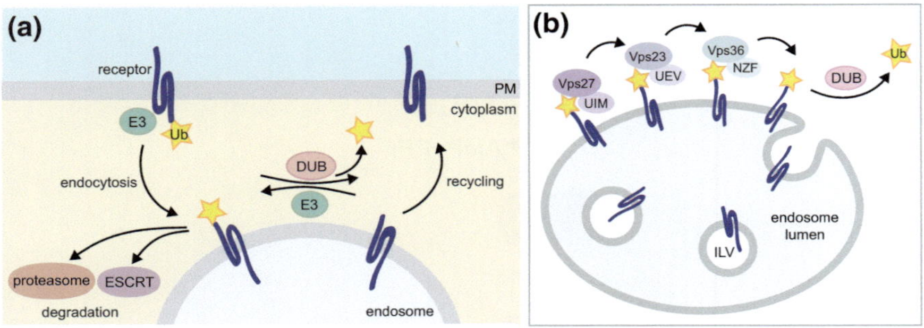

Fig. 4.2 Ubiquitination and deubiquitination of proteins a Ubiquitination and deubiquitination reactions at the plasma membrane and on endosomes. **b** Ubiquitination of membrane proteins induces ESCRT-dependent sorting into intraluminal vesicles (ILV) on endosomes. ESCRT subunits interact directly via their ubiquitin-binding domains. Deubiquitination takes place prior to ILV formation. Ubiquitin-interacting motiv (UIM), ubiquitin E2 variant (UEV), novel zink finger (NZF), plasma membrane (PM)

The flexible tails and the different ubiquitination modes provide a code that can be read by different ubiquitin-binding domains (UBDs) and thus determine the fate of the protein. UBDs recognize different forms of ubiquitination (mono- versus polyubiquitin chains) and/or different surface regions within the ubiquitin moiety and thus bind to ubiquitinated proteins with different affinities (Di Fiore et al. 2003; Schnell and Hicke 2003). Some examples of UBDs are: ubiquitin-interacting motif (UIM), ubiquitin-associated domain (UBA), ubiquitin-conjugating enzyme-like (UBC), ubiquitin E2 variant (UEV), Cue1-homologous (CUE), polyubiquitin-associated zinc finger (PAZ) and novel zinc finger (NZF).

Ubiquitination is essential for many cellular processes. In this chapter, we focus on the ubiquitination of plasma membrane proteins and the subsequent endocytosis and lysosomal degradation via the MVB pathway. We also briefly discuss some of the involved ubiquitin ligases and DUBs.

4.1.1 Ubiquitination of Membrane Proteins

The ubiquitination of membrane proteins also involves E1 and E2 enzymes and E3 ubiquitin ligases. Two major types of E3 ligation enzymes are involved: The HECT-type Nedd4 (neural precursor cell expressed developmentally down-regulated protein 4) family and the RING-type MARCH ligases (membrane associated RING-CH ligases).

The yeast Nedd4 homologue Rsp5 is one of the best-studied E3 ligases (Rotin and Kumar 2009; Staub et al. 2000; Belgareh-Touzé et al. 2008; Dupré et al. 2004).

The human genome encodes nine Nedd4 family members. Nedd4/Rsp5 have a characteristic domain organization with an N-terminal C2 domain, a variable number of WW domains (protein–protein interaction domains) and the HECT domain.

The C2 domain spans approximately 130 amino acids and functions as a protein–lipid interaction module (Nalefski and Falke 1996; Hurley and Misra 2000).

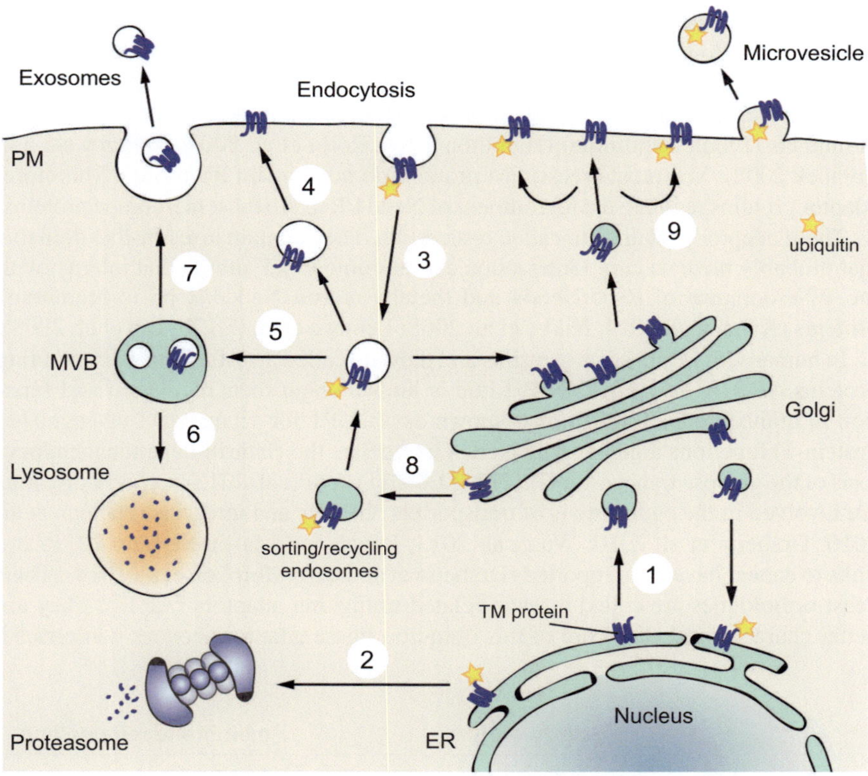

Fig. 4.3 Regulation of membrane proteins Ubiquitination influences the lifecycle and destination of transmembrane proteins. Membrane proteins are synthesized at the endoplasmatic reticulum (ER) and undergo quality control (ERQC). (1) Membrane proteins are transported to the Golgi, for modification and quality control. (2) Misfolded proteins are recognized by the ER-associated degradation (ERAD), ubiquitinated and a target for proteasomal degradation. (3) Ubiquitinated plasma membrane proteins undergo endocytosis. (4) Endocytosed proteins are either recycled back to the plasma membrane or to the Golgi. (5) Endosomes can mature via the ESCRT mediated multivesicular body (MVB) pathway, which (6) targets ubiquitinated membrane proteins for lysosomal degradation. (7) MVBs can fuse with the plasma membrane and release their vesicles as exosomes. (8) Ubiquitination at the Golgi contributes to the endosomal sorting or recycling. Proof read proteins are transported from the Golgi to the plasma membrane (PM). (9) Ubiquitination can be a signal for microvesicle formation

The C2 of Rsp5 interacts with phosphoinositides in membranes a function that is important for the ubiquitination of membrane proteins (Dunn et al. 2004).

The WW domains are about 40 amino acids long and bind to proline-rich sequences, so-called PY motifs (Kaliszewski and Zoładek 2008; Bedford et al. 2000). Group I binds to PXY, LPXY and PPXY motifs. group II binds to PPLP, group III binds PPR motifs, and group IV binds short sequences containing phospho-serine or phospho-threonine followed by proline. Some membrane proteins such as the epithelial sodium channel, ENaC, contain PPxY motifs and thus permit direct interaction with the WW domain of Nedd4-2, allowing its ubiquitination and endocytosis (Lu et al. 2007; Staub et al. 2000). The down-regulation of ENaC controls blood volume and pressure, and the defects in this process result in hereditary hypertension in humans (Liddle syndrome) (Hamilton 2014; Rotin et al. 2000; Kaliszewski and Zoładek 2008). Yet, most (yeast) PM proteins do not contain PY motifs. Therefore, adaptor proteins regulate the recruitment of Nedd4/Rsp5 to most membrane proteins.

These adaptor proteins are called α-arrestins. They contain arrestin-like domains that probably mediate cargo interaction and multiple PPxY motifs that interact with the WW domains of Rsp5/Nedd4 and thereby recruit Nedd4/Rsp5 to membrane proteins (Kang et al. 2014; Nikko et al. 2008; Becuwe et al. 2012b; Lin et al. 2008).

In humans, there are six α-arrestins and they are called arrestin domain containing proteins ARRDC1-5 and TXNIP. Little is known about their regulation and function in human cells. TXNIP (also known as VDUP1 for vitamin D3 up-regulated protein-1) functions together with Nedd4 to mediate the clathrin-dependent endocytosis of the glucose transporter GLUT1 (SLC2A1) (Wu et al. 2013). Other α-arrestins are involved in the endocytosis of transporters, GPCRs and integrins (Nabhan et al. 2010; Draheim et al. 2010; Wu et al. 2013; Rauch and Martin-Serrano 2011) and links to cancer have been reported (Draheim et al. 2010; Morrison et al. 2014). Their yeast orthologues are called arrestin-related trafficking adaptors (ARTs). They are better characterized. The core of this ubiquitin ligase adaptor network comprises a

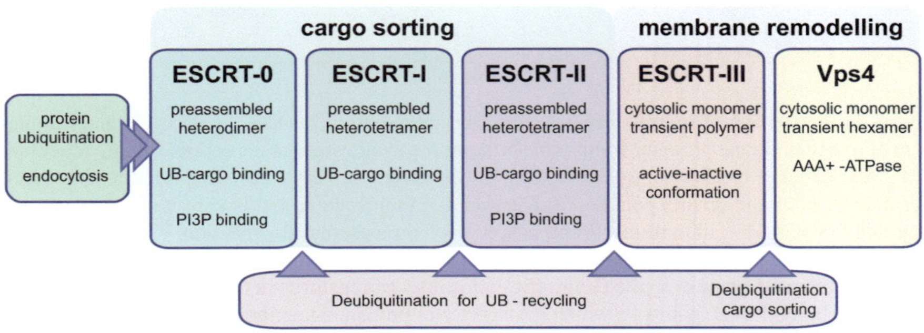

Fig. 4.4 Ubiquitination—deubiquitination and ESCRT complexes Overview on the most important features of the five ESCRT complexes. The ESCRT-0, -I and -II complexes bind to ubiquitinated cargo and PI3P. The ESCRT-III and Vps4 mediate membrane budding and scission. Deubiquitination can take place at different steps

family of 10 ART proteins (Art1-10), and Bul1 and Bul2. Art2/Art8 (Ecm21/Csr2), Art3/Art6 (Aly2/Aly1), Art4/Art7 (Rod1/Rog3) and Bul1/Bul2 are considered gene duplications.

The adaptor function for most yeast ART proteins was demonstrated in recent years. Under different growth conditions or under stress (mainly nutrient excess) ARTs regulate, in a partially redundant manner, several different transporters and thereby enable ubiquitin-dependent nutrient transporter endocytosis (Suzuki et al. 2013; O'Donnell et al. 2010, 2013; Alvaro et al. 2014; Marqués and Zamarbide-Forés 2015; O'Donnell et al. 2015; Becuwe et al. 2012a; Hatakeyama et al. 2010; Smardon and Kane 2014; Crapeau et al. 2014; Merhi and Andre 2012; Ghaddar et al. 2014; Nikko and Pelham 2009; Nikko et al. 2008; Lin et al. 2008; MacGurn et al. 2011; Zhao et al. 2013). This is exemplified by the regulation of Lyp1 (lysine transporter) endocytosis. Art1/Ldb19 is required for Lyp1 endocytosis in response to excess lysine, whereas Art2/Ecm21 is required for Lyp1 internalization in response to cycloheximide (CHX causes TORC1 hyper-activation) treatment. These findings suggested that ART function and transporter endocytosis are controlled by different stimuli (Lin et al. 2008). The regulation of ARTs is controlled by post-translation modifications (PTMs) (Becuwe et al. 2012b). One common regulatory concept is the ubiquitination of the ARTs by Rsp5. In addition, it is becoming clear that phosphorylation inhibits ART function whereas de-phosphorylation activates ART function. Some examples are described below: Under energy stress, mammalian cells activate the 5'AMP-activated protein kinase, AMPK. AMPK has many targets; among others, it phosphorylates the α-arrestin TXNIP, leading to its degradation. Low levels of TXNIP can no longer efficiently down-regulate the glucose transporter GLUT1 (SLC2A1), which in turn allows the influx of glucose to restore ADP/ATP homeostasis (Wu et al. 2013).

When yeast cells experience glucose starvation and grow in the presence of lactate, they also activate AMPK, Snf1. In turn, Snf1 phosphorylates Art4 (Rod1). 14-3-3 proteins bind to phosphorylated Art4 and inactivate it (Becuwe et al. 2012a). Thereby, Snf1-dependent inactivation of Art4 prevents endocytosis of the lactate transporter Jen1 and thus stimulates lactate influx. Upon addition of glucose, a preferred carbon source, Art4 is dephosphorylated by the PP1 phosphatase, Glc7, which releases the 14-3-3 proteins. Now, Art4-Rsp5 complexes become active and ubiquitinate Jen1, which results in endocytosis and degradation of Jen1 (Becuwe et al. 2012b). Conversely, glucose deprivation regulates the ART protein Csr2/Art8 at multiple levels to trigger high-affinity glucose transporter endocytosis (Hovsepian et al. 2017). Also, amino acid starvation triggers endocytosis (Muller et al. 2015; Jones et al. 2012), but how the ART proteins are regulated in response to nitrogen deprivation is not clear.

The nutrient-sensitive kinase complex—target of rapamycin complex 1 (TORC1)—also controls the ART-Rsp5 network via phospho-inhibition of the kinase Npr1. Inactivation of TORC1 by rapamycin treatment or by amino acid/nitrogen starvation activates Npr1. Activated Npr1 phosphorylates Art1 and Bul1/2 and thereby inhibits their function (Becuwe et al. 2012b; MacGurn et al. 2011; Merhi and Andre 2012; Crapeau et al. 2014). Phosphorylated Bul1/2 is bound by 14-3-3 proteins and inactivated (Merhi and Andre 2012). Thus, inactivation of TORC1 should inhibit endocytosis of Art1- and Bul1/2-dependent PM proteins. Conversely, activation of TORC1 by nutrient excess or CHX treatment inactivates Npr1, which in turn stimulates endocytosis of the Art1 cargo, Can1 (arginine transporter) and the Bul1/2 cargo, general amino acid permease (Gap1). Activation of Bul1/2 and subsequent endocytosis of Gap1 additionally depends on Sit4 phosphates activity (Merhi and Andre 2012).

Interestingly, Gap1 is also down-regulated in rapamycin-treated cells, which shuts off TORC1 signalling. The rapamycin-induced endocytosis of Gap1 still requires Bul1/2 but additionally depends on Art3/6 (Aly2/1) (Crapeau et al. 2014). An additional layer of complexity may be added by stimulus-induced transporter phosphorylation, akin to the recognition of GPCRs by β-arrestins in mammalian cells (Shukla et al. 2014; Nikko et al. 2008). It seems that ART-Rsp5-mediated ubiquitin-dependent nutrient transporter endocytosis is controlled by central signalling networks (TORC1 and AMPK) to adjust nutrient influx with cellular metabolism and growth.

The second major family of membrane protein E3 ligases is represented by RING ligases mostly of the MARCH family. There are 11 MARCH proteins and they were originally discovered as structural homologues to the viral E3 ligases, K3 and K5 from Kaposi's sarcoma-associated herpesvirus (KSHV) that play an important role in immune evasion (Samji et al. 2014).

Most of the MARCH ligases have two transmembrane domains with some having up to 14 putative transmembrane domains (Nathan and Lehner 2009; Piper and Lehner 2011). The best-characterized member is MARCH1, which ubiquitinates MHC-II on the plasma membrane of antigen presenting cells, initiates its endocytosis, and thereby influences immune and inflammatory responses (Oh et al. 2013; Cho et al. 2015). Other surface receptors known to be down-regulated by MARCH are CD44, CD81, CD4, CD86, transferrin receptor and others mostly important for cell differentiation (Metzger et al. 2012; Nakamura 2011; Samji et al. 2014). Overall, it is becoming clear that MARCH proteins have a wide variety of substrates and binding partners, including other membrane-bound proteins, SNAREs and even cytoskeletal elements, like microtubules. Despite increasing evidence that MARCH proteins are important regulators of cellular physiology, they are currently understudied.

Another prominent family of the RING ligases involved in plasma membrane protein ubiquitination are the three CBL proteins (casitas B-lineage Lymphoma). c-CBL, CBL-b, CBL-3 contain a PTB domain (phosphotyrosine-binding domain or tyrosine kinase domain) and a RING finger domain. c-CBL and CBL-b additionally contain a proline-rich region and a C-terminal ubiquitin-associated domain (UBA) (Schmidt and Dikic 2005; Kozlov et al. 2007; Gay et al. 2008). CBL ubiquitinates numerous target proteins, including the activated epidermal growth factor receptor (EGFR) (Haglund 2003; Levkowitz et al. 1998, 1999; Sigismund et al. 2005, 2008; Yokouchi et al. 1999; Deshaies and Joazeiro 2009; Thien and Langdon 2001; Wee and Wang 2017; Mohapatra et al. 2013), which results in EGFR endocytosis and subsequent lysosomal degradation via the MVB pathway. This is most likely the best-studied ubiquitination of a membrane protein and has been subject to recent reviews (Haglund and Dikic 2012; Cruickshanks 2014).

So far, we have mainly discussed how membrane proteins are ubiquitinated once they have been properly folded and exported from the ER (Fig. 4.3). Early in their lifetime—still in the ER—membrane proteins undergo a ubiquitin-dependent quality control process that ensures, that misfolded proteins are not efficiently exported from the ER. Misfolded membrane proteins are retro-translocated into the cytoplasm, ubiquitinated and degraded by the proteasome. This ER-associated degradation pathway (ERAD) has been extensively studied in yeast. Two RING E3 ligases, Doa10p and Hrd1, have been discovered to target almost all substrates: Doa10 targeting cytosolic proteins and Hrd1 is ubiquitinating misfolded membrane proteins (Christianson and Ye 2014; Bays et al. 2001; Carvalho et al. 2006; Vashist and Ng 2004). Sometimes, misfolded proteins escape the ERAD system and arrive at the Golgi. The ubiquitin quality control at the Golgi and the E3 ligases involved is not yet characterized well. In fission yeast and in budding yeast, a transmembrane ubiquitin ligase complex called Dsc operates at the Golgi and can ubiquitinate misfolded membrane proteins which ensure that they are degraded in lysosomes (Reggiori and Pelham 2002). In addition, the Dsc complex is also required for the hypoxia-induced activation of yeast SREBP cleavage (Stewart et al. 2011). In budding yeast the E3 ligase of the Dsc complex, Tul1 also functions together with Rsp5 on the surface of the vacuole and there they ubiquitinate different membrane proteins and thereby target them for lysosomal degradation (Li et al. 2015a, b). Overall it is becoming clear that ubiquitin ligases target membrane proteins at the ER, the Golgi, at the PM and on endosomes, and even on lysosomes. In almost all cases, the ubiquitination of membrane proteins targets them for degradation (Foot et al. 2017) (Fig. 4.3).

Ubiquitination is reversed by approximately 100 DUBs in human cells and 20 DUBs in yeast. There are six structurally distinct DUB families that can deal with the vast complexity of the ubiquitination code and additionally contribute to the complexity of the system. Their structural features and functional properties have been summarized recently (Mevissen and Komander 2016). Later, we will briefly discuss DUBs that are involved in the degradation of membrane proteins (please see Sect. 4.3.1).

4.1.2 Endocytosis—As Consequence of Ubiquitination

Adjusting the cell surface abundance of ion channels, growth factor receptors, GPCRs and nutrient transporters is essential to control cell proliferation, migration, electrolyte balance, metabolism and differentiation (Lemmon and Schlessinger 2010). Frequently ubiquitination serves as signal for endocytosis, but there is also evidence that ubiquitin-independent endocytic pathways exist (Galan et al. 1996; Hicke and Riezman 1996; MacGurn et al. 2012). During endocytosis, the target plasma membrane proteins are sequestered in vesicles, which bud into the cytosol to form primary endosomes (Fig. 4.3).

The role of ubiquitin in endocytosis is complex and involves numerous endocytic adaptors. Such adaptors are the Epsin proteins. They bind to ubiquitinated membrane proteins and to the clathrin machinery at the same time. The proteins are called Epsin1/2/3 (Ent1, Ent2 in yeast) and the Eps15 and Eps15R (Ede1, Ede3 in yeast). Their binding to ubiquitinated membrane proteins at the plasma membrane leads to the recruitment of the clathrin effector proteins and the actin cytoskeleton, and together with many other endocytic proteins, this ultimately results in endocytosis (Polo et al. 2002; Shih et al. 2002; Godlee and Kaksonen 2013).

4.2 Ubiquitin-Mediated Protein Sorting on Endosomes

Once ubiquitinated plasma membrane proteins have been internalized, they enter the endosomal network. From there, membrane proteins can either recycle back to the plasma membrane or are transported to the lumen of lysosomes for degradation via the MVB pathway. MVB formation requires the budding of the limiting endosomal membrane into the lumen of the organelle to generate intraluminal vesicles (ILVs). Membrane proteins—cargo—that are destined for degradation in lysosomes must be sorted into the growing ILVs. It is assumed that cargo molecules are typically ubiquitinated. In some case, cargo sorting along the MVB may also occur independent from cargo ubiquitination (e.g. for P2Y1 purinergic receptor) (Dores et al. 2012). Tetraspanins can support the sorting of non-ubiqutinated cargo molecules by providing a ubiquitination signal in trans (MacDonald et al. 2015). The formation of ILVs can be driven by lipids only or by the Endosomal Sorting Complexes Required for Transport (ESCRT) (for a review see Babst (2011).

The ESCRT machinery (Fig. 4.4) couples the sorting of ubiquitinated membrane proteins to the biogenesis of ILVs. ESCRT complexes bind directly to ubiquitinated membrane proteins, induce cargo crowing, ILV budding, cargo deubiquitination and finally scission of the cargo laden ILV from the limiting MVB membrane. The interplay between cargo, endosomal phospho-lipids and five different ESCRT complexes (ESCRT-0, -I, -II, -III and Vps4) ultimately leads to the biogenesis of MVBs.

4.3 The ESCRT Complexes—Domains, Conformation and Interaction

ESCRT-0, -I and -II contain specific ubiquitin-binding domains to recognize the ubiquitinated cargo of endosomes (Schmidt and Teis 2012) (for details see below) (Table 4.1).

ESCRT-0 consists of a 1:1 heterodimer consisting of Vps27/Hse1 (Hrs1/STAM in mammalian) (Ren et al. 2009). It is recruited to phosphatidylinositol 3-phosphate on endosomes via the FYVE zinc finger domain of HRS/Vps27, (Myromslien et al. 2006; Raiborg et al. 2001a, b; Wollert and Hurley 2010). The ESCRT-0 complex contains multiple ubiquitin-binding domains (UBDs). STAM/Hse1 contains one VHS (Vps27,

Table 4.1 ESCRT complexes on endosomes, co-factors, ubiquitin ligases and deubiquitinating enzymes

ESCRT protein	Yeast/human homologue		Domains and motives	Modification	Function
ESCRT-0	Vps27	HRS	UIM, FYVE, VHS, clathrin binding	Ubiquitinated	Binds to ubiquitinated membrane proteins and PI3P
	Hse1	STAM1, STAM2	VHS, UIM, GAT, coiled coil, CB	Ubiquitinated	Interaction with Hua1 and Rsp5
ESCRT-I	Vps23	TSG101	UEV, PRD, coiled coil	Ubiquitinated	Cargo and Vps27 interaction
	Vps28	VPS28	CTD		Binding to ESCRT-II (Vps36)
	Vps37	VPS37 A, B, C, D	Coiled coil, MOD(r)		
	Mvb12		MABP, UMA, UBD	Ubiquitin	Ubiquitin binding
ESCRT-II	Vps22	EAP30, SNF8	Coiled coil		
	Vps25	EAP20			Binding to ESCRT-III (Vps20)
	Vps36	EAP45	GLUE		Cargo and PI3P interaction; interaction with Vps28
ESCRT-III	Vps20	CHMP6	MIM2	Myristoylated	Binding to ESCRT-II (Vps25)
	Vps32/Snf7	CHMP4A, B, C	MIM1, MIM2		Membrane deformation; membrane invagination
	Vps2/Did4	CHMP2A, B	MIM1	Ubiquitinated	Vps4 recruitment
	Vps24	CHMP3	MIM1		

(continued)

Table 4.1 (continued)

ESCRT protein	Yeast/human homologue		Domains and motives	Modification	Function
	Ist1	IST1	MIM1, MIM2		Interacts with Did2 and Vps4
	Vps60/Mos10	CHMP5	MIM1, coiled coil		ESCRT-III like
	Vps46/Did2	CHMP1A, B	MIM1, coiled coil		ESCRT-III like
AAA+-ATPase	Vps4	VPS4A, B/SKD1	MIT, AAA+		ESCRT disassembly and recycling
	Vta1	LIP5	MIT, VSL, Coiled coil		Positive regulator of Vps4
Modulators and adaptors	Vps31/Bro1	ALIX/AIP1	Coiled coil, PRD, PTAP-like	Ubiquitin	Doa4 recruitment; ESCRT-III interaction
PI3 Kinase complex	Vps34	VPS34	C2, PIK		PI(3)P synthesis
	Vps15	PIK3R4(p150)	PKinase, HEAT, WD40		
Ubiquitin ligase	Rsp5	Nedd4/Cbl and many others	C2, WW, HECT		Cargo ubiquitination
De-ubiquitinase	Doa4	UBPY/USP8	Rhod, UBP		Cargo deubiquitination
	Ubp7	AMSH	MIT, JAMM		Cargo deubiquitination
	Hua1	Plant homologues			Links Rsp5 to Hse1
	Rup1	Plant homologues			Complex with Rsp5 and Ubp2

Abbreviations AAA+(ATPase associated with diverse cellular activities; CB (clathrin-box motive); CHMP (charged multivesicular body protein); CTD (C-terminal domain); FYVE (Fab1p/YOTB/Vac1p/EEA1 domain); GAT (GGA/TOM1 domain); GLUE (GRAM-like ubiquitin in EAP45 domain); HRS (hepatocyte growth factor-regulated tyrosine kinase substrate); JAMM (JAB1/MPN/MOV34 metalloenzyme); MABP (MVB12-associated β-prism domain); MIT-interacting motif (MIM); microtubule-interacting and transport (MIT) domain; MOD(r) (modifier of rudimentary domain); PI3P (phosphatidylinositol 3-phosphate); PIK (Phosphoinositide 3-kinas); PRD (proline-rich domain); Rhod (Rhodanese Homology Domain); STAM (signal-transducing adaptor molecule); ubiquitin-binding domain (UBD); ubiquitin-conjugated enzyme E2 variant (UEV); ubiquitin-interacting motif (UIM); UMA (UBAP1/MVB12-associated domain); VHS (Vps27/HRS/STAM domain); VSL (Vta1/SBP1/LIP1 domain)

HRS and STAM) and one ubiquitin interaction motif (UIM) domain. HRS/Vps27 contains a VHS domain plus a double UIM (Zakalskiy et al. 2002) or two UIMs (Vps27) (Babst 2005; Bilodeau et al. 2003; Hicke et al. 2005). Hence, a single ESCRT-0 complex could bind up to five different ubiquitinated membrane proteins or multiple ubiquitin moieties of poly-ubiquitinated cargo to sort these away from other membrane proteins (Prag et al. 2007; Raiborg and Stenmark 2009; Wollert and Hurley 2010). HRS of ESCRT-0 can also interact with flat clathrin coats on endosomes, which may further promote the ubiquitinated cargo crowding (Bilodeau and Urbanowski 2002; Clague 2002; Raiborg et al. 2002; Shih et al. 2002). The C-terminal region of HRS/Vps27 domains of both proteins contains PTAP motifs (proline, threonine, alanine, proline) required for the binding to ESCRT-I (Katzmann et al. 2003; Chu et al. 2006). STAM2 also interacts with ESCRT accessory protein HD-PTP (cytosolic protein tyrosine phosphatase). The N-terminal Bro1 domain of HD-PTP in turn interacts with ESCRT-III protein CHMP4B. In yeast, Bro1 binds ESCRT-I and -III and recruits and catalytically activates Doa4. HD-PTP might fulfil a similar role in establishing a connection between ubiquitinated cargo, ESCRT-III subunits and the coordination of deubiquitinating enzymes (Ali et al. 2013; Lee et al. 2016).

ESCRT-I is an elongated heterotetramer that spans approx. 18 nm. It consists of four subunits Vps23 (Tsg101), Vps28 (VPS28), Vps37 (VPS37A, B) and Mvb12 (MVB12A, B) or UBAP1 (Katzmann et al. 2001; Chu et al. 2006; Oestreich et al. 2007; Audhya et al. 2007; Curtiss et al. 2007; Wunderley et al. 2014). Together they form a soluble complex that is transiently recruited from the cytoplasm to the surface of endosomes. Different structural conformations of the ESCRT-I complex in solution have been resolved (Kostelansky et al. 2007; Boura et al. 2011). Together Vps23, Vps37 and Mvb12 form a long intertwined coiled–coiled stalk. The headpiece consists of Vps23, Vps28 and Vps37 (which binds to ESCRT-II). On its one end, the ubiquitin-conjugating enzyme E2 variant (UEV) domain of TSG101/Vps23 interacts with ESCRT-0 and with ubiquitinated membrane proteins. UBAP1 and yeast Mvb12 also bind ubiquitin. Thus, ESCRT-I could bind two ubiquitin moieties on endosomes. At the opposite end of the ESCRT-I rod, Vps28 binds to the GRAM-like ubiquitin binding in EAP45 (GLUE) domain of ESCRT-II (Gill et al. 2007).

Vps37 carries N-terminal helices, which interacts with the endosomal membrane and contributes to membrane association, which could be additionally enhanced by the β-prism fold—a hydrophobic membrane-anchoring loop and an electropositive phosphoinositide-binding patch of human MVB12b (Kostelansky et al. 2007; Boura and Hurley 2012). While Vps23 can interact with ESCRT-0, Vps28 interacts with the downstream ESCRT-II complex, thus potentially linking ESCRT-0 and ESCRT-II together. ESCRT-I can also interact with HD-PTP and it recruits the DUB UBPY/USP. This may facilitate the transfer of EGFR to the further ESCRT complexes (Ali et al. 2013; Doyotte et al. 2008; Stefani et al. 2011; Alwan and van Leeuwen 2007; Bowers et al. 2006; Pareja et al. 2012; Row et al. 2006; Wunderley et al. 2014). Overall, ESCRT-I recognizes ubiquitinated cargo and it may play a central role in the formation of a sorting domain by binding to ESCRT-0 and ESCRT-II.

ESCRT-II is a stable heterotetramer formed of Vps22 (EAP30), Vps36 (EAP45) and two copies of Vps25 (EAP20) (Hierro et al. 2004; Langelier et al. 2006; Teo et al. 2004). Vps22 and Vps36 constitute the main body of the Y-shaped structure, while the two copies of Vps25 form the flexible arms. The interaction with ESCRT-I (Vps28) and recruitment of ESCRT-II is mediated via the N-terminal GLUE domain of Vps36. The GLUE domain functions as a hub that also binds to PI3P in the endosomal membrane and it interacts with ubiquitin via its NZF (Np14-type zinc finger domains) (Teo et al. 2006). Membrane binding of ESCRT-II is additionally stabilized by the first helix of Vps22 (Im and Hurley 2008). Most likely ESCRT-0, -I and -II could interact simultaneously with up to eight different ubiquitinated membrane proteins and thereby generate a sorting domain on the surface of endosomes that matures into a site of MVB formation. While they collect cargo, ESCRT-I and ESCRT-II can form a super-complex that can adopt different shapes which may help to drive an initial phase of membrane budding (Wollert and Hurley 2010; Boura and Hurley 2012). At the same time, both Vps25 arms could already interact with one copy of the first ESCRT-III subunit, charged multivesicular body protein 6 (CHMP6)/Vps20, and convert them into an active nucleator for ESCRT-III assembly on endosomes (Im et al. 2009; Teis et al. 2010). In parallel, ESCRT-III assembly could be stimulated by ESCRT-0/Bro1 (Tang et al. 2016).

ESCRT-III consists of four core subunits: Vps20 (CHMP6), Snf7 (CHMP4A, B, C), Vps24 (CHMP3) and Vps2 (CHMP2A, B) and the associated subunits Did2 (Chmp1A, B), Vps60 (Chmp5) and the ESCRT-III-related protein Ist1 (hIst1) (Adell et al. 2016).

Unlike ESCRT-0, -I, -II, ESCRT-III subunits are monomers in the cytoplasm and only transiently assembly into poorly defined polymers, once recruited by upstream activators. So far, structural data on ESCRT-III polymers is available only from in vitro data and upon strong overexpression or Vps4 inhibition in situ. ESCRT-III polymers can adopt a variety of shapes, reaching from filaments, rings, spirals, helical filaments sheets and dome-shaped structures in vitro (Bajorek et al. 2009a; Ghazi-Tabatabai et al. 2008; Lata et al. 2008; Chiaruttini et al. 2015; Henne et al. 2012; Schöneberg et al. 2017). The architecture of the membrane scission competent ESCRT-III complex has not been visualized.

ESCRT-III subunits have a similar size, structure and domain organization, including a positively charged N-terminus and an acidic domain at their C-terminus (Muziol et al. 2006). The crystal structure of Vps24 for example revealed an antiparallel four-helix bundle (α1 to α4). In addition, the C-terminal helices α5 and α6 can fold back (and mediate autoinhibition in the cytoplasm or open up to allow polymerization on endosomes (Bajorek et al. 2009b; Shim et al. 2007; Zamborlini et al. 2006). In yeast, Snf7 is the most abundant subunit (Teis et al. 2008) and it undergoes even more dramatic conformational rearrangements that are required to promote the assembly of Snf7 arrays with ~30 Å periodicity into a membrane-sculpting filament (Tang et al. 2015). This involves the release of autoinhibition and the repositioning of α3 and α4 to mediate membrane binding and polymerization. Snf7 can also insert the terminal helix into the membrane (Buchkovich et al. 2013; Tang et al. 2015) a property that appears to be essential for membrane budding. In vitro Snf7 has been shown to polymerize in flexible rings spirals (Henne et al. 2012; Shen et al. 2014), which may accumulate elastic energy to spring and deform membranes as a consequence of tension release (Chiaruttini et al. 2015).

The binding of Vps24 and Vps2 to Snf7 polymers changes the architecture of ESCRT-III from flat spirals into 3D helices in vitro and stimulates the recruitment of the AAA-ATPase Vps4 (Babst et al. 2002; Im et al. 2009; Teis et al. 2008, 2010; Adell et al. 2016; Henne et al. 2012). Binding of Vps4 is mediated by the microtubule-interacting and transport (MIT)-interacting motif (MIM) at the C-terminus of Vps2. The recruitment of Vps4 may additionally be regulated by Did2 (CHMP1A, B), Vps60 (CHMP5) and Ist1 (IST1). Did2 has been shown to form a complex with Ist1, promoting Vps4 binding and disassembly of ESCRT-III. Additionally, Vps60 interacts with Vta1 a co-factor that stabilizes Vps4 and stimulates the ATPase activity of Vps4 (Amerik et al. 2000b; Nickerson et al. 2006, 2010; Rue et al. 2008; Zhongzheng et al. 2012). On yeast endosomes, the ESCRT-III assemblies accumulated 75–200 Snf7 and 15–50 Vps24 molecules and more than 4 Vps4 hexamers during their 3–45s lifetimes (Adell et al. 2017). Instead of growing into stable filaments, the dynamic and largely stochastic exchange of ESCRT-III subunits and of Vps4 is one of the prerequisites for successful membrane budding (Adell et al. 2017; Mierzwa et al. 2017). Vps4 is recruited by ESCRT-III in the first second—and not at the end of the pathway—and is present for the entire lifetime until an all-or-none step led to final release of ESCRT-III and Vps4 from endosomes (Adell et al. 2017). How these dynamic properties of ESCRT-III and Vps4 assemblies drive membrane bulging and scission, is a major open question.

The AAA-ATPase **Vps4** is the only thermodynamic driving force of the ESCRT machinery (Babst et al. 1998). Vps4 belongs to the class of type I AAA+ (ATPase associated with a variety of cellular activities). Vps4 is largely a monomer in the cytoplasm.

Each Vps4 monomer contains an N-terminal MIT domain (microtubule-interacting and trafficking), which is crucial for the interaction with the c-terminal MIMs of ESCRT-III subunits in the membrane-bound polymer, the MIM of Vps2 being most critical. This MIT domain is connected by a ~40 residue flexible linker with the ATPase cassette (Gonciarz et al. 2008; Scott et al. 2005; Xiao et al. 2007). Vfa1, a recently identified regulator of Vps4 binds to the N-terminal MIT domain of Vps4 and has been shown to stimulate the ATPase activity of Vps4 (Arlt et al. 2011; Vild and Xu, 2014; Kojima et al., 2016). Once recruited to ESCRT-III, it most likely assembles into an active hexamer (Inoue et al. 2008; Xiao et al. 2007; Yang et al. 2015; Monroe et al. 2017; Su et al. 2017), that can bind to and hydrolyse ATP. The ATPase cassette typically consists of a large ATPase domain, a small ATPase domain and an β-domain insertion which is specific for Vps4 (Gonciarz et al. 2008; Inoue et al. 2008; Xiao et al. 2007).

The β-domain mediates the interaction of Vps4 with its co-factor Vta1. Vta1 promotes the oligomerization and hence the ATPase activity of Vps4 and it facilitates ESCRT-III binding, via its MIT domains (Azmi et al. 2006; Lottridge et al. 2006; Shiflett et al. 2004), (Yeo et al. 2003), ATP hydrolysis of Vps4 protomers leads to conformational changes within the hexamer, most likely leading to the translocation of single ESCRT-III subunits trough the central pore of Vps4 (Monroe et al. 2017; Su et al. 2017; Adell et al. 2014; Landsberg et al. 2009; Scott et al. 2005; Yang et al. 2015; Yamada-Inagawa et al. 2003).

It is becoming clear that Vps4 is not simply an ESCRT-III disassembly factor that works at the end of each functional ESCRT cycle. The coordinated binding of Vps4 hexamers to Vps2 and Snf7 drives neck constriction (Adell et al. 2014) and Vps4 is immediately recruited to ESCRT and just at the end (Adell et al. 2017). How Vps4 and ESCRT-III function together during membrane budding is unclear but it may be that Vps4 constricts ESCRT-III filaments by pulling individual ESCRT-III subunits through the central pore of the Vps4 hexamer (Monroe et al. 2017; Su et al. 2017; Ghazi-Tabatabai et al. 2008; Yang et al. 2015; Shen et al. 2015; Babst et al. 1998; Shestakova et al. 2010; Adell et al. 2014). Repeating this process may help to concentrate cargo, buckle membrane and ultimately pull membranes together to drive membrane scission.

4.3.1 Deubiquitinating Enzymes and Co-factors on Endosomes

Degradation of alpha 4 (Doa4), or its mammalian orthologue ubiquitin-binding protein Y (UBPY), also known as ubiquitin-specific protease 8 (USP8) is required for the deubiquitination of membrane proteins during the ESCRT-dependent sorting process (Amerik et al. 2000a, b). UBPY and AMSH (associated molecule with the SH3 domain of STAM) are recruited to early endosome. They interact via the SH domain with STAM and via their MIT domains with ESCRT-III subunits (Kato et al. 2000; Agromayor and Martin-Serrano 2006; Row et al. 2007; McCullough et al. 2004; Nakamura et al. 2006). AMSH, also known as STAM-binding protein (STAMBP), is an endosome-associated isopeptidase (McCullough et al. 2004; Clague and Urbe 2006), which together with USP8 negatively regulates the degradation of proteins. They may stimulate early deubiquitination of cargos such as EGFR and therefore promote cargo recycling instead of degradation (Niendorf et al. 2007; Pareja et al. 2012; Clague and Urbe 2006).

But cargo deubiquitination can also occur later. In yeast, Doa4 is specifically recruited to ESCRT-III polymers by Bro1 (BCK1-like Resistance to Osmotic shock). Bro1 itself binds to Snf7 in ESCRT-III filaments and it is also directly involved in membrane budding, possible by delaying Vps4 mediated disassembly of ESCRT-III (Tanaka et al. 1999; Luhtala et al. 2004; McCullough et al. 2008; Richter et al. 2007; Wemmer et al. 2011). These mechanisms might help to ensure, that membrane proteins are de-ubiquitinated only when ESCRT-III filaments are already established. This may prevent their escape into recycling pathways.

Alix (formerly apoptosis-linked gene 2-interacting protein X; official name programmed cell death 6-interacting protein (PDCD6IP) also contains a Bro domain and interacts with Chmp4. Both, Bro1 and Alix can bind to ESCRT-0 or ESCRT-I and stimulate parallel ESCRT-III assembly pathways that play a key role in ESCRT-III-mediated membrane budding events (Clague and Urbe 2006; Bissig and Gruenberg 2014; Tang et al. 2016). Cargo deubiquitination may contribute to enhance the plasticity and flexibility of the sorting system. It has also been proposed, that this early deubiquitination of cargo proteins may provide an additional proofreading and assure the specificity of the lysosomal degradation pathway (Piper and Lehner 2011; Wright et al. 2011). How cargo deubiquitation process is regulated (early or later) and how deubiquitination events are timed to allow efficient recycling or sorting into ILVs is not entirely clear.

Most importantly, cargo deubiquitination prevents ubiquitin wasting and thereby contributes to the recycling of ubiquitin, in analogy to deubiquitination prior to proteasomal degradation (Papa et al. 1999). This also highlights the important interplay between ubiquitinating and deubiquitinating enzymes, to maintain essential cellular processes.

4.4 Reverse Membrane Remodelling—Driven by ESCRT-III and Vps4

Additional cellular processes at the PM and the Nuclear envelop/ER require ESCRT-III and Vps4-mediated reverse membrane budding (Adell et al. 2016; Alfred and Vaccari 2016; Campsteijn et al. 2016; Christ et al. 2017; Schöneberg et al. 2017).

At the plasma membrane, the ESCRT system is required for repair, neuronal pruning, exosome and microvesicle shedding. The ESCRT machinery is also hijacked by retrovirus (e.g. HIV/Ebola) to cut the membrane stalk connecting the budding virus with the host cell during the release of retroviral particles (Zhang et al. 2014; Loncle et al. 2015; Jimenez et al. 2014; Scheffer et al. 2014; Nabhan et al. 2012; Colombo et al. 2013; Géminard et al. 2004; Baietti et al. 2012; Roucourt et al. 2015; Wehman et al. 2011; Martin-Serrano et al. 2005; Martin-Serrano and Neil 2011; Garrus et al. 2001; von Schwedler et al. 2003; Pornillos et al. 2003; Strack et al. 2003; Morita et al. 2011; Effantin et al. 2013; Cashikar et al. 2014; Prescher et al. 2015; Jouvenet et al. 2011; Baumgärtel et al. 2011).

Another ESCRT-dependent process at the PM is cytokinesis (Carlton and Martin-Serrano 2007; Morita et al. 2007; Lee et al. 2008; Morita et al. 2010; Carlton et al. 2008; Guizetti et al. 2011; Bajorek et al. 2009a; Yang 2008; Reid et al. 2005; Elia et al. 2011, 2012; Caballe et al. 2015; Capalbo et al. 2012). At the nuclear envelop, ESCRTs are involved in the removal of defective nuclear pore complexes and in the resealing of the nuclear envelop after mitosis or after rupture in interphase cells (Vietri et al. 2015; Olmos et al. 2015; Webster et al. 2014; Bauer et al. 2015; Denais et al. 2016; Raab et al. 2016). Not only damage of the nuclear envelop or the PM, but also lysosomal damage triggers the transient recruitment of the ESCRT machinery for repair (Skowyra et al. 2018). ESCRTs are also required for the biogenesis of double membrane compartments, so-called compartment of unconventional protein secretion (CUPS), that mediate a conventional secretion pathway in yeast (Curwin et al. 2016). Remarkably, this process was largely independent of Vps4.

On endosomes, the ubiquitination signal on cargo proteins are the main recruiters of the ESCRT machinery, while all other ESCRTs depended processes require specific adaptor proteins. Some have been discovered such as Cep55 (Centrosomal protein of 55 kDa) (Carlton and Martin-Serrano 2007), that recruits ESCRT-III during cytokinesis, or Helix-Extension-Helix domain 2 (Heh2), Charged multivesicular body protein 7 (Chmp7), Lap2-emerin-man1 (Lem2) (Bauer et al. 2015; Olmos et al. 2015; Vietri et al. 2015; Webster et al. 2014; Gu et al. 2017; Webster et al. 2016; Olmos et al. 2016), which recruit ESCRT-III and Vps4 to the nuclear envelop. HIV-1 structural proteins such as GAG (Garrus et al. 2001; VerPlank et al. 2001; Martin-Serrano et al. 2001) hijack the hosts ESCRT proteins for budding. Many of the recruiting factors still need to be investigated, as well as the dynamic assembly and disassembly of ESCRT components at the distinct sides. Along with the open question, how ESCRT-III and Vps4 are capable of executing membrane budding in such a broad variety of cellular processes.

Summary

The basic concepts for membrane protein ubiquitination and subsequent degradation are emerging but many questions are not answered yet. Membrane proteins can constitute up to 20% of eukaryotic genomes but for the vast majority, it is completely unclear how they are ubiquitinated. What are the signals that mediate their ubiquitination and which ubiquitin ligase complexes are unsolved. How the ESCRT machinery sorts cargo proteins and coordinates this process with reverse membrane budding is mechanistically largely unresolved.

Answering these questions will provide a better understanding of how cells can adapt their repertoire of membrane proteins and thereby control cell growth and cellular homeostasis.

References

Adell MA, Vogel GF, Pakdel M, Muller M, Lindner H, Hess MW, Teis D (2014) Coordinated binding of Vps4 to ESCRT-III drives membrane neck constriction during MVB vesicle formation. J Cell Biol 205:33–49

Adell M, Migliano SM, Teis D (2016) ESCRT-III and Vps4: a dynamic multipurpose tool for membrane budding and scission. FEBS J

Adell MAY, Migliano SM, Upadhyayula S, Bykov YS, Sprenger S, Pakdel M, Vogel GF, Jih G, Skillern W, Behrouzi R, Babst M, Schmidt O, Hess MW, Briggs JA, Kirchhausen T, Teis D (2017) Recruitment dynamics of ESCRT-III and Vps4 to endosomes and implications for reverse membrane budding. Elife 6

Agromayor M, Martin-Serrano J (2006) Interaction of AMSH with ESCRT-III and deubiquitination of endosomal cargo. J Biol Chem 281:23083–23091

Alfred V, Vaccari T (2016) When membranes need an ESCRT: endosomal sorting and membrane remodelling in health and disease, Swiss Med Wkly

Ali N, Zhang L, Taylor S, Mironov A, Urbé S, Woodman P (2013) Recruitment of UBPY and ESCRT exchange drive HD-PTP-dependent sorting of EGFR to the MVB. Curr Biol 23:453–461

Alvaro CG, O'Donnell AF, Prosser DC, Augustine AA, Goldman A, Brodsky JL, Cyert MS, Wendland B, Thorner J (2014) Specific alpha-arrestins negatively regulate *Saccharomyces cerevisiae* pheromone response by down-modulating the G-protein-coupled receptor Ste2. Mol Cell Biol 34:2660–2681

Alwan HA, van Leeuwen JE (2007) UBPY-mediated epidermal growth factor receptor (EGFR) de-ubiquitination promotes EGFR degradation. J Biol Chem 282:1658–1669

Amerik AY, Li SJ, Hochstrasser M (2000a) Analysis of the deubiquitinating enzymes of the yeast *Saccharomyces cerevisiae*. Biol Chem 381:981–992

Amerik AY, Nowak J, Swaminathan S, Hochstrasser M (2000b) The Doa4 deubiquitinating enzyme is functionally linked to the vacuolar protein-sorting and endocytic pathways. Mol Biol Cell 11:3365–3380

Arlt H, Perz A, Ungermann C (2011) An overexpression screen in *Saccharomyces cerevisiae* identifies novel genes that affect endocytic protein trafficking. Traffic 12:1592–1603

Audhya A, Desai A, Oegema K (2007) A role for Rab5 in structuring the endoplasmic reticulum. J Cell Biol 178:43–56

Azmi I, Davies B, Dimaano C, Payne J, Eckert D, Babst M, Katzmann DJ (2006) Recycling of ESCRTs by the AAA-ATPase Vps4 is regulated by a conserved VSL region in Vta1. J Cell Biol 172:705–717

Babst M (2005) A protein's final ESCRT. Traffic 6:2–9

Babst M (2011) MVB vesicle formation: ESCRT-dependent, ESCRT-independent and everything in between. Curr Opin Cell Biol 23:452–457

Babst M, Wendland B, Estepa EJ, Emr SD (1998) The Vps4p AAA ATPase regulates membrane association of a Vps protein complex required for normal endosome function. EMBO J 17:2982–2993

Babst M, Katzmann DJ, Estepa-Sabal EJ, Meerloo T, Emr SD (2002) Escrt-III: an endosome-associated heterooligomeric protein complex required for MVB sorting. Dev Cell 3:271–282

Baietti MF, Zhang Z, Mortier E, Melchior A (2012) Syndecan-syntenin-ALIX regulates the biogenesis of exosomes. Nat Cell

Bajorek M, Morita E, Skalicky JJ, Morham SG, Babst M, Sundquist WI (2009a) Biochemical analyses of human IST1 and its function in cytokinesis. Mol Biol Cell 20:1360–1373

Bajorek M, Schubert HL, Mccullough J, Langelier C, Eckert DM, Stubblefield WM, Uter NT, Myszka DG, Hill CP, Sundquist WI (2009b). Structural basis for ESCRT-III protein autoinhibition. Nat Struct Mol Biol

Bauer I, Brune T, Preiss R, Kölling R (2015) Evidence for a nonendosomal function of the *Saccharomyces cerevisiae* ESCRT-III-like protein Chm7. Genetics 201:1439–1452

Baumgärtel V, Ivanchenko S, Dupont A (2011) Live-cell visualization of dynamics of HIV budding site interactions with an ESCRT component. Nat cell

Bays NW, Gardner RG, Seelig LP (2001) Hrd1p/Der3p is a membrane-anchored ubiquitin ligase required for ER-associated degradation. Nat Cell

Becuwe M, Herrador A, Haguenauer-Tsapis R, Vincent O, Leon S (2012a) Ubiquitin-mediated regulation of endocytosis by proteins of the arrestin family. Biochem Res Int 2012:242764

Becuwe M, Vieira N, Lara D, Gomes-Rezende J, Soares-Cunha C, Casal M, Haguenauer-Tsapis R, Vincent O, Paiva S, Leon S (2012b) A molecular switch on an arrestin-like protein relays glucose signaling to transporter endocytosis. J Cell Biol 196:247–259

Bedford MT, Frankel A, Yaffe MB, Clarke S, Leder P, Richard S (2000) Arginine methylation inhibits the binding of proline-rich ligands to Src homology 3, but not WW, domains. J Biol Chem 275:16030–16036

Belgareh-Touzé N, Léon S, Erpapazoglou Z, Stawiecka-Mirota M, Urban-Grimal D, Haguenauer-Tsapis R (2008) Versatile role of the yeast ubiquitin ligase Rsp5p in intracellular trafficking. Biochem Soc Trans 36:791–796

Bilodeau PS, Urbanowski JL (2002) The Vps27p–Hse1p complex binds ubiquitin and mediates endosomal protein sorting. Nat Cell

Bilodeau PS, Winistorfer SC, Kearney WR, Robertson AD, Piper RC (2003) Vps27-Hse1 and ESCRT-I complexes cooperate to increase efficiency of sorting ubiquitinated proteins at the endosome. J Cell Biol 163:237–243
Bissig C, Gruenberg J (2014) ALIX and the multivesicular endosome: ALIX in Wonderland. Trends Cell Biol 24:19–25
Boura E, Hurley JH (2012) Structural basis for membrane targeting by the MVB12-associated β-prism domain of the human ESCRT-I MVB12 subunit. Proc Natl Acad Sci USA 109:1901–1906
Boura E, Różycki B, Herrick DZ, Chung HS, Vecer J, Eaton WA, Cafiso DS, Hummer G, Hurley JH (2011) Solution structure of the ESCRT-I complex by small-angle X-ray scattering, EPR, and FRET spectroscopy. Proc Natl Acad Sci USA 108:9437–9442
Bowers K, Piper SC, Edeling MA, Gray SR, Owen DJ, Lehner PJ, Luzio JP (2006) Degradation of endocytosed epidermal growth factor and virally ubiquitinated major histocompatibility complex class I is independent of mammalian ESCRTII. J Biol Chem 281:5094–5105
Buchkovich NJ, Henne WM, Shaogeng T, Emr SD (2013) Essential N-terminal insertion motif anchors the ESCRT-III filament during MVB vesicle formation. Dev Cell 27:201–214
Caballe A, Wenzel DM, Agromayor M, Alam SL, Skalicky JJ, Kloc M, Carlton JG, Labrador L, Sundquist WI, Martin-Serrano J (2015) ULK3 regulates cytokinetic abscission by phosphorylating ESCRT-III proteins. Elife 4
Campsteijn C, Vietri M, Stenmark H (2016) Novel ESCRT functions in cell biology: spiraling out of control? Curr Opin Cell Biol 41:1–8
Capalbo L, Montembault E, Takeda T, Bassi ZI, Glover DM, D'Avino PP (2012) The chromosomal passenger complex controls the function of endosomal sorting complex required for transport-III Snf7 proteins during cytokinesis. Open Biol 2:120070
Carlton JG, Martin-Serrano J (2007) Parallels between cytokinesis and retroviral budding: a role for the ESCRT machinery. Science 316:1908–1912
Carlton JG, Agromayor M, Martin-Serrano J (2008) Differential requirements for Alix and ESCRT-III in cytokinesis and HIV-1 release. Proc Natl Acad Sci USA 105:10541–10546
Carvalho P, Goder V, Rapoport TA (2006) Distinct ubiquitin-ligase complexes define convergent pathways for the degradation of ER proteins. Cell 126:361–373
Cashikar AG, Shim S, Roth R, Maldazys MR, Heuser JE, Hanson PI (2014) Structure of cellular ESCRT-III spirals and their relationship to HIV budding. Elife 3
Chiaruttini N, Redondo-Morata L, Colom A, Humbert F, Lenz M, Scheuring S, Roux A (2015) Relaxation of loaded ESCRT-III spiral springs drives membrane deformation. Cell 163:866–879
Cho KJ, Walseng E, Ishido S, Roche PA (2015) Ubiquitination by March-I prevents MHC class II recycling and promotes MHC class II turnover in antigen-presenting cells. Proc Natl Acad Sci USA 112:10449–10454
Christ L, Raiborg C, Wenzel EM, Campsteijn C, Stenmark H (2017) Cellular functions and molecular mechanisms of the ESCRT membrane-scission machinery. Trends Biochem Sci 42:42–56
Christianson JC, Ye Y (2014) Cleaning up in the endoplasmic reticulum: ubiquitin in charge. Nat Struct Mol Biol
Chu T, Sun J, Saksena S, Emr SD (2006) New component of ESCRT-I regulates endosomal sorting complex assembly. J Cell Biol 175:815–823
Clague MJ (2002) Membrane transport: a coat for ubiquitin. Curr Biol 12:R529–R531
Clague MJ, Urbe S (2006) Endocytosis: the DUB version. Trends Cell Biol 16:551–559
Colombo M, Moita C, van Niel G, Kowal J, Vigneron J, Benaroch P, Manel N, Moita LF, Thery C, Raposo G (2013) Analysis of ESCRT functions in exosomes biogenesis, composition and secretion highlights the heterogeneity of extracellular vesicles. J Cell Sci 126:5553–5565
Crapeau M, Merhi A, Andre B (2014) Stress conditions promote yeast Gap1 permease ubiquitylation and down-regulation via the arrestin-like Bul and Aly proteins. J Biol Chem 289:22103–22116
Cruickshanks N (2014) When to screen and not to screen. Cancer Biol Ther 15:147–148
Curtiss M, Jones C, Babst M (2007) Efficient cargo sorting by ESCRT-I and the subsequent release of ESCRT-I from multivesicular bodies requires the subunit Mvb12. Mol Biol Cell 18:636–645

Curwin AJ, Brouwers N, Adell MAY, Teis D, Turacchio G, Parashuraman S, Ronchi P, Malhotra V, Schekman R (2016) ESCRT-III drives the final stages of CUPS maturation for unconventional protein secretion. Elife 5:e16299

Denais CM, Gilbert RM, Isermann P, McGregor AL, te Lindert M, Weigelin B, Davidson PM, Friedl P, Wolf K, Lammerding J (2016) Nuclear envelope rupture and repair during cancer cell migration. Science 352:353–358

Deshaies RJ, Joazeiro CAP (2009) RING domain E3 ubiquitin ligases 78:399–434

Di Fiore PP, Polo S, Hofmann K (2003) When ubiquitin meets ubiquitin receptors: a signalling connection. Nat Rev Mol Cell Biol

Dikic I, Robertson M (2012) Ubiquitin ligases and beyond. BMC Biol 10:22

Dores MR, Chen B, Lin H, Soh UJ, Paing MM, Montagne WA, Meerloo T, Trejo J (2012) ALIX binds a YPX(3)L motif of the GPCR PAR1 and mediates ubiquitin-independent ESCRT-III/MVB sorting. J Cell Biol 197:407–419

Doyotte A, Mironov A, McKenzie E, Woodman P (2008) The Bro1-related protein HD-PTP/PTPN23 is required for endosomal cargo sorting and multivesicular body morphogenesis. Proc Natl Acad Sci USA 105:6308–6313

Draheim KM, Chen HB, Tao Q, Moore N, Roche M, Lyle S (2010) ARRDC3 suppresses breast cancer progression by negatively regulating integrin beta4. Oncogene 29:5032–5047

Dunn R, Klos DA, Adler AS, Hicke L (2004) The C2 domain of the Rsp5 ubiquitin ligase binds membrane phosphoinositides and directs ubiquitination of endosomal cargo. J Cell Biol 165:135–144

Dupré S, Urban-Grimal D, Haguenauer-Tsapis R (2004) Ubiquitin and endocytic internalization in yeast and animal cells. Biochim Biophys Acta 1695:89–111

Effantin G, Dordor A, Sandrin V, Martinelli N, Sundquist WI, Schoehn G, Weissenhorn W (2013) ESCRT-III CHMP2A and CHMP3 form variable helical polymers in vitro and act synergistically during HIV-1 budding. Cell Microbiol 15:213–226

Elia N, Sougrat R, Spurlin T, Hurley J, Lippincott-Schwartz J (2011) Dynamics of endosomal sorting complex required for transport (ESCRT) machinery during cytokinesis and its role in abscission. PNAS 108:4846–4851

Elia N, Fabrikant G, Kozlov MM, Lippincott-Schwartz J (2012) Computational model of cytokinetic abscission driven by ESCRT-III polymerization and remodeling. Biophys J 102:2309–2320

Foot N, Henshall T, Kumar S (2017) Ubiquitination and the regulation of membrane proteins. Physiol Rev 97:253–281

Galan JM, Moreau V, Andre B, Volland C, Haguenauer-Tsapis R (1996) Ubiquitination mediated by the Npi1p/Rsp5p ubiquitin-protein ligase is required for endocytosis of the yeast uracil permease. J Biol Chem 271:10946–10952

Garrus JE, von Schwedler UK, Pornillos OW, Morham SG, Zavitz KH, Wang HE, Wettstein DA, Stray KM, Cote M, Rich RL, Myszka DG, Sundquist WI (2001) Tsg101 and the vacuolar protein sorting pathway are essential for HIV-1 budding. Cell 107:55–65

Gay DL, Ramón H, Oliver PM (2008) Cbl- and Nedd4-family ubiquitin ligases: balancing tolerance and immunity. Immunol Res 42:51–64

Géminard C, de Gassart A, Blanc L, Vidal M (2004) Degradation of AP2 during reticulocyte maturation enhances binding of Hsc70 and Alix to a common site on TfR for sorting into exosomes. Traffic 5:181–193

Ghaddar K, Merhi A, Saliba E, Krammer EM, Prevost M, Andre B (2014) Substrate-induced ubiquitylation and endocytosis of yeast amino acid permeases. Mol Cell Biol 34:4447–4463

Ghazi-Tabatabai S, Saksena S, Short JM, Pobbati AV, Veprintsev DB, Crowther RA, Emr SD, Egelman EH, Williams RL (2008) Structure and disassembly of filaments formed by the ESCRT-III subunit Vps24. Structure 16:1345–1356

Gill DJ, Teo H, Sun J, Perisic O, Veprintsev DB, Emr SD, Williams RL (2007) Structural insight into the ESCRT-I/-II link and its role in MVB trafficking. EMBO J 26:600–612

Godlee C, Kaksonen M (2013) From uncertain beginnings: initiation mechanisms of clathrin-mediated endocytosis. J Cell Biol 203:717–725

Gonciarz MD, Whitby FG, Eckert DM, Kieffer C, Heroux A, Sundquist WI, Hill CP (2008) Biochemical and structural studies of yeast Vps4 oligomerization. J Mol Biol 384:878–895

Gu M, Lajoie D, Chen OS, von Appen A, Ladinsky MS, Redd MJ, Nikolova L, Bjorkman PJ, Sundquist WI, Ullman KS, Frost A (2017) LEM2 recruits CHMP7 for ESCRT-mediated nuclear envelope closure in fission yeast and human cells. Proc Natl Acad Sci USA 114:E2166–E2175

Guizetti J, Schermelleh L, Mäntler J, Maar S, Poser I, Leonhardt H, Müller-Reichert T, Gerlich WD (2011) Cortical constriction during abscission involves helices Of ESCRT-III-dependent filaments. Science 331:1616–1620

Haglund K (2003) Distinct monoubiquitin signals in receptor endocytosis. Trends Biochem Sci 28:598–604

Haglund K, Dikic I (2012) The role of ubiquitylation in receptor endocytosis and endosomal sorting. J Cell Sci 125:265–275

Hamilton KL (2014) Cleavage: what's up with prostasin and ENaC these days? Am J Physiol Renal Physiol 307:F1196–F1197

Hatakeyama R, Kamiya M, Takahara T, Maeda T (2010) Endocytosis of the aspartic acid/glutamic acid transporter Dip5 is triggered by substrate-dependent recruitment of the Rsp5 ubiquitin ligase via the arrestin-like protein Aly2. Mol Cell Biol 30:5598–5607

Henne WM, Buchkovich NJ, Yingying Z, Emr SD (2012) The endosomal sorting complex ESCRT-II mediates the assembly and architecture of ESCRT-III helices. Cell 151:356–371

Hershko A, Ciechanover A (1998) The ubiquitin system. Annu Rev Biochem 67:425–479

Hicke L, Riezman H (1996) Ubiquitination of a yeast plasma membrane receptor signals its ligand-stimulated endocytosis. Cell 84:277–287

Hicke L, Schubert HL, Hill CP (2005) Ubiquitin-binding domains. Nat Rev Mol Cell Biol 6:610–621

Hierro A, Sun J, Rusnak AS, Kim J, Prag G, Emr SD, Hurley JH (2004) Structure of the ESCRT-II endosomal trafficking complex. Nature 431:221–225

Hofmann RM, Pickart CM (2001) In vitro assembly and recognition of Lys-63 polyubiquitin chains. J Biol Chem 276:27936–27943

Hovsepian J, Defenouillere Q, Albanese V, Vachova L, Garcia C, Palkova Z, Leon S (2017) Multilevel regulation of an alpha-arrestin by glucose depletion controls hexose transporter endocytosis. J Cell Biol

Huang L, Kinnucan E, Wang G, Beaudenon S, Howley PM, Huibregtse JM, Pavletich NP (1999) Structure of an E6AP-UbcH7 complex: insights into ubiquitination by the E2-E3 enzyme cascade. Science 286:1321–1326

Hurley JH, Misra S (2000) Signaling and subcellular targeting by membrane-binding domains. Annu Rev Biophys Biomol Struct 29:49–79

Im YJ, Hurley JH (2008) Integrated structural model and membrane targeting mechanism of the human ESCRT-II complex. Dev Cell 14:902–913

Im YJ, Wollert T, Boura E, Hurley JH (2009) Structure and function of the ESCRT-II-III interface in multivesicular body biogenesis. Dev Cell 17:234–243

Inoue M, Kamikubo H, Kataoka M, Kato R, Yoshimori T, Wakatsuki S, Kawasaki M (2008) Nucleotide-dependent conformational changes and assembly of the AAA ATPase SKD1/VPS4b. Traffic 9:2180–2189

Jimenez AJ, Maiuri P, Lafaurie-Janvore J, Divoux S, Piel M, Perez F (2014) ESCRT machinery is required for plasma membrane repair. Science 343:1247136

Jones CB, Ott EM, Keener JM, Curtiss M, Sandrin V, Babst M (2012) Regulation of membrane protein degradation by starvation-response pathways. Traffic 13:468–482

Jouvenet N, Simon SM, Bieniasz PD (2011) Visualizing HIV-1 assembly. J Mol Biol 410:501–511

Kaliszewski P, Zoładek T (2008) The role of Rsp5 ubiquitin ligase in regulation of diverse processes in yeast cells. Acta Biochim Pol 55:649–662

Kang DS, Tian X, Benovic JL (2014) Role of β-arrestins and arrestin domain-containing proteins in G protein-coupled receptor trafficking. Curr Opin Cell Biol 27:63–71

Kato M, Miyazawa K, Kitamura N (2000) A deubiquitinating enzyme UBPY interacts with the Src homology 3 domain of Hrs-binding protein via a novel binding motif PX(V/I)(D/N)RXXKP. J Biol Chem 275:37481–37487

Katzmann DJ, Babst M, Emr SD (2001) Ubiquitin-dependent sorting into the multivesicular body pathway requires the function of a conserved endosomal protein sorting complex, ESCRT-I. Cell 106:145–155

Katzmann DJ, Stefan CJ, Babst M, Emr SD (2003) Vps27 recruits ESCRT machinery to endosomes during MVB sorting. J Cell Biol 162:413–423

Kerscher O, Felberbaum R, Hochstrasser M (2006) Modification of proteins by ubiquitin and ubiquitin-like proteins. Annu Rev Cell Dev Biol 22:159–180

Kojima R, Obita T, Onoue K, Mizuguchi M (2016) Structural fine-tuning of MIT-interacting motif 2 (MIM2) and allosteric regulation of ESCRT-III by Vps4 in yeast. J Mol Biol 428:2392–2404

Komander D (2009) The emerging complexity of protein ubiquitination. Biochem Soc Trans 37:937–953

Kostelansky MS, Schluter C, Tam YY, Lee S, Ghirlando R, Beach B, Conibear E, Hurley JH (2007) Molecular architecture and functional model of the complete yeast ESCRT-I heterotetramer. Cell 129:485–498

Kozlov G, Nguyen L, Lin T, de Crescenzo G, Park M, Gehring K (2007) Structural basis of ubiquitin recognition by the ubiquitin-associated (UBA) domain of the ubiquitin ligase EDD. J Biol Chem 282:35787–35795

Landsberg MJ, Vajjhala PR, Rothnagel R, Munn AL, Hankamer B (2009) Three-dimensional structure of AAA ATPase Vps4: advancing structural insights into the mechanisms of endosomal sorting and enveloped virus budding. Structure 17:427–437

Langelier C, von Schwedler UK, Fisher RD, de Domenico I, White PL, Hill CP, Kaplan J, Ward D, Sundquist WI (2006) Human ESCRT-II complex and its role in human immunodeficiency virus type 1 release. J Virol 80:9465–9480

Lata S, Schoehn G, Jain A, Pires R, Piehler J, Gottlinger HG, Weissenhorn W (2008) Helical structures of ESCRT-III are disassembled by VPS4. Science 321:1354–1357

Lee HH, Elia N, Ghirlando R, Lippincott-Schwartz J, Hurley JH (2008) Midbody targeting of the ESCRT machinery by a noncanonical coiled coil in CEP55. Science 322:576–580

Lee J, Oh K-J, Lee D, Kim BY, Choi JS, Ku B, Kim SJ (2016) Structural study of the HD-PTP Bro1 domain in a complex with the core region of STAM2, a subunit of ESCRT-0. PLoS ONE 11:e0149113

Lemmon MA, Schlessinger J (2010) Cell signaling by receptor tyrosine kinases. Cell 141:1117–1134

Levkowitz G, Waterman H, Zamir E, Kam Z, Oved S, Langdon WY, Beguinot L, Geiger B, Yarden Y (1998) c-Cbl/Sli-1 regulates endocytic sorting and ubiquitination of the epidermal growth factor receptor. Genes Dev 12:3663–3674

Levkowitz G, Waterman H, Ettenberg SA, Katz M, Tsygankov AY, Alroy I, Lavi S, Iwai K, Reiss Y, Ciechanover A, Lipkowitz S, Yarden Y (1999) Ubiquitin ligase activity and tyrosine phosphorylation underlie suppression of growth factor signaling by c-Cbl/Sli-1. Mol Cell 4:1029–1040

Li M, Koshi T, Emr SD (2015a) Membrane-anchored ubiquitin ligase complex is required for the turnover of lysosomal membrane proteins. J Cell Biol 211:jcb.201505062-652

Li M, Rong Y, Chuang Y-S, Peng D, Emr SD (2015b) Ubiquitin-dependent lysosomal membrane protein sorting and degradation. Mol Cell 57:467–478

Lin CH, Macgurn JA, Chu T, Stefan CJ, Emr SD (2008) Arrestin-related ubiquitin-ligase adaptors regulate endocytosis and protein turnover at the cell surface. Cell 135:714–725

Loncle N, Agromayor M, Martin-Serrano J, Williams DW (2015) An ESCRT module is required for neuron pruning. Sci Rep 5:127

Lorick KL, Jensen JP, Fang S, Ong AM, Hatakeyama S, Weissman AM (1999) RING fingers mediate ubiquitin-conjugating enzyme (E2)-dependent ubiquitination. Proc Natl Acad Sci 96:11364–11369

Lottridge JM, Flannery AR, Vincelli JL, Stevens TH (2006) Vta1p and Vps46p regulate the membrane association and ATPase activity of Vps4p at the yeast multivesicular body. Proc Natl Acad Sci 103:6202–6207

Lu Y, Li X, Liang K, Luwor R, Siddik ZH, Mills GB, Mendelsohn J, Fan Z (2007) Epidermal growth factor receptor (EGFR) ubiquitination as a mechanism of acquired resistance escaping treatment by the anti-EGFR monoclonal antibody cetuximab. Cancer Res 67:8240–8247

Luhtala N, Odorizzi G (2004) Bro1 coordinates deubiquitination in the multivesicular body pathway by recruiting Doa4 to endosomes. J Cell Biol 166:717–729

Macdonald C, Payne JA, Aboian M, Smith W, Katzmann DJ, Piper RC (2015) A family of tetraspans organizes cargo for sorting into multivesicular bodies. Dev Cell 33:328–342

Macgurn JA, Hsu P-C, Smolka MB, Emr SD (2011) TORC1 regulates endocytosis via Npr1-mediated phosphoinhibition of a ubiquitin ligase adaptor. Cell 147:1104–1117

Macgurn JA, Hsu PC, Emr SD (2012) Ubiquitin and membrane protein turnover: from cradle to grave. Annu Rev Biochem 81:231–259

Marqués MC, Zamarbide-Forés S (2015) A functional Rim101 complex is required for proper accumulation of the Ena1 Na+-ATPase protein in response to salt stress in *Saccharomyces cerevisiae*. FEMS Yeast Res

Martin-Serrano J, Neil SJ (2011) Host factors involved in retroviral budding and release. Nat Rev Microbiol 9:519–531

Martin-Serrano J, Zang T, Bieniasz PD (2001) HIV-1 and Ebola virus encode small peptide motifs that recruit Tsg101 to sites of particle assembly to facilitate egress. Nat Med 7:1313–1319

Martin-Serrano J, Eastman SW, Chung W, Bieniasz PD (2005) HECT ubiquitin ligases link viral and cellular PPXY motifs to the vacuolar protein-sorting pathway. J Cell Biol 168:89–101

McCullough J, Clague MJ, Urbe S (2004) AMSH is an endosome-associated ubiquitin isopeptidase. J Cell Biol 166:487–492

McCullough J, Fisher RD, Whitby FG, Sundquist WI, Hill CP (2008) ALIX-CHMP4 interactions in the human ESCRT pathway. Proc Natl Acad Sci USA 105:7687–7691

Merhi A, Andre B (2012) Internal amino acids promote Gap1 permease ubiquitylation via TORC1/Npr1/14-3-3-dependent control of the Bul arrestin-like adaptors. Mol Cell Biol 32:4510–4522

Metzger MB, Hristova VA, Weissman AM (2012) HECT and RING finger families of E3 ubiquitin ligases at a glance. J Cell Sci 125:531–537

Mevissen TET, Komander D (2016) Mechanisms of deubiquitinase specificity and regulation 86:annurev-biochem-061516-044916

Mierzwa BE, Chiaruttini N, Redondo-Morata L, Moser Von Filseck J, Konig J, Larios J, Poser I, Muller-Reichert T, Scheuring S, Roux A, Gerlich DW (2017) Dynamic subunit turnover in ESCRT-III assemblies is regulated by Vps4 to mediate membrane remodelling during cytokinesis. Nat Cell Biol

Mohapatra B, Ahmad G, Nadeau S, Zutshi N, An W, Scheffe S, Dong L, Feng D, Goetz B, Arya P, Bailey TA, Palermo N, Borgstahl GEO, Natarajan A, Raja SM, Naramura M, Band V, Band H (2013) Protein tyrosine kinase regulation by ubiquitination: Critical roles of Cbl-family ubiquitin ligases. Biochim Biophys Acta (BBA) Mol Cell Res 1833:122–139

Monroe N, Han H, Shen PS, Sundquist WI, Hill CP, Subramaniam S (2017) Structural basis of protein translocation by the Vps4-Vta1 AAA ATPase. Elife 6:e24487

Morita E, Sandrin V, Chung HY, Morham SG, Gygi SP, Rodesch CK, Sundquist WI (2007) Human ESCRT and ALIX proteins interact with proteins of the midbody and function in cytokinesis. EMBO J 26:4215–4227

Morita E, Colf LA, Karren MA, Sandrin V, Rodesch CK, Sundquist WI (2010) Human ESCRT-III and VPS4 proteins are required for centrosome and spindle maintenance. Proc Natl Acad Sci USA 107:12889–12894

Morita E, Sandrin V, McCullough J, Katsuyama A, Baci Hamilton I, Sundquist WI (2011) ESCRT-III protein requirements for HIV-1 budding. Cell Host Microbe 9:235–242

Morrison JA, Pike LA, Sams SB, Sharma V, Zhou Q, Severson JJ, Tan AC, Wood WM, Haugen BR (2014) Thioredoxin interacting protein (TXNIP) is a novel tumor suppressor in thyroid cancer. Mol Cancer 13:62

Muller M, Schmidt O, Angelova M, Faserl K, Weys S, Kremser L, Pfaffenwimmer T, Dalik T, Kraft C, Trajanoski Z, Lindner H, Teis D (2015) The coordinated action of the MVB pathway and autophagy ensures cell survival during starvation. Elife 4

Muziol T, Pineda-Molina E, Ravelli RB, Zamborlini A, Usami Y, Gottlinger H, Weissenhorn W (2006) Structural basis for budding by the ESCRT-III factor CHMP3. Dev Cell 10:821–830

Myromslien FD, Grovdal LM, Raiborg C, Stenmark H, Madshus IH, Stang E (2006) Both clathrin-positive and -negative coats are involved in endosomal sorting of the EGF receptor. Exp Cell Res 312:3036–3048

Nabhan JF, Pan H, Lu Q (2010) Arrestin domain-containing protein 3 recruits the NEDD4 E3 ligase to mediate ubiquitination of the β2-adrenergic receptor. EMBO Rep 11:605–611

Nabhan JF, Hu R, Oh RS, Cohen SN, Lu Q (2012) Formation and release of arrestin domain-containing protein 1-mediated microvesicles (ARMMs) at plasma membrane by recruitment of TSG101 protein. Proc Natl Acad Sci USA 109:4146–4151

Nakamura N (2011) The role of the transmembrane RING finger proteins in cellular and organelle function. Membranes 1:354–393

Nakamura M, Tanaka N, Kitamura N, Komada M (2006) Clathrin anchors deubiquitinating enzymes, AMSH and AMSH-like protein, on early endosomes. Genes Cells: Devoted Mol Cell Mech 11:593–606

Nalefski EA, Falke JJ (1996) The C2 domain calcium-binding motif: structural and functional diversity. Protein Sci: Publ Protein Soc 5:2375–2390

Nathan JA, Lehner PJ (2009) The trafficking and regulation of membrane receptors by the RING-CH ubiquitin E3 ligases. Exp Cell Res 315:1593–1600

Nickerson DP, West M, Odorizzi G (2006) Did2 coordinates Vps4-mediated dissociation of ESCRT-III from endosomes. J Cell Biol 175:715–720

Nickerson DP, West M, Henry R, Odorizzi G (2010) Regulators of Vps4 ATPase activity at endosomes differentially influence the size and rate of formation of intralumenal vesicles. Mol Biol Cell 21:1023–1032

Niendorf S, Oksche A, Kisser A, Lohler J, Prinz M, Schorle H, Feller S, Lewitzky M, Horak I, Knobeloch KP (2007) Essential role of ubiquitin-specific protease 8 for receptor tyrosine kinase stability and endocytic trafficking in vivo. Mol Cell Biol 27:5029–5039

Nikko E, Pelham HRB (2009) Arrestin-mediated endocytosis of yeast plasma membrane transporters. Traffic 10:1856–1867

Nikko E, Sullivan JA, Pelham HR (2008) Arrestin-like proteins mediate ubiquitination and endocytosis of the yeast metal transporter Smf1. EMBO Rep 9:1216–1221

O'Donnell AF, Apffel A, Gardner RG, Cyert MS (2010) Alpha-arrestins Aly1 and Aly2 regulate intracellular trafficking in response to nutrient signaling. Mol Biol Cell 21:3552–3566

O'Donnell AF, Huang L, Thorner J, Cyert MS (2013) A calcineurin-dependent switch controls the trafficking function of alpha-arrestin Aly1/Art6. J Biol Chem 288:24063–24080

O'Donnell AF, McCartney RR, Chandrashekarappa DG, Zhang BB, Thorner J, Schmidt MC (2015) 2-Deoxyglucose impairs *Saccharomyces cerevisiae* growth by stimulating Snf1-regulated and alpha-arrestin-mediated trafficking of hexose transporters 1 and 3. Mol Cell Biol 35:939–955

Oestreich AJ, Davies BA, Payne JA, Katzmann DJ (2007) Mvb12 is a novel member of ESCRT-I involved in cargo selection by the multivesicular body pathway. Mol Biol Cell 18:646–657

Oh J, Wu N, Baravalle G, Cohn B, Ma J, Lo B, Mellman I, Ishido S, Anderson M, Shin JS (2013) MARCH1-mediated MHCII ubiquitination promotes dendritic cell selection of natural regulatory T cells. J Exp Med 210:1069–1077

Olmos Y, Hodgson L, Mantell J, Verkade P, Carlton JG (2015) ESCRT-III controls nuclear envelope reformation. Nature 522:236–239

Olmos Y, Perdrix-Rosell A, Carlton JG (2016) Membrane binding by CHMP7 coordinates ESCRT-III-dependent nuclear envelope reformation. Curr Biol 26:2635–2641

Papa FR, Amerik AY, Hochstrasser M (1999) Interaction of the Doa4 deubiquitinating enzyme with the yeast 26S proteasome. Mol Biol Cell 10:741–756
Pareja F, Ferraro DA, Rubin C, Cohen-Dvashi H, Zhang F, Aulmann S, Ben-Chetrit N, Pines G, Navon R, Crosetto N, Köstler W, Carvalho S, Lavi S, Schmitt F, Dikic I, Yakhini Z, Sinn P, Mills GB, Yarden Y (2012) Deubiquitination of EGFR by Cezanne-1 contributes to cancer progression. Oncogene 31:4599–4608
Piper RC, Lehner PJ (2011) Endosomal transport via ubiquitination. Trends Cell Biol 21:647–655
Polo S, Sigismund S, Faretta M, Guidi M, Capua MR, Bossi G, Chen H, de Camilli P, di Fiore PP (2002) A single motif responsible for ubiquitin recognition and monoubiquitination in endocytic proteins. Nature 416:451–455
Pornillos O, Higginson DS, Stray KM, Fisher RD, Garrus JE, Payne M, He G-P, Wang HE, Morham SG, Sundquist WI (2003) HIV Gag mimics the Tsg101-recruiting activity of the human Hrs protein. J Cell Biol 162:425–434
Prag G, Watson H, Kim YC, Beach BM, Ghirlando R, Hummer G, Bonifacino JS, Hurley JH (2007) The Vps27/Hse1 complex is a GAT domain-based scaffold for ubiquitin-dependent sorting. Dev Cell 12:973–986
Prescher J, Baumgartel V, Ivanchenko S, Torrano AA, Brauchle C, Muller B, Lamb DC (2015) Super-resolution imaging of ESCRT-proteins at HIV-1 assembly sites. PLoS Pathog 11:e1004677
Raab M, Gentili M, de Belly H, Thiam HR, Vargas P, Jimenez AJ, Lautenschlaeger F, Voituriez R, Lennon-Dumenil AM, Manel N, Piel M (2016) ESCRT III repairs nuclear envelope ruptures during cell migration to limit DNA damage and cell death. Science 352:359–362
Raiborg C, Stenmark H (2009) The ESCRT machinery in endosomal sorting of ubiquitylated membrane proteins. Nature 458:445–452
Raiborg C, Bache KG, Mehlum A, Stenmark H (2001a) Function of Hrs in endocytic trafficking and signalling. Biochem Soc Trans 29:472–475
Raiborg C, Bremnes B, Mehlum A, Gillooly DJ, D'Arrigo A, Stang E, Stenmark H (2001b) FYVE and coiled-coil domains determine the specific localisation of Hrs to early endosomes. J Cell Sci 114:2255–2263
Raiborg C, Bache KG, Gillooly DJ, Madshus IH, Stang E, Stenmark H (2002) Hrs sorts ubiquitinated proteins into clathrin-coated microdomains of early endosomes. Nat Cell Biol 4:394–398
Rauch S, Martin-Serrano J (2011) Multiple interactions between the ESCRT machinery and arrestin-related proteins: implications for PPXY-dependent budding. J Virol 85:3546–3556
Ravid T, Hochstrasser M (2007) Autoregulation of an E2 enzyme by ubiquitin-chain assembly on its catalytic residue. Nat Cell Biol 9:422–427
Reggiori F, Pelham HRB (2002) A transmembrane ubiquitin ligase required to sort membrane proteins into multivesicular bodies. Nat Cell Biol 4:117–123
Reid E, Connell J, Edwards TL, Duley S, Brown SE, Sanderson CM (2005) The hereditary spastic paraplegia protein spastin interacts with the ESCRT-III complex-associated endosomal protein CHMP1B. Hum Mol Genet 14:19–38
Ren X, Kloer DP, Kim YC, Ghirlando R, Saidi LF, Hummer G, Hurley JH (2009) Hybrid structural model of the complete human ESCRT-0 complex. Structure 17:406–416
Richter C, West M, Odorizzi G (2007) Dual mechanisms specify Doa4-mediated deubiquitination at multivesicular bodies. EMBO J 26:2454–2464
Rotin D, Kumar S (2009) Physiological functions of the HECT family of ubiquitin ligases. Nat Rev Mol Cell Biol 10:398–409
Rotin D, Staub O, Haguenauer-Tsapis R (2000) Ubiquitination and endocytosis of plasma membrane proteins: role of Nedd4/Rsp5p family of ubiquitin-protein ligases. J Membr Biol 176:1–17
Roucourt B, Meeussen S, Bao J, Zimmermann P, David G (2015) Heparanase activates the syndecan-syntenin-ALIX exosome pathway. Cell Res 25:412–428
Row PE, Prior IA, McCullough J, Clague MJ, Urbe S (2006) The ubiquitin isopeptidase UBPY regulates endosomal ubiquitin dynamics and is essential for receptor down-regulation. J Biol Chem 281:12618–12624

Row PE, Liu H, Hayes S, Welchman R, Charalabous P, Hofmann K, Clague MJ, Sanderson CM, Urbé S (2007) The MIT domain of UBPY constitutes a CHMP binding and endosomal localization signal required for efficient epidermal growth factor receptor degradation. J Biol Chem 282:30929–30937

Rue SM, Mattei S, Saksena S, Emr SD (2008) Novel Ist1-Did2 complex functions at a late step in multivesicular body sorting. Mol Biol Cell 19:475–484

Samji T, Hong S, Means RE (2014) The membrane associated RING-CH proteins: a family of E3 ligases with diverse roles through the cell. Int Sch Res Not 2014:637295–637323

Scheffer LL, Sreetama SC, Sharma N, Medikayala S, Brown KJ, Defour A, Jaiswal JK (2014) Mechanism of Ca^{2+}-triggered ESCRT assembly and regulation of cell membrane repair. Nat Commun 5:5646

Schmidt MHH, Dikic I (2005) The Cbl interactome and its functions. Nat Rev Mol Cell Biol 6:907–918

Schmidt O, Teis D (2012) The ESCRT machinery. Curr Biol: CB 22:R116–R120

Schnell JD, Hicke L (2003) Non-traditional functions of ubiquitin and ubiquitin-binding proteins. J Biol Chem 278:35857–35860

Schöneberg J, Lee I-H, Iwasa JH, Hurley JH (2017) Reverse-topology membrane scission by the ESCRT proteins. Nat Rev Mol Cell Biol 18:5–17

Scott A, Chung HY, Gonciarz-Swiatek M, Hill GC, Whitby FG, Gaspar J, Holton JM, Viswanathan R, Ghaffarian S, Hill CP, Sundquist WI (2005) Structural and mechanistic studies of VPS4 proteins. EMBO J 24:3658–3669

Shen QT, Schuh AL, Zheng Y, Quinney K, Wang L, Hanna M, Mitchell JC, Otegui MS, Ahlquist P, Cui Q, Audhya A (2014) Structural analysis and modeling reveals new mechanisms governing ESCRT-III spiral filament assembly. J Cell Biol 206:763–777

Shen Q-T, Ren X, Zhang R, Lee I-H, Hurley JH (2015) HIV-1 Nef hijacks clathrin coats by stabilizing AP-1:Arf1 polygons. Science 350:aac5137

Shestakova A, Hanono A, Drosner S, Curtiss M, Davies BA, Katzmann DJ, Babst M (2010) Assembly of the AAA ATPase Vps4 on ESCRT-III. Mol Biol Cell 21:1059–1071

Shiflett SL, Ward DM, Huynh D, Vaughn MB, Simmons JC, Kaplan J (2004) Characterization of Vta1p, a class E Vps protein in *Saccharomyces cerevisiae*. J Biol Chem 279:10982–10990

Shih SC, Katzmann DJ, Schnell JD, Sutanto M, Emr SD, Hicke L (2002) Epsins and Vps27p/Hrs contain ubiquitin-binding domains that function in receptor endocytosis. Nat Cell Biol 4:389–393

Shim S, Kimpler LA, Hanson PI (2007) Structure/function analysis of four core ESCRT-III proteins reveals common regulatory role for extreme C-terminal domain. Traffic 8:1068–1079

Shukla AK, Westfield GH, Xiao K, Reis RI, Huang LY, Tripathi-Shukla P, Qian J, Li S, Blanc A, Oleskie AN, Dosey AM, Su M, Liang CR, Gu LL, Shan JM, Chen X, Hanna R, Choi M, Yao XJ, Klink BU, Kahsai AW, Sidhu SS, Koide S, Penczek PA, Kossiakoff AA, Woods VL Jr, Kobilka BK, Skiniotis G, Lefkowitz RJ (2014) Visualization of arrestin recruitment by a G-protein-coupled receptor. Nature 512:218–222

Sigismund S, Woelk T, Puri C, Maspero E, Tacchetti C, Transidico P, di Fiore PP, Polo S (2005) Clathrin-independent endocytosis of ubiquitinated cargos. Proc Natl Acad Sci 102:2760–2765

Sigismund S, Argenzio E, Tosoni D, Cavallaro E, Polo S, di Fiore PP (2008) Clathrin-mediated internalization is essential for sustained EGFR signaling but dispensable for degradation. Dev Cell 15:209–219

Skowyra ML, Schlesinger PH, Naismith TV, Hanson PI (2018) Triggered recruitment of ESCRT machinery promotes endolysosomal repair. Science 360

Smardon AM, Kane PM (2014) Loss of vacuolar H+-ATPase activity in organelles signals ubiquitination and endocytosis of the yeast plasma membrane proton pump Pma1p. J Biol Chem 289:32316–32326

Staub O, Abriel H, Plant P, Ishikawa T, Kanelis V, Saleki R, Horisberger JD, Schild L, Rotin D (2000) Regulation of the epithelial Na+ channel by Nedd4 and ubiquitination. Kidney Int 57:809–815

Stefani F, Zhang L, Taylor S, Donovan J, Rollinson S, Doyotte A, Brownhill K, Bennion J, Pickering-Brown S, Woodman P (2011) UBAP1 is a component of an endosome-specific ESCRT-I complex that is essential for MVB sorting. Curr Biol: CB 21:1245–1250

Stewart EV, Nwosu CC, Tong Z, Roguev A, Cummins TD, Kim D-U, Hayles J, Park H-O, Hoe K-L, Powell DW, Krogan NJ, Espenshade PJ (2011) Yeast SREBP cleavage activation requires the Golgi Dsc E3 ligase complex. Mol Cell 42:160–171

Strack B, Calistri A, Craig S, Popova E, Gottlinger HG (2003) AIP1/ALIX is a binding partner for HIV-1 p6 and EIAV p9 functioning in virus budding. Cell 114:689–699

Su M, Guo EZ, Ding X, Li Y, Tarrasch JT, Brooks CL, Xu Z, Skiniotis G (2017) Mechanism of Vps4 hexamer function revealed by cryo-EM. Sci Adv 3:e1700325

Suzuki A, Mochizuki T, Uemura S, Hiraki T, Abe F (2013) Pressure-induced endocytic degradation of the *Saccharomyces cerevisiae* low-affinity tryptophan permease Tat1 is mediated by Rsp5 ubiquitin ligase and functionally redundant PPxY motif proteins. Eukaryot Cell 12:990–997

Swatek KN, Komander D (2016) Ubiquitin modifications. Cell Res 26:399–422

Tanaka N, Kaneko K, Asao H, Kasai H, Endo Y, Fujita T, Takeshita T, Sugamura K (1999) Possible involvement of a novel STAM-associated molecule "AMSH" in intracellular signal transduction mediated by cytokines. J Biol Chem 274:19129–19135

Tang G, Zhang Z, Qian H, Chen J, Wang Y, Chen X, Chen B, Chen Y (2015) (-)-Epigallocatechin-3-gallate inhibits osteosarcoma cell invasiveness by inhibiting the MEK/ERK signaling pathway in human osteosarcoma cells. J Environ Pathol Toxicol Oncol 34:85–93

Tang S, Buchkovich NJ, Henne WM, Banjade S, Kim YJ, Emr SD (2016) ESCRT-III activation by parallel action of ESCRT-I/II and ESCRT-0/Bro1 during MVB biogenesis. Elife 5

Teis D, Saksena S, Emr SD (2008) Ordered assembly of the ESCRT-III complex on endosomes is required to sequester cargo during MVB formation. Dev Cell 15:578–589

Teis D, Saksena S, Judson B, Emr S (2010) ESCRT-II coordinates the assembly of ESCRT-III filaments for cargo sorting and multivesicular body vesicle formation. EMBO J 29(5):871–883

Teo H, Perisic O, Gonzalez B, Williams RL (2004) ESCRT-II, an endosome-associated complex required for protein sorting: crystal structure and interactions with ESCRT-III and membranes. Dev Cell 7:559–569

Teo H, Gill DJ, Sun J, Perisic O, Veprintsev DB, Vallis Y, Emr SD, Williams RL (2006) ESCRT-I core and ESCRT-II GLUE domain structures reveal role for GLUE in linking to ESCRT-I and membranes. Cell 125:99–111

Thien CB, Langdon WY (2001) Cbl: many adaptations to regulate protein tyrosine kinases. Nat Rev Mol Cell Biol 2:294–307

Vashist S, Ng DTW (2004) Misfolded proteins are sorted by a sequential checkpoint mechanism of ER quality control. J Cell Biol 165:41–52

Verplank L, Bouamr F, Lagrassa TJ, Agresta B, Kikonyogo A, Leis J, Carter CA (2001) Tsg101, a homologue of ubiquitin-conjugating (E2) enzymes, binds the L domain in HIV type 1 Pr55(Gag). Proc Natl Acad Sci USA 98:7724–7729

Vietri M, Schink KO, Campsteijn C, Wegner CS, Schultz SW, Christ L, Thoresen SB, Brech A, Raiborg C, Stenmark H (2015) Spastin and ESCRT-III coordinate mitotic spindle disassembly and nuclear envelope sealing. Nature 522:231–235

Vild CJ, Xu Z (2014) Vfa1 binds to the N-terminal microtubule-interacting and trafficking (MIT) domain of Vps4 and stimulates its ATPase activity. J Biol Chem 289:10378–10386

von Schwedler UK, Stuchell M, Müller B, Ward DM, Chung HY, Morita E, Wang HE, Davis T, He G-P, Cimbora DM, Scott A, Kräusslich H-G, Kaplan J, Morham SG, Sundquist WI (2003) The protein network of HIV budding. Cell 114:701–713

Wang M, Pickart CM (2005) Different HECT domain ubiquitin ligases employ distinct mechanisms of polyubiquitin chain synthesis. EMBO J 24:4324–4333

Webster BM, Colombi P, Jager J, Lusk CP (2014) Surveillance of nuclear pore complex assembly by ESCRT-III/Vps4. Cell 159:388–401

Webster BM, Thaller DJ, Jager J, Ochmann SE, Borah S, Lusk CP (2016) Chm7 and Heh1 collaborate to link nuclear pore complex quality control with nuclear envelope sealing. EMBO J 35:2447–2467

Wee P, Wang Z (2017) Epidermal growth factor receptor cell proliferation signaling pathways. Cancers 9:52

Wehman AM, Poggioli C, Schweinsberg P, Grant BD, Nance J (2011) The P4-ATPase TAT-5 inhibits the budding of extracellular vesicles in C. elegans embryos. Curr Biol: CB 21:1951–1959

Wemmer M, Azmi I, West M, Davies B, Katzmann D, Odorizzi G (2011) Bro1 binding to Snf7 regulates ESCRT-III membrane scission activity in yeast. J Cell Biol 192(2):295–306

Wollert T, Hurley JH (2010) Molecular mechanism of multivesicular body biogenesis by ESCRT complexes. Nature 464:864–869

Wright MH, Berlin I, Nash PD (2011) Regulation of endocytic sorting by ESCRT-DUB-mediated deubiquitination. Cell Biochem Biophys 60:39–46

Wu N, Zheng B, Shaywitz A, Dagon Y, Tower C, Bellinger G, Shen C-H, Wen J, Asara J, McGraw TE, Kahn BB, Cantley LC (2013) AMPK-dependent degradation of TXNIP upon energy stress leads to enhanced glucose uptake via GLUT1. Mol Cell 49:1167–1175

Wunderley L, Brownhill K, Stefani F, Tabernero L, Woodman P (2014) The molecular basis for selective assembly of the UBAP1-containing endosome-specific ESCRT-I complex. J Cell Sci 127:663–672

Xiao J, Xia H, Yoshino-Koh K, Zhou J, Xu Z (2007) Structural characterization of the ATPase reaction cycle of endosomal AAA protein Vps4. J Mol Biol 374:655–670

Yamada-Inagawa T, Okuno T, Karata K, Yamanaka K, Ogura T (2003) Conserved pore residues in the AAA protease FtsH are important for proteolysis and its coupling to ATP hydrolysis. J Biol Chem 278:50182–50187

Yang DEA (2008) Structural basis for midbody targeting of spastin by the ESCRT-III protein CHMP1B. Nat Struct Mol Biol 15:1278–1286

Yang B, Stjepanovic G, Shen Q, Martin A, Hurley JH (2015) Vps4 disassembles an ESCRT-III filament by global unfolding and processive translocation. Nat Struct Mol Biol 22:492–498

Yeo SC, Xu L, Ren J, Boulton VJ, Wagle MD, Liu C, Ren G, Wong P, Zahn R, Sasajala P, Yang H, Piper RC, Munn AL (2003) Vps20p and Vta1p interact with Vps4p and function in multivesicular body sorting and endosomal transport in *Saccharomyces cerevisiae*. J Cell Sci 116:3957–3970

Yokouchi M, Kondo T, Houghton A, Bartkiewicz M, Horne WC, Zhang H, Yoshimura A, Baron R (1999) Ligand-induced ubiquitination of the epidermal growth factor receptor involves the interaction of the c-Cbl RING finger and UbcH7. J Biol Chem 274:31707–31712

Zakalskiy A, Hogenauer G, Ishikawa T, Wehrschutz-Sigl E, Wendler F, Teis D, Zisser G, Steven AC, Bergler H (2002) Structural and enzymatic properties of the AAA protein Drg1p from *Saccharomyces cerevisiae*. Decoupling of intracellular function from ATPase activity and hexamerization. J Biol Chem 277:26788–26795

Zamborlini A, Usami Y, Radoshitzky SR, Popova E, Palu G, Gottlinger H (2006) Release of autoinhibition converts ESCRT-III components into potent inhibitors of HIV-1 budding. Proc Natl Acad Sci USA 103:19140–19145

Zhang H, Wang Y, Wong JJ, Lim KL, Liou YC, Wang H, Yu F (2014) Endocytic pathways downregulate the L1-type cell adhesion molecule neuroglian to promote dendrite pruning in Drosophila. Dev Cell 30:463–478

Zhao Y, Macgurn JA, Liu M, Emr S (2013) The ART-Rsp5 ubiquitin ligase network comprises a plasma membrane quality control system that protects yeast cells from proteotoxic stress. Elife 2:e00459

Zhongzheng Y, Cody V, Jiaying J, Xu Z, Jianping L, Jie S, Bin Z, Wenxian L, Fuchung G, Maili L, Chunyang C, Zhaohui X (2012) Structural basis of molecular recognition between ESCRT-III like protein Vps60 and AAA-ATPase regulator Vta1 in the multi-vesicular body pathway. JBC 287:43899–43908

Chapter 5
Retromer and Its Role in Regulating Signaling at Endosomes

Matthew N. J. Seaman

Contents

Abstract The retromer complex is a key element of the endosomal protein sorting machinery being involved in trafficking of proteins from endosomes to the Golgi and also endosomes to the cell surface. There is now accumulating evidence that retromer also has a prominent role in regulating the activity of many diverse signaling proteins that traffic through endosomes and this activity has profound implications for the functioning of many different cell and tissue types from neuronal cells to cells of the immune system to specialized polarized epithelial cells of the retina. In this review, the protein composition of the retromer complex will be described along with many of the accessory factors that facilitate retromer-mediated endosomal protein sorting to detail how retromer activity contributes to the regulation of several distinct signaling pathways.

M. N. J. Seaman (✉)
Cambridge Institute for Medical Research, University of Cambridge, Wellcome Trust/MRC Building, Addenbrookes Hospital, Cambridge CB2 0XY, UK
e-mail: mnjs100@cam.ac.uk

C. Lamaze and I. Prior (eds.), *Endocytosis and Signaling*, Progress in Molecular and Subcellular Biology 57, https://doi.org/10.1007/978-3-319-96704-2_5

5.1 Introduction—Retromer and Endosomal Protein Sorting

Signaling events initiated at the plasma membrane by a receptor binding its respective ligand may be modulated by the receptor being endocytosed and then delivered to an endosome. A classical example of this process is the downregulation of the epidermal growth factor receptor (EGFR) after binding of ligand, ubiquitylation, and uptake into clathrin-coated vesicles. Following endocytosis and arrival at a structure termed a signaling endosome, the activation of the tyrosine kinase domain of the EGFR can occur. Once in an endosome, the activated EGFR is sorted by the endosome sorting complex required for transport (ESCRT) machinery into nascent intralumenal vesicles which results in the EGFR being effectively removed from the cytoplasm of the cell silencing the active tyrosine kinase in the cytoplasmic tail of the EGFR. The subsequent proteolytic destruction of the EGFR in a lysosome finishes the task of modulating the signaling cascade initiated by ligand binding at the cell surface (for review, see Schuh and Audhya 2014). It is noteworthy that some cancers have been linked to mutations in the ESCRT-1 complex that operates to recognize the ubiquitin tag added to the activated EGFR underscoring the biological importance of endosomally localized sorting machinery in signal modulation. It is also now well established that other receptor tyrosine kinases (RTKs) such as the platelet-derived growth factor receptor (PDGFR) and insulin receptor (IR) are similarly silenced by ESCRT-mediated sorting into endosomal intralumenal vesicles.

While the role of ESCRT proteins in modulating signaling is now well established and has been covered in depth in other reviews, there is now growing evidence of the importance of additional endosomally localized sorting machinery—namely the retromer complex—in the modulation of signaling events. The retromer complex was first identified several years ago from studies in yeast focused on the endosome-to-Golgi pathway. The yeast retromer complex was shown to mediate the retrieval of a vacuolar hydrolase receptor protein, Vps10p from endosomes to the Golgi complex. The yeast retromer complex was revealed to comprise five proteins, all encoded by vacuolar protein sorting (VPS) genes, that assemble to form a stable heteropentamer (Seaman et al. 1997, 1998). The five retromer proteins are: Vps35p, Vps29p, Vps26p, Vps17p, and Vps5p. It has been shown that Vps35p forms a stable trimer with Vps29p and Vps26p which functions in selecting membrane proteins, e.g., Vps10p (cargo) for retrieval to the Golgi. The Vps5p and Vps17p are both members of the sorting nexin (SNX) family and form a stable dimer that can assemble on membranes to drive tubulation (Horazdovsky et al. 1997). The ability of the Vps5p–Vps17p dimer to tubulate membranes is derived from their carboxy-terminal Bin–amphiphysin–Rvs (BAR) domains, and hence Vps5p and Vps17p are referred to as SNX-BAR proteins (reviewed in Van Weering and Cullen 2014). Loss of any of the retromer components results in a defect in endosome-to-Golgi retrieval and can also lead to instability and rapid degradation of the remaining proteins.

The retromer complex is conserved in all eukaryotes, and in mammalian cells a cargo-selective trimer of VPS35, VPS29, and VPS26 operates with a sorting nexin dimer (containing SNX-BAR proteins) of SNX1 or SNX2 paired with either SNX5 or SNX6 to mediate the retrieval of various proteins from endosomes to the Golgi including lysosomal hydrolase receptors such as the cation-independent mannose 6-phosphate receptor (CIMPR) and the Vps10p homologue, sortilin (Haft et al. 2000; Seaman 2004; Arighi et al. 2004; Carlton et al. 2004). It is worth noting, however, that in higher eukaryotes such as humans, the cargo-selective trimer forms only a loose association with the sorting nexin dimer (Swarbrick et al. 2011; Harbour and Seaman 2011) and thus the mechanisms that govern endosome-to-Golgi retrieval may be subtly different from yeast to humans. One major difference between yeast and humans is that, in humans and other higher eukaryotes, retromer has a prominent role to play in the endosome-to-cell surface retrieval pathway. This is because the mammalian retromer complex is able to associate with a host of additional 'accessory' factors that play a key role in endosomal protein sorting, especially the endosome-to-cell surface pathway (reviewed in Seaman 2012). This means that the retromer complex can regulate the cell surface levels of many different proteins, some of which function as signaling receptors. Recent data have also implicated retromer in directly regulating the activity of signaling receptors by displacing key signaling intermediates from the cytoplasmic domains of signaling receptors.

5.2 VPS26—An α-Arrestin

The first clues that retromer may play a direct role in regulating signaling events came from the determination of the structure of the VPS26 protein. Once the crystal structure of VPS26 was solved through X-ray diffraction, it was realized that VPS26 was structurally highly related to the β-arrestin family of proteins (Shi et al. 2006; Collins et al. 2008). The similarity of VPS26 to the β-arrestin family of proteins was such that VPS26 has been termed an α-arrestin (for reviews, see Aubry and Klein 2013; Gurevich and Gurevich 2014). The β-arrestin proteins are primarily involved in the uptake of activated G protein-coupled receptors (GPCRs) by linking GPCRs to the endocytic machinery that forms clathrin-coated pits at the cell surface. Once GPCRs have reached an endosome, their signaling activity is downregulated and they are then recycled to the cell surface. The recycling of the β2-adrenergic receptor—a classical GPCR—from endosomes to the cell surface requires the function of a sorting nexin protein, specifically SNX27 in conjunction with retromer (Temkin et al. 2011). Importantly, SNX27 differs from the SNX-BAR proteins (e.g., SNX1 and SNX2) by lacking a BAR domain and thus is not believed to mediate membrane tubulation. Although lacking a BAR domain, SNX27 does possess other functional domains, namely a PDZ (PSD95, Dlg1, zo-1)-binding domain and a FERM (4.1, ezrin, radixin, moesin) domain—both of which have important roles in how SNX27 links retromer to signaling proteins.

5.3 Retromer Accessory Proteins: The WASH Complex and SNX27

The SNX27 protein associates with the retromer cargo-selective trimer by binding to VPS26 (Gallon et al. 2014). In addition to binding to VPS26, the SNX27 protein also interacts with a protein called FAM21 (also now known as WASHC2) which is a component of the WASH complex (Temkin et al. 2011; Freeman et al. 2014). Like retromer, the WASH complex is comprised of five proteins, namely WASH1, KIAA1033 (now WASHC3), strumpellin (now WASHC4), CCDC53 (now WASHC5), and FAM21 (Derivery et al. 2009), but unlike retromer the WASH complex is not universally conserved being absent in many organisms including yeast. The WASH complex mediates formation of branched filamentous (F-) actin on endosomes and is required for sorting many endosomal proteins to their destination but is generally more prominently required for endosome-to-cell surface recycling [see Seaman et al. (2013) for review]. Indeed, it has been reported that FAM21 can prevent the endosome-to-Golgi retrieval of proteins such as the Glut-1 glucose transporter and direct it into a SNX27-retromer-mediated endosome-to-cell surface pathway (Lee et al. 2016). In mammals, the WASH complex is recruited to endosomes by the direct interaction of FAM21 with VPS35 (Harbour et al. 2012; Jia et al. 2012; Helfer et al. 2013).

Thus, there is a complex and dynamic network of protein–protein interactions that underlie the mechanism of endosome-to-cell surface recycling and revolve around the cargo-selective retromer trimer of VPS35-VPS29-VPS26 (see Fig. 5.1). One way retromer may, through the WASH complex, contribute to regulating signaling events on endosomes is through the creation of actin-stabilized microdomains that potentiate signaling initiated by specific receptors. This suggestion, while conceptually attractive, currently lacks strong experimental evidence.

The importance of the retromer trimer in regulating GPCR signaling goes beyond that of providing binding sites for the WASH complex and SNX27. The VPS26 protein can, due to its similarity to β-arrestin, displace β-arrestin from certain GPCRs and thereby modulate the signaling activity. For the parathyroid hormone receptor (PTHR), activation by ligand binding while at the cell surface triggers the production of cyclic AMP (cAMP) and results in binding of the PTHR by β-arrestin. Following endocytosis of the PTHR with β-arrestin, the production of cyclic adenosine monophosphate (cAMP) continues until the β-arrestin is displaced from the cytoplasmic domain of the PTHR by retromer—an action believed to be driven by VPS26 (Feinstein et al. 2011). Subsequently, the PTHR is recycled back to the cell surface through an association with SNX27 mediated through the PDZ domain of SNX27 binding to a PDZ ligand in the cytoplasmic domain of the PTHR (Chan et al. 2016; McGarvey et al. 2016). This process effectively silences the signaling activity of the PTHR, 'resets' the system ready for another round of ligand binding, receptor activation, and signaling, and thus is quite different to the downregulation of, for example, the EGFR by the ESCRT complex.

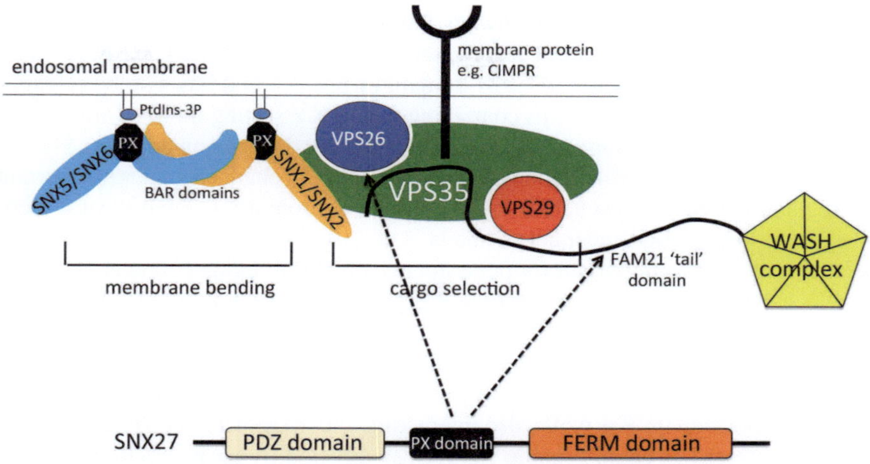

Fig. 5.1 Schematic of the retromer complex and key accessory proteins. The retromer complex comprises two functional units: a cargo-selective trimer of VPS35, VPS29, and VPS26 (often referred to as the cargo-selective complex—CSC) and a membrane bending/tubulating dimer of SNX1 or SNX2 with either SNX5 or SNX6. The VPS26 protein is structurally related to β-arrestin and can modulate GPCR signaling at endosomes. SNX27 interacts with both VPS26 and the FAM21 subunits of the WASH complex that promotes F-actin formation on endosomes. The PDZ and FERM domains of SNX27 both play important roles in trafficking membrane proteins (including those involved in signaling) from endosomes to the cell surface. The SNX27 PX domain binds to phosphatidylinositol 3-phosphate (PtdIns3P) in endosomal membranes

The role of retromer and its associated proteins in regulating the activity and localization of GPCRs such as the β2-adrenergic receptor and the PTHR is now well established. How the control of signaling and localization of these GPCRs precisely affects the operation of tissues and organs, however, remains to be determined. There is now an example where retromer function in regulating a specific GPCR does have profound implications for the operation of a specific tissue—namely the retina of the eye. The rhodopsin1 protein is a GPCR that functions as a light sensor in photoreceptor cells which is endocytosed upon activation by light. Studies in drosophila have shown that, following its endocytosis, the retromer complex is required to mediate the endosome-to-cell surface recycling of rhodopsin1 (Wang et al. 2014). Loss of retromer function results in mistrafficking of activated rhodopsin1 to a lysosome resulting in its destruction. This, in turn, leads to light-induced degeneration of photoreceptor cells, although mechanistically it is not clear why this happens. Additionally, whether the recycling of rhodopsin1 by retromer also involves a drosophila homologue of SNX27 also remains to be determined, but it appears that there is homologue of SNX27 in flies encoded by the CG32758 gene.

In addition to binding to both VPS26 and also the FAM21 subunit of the WASH complex, the SNX27 protein also has the ability to interact with Ras (Ghai et al. 2011). Ras can be considered to be the archetypal small GTPase and is associated with multiple signaling events/pathways that have long been implicated in oncogenic signaling mechanisms (for review, see Herrero et al. 2016). The SNX27–Ras interaction occurs through the FERM domain of SNX27 that has also been implicated in binding to cargo proteins containing the NPXY sorting motif (Burden et al. 2004; Ghai et al. 2011). One of the cargo proteins that depends on an intact SNX27 FERM domain for its localization is a class of potassium channels that are important for regulating neuronal excitability—namely the G protein-regulated inward rectifying potassium (GIRK) channels (Balana et al. 2013)—although other domains in SNX27 may also be important for potassium channel trafficking (Lunn et al. 2007). In cells expressing a dominant negative Ras protein that cannot bind GTP, the ability of SNX27 to regulate cell surface levels of the GIRK potassium channels was compromised (Balana et al. 2013).

These findings raise some intriguing possibilities and questions: For example, does SNX27 associate with Ras while also interacting with retromer (via VPS26) and the WASH complex? Is cargo-binding by SNX27 affected by an interaction between SNX27 and Ras? Does the SNX27–Ras interaction modulate the role and/or function of Ras in various signaling pathways? Answers to these questions have yet to be reported, but there is indeed a rich vein of research to be mined for functional links between the retromer-centered endosomal sorting machinery and the Ras protein. Adding to the questions posed by the SNX27–Ras association is the recent identification of an interaction between the VPS35 component of retromer and N-Ras, a version of the Ras protein that is farnesylated rather than being palmitoylated and is generally found to be cytoplasmic (Zhou et al. 2016). The interaction between retromer and N-Ras appears to be restricted to cytoplasmic retromer which would not be regarded as functional as retromer must be localized to endosomes to operate in endosomal protein sorting. Thus, it is possible that the retromer–N-Ras interaction may represent a means of regulating the activity of retromer, while it is cytoplasmic although this hypothesis has yet to be proven.

5.4 Retromer-Mediated Regulation of Neuronal Signaling

SNX27 plays a key role in regulating the trafficking of several membrane proteins that can function as receptors (e.g., rhodopsin, a light receptor) or channels (e.g., the GIRK channels) and does indeed seem to have a preference for multi-membrane spanning proteins as cargo, another notable example being the Glut-1, glucose transporter protein (Steinberg et al. 2013). Indeed, it has also been reported that SNX27 mediates the endosome-to-cell surface recycling of the AMPA (α-amino-3-hydroxy-5-methyl-4-isoxazolepropionic acid) receptor (Wang et al. 2013; Loo et al. 2014; Hussain et al. 2014)—a receptor for glutamate in neuronal cells that is implicated in long-term potentiation (LTP), a process intricately linked with memory and learning. In a study of AMPA receptor trafficking, it was shown that following stimulation of LTP, K-Ras is recruited to endosomes that are enriched in SNX27 and that this is necessary for synaptic delivery of GluA1, a glutamate receptor. Studies in neuronal cells have revealed a role for retromer in trafficking proteins to dendritic spines (Choy et al. 2014), and interestingly, localization and trafficking of GluA1 receptors have been shown to be inhibited in cells expressing a mutant of VPS35 that is responsible for a rare form of inherited Parkinson's disease (PD). The PD-causing allele of VPS35 is a mutation of aspartate at position 620 to glutamate and results in reduced binding of the WASH complex by VPS35 (Zavodszky et al. 2014; McGough et al. 2014) which, in turn, leads to mistrafficking of many WASH complex-dependent cargo proteins including the Glut-1 glucose transporter and the GluA1 glutamate receptor (Munsie et al. 2015; Tian et al. 2015). It is currently not known how the PD-causing VPS35 D620N mutant might affect the signaling pathways that depend upon the retromer–WASH complex association, but it seems likely that several distinct pathways may be affected. Not only does mutation of VPS35 affect transport of AMPA receptors such as GluA1, but loss of SNX27 can also lead to deficient transport of excitatory receptors. In a small number of patients, mutations in SNX27 cause a form of epilepsy associated with additional neurological abnormalities (Damseh et al. 2015). These observations demonstrate the key role that retromer, along with its accessory proteins such as the WASH complex and SNX27, plays in mediating the localization of neuronally important proteins such as the AMPA receptors and thereby illustrates how retromer contributes to signaling events that are key to the function of the brain such as the establishment of long-term potentiation.

5.5 A Role for Retromer in Regulating Bone Remodeling

The trafficking of glutamate receptors in neuronal cells is an example of how retromer function can influence relatively rapid signaling events that may, in some circumstances, lead to longer-term alterations in cell–cell contacts which can influence processes such as synaptic plasticity. Another example of how retromer can influence signaling that controls large-scale changes is the regulating of the localization and trafficking of the RANK protein by retromer. The receptor activator of NF-κB (RANK) protein functions in bone remodeling—osteoclastogenesis—and as such is important in maintaining bone density and structure. Loss of VPS35 function results in mislocalization of RANK which causes dysregulation of signaling from the RANK ligand and leads to decreased bone formation and a condition similar to osteoporosis (Xia et al. 2013). Precisely, how VPS35 (and presumably retromer) contributes to RANK localization has yet to be determined, but these data illustrate how important it is for normal signaling to be maintained and the key role that retromer plays in this form of homeostasis.

5.6 Retromer Regulation of Wnt-Mediated Morphogenic Signaling

Another example of how retromer function can influence signaling events in the longer term or over longer distances is the requirement for retromer to control the localization of the Wntless protein. The secretion of the Wnt morphogen that regulates some key developmental processes requires the Wntless protein—a multi-pass membrane protein that traffics between the cell surface and the TGN. While in the TGN, Wntless can bind to Wnt and then facilitate the secretion of Wnt at the cell surface. Wntless is subsequently endocytosed and then recycled back to the TGN via retromer-mediated endosome-to-Golgi retrieval (Franch-Marro et al. 2008; Port et al. 2008; Belenkaya et al. 2008; Yang et al. 2008; Pan et al. 2008). In this instance, retromer operates with another sorting nexin that, like SNX27, is not a SNX-BAR protein involved in membrane tubule formation. Here, retromer operates with SNX3, a small sorting nexin that comprises a phosphatidylinositol 3-phosphate binding domain and little more (Harterink et al. 2011; Zhang et al. 2011). SNX3 is conserved and functions with retromer in endosome-to-Golgi retrieval in yeast, although it is known as Grd19p in yeast (Strochlic et al. 2007). Failure of recycling of Wntless by retromer leads to reduced Wnt secretion due to a lack of Wntless in the TGN much like loss of retromer leading to reduced TGN localization of the CIMPR which in turn results in a failure of delivery of lysosomal hydrolases (e.g., Cathepsin D) to the lysosome (see Fig. 5.2). The reduction in Wnt secretion following inhibition of retromer function results in developmental abnormalities.

5.7 Retromer-Mediated Regulation of Signaling in the Immune System

One of the tissue types often associated with complicated signaling pathways is the immune system—specifically signaling in T cells in response to the various cytokines. Recently, it has been reported that retromer can modulate the signaling response initiated by binding of type I interferons to their receptor. The type I family of interferons is typically associated with the response to bacterial or viral infections. Once ligand (i.e., the type I interferon) has bound to a receptor comprising two subunits (IFNAR1 and IFNAR2), the receptor is internalized through clathrin-mediated endocytosis and delivered to an endosome where signaling via JAK/STAT is initiated. The two receptor subunits then dissociate with the IFNAR2 being recycled to the cell surface, while the IFNAR1 is directed to the lysosome for degradation. Recycling of IFNAR2 to the cell surface requires retromer (Chmiest et al. 2016). Interestingly, retromer association with IFNAR2 may facilitate the dissociation of the interferon receptor and thereby downregulate the signaling by JAK/STAT as loss of VPS35 leads to increased localization of the IFNAR1 and IFNAR2 proteins at the endosome and prolonged activation of JAK/STAT with increased downstream events as well such as transcription of specific genes associated with the immune response.

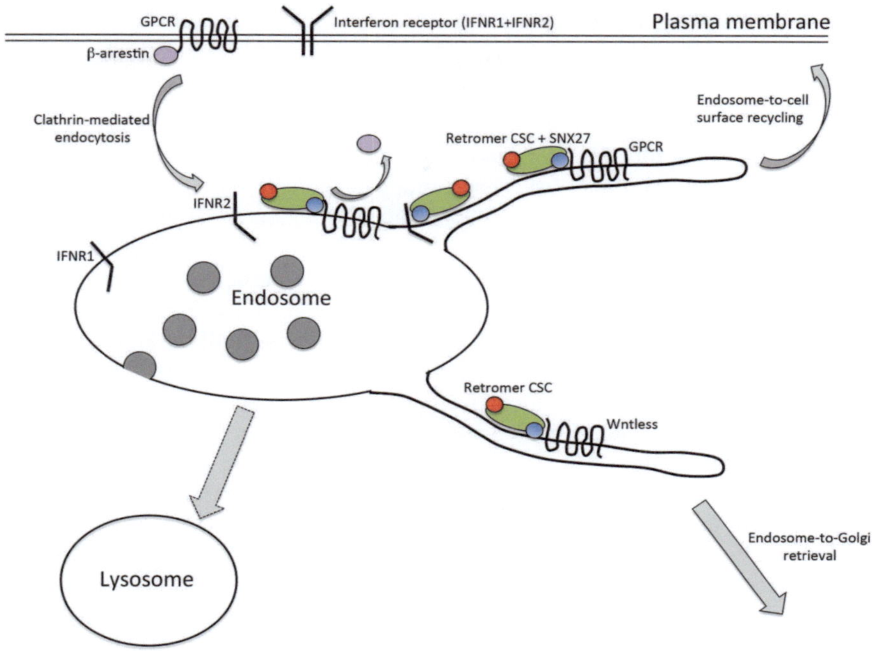

Fig. 5.2 Involvement of retromer and accessory proteins in trafficking pathways from endosomes. The retromer cargo-selective complex (CSC) along with SNX27 functions to sort proteins into tubular carriers for the endosome-to-cell surface recycling pathway. Cargo proteins sorted by retromer with SNX27 include GPCRs where the VPS26 subunit of retromer can displace β-arrestin from the cytoplasmic tail of the GPCR. Other membrane proteins involved in signaling that rely on retromer include the interferon receptor subunit, IFNAR2. At the endosome, retromer causes the IFNAR1 and IFNAR2 proteins to dissociate and directs the IFNAR2 protein back to the cell surface, while the IFNAR1 will be degraded in the lysosome. Retromer also functions in endosome-to-Golgi retrieval and maintains a pool of receptors in the TGN including Wntless, the morphogen receptor required for Wnt secretion

Thus, although at the beginning of this review I suggested that retromer can modulate signaling pathways in very different ways to the mode of action of the ESCRT machinery, there are some similarities depending on the mechanism of action of the proteins that initiate the signaling event. For GPCRs (e.g., the parathyroid hormone receptor), retromer can modulate signaling by displacing the β-arrestin protein from the activated GPCR, thereby silencing the signaling prior to directing the GPCR into an endosome-to-cell surface recycling pathway. But for activated interferon receptors, retromer downregulates the receptor by causing the two subunits to dissociate: One is then degraded and the other recycled.

5.8 Concluding Remarks

In conclusion, endosomal protein sorting mediated by the retromer complex along with associated proteins such as the WASH complex and SNX27 has been clearly shown to be critical for trafficking and recycling of membrane proteins from endosomes to the Golgi and also from endosomes to the cell surface. There are now numerous examples of how retromer contributes to regulating signaling events by controlling the trafficking and localization of membrane proteins that initiate signaling events. In some instances, retromer can directly participate in the signaling activity of a pathway, for example when the VPS26 subunit of retromer displaces the β-arrestin from an activated GPCR thereby bringing to a halt the signaling pathway initiated when the GPCR was activated. In other reports, retromer has been shown to control the cell surface localization of important neuronal receptors, (e.g., GluA1) and hence retromer function contributes to signaling events that are instrumental in neuronal processes such as long-term potentiation. Retromer may also indirectly affect signaling pathways by regulating the localization of a protein that is involved in secretion of a morphogen that operates over relatively long distances and timescale—as is the case for the Wnt morphogen and the influence that retromer plays in its secretion via the Wntless protein.

It seems likely that further examples of how retromer function contributes to the regulation of signaling pathways and events that may be initiated at the cell surface but are modulated at endosomal membrane will be reported in the near future. For cell biologists interested in understanding how protein localization contributes to signaling, these are interesting and exciting times.

Acknowledgements I would like to thank Aamir Mukadam for critical reading of the manuscript. MNJS is funded by the Medical Research Council.

References

Arighi CN, Hartnell LM, Aguilar RC, Haft CR, Bonifacino JS (2004) Role of the mammalian retromer in sorting of the cation-independent mannose 6-phosphate receptor. J Cell Biol 165(1):123–133

Aubry L, Klein G (2013) True arrestins and arrestin-fold proteins: a structure-based appraisal. Prog Mol Biol Transl Sci. 118:21–56

Balana B, Bahima L, Bodhinathan K, Taura JJ, Taylor NM, Nettleton MY, Ciruela F, Slesinger PA (2013) Ras-association domain of sorting Nexin 27 is critical for regulating expression of GIRK potassium channels. PLoS ONE 8(3):e59800

Belenkaya TY, Wu Y, Tang X, Zhou B, Cheng L, Sharma YV, Yan D, Selva EM, Lin X (2008) The retromer complex influences Wnt secretion by recycling wntless from endosomes to the trans-Golgi network. Dev Cell 14(1):120–131

Burden JJ, Sun XM, García AB, Soutar AK (2004) Sorting motifs in the intracellular domain of the low density lipoprotein receptor interact with a novel domain of sorting nexin-17. J Biol Chem 279(16):16237–16245

Carlton J, Bujny M, Peter BJ, Oorschot VM, Rutherford A, Mellor H, Klumperman J, McMahon HT, Cullen PJ (2004) Sorting nexin-1 mediates tubular endosome-to-TGN transport through coincidence sensing of high- curvature membranes and 3-phosphoinositides. Curr Biol 14(20):1791–800

Chan AS, Clairfeuille T, Landao-Bassonga E, Kinna G, Ng PY, Loo LS, Cheng TS, Zheng M, Hong W, Teasdale RD, Collins BM, Pavlos NJ. (2016) Sorting nexin 27 couples PTHR trafficking to retromer for signal regulation in osteoblasts during bone growth. Mol Biol Cell 27(8):1367–82

Chmiest D, Sharma N, Zanin N, Viaris de Lesegno C, Shafaq-Zadah M, Sibut V, Dingli F, Hupé P, Wilmes S, Piehler J, Loew D, Johannes L, Schreiber G, Lamaze C (2016) Spatiotemporal control of interferon-induced JAK/STAT signaling and gene transcription by the retromer complex. Nat Commun 7:13476

Choy RW, Park M, Temkin P, Herring BE, Marley A, Nicoll RA, von Zastrow M (2014) Retromer mediates a discrete route of local membrane delivery to dendrites. Neuron 82(1):55–62

Collins BM, Norwood SJ, Kerr MC, Mahony D, Seaman MN, Teasdale RD, Owen DJ (2008) Structure of Vps26B and mapping of its interaction with the retromer protein complex. Traffic 9(3):366–379

Damseh N, Danson CM, Al-Ashhab M, Abu-Libdeh B, Gallon M, Sharma K, Yaacov B, Coulthard E, Caldwell MA, Edvardson S, Cullen PJ, Elpeleg O (2015) A defect in the retromer accessory protein, SNX27, manifests by infantile myoclonic epilepsy and neurodegeneration. Neurogenetics 16(3):215–221

Derivery E, Sousa C, Gautier JJ, Lombard B, Loew D, Gautreau A (2009) The Arp2/3 activator WASH controls the fission of endosomes through a large multiprotein complex. Dev Cell 17(5):712–723

Feinstein TN, Wehbi VL, Ardura JA, Wheeler DS, Ferrandon S, Gardella TJ, Vilardaga JP (2011) Retromer terminates the generation of cAMP by internalized PTH receptors. Nat Chem Biol 7(5):278–284

Franch-Marro X, Wendler F, Guidato S, Griffith J, Baena-Lopez A, Itasaki N, Maurice MM, Vincent JP (2008) Wingless secretion requires endosome-to-Golgi retrieval of Wntless/Evi/Sprinter by the retromer complex. Nat Cell Biol 10(2):170–177

Freeman CL, Hesketh G, Seaman MN (2014) RME-8 coordinates the activity of the WASH complex with the function of the retromer SNX dimer to control endosomal tubulation. J Cell Sci 127(Pt 9):2053–2070

Gallon M, Clairfeuille T, Steinberg F, Mas C, Ghai R, Sessions RB, Teasdale RD, Collins BM, Cullen PJ (2014) A unique PDZ domain and arrestin-like fold interaction reveals mechanistic details of endocytic recycling by SNX27-retromer. Proc Natl Acad Sci U S A 111(35):E3604–E3613

Ghai R, Mobli M, Norwood SJ, Bugarcic A, Teasdale RD, King GF, Collins BM (2011) Phox homology band 4.1/ezrin/radixin/moesin-like proteins function as molecular scaffolds that interact with cargo receptors and Ras GTPases. Proc Natl Acad Sci U S A 108(19):7763–7768

Gurevich VV, Gurevich EV (2014) Overview of different mechanisms of arrestin-mediated signaling. Curr Protoc Pharmacol 1:67

Haft CR, de la Luz Sierra M, Bafford R, Lesniak MA, Barr VA, Taylor SI (2000) Human orthologs of yeast vacuolar protein sorting proteins Vps 26, 29, and 35: assembly into multimeric complexes. Mol Biol Cell 11(12):4105–4116

Harbour ME, Breusegem SY, Seaman MN (2012) Recruitment of the endosomal WASH complex is mediated by the extended 'tail' of Fam21 binding to the retromer protein Vps35. Biochem J 442(1):209–220

Harbour ME, Seaman MN (2011) Evolutionary variations of VPS29, and their implications for the heteropentameric model of retromer. Commun Integr Biol 4(5):619–622

Harterink M, Port F, Lorenowicz MJ, McGough IJ, Silhankova M, Betist MC, van Weering JR, van Heesbeen RG, Middelkoop TC, Basler K, Cullen PJ, Korswagen HC (2011) A SNX3-dependent retromer pathway mediates retrograde transport of the Wnt sorting receptor Wntless and is required for Wnt secretion. Nat Cell Biol 13(8):914–923

Helfer E, Harbour ME, Henriot V, Lakisic G, Sousa-Blin C, Volceanov L, Seaman MN, Gautreau A (2013) Endosomal recruitment of the WASH complex: active sequences and mutations impairing interaction with the retromer. Biol Cell 105(5):191–207

Herrero A, Matallanas D, Kolch W (2016) The spatiotemporal regulation of RAS signaling. Biochem Soc Trans 44(5):1517–1522

Horazdovsky BF, Davies B, Seaman MNJ, McLauglin SA, Yoon S-K, Emr SD (1997) A Yeast homolog of Snx1, Vps5p forms a membrane associated complex with Vps17p and is required for recycling of the vacuolar sorting receptor, Vps10p. Mol Biol Cell 8:1529–1541

Hussain NK, Diering GH, Sole J, Anggono V, Huganir RL (2014) Sorting Nexin 27 regulates basal and activity-dependent trafficking of AMPARs. Proc Natl Acad Sci U S A 111(32):11840–11845

Jia D, Gomez TS, Billadeau DD, Rosen MK (2012) Multiple repeat elements within the FAM21 tail link the WASH actin regulatory complex to the retromer. Mol Biol Cell 23(12):2352–2361

Lee S, Chang J, Blackstone C (2016) FAM21 directs SNX27-retromer cargoes to the plasma membrane by preventing transport to the Golgi apparatus. Nat Commun 9(7):10939

Loo LS, Tang N, Al-Haddawi M, Dawe GS, Hong W (2014) A role for sorting nexin 27 in AMPA receptor trafficking. Nat Commun 5:3176

Lunn ML, Nassirpour R, Arrabit C, Tan J, McLeod I, Arias CM, Sawchenko PE, Yates JR 3rd, Slesinger PA (2007) A unique sorting nexin regulates trafficking of potassium channels via a PDZ domain interaction. Nat Neurosci 10(10):1249–1259

McGarvey JC, Xiao K, Bowman SL, Mamonova T, Zhang Q, Bisello A, Sneddon WB, Ardura JA, Jean-Alphonse F, Vilardaga JP, Puthenveedu MA, Friedman PA (2016) Actin-Sorting Nexin 27 (SNX27)-retromer complex mediates rapid parathyroid hormone receptor recycling. J Biol Chem 291(21):10986–101002

McGough IJ, Steinberg F, Jia D, Barbuti PA, McMillan KJ, Heesom KJ, Whone AL, Caldwell MA, Billadeau DD, Rosen MK, Cullen PJ (2014) Retromer binding to FAM21 and the WASH complex is perturbed by the Parkinson disease-linked VPS35(D620 N) mutation. Curr Biol 24(14):1670–1676

Munsie LN, Milnerwood AJ, Seibler P, Beccano-Kelly DA, Tatarnikov I, Khinda J, Volta M, Kadgien C, Cao LP, Tapia L, Klein C, Farrer MJ (2015) Retromer-dependent neurotransmitter receptor trafficking to synapses is altered by the Parkinson's disease VPS35 mutation p. D620N. Hum Mol Genet 24(6):1691–1703

Pan CL, Baum PD, Gu M, Jorgensen EM, Clark SG, Garriga G (2008) C. elegans AP-2 and retromer control Wnt signaling by regulating mig-14/Wntless. Dev Cell 14(1):132–139

Port F, Kuster M, Herr P, Furger E, Bänziger C, Hausmann G, Basler K (2008) Wingless secretion promotes and requires retromer-dependent cycling of Wntless. Nat Cell Biol 10(2):178–185

Schuh AL, Audhya A (2014) The ESCRT machinery: from the plasma membrane to endosomes and back again. Crit Rev Biochem Mol Biol. 49(3):242–261

Seaman MN (2004) Cargo-selective endosomal sorting for retrieval to the Golgi requires retromer. J Cell Biol 165(1):111–122

Seaman MN (2012) The retromer complex—endosomal protein recycling and beyond. J Cell Sci 125(Pt 20):4693–4702

Seaman MN, Gautreau A, Billadeau DD (2013) Retromer-mediated endosomal protein sorting: all WASHed up! Trends Cell Biol 23(11):522–528

Seaman MN, Marcusson EG, Cereghino JL, Emr SD (1997) Endosome to Golgi retrieval of the vacuolar protein sorting receptor, Vps10p, requires the function of the VPS29, VPS30, and VPS35 gene products. J Cell Biol 137(1):79–92

Seaman MN, McCaffery JM, Emr SD (1998) A membrane coat complex essential for endosome-to-Golgi retrograde transport in yeast. J Cell Biol 142(3):665–681

Shi H, Rojas R, Bonifacino JS, Hurley JH (2006) The retromer subunit Vps26 has an arrestin fold and binds Vps35 through its C-terminal domain. Nat Struct Mol Biol 13(6):540–548

Steinberg F, Gallon M, Winfield M, Thomas EC, Bell AJ, Heesom KJ, Tavaré JM, Cullen PJ (2013) A global analysis of SNX27-retromer assembly and cargo specificity reveals a function in glucose and metal ion transport. Nat Cell Biol 15(5):461–471

Strochlic TI, Setty TG, Sitaram A, Burd CG (2007) Grd19/Snx3p functions as a cargo-specific adapter for retromer-dependent endocytic recycling. J Cell Biol 177(1):115–125

Swarbrick JD, Shaw DJ, Chhabra S, Ghai R, Valkov E, Norwood SJ, Seaman MN, Collins BM (2011) VPS29 is not an active metallo-phosphatase but is a rigid scaffold required for retromer interaction with accessory proteins. PLoS ONE 6(5):e20420

Temkin P, Lauffer B, Jäger S, Cimermancic P, Krogan NJ, von Zastrow M (2011) SNX27 mediates retromer tubule entry and endosome-to-plasma membrane trafficking of signaling receptors. Nat Cell Biol 13(6):715–721

Tian Y, Tang FL, Sun X, Wen L, Mei L, Tang BS, Xiong WC (2015) VPS35-deficiency results in an impaired AMPA receptor trafficking and decreased dendritic spine maturation. Mol Brain 8(1):70

van Weering JR, Cullen PJ (2014) Membrane-associated cargo recycling by tubule-based endosomal sorting. Semin Cell Dev Biol 31:40–47

Wang S, Tan KL, Agosto MA, Xiong B, Yamamoto S, Sandoval H, Jaiswal M, Bayat V, Zhang K, Charng WL, David G, Duraine L, Venkatachalam K, Wensel TG, Bellen HJ (2014) The retromer complex is required for rhodopsin recycling and its loss leads to photoreceptor degeneration. PLoS Biol 12(4):e1001847

Wang X, Zhao Y, Zhang X, Badie H, Zhou Y, Mu Y, Loo LS, Cai L, Thompson RC, Yang B, Chen Y, Johnson PF, Wu C, Bu G, Mobley WC, Zhang D, Gage FH, Ranscht B, Zhang YW, Lipton SA, Hong W, Xu H (2013) Loss of sorting nexin 27 contributes to excitatory synaptic dysfunction by modulating glutamate receptor recycling in Down's syndrome. Nat Med 19(4):473–480

Xia WF, Tang FL, Xiong L, Xiong S, Jung JU, Lee DH, Li XS, Feng X, Mei L, Xiong WC (2013) Vps35 loss promotes hyperresorptive osteoclastogenesis and osteoporosis via sustained RANKL signaling. J Cell Biol 200(6):821–837

Yang PT, Lorenowicz MJ, Silhankova M, Coudreuse DY, Betist MC, Korswagen HC (2008) Wnt signaling requires retromer-dependent recycling of MIG-14/Wntless in Wnt-producing cells. Dev Cell 14(1):140–147

Zavodszky E, Seaman MN, Moreau K, Jimenez-Sanchez M, Breusegem SY, Harbour ME, Rubinsztein DC (2014) Mutation in VPS35 associated with Parkinson's disease impairs WASH complex association and inhibits autophagy. Nat Commun 5:3828

Zhang P, Wu Y, Belenkaya TY, Lin X (2011) SNX3 controls Wingless/Wnt secretion through regulating retromer-dependent recycling of Wntless. Cell Res 21(12):1677–1690

Zhou M, Wiener H, Su W, Zhou Y, Liot C, Ahearn I, Hancock JF, Philips MR (2016) VPS35 binds farnesylated N-Ras in the cytosol to regulate N-Ras trafficking. J Cell Biol 214(4):445–458

Chapter 6
The Lysosome and Intracellular Signalling

Geoffrey G. Hesketh, Lena Wartosch, Luther J. Davis, Nicholas A. Bright and J. Paul Luzio

Contents

Abstract In addition to being the terminal degradative compartment of the cell's endocytic and autophagic pathways, the lysosome is a multifunctional signalling hub integrating the cell's response to nutrient status and growth factor/hormone signalling. The cytosolic surface of the limiting membrane of the lysosome is the site of activation of the multiprotein complex mammalian target of rapamycin complex 1 (mTORC1), which phosphorylates numerous cell growth-related substrates, including transcription factor EB (TFEB). Under conditions in which mTORC1 is inhibited including starvation, TFEB becomes dephosphorylated and translocates to

G. G. Hesketh
Lunenfeld-Tanenbaum Research Institute, Sinai Health System, Toronto, ON M5G 1X5, Canada

L. Wartosch · L. J. Davis · N. A. Bright · J. P. Luzio (✉)
Department of Clinical Biochemistry and Cambridge Institute for Medical Research, School of Clinical Medicine, Wellcome Trust/MRC Building, University of Cambridge, Cambridge Biomedical Campus, Hills Road, Cambridge CB2 0XY, UK
e-mail: jpl10@cam.ac.uk

C. Lamaze and I. Prior (eds.), *Endocytosis and Signaling*, Progress in Molecular and Subcellular Biology 57, https://doi.org/10.1007/978-3-319-96704-2_6

the nucleus where it functions as a master regulator of lysosome biogenesis. The signalling role of lysosomes is not limited to this pathway. They act as an intracellular Ca^{2+} store, which can release Ca^{2+} into the cytosol for both local effects on membrane fusion and pleiotropic effects within the cell. The relationship and crosstalk between the lysosomal and endoplasmic reticulum (ER) Ca^{2+} stores play a role in shaping intracellular Ca^{2+} signalling. Lysosomes also perform other signalling functions, which are discussed. Current views of the lysosomal compartment recognize its dynamic nature. It includes endolysosomes, autolysosome and storage lysosomes that are constantly engaged in fusion/fission events and lysosome regeneration. How signalling is affected by individual lysosomal organelles being at different stages of these processes and/or at different sites within the cell is poorly understood, but is discussed.

6.1 The Discovery of the Lysosomal Membrane as a Signalling Hub

Less than ten years ago, a description of the role of the lysosome in intracellular signalling would likely have focused mainly on its function in the down-regulation, by degradation, of endocytosed cell surface receptors. An extensive literature had built up on the endocytic pathways taken by different receptors, including the role of endosomes as signalling platforms, the mechanisms of sorting into recycling pathways, the loss of signalling when they were sorted into the intraluminal vesicles (ILVs) of multivesicular bodies (MVBs) and delivery to lysosomes for degradation by proteases [for reviews, see (Platta and Stenmark 2011; Alonso and Friedman 2013; Goh and Sorkin 2013; Bowman et al. 2016)]. However, two papers in *Science* published at the end of the last decade radically changed the perception of the lysosome's role in intracellular signalling. David Sabatini's group found that whereas in starved, cultured mammalian cells, mammalian/mechanistic target of rapamycin (mTOR) was distributed throughout the cytoplasm, amino acid feeding resulted in a substantial fraction translocating to late endosomal and lysosomal compartments (Sancak et al. 2008). Initially using a systems biology approach, Andrea Ballabio's group discovered that many lysosomal genes exhibited coordinated transcriptional behaviour and are regulated by transcription factor EB (TFEB), which translocates from the cytoplasm to the nucleus under various conditions of lysosomal stress and acts as a master regulator of lysosome biogenesis (Sardiello et al. 2009). The Sabatini and Ballabio research groups went on to show that TFEB and the multiprotein complex mTORC1 [mTOR complex 1, reviewed in (Eltschinger and Loewith 2016)] co-localize on the lysosome membrane where mTORC1 phosphorylates TFEB (Settembre et al. 2012). Under conditions in which mTORC1 is inhibited, they observed that TFEB becomes dephosphorylated and translocates to the nucleus. The discovery of a lysosome-to-nucleus signalling mechanism involving mTORC1 and TFEB, together with subsequent work showing that multiple signals are sensed and

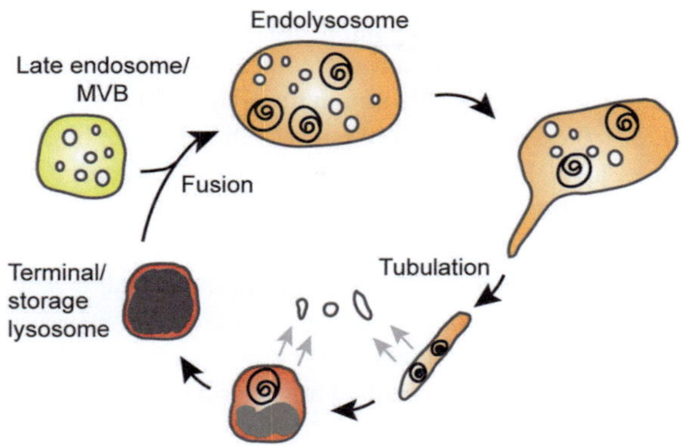

Fig. 6.1 Lysosome fusion/regeneration cycle. Electron-dense, terminal storage lysosomes fuse with late endosomes/MVBs to form catalytically active, acidic endolysosomes. Subsequent tubulation, maturation and content condensation steps are involved in the regeneration/re-formation of storage lysosomes. A similar cycle occurs for the fusion of lysosomes with autophagosomes to form autolysosomes from which lysosomes are regenerated. Electron-dense material is shaded in dark grey

integrated on the lysosomal surface to regulate the activation state of mTORC1, has led to the current consensus that the lysosomal membrane is a major hub of intracellular signalling, which regulates cell metabolism and growth (Perera and Zoncu 2016).

In addition to these major discoveries about the role of the lysosomal membrane in mTORC1 and TFEB signalling, there has also been a growing realization over the past two decades that the mammalian lysosomal compartment is heterogeneous and is made up of less acidic/neutral lysosomes, acidic endolysosomes and autolysosomes, formed respectively by fusion of late endosomes or autophagosomes with lysosomes, as well as organelles that are at different stages of maturation on lysosome regeneration pathways (Fig. 6.1). As more and more components of the molecular machinery of lysosome fusion and re-formation are discovered, there is also an increasing emphasis on how these processes are regulated and how they contribute to or are affected by intracellular signalling.

6.2 The Lysosomal Compartment and the Lysosome Fusion/Regeneration Cycle

Lysosomes were originally discovered in the mid-twentieth century as membrane-bound organelles containing acid hydrolases and were rapidly recognized as the terminal degradative compartment of the endocytic and autophagic pathways [reviewed in (de Duve 2005)]. Nowadays, we are more aware of the dynamic nature of the endocytic pathway and the role of the lysosome fusion/regeneration cycle in determining the complexity of the late endosomal and lysosomal compartments (Fig. 6.1). The endocytic pathway taken by macromolecules en route to lysosomes has been well described (Luzio et al. 2007; Woodman and Futter 2008; Huotari and Helenius 2011; Luzio et al. 2014), as has the autophagic (i.e. macroautophagic) pathway (Feng et al. 2014; Bento et al. 2016). The delivery of endocytosed cargo to lysosomal acid hydrolases is achieved by kiss-and-run as well as full-fusion events between late endosomes, also known as multivesicular bodies (MVBs), and lysosomes (Bright et al. 2005, 2016). These events result in the formation of endolysosomes, which are hybrid organelles with characteristics of both late endosomes and lysosomes. It is in endolysosomes that hydrolytic degradation commences and these organelles can undertake further, multiple, fusion events with other late endosomes and lysosomes.

6.2.1 *Lysosome Fusion with Endosomes and Autophagosomes—Core Machinery and Regulation*

The molecular mechanism of late endosome–lysosome fusion is broadly understood and, like other fusion events in secretory and endocytic membrane traffic pathways, involves tethering, docking and phospholipid bilayer fusion steps (Luzio et al. 2007). Tethering requires the heterohexameric HOPS (homotypic fusion and vacuole protein sorting) complex (Wartosch et al. 2015). In the yeast *Saccharomyces cerevisiae,* HOPS, which structurally is a three-legged filamentous complex (Chou et al. 2016), is recruited directly to the limiting membrane of the vacuole (the yeast equivalent of the mammalian lysosome) by Ypt7p, the ortholog of the mammalian small GTPase Rab7 (Hickey and Wickner 2010). However, in mammalian cells, the Rab7 effector RILP (Lin et al. 2014; van der Kant et al. 2015) and the small GTPase Arl8b are needed for HOPS recruitment to late endosomal/lysosomal membranes (Khatter et al. 2015). Following tethering, the late endosome–lysosome fusion process requires the formation of a trans-SNARE (soluble N-ethylmaleimide-sensitive factor attachment protein receptor) complex consisting of three Q-SNAREs, namely syntaxin 7 (a Qa-SNARE), Vti1b (a Qb-SNARE), syntaxin 8 (a Qc-SNARE) and VAMP7 (an R-SNARE), which interact to form a parallel four-helix bundle that brings the two phospholipid bilayers together to enable membrane fusion (Pryor et al. 2004). It has been suggested that in vivo, the requirement for VAMP7 may be complemented by another R-SNARE, VAMP8 (Pols et al. 2013).

Components of the molecular machinery required for autophagosome–lysosome fusion have also been discovered. The HOPS complex is required (Jiang et al. 2014), but an additional factor involved in tethering is the Rab7 effector and HOPS-interacting protein, PLEKHM1 (pleckstrin homology domain-containing protein family member 1), which functions as a central hub in integrating endocytic and autophagic pathways at the lysosome (McEwan et al. 2015; McEwan and Dikic 2015; Marwaha et al. 2017). PLEKHM1 contains an LC3 (microtubule-associated protein

1 light chain 3)-interacting region that mediates its binding to autophagosomes via LC3 or GABARAP (Gamma-aminobutyric acid type A receptor-associated protein) members of the ATG8 (autophagy-related 8) family of proteins, which are essential for autophagosome–lysosome fusion (Nguyen et al. 2016). The recruitment of PLEKHM1 and HOPS may also be regulated by the cholesterol-sensing Rab7 effector ORP1L which localizes to autophagosomes and under low cholesterol conditions is involved in the formation of autophagosome-ER (endoplasmic reticulum) membrane contact sites (Wijdeven et al. 2016). The trans-SNARE complex required for autophagosome–lysosome fusion has been identified as consisting of syntaxin 17 as the Qa-SNARE, SNAP-29 as the Qbc-SNARE and VAMP8 as the R-SNARE (Itakura et al. 2012). A role for an alternative R-SNARE, VAMP7, has been proposed (Fader et al. 2009).

What is much less clear than our knowledge of the minimal protein machinery required for lysosome fusion with either late endosomes or autophagosomes is our understanding of how these processes are regulated. Relatively little is known about how flux through the late endocytic pathway or the route taken are regulated, although it is clear that regulation does occur. For example, whereas we have a good understanding of how epidermal growth factor (EGF) binding to its receptor at the plasma membrane leads to receptor phosphorylation, ubiquitination, endocytosis and sorting into endosomal ILVs, as well as intracellular signalling (Sorkin and Goh 2008), the regulation of its passage through late endocytic compartments is less well understood. EGF stimulates both the number of MVBs per unit of cytoplasm and the number of ILVs per MVB, an effect specific to the MVBs containing the EGF receptor (White et al. 2006), but the mechanism is unclear. Even within a single MVB, ILVs are of different sizes and are formed by more than one mechanism, with evidence that a competitive relationship exists between the well-described endocytic sorting complex required for transport (ESCRT) pathway for ILV formation and sorting of ubiquitinated cargo such as the EGF receptor and an ESCRT-independent pathway, which requires the tetraspanin membrane protein CD63 (Edgar et al. 2014). Details of the latter pathway are very sketchy, although recent work on the role of tetraspanins in MVB formation in yeast may provide some clues (MacDonald et al. 2015).

Whilst it is also known that lysosomes fuse with late endosomes much more efficiently than they do with early endosomes, how fusion is signalled is only poorly understood. A partial explanation is the Rab5/Rab7 switch that occurs as the endosome matures allowing different Rab effectors to be recruited [reviewed in (Huotari and Helenius 2011)]. However, there is also a need to ensure that the MVBs that fuse have been depleted of recycling membrane proteins such as the mannose 6-phosphate receptors and that formation of ILVs, as well as sorting of ubiquitinated cargo into ILVs by the ESCRT pathway, is complete. Although it is not known how a mature MVB is identified for fusion, it could be as simple as the loss of ubiquitinated cargo from the limiting membrane. There may also be a role for some components of the ESCRT pathway, since an involvement of ESCRT-III proteins in fusion of late endosomes with lysosomes, separate to the requirement for ILV formation, has been described [reviewed in (Metcalf and Isaacs 2010)].

Efficient fusion of lysosomes with late endosomes requires release of luminal Ca^{2+} at a late stage in the fusion process (Pryor et al. 2000). Although the mechanism by which this aids membrane fusion is not resolved, it may well underlie the effects of other perturbations of lysosome fusion. For example, Niemann–Pick type-C1-deficient cells have a defect in fusion of lysosomes with endosomes, which is likely to be a consequence of altered luminal Ca^{2+} content (Lloyd-Evans et al. 2008). It is also unclear which Ca^{2+} channel(s) in the lysosome membrane is most important in release of this luminal Ca^{2+} with evidence that mucolipins [especially transient receptor potential mucolipin 1, MCOLN1 also known as TRPML1 (Dong et al. 2010)], two-pore channels [especially TPC2 (Grimm et al. 2014)], a voltage-gated calcium channel (Tian et al. 2015) and/or the purinergic receptor, P2X4 (Cao et al. 2015), may be involved. Intriguingly, cation transport through both mucolipin1 and TPC2 is activated by phosphatidylinositol 3,5-bisphosphate [PI(3,5)P2], which, in mammalian cells, is formed by the action of the enzyme PIKfyve on phosphatidylinositol 3-phosphate [PI(3)P] in endosomal membranes (Jin et al. 2016). It has been reported that a rapid increase in the concentration of PI(3,5)P2 occurs on the membranes of late endocytic organelles immediately before they fuse (Li et al. 2013), although the specificity of the PI(3,5)P2 probe used in the studies leading to this conclusion has been questioned (Hammond et al. 2015). In addition, it has been argued that PI(3,5)P2 may be more important in regulating fission rather than fusion (see below). Lysosomal Ca^{2+} release via mucolipin1 is also thought to be important for the fusion of autophagosomes with lysosomes, as are appropriate levels of PI(3,5)P2 (Martens et al. 2016). Unlike PI(3,5)P2, PI(4,5)P2 is mostly found in the plasma membrane, some can be generated at the lysosomal membrane where it can act as an inhibitor of Ca^{2+} transport through mucolipin1. Part of the cellular complement of the PI(4,5)P2 5-phosphatase OCRL (oculocerebrorenal syndrome of Lowe) is associated with mucolipin1 and maintains PI(4,5)P2-free microdomains around this Ca^{2+} channel (De Leo et al. 2016). Thus, when OCRL activity is absent either as a result of experimental manipulation or in the disease Lowe syndrome, an unrestricted accumulation of PI(4,5)P2 occurs on the lysosomal membrane with consequent inhibition of Ca^{2+} flux through mucolipin1 and inhibition of lysosome fusion with autophagosomes. This has been proposed as the reason why autophagosomes accumulate in cells from Lowe syndrome patients. Local generation of PI(4,5)P2 is also thought to play a role in lysosome re-formation from autolysosomes (see below), emphasizing the likely importance of spatiotemporal regulation of the concentration of phosphoinositides on the lysosomal membrane. A role for luminal acidity has also been proposed for fusion events undertaken by lysosomes and yeast vacuoles, but remains controversial as does the requirement for the V-ATPase (vacuolar H^+ ATPase), which may involve an unconventional role of the V_O sector interacting with SNARE proteins and contributing physically to membrane fusion (Coonrod et al. 2013; Mauvezin et al. 2015; Desfougeres et al. 2016).

Regulatory cytosolic factors are also implicated in lysosome fusion with endosomes and/or autophagosomes and include the VAMP7-binding protein, VARP (vacuolar protein sorting 9 ankyrin repeat protein) (Schafer et al. 2012), ATG14 (Diao et al. 2015) and EPG5 (ectopic P-granules autophagy protein 5) (Wang et al. 2016), which have effects on SNARE complex assembly and/or stabilization. The leucine-rich repeat kinases LRRK1 and LRRK2 have also been implicated in defective traffic to lysosomes via the endocytic and autophagy pathways. LRRK2 is highly expressed in brain, kidney and some immune cells. It localizes to several membrane-bound organelles on post-Golgi membrane traffic pathways and regulates a subset of Rab GTPases (Steger et al. 2016). Autosomal dominant mutations in the gene encoding LRRK2 are the most common cause of familial forms of the neurodegenerative disorder, Parkinson's disease and are also associated with sporadic forms of the disease. LRRK1 is widely expressed, but less highly in brain than LRRK2. LRRK1 indirectly regulates Rab7 activity and is recruited to lysosomes by VAMP7 (Toyofuku et al. 2015). Lysosome fusion with both autophagosomes and endosomes is negatively regulated by RUBICON (Run domain BECLIN-1-interacting and cysteine-rich domain-containing protein), which binds to UVRAG (UV radiation resistance-associated gene product), when this component of VPS34 complex II (consisting of the phosphatidylinositol-3-kinase VPS34, VPS15, BECLIN-1 and UVRAG) is phosphorylated by mTORC1 (Kim et al. 2015). This results in inhibition of Rab7 and HOPS activity and has led to the suggestion that the mTORC1-UVRAG pathway is an important regulatory axis through which cells coordinate autophagy and the endosome–lysosomal degradation pathway (Kim et al. 2015). In addition to possible nutritional regulation via mTORC1 (Antonioli et al. 2016), lysosome–autophagosome fusion is reduced when cells are incubated in high glucose as a result of increased addition of *O*-linked-N-acetylglucosamine to SNAP29, which inhibits the formation of trans-SNARE complexes. Conversely, glucose starvation results in less modification of SNAP29 with *O*-linked-N-acetylglucosamine and increased formation of autolysosomes (Guo et al. 2014). Also implicated in regulating the efficiency of fusion with lysosomes are motor proteins, which are bound to organelles destined for fusion and move them along microtubules and/or actin filaments (van der Kant et al. 2013; Kruppa et al. 2016). Finally, it should be noted that inefficient degradation of macromolecules may itself reduce the efficiency of fusion via effects on membrane cholesterol (Fraldi et al. 2010) and also prevent lysosome re-formation (Bright et al. 1997, 2016; Schmid et al. 1999).

6.2.2 Other Lysosomal Fusion Events

In addition to fusing with late endosomes and autophagosomes, lysosomes can also fuse with phagosomes, macropinosomes and the plasma membrane [reviewed in (Luzio et al. 2007)]. Lysosome fusion with the plasma membrane is triggered by an increase in cytosolic Ca^{2+} concentration and is regulated by the Ca^{2+} sensor synaptotagmin VII, which restricts both the kinetics and extent of fusion and interacts with the core fusion machinery comprising the SNARE proteins syntaxin 4, SNAP23 and VAMP7 (Rao et al. 2004). The increased cytosolic Ca^{2+} concentration can occur as a result of plasma membrane damage causing Ca^{2+} influx. The resultant exocytosis of a peripheral population of lysosomes starts a plasma membrane sealing and re-modelling process that is essential for the survival of cells wounded by mechanical stress or attacked by some pathogens (Castro-Gomes et al. 2016). Lysosomal secretion is also regulated by the transcription factor TFEB, which can cause translocation of lysosomes to the plasma membrane and increased secretion, but the molecular mechanisms are unclear (Medina et al. 2011). Over-expression of TFEB in some cellular models of lysosomal storage disease results in increased lysosomal exocytosis and clearance of accumulated metabolites, suggesting TFEB as a therapeutic target in these diseases (Medina et al. 2011; Spampanato et al. 2013). Although all cell types seem to have the ability to fuse lysosomes with the plasma membrane, some cells have specialized "secretory lysosomes" or "lysosome-related organelles" (LROs). The variety of LROs and signalling pathways to trigger their secretion are reviewed in (Luzio et al. 2014).

6.2.3 Lysosome Re-formation and Its Regulation

In contrast to the molecular machinery of fusion, less is known about the mechanism(s) by which lysosomes are re-formed from endolysosomes and autolysosomes. In both cases, tubulation and fission events have been suggested to occur. Some of these events are likely concerned with the recycling of membrane components, e.g. SNAREs that should not be present on the re-formed lysosomes. However, others are necessary in the formation and scission, along their length, of protolysosomal tubules from which mature re-formed lysosomes are generated. In the re-formation of lysosomes from autolysosomes, the formation of protolysosomal tubules is regulated by mTORC1 (Yu et al. 2010; Rong et al. 2011) and their scission/vesiculation is mediated by the GTPase dynamin2 (Schulze et al. 2013). Currently, the best model for re-formation of lysosomes from autolysosomes comes from the results of experimental manipulations, suggesting that localized phosphoinositide generation on the autolysosome membrane causes the recruitment of the sorting adaptor AP-2, clathrin and the kinesin motor KIF5B to microdomains enriched in PI(4,5)P2, which results in the formation and extension of protolysosomal tubules along microtubules (Rong et al. 2012; Du et al. 2016). One note of caution about this model is that much earlier experiments showed how easy it was to mis-target AP-2 and clathrin to intracellular compartments and away from the plasma membrane, where their recruitment and function are well understood, simply by adding GTPγS or excess Ca^{2+} (Seaman et al. 1993). Recently, it has been suggested that it is not alterations in mTORC1 activity per se that induces lysosome re-formation, but the delivery to the autolysosome of mitochondrial DNA, which binds to TLR9 (toll-like receptor 9). This triggers an increase in local PI(4,5)P2 concentration, resulting in the recruitment of AP-2 and clathrin (De Leo et al. 2016).

Additional clues about the machinery of lysosome re-formation have come from the study of cells from patients with lysosomal storage diseases. These are rare, inherited genetic defects, in many cases causing deficiencies in specific lysosomal acid hydrolases, but in others resulting in defects in lysosomal membrane proteins or nonenzymatic soluble lysosomal proteins. Cells from such patients contain membrane-bound, heterogeneous storage lesions, most probably abnormal endolysosomes/autolysosomes, filled with different contents in different diseases (Platt et al. 2012). Amongst lysosomal disease-associated proteins implicated in lysosome re-formation are Niemann–Pick type-C2 (Goldman and Krise 2010), lysosomal trafficking regulator (LYST) (Holland et al. 2014), the sorting adaptors AP-4 (Rong et al. 2012) and AP-5, along with its associated proteins spatacsin and spastizin (Hirst et al. 2015), as well as mucolipin1 (Miller et al. 2015), but molecular mechanisms remain elusive. In the case of mucolipin1, it has been proposed that this cation channel is responsible for the release of luminal Ca^{2+} and earlier in vitro experiments had shown that luminal Ca^{2+} is necessary for the formation of dense core lysosomes from endolysosomes (Pryor et al. 2000). As described above, PI(3,5)P2 is an activator and PI(4,5)P2 an inhibitor of this channel, which suggests at the very least tight spatiotemporal control of the concentrations of these phosphoinositides on the lysosomal membrane if fusion events and re-formation events are to be properly coordinated. Certainly, depletion or pharmacological inhibition of PIKfyve, the enzyme synthesizing PI(3,5)P2, results in the formation of enlarged endocytic compartments with many characteristics of endolysosomes [reviewed in (Dove et al. 2009)] and small molecule activators of mucolipin1 can reverse the enlarged endolysosomal phenotype observed when a protein acting as a scaffold for PIKfyve is depleted (Zou et al. 2015). The Ca^{2+} released through mucolipin1 may be required for the extension and/or scission of the membrane bridges connecting endolysosomes to nascent lysosomes in the protolysosomal tubules (Miller et al. 2015). It has been argued that a good candidate for a Ca^{2+}-regulated target is actin (Miller et al. 2015), the polymerization state of which can also be affected by PI(4,5)P2 (Saarikangas et al. 2010).

One of the most interesting aspects of lysosome re-formation is that it may be associated with an alteration in luminal acidity. For a long time, it has been widely assumed that the lumen of all lysosomes is acidic ($pH \leq 5$), despite evidence from early ratiometric imaging experiments in the 1980s (Yamashiro and Maxfield 1987) and experiments on lysosomal enzyme function and acidophilic dye accumulation in the 1990s (Butor et al. 1995), suggesting that lysosomes exhibit a wide range of pH. This has been reinforced by recent studies, suggesting that a proportion of lysosomes (up to ~25%) have a luminal pH that is closer to neutral (>pH 6.5) (Johnson et al. 2016; Bright et al. 2016). These less acidic lysosomes are preferentially distributed closer to the cell periphery in cultured mammalian cells (Johnson et al. 2016) (Fig. 6.2). The subcellular localization of lysosomes is determined by the balance between the small GTPases Rab7 and Arl8b, which interact with kinesin and dynein microtubule motors via different effectors (Jordens et al. 2001; Rosa-Ferreira and Munro 2011; Pu et al. 2015; Guardia et al. 2016; Fujiwara et al. 2016), as well as an ER-located ubiquitin ligase system that contributes to their immobilisation in the perinuclear region (Jongsma et al. 2016). Over-expression of the GTPases and/or their effectors can alter lysosome distribution within the cell, as can the use of motor inhibitors. It was found that if cells were experimentally manipulated to drive more lysosomes to the periphery, there was a reduction in their acidity (Johnson et al.

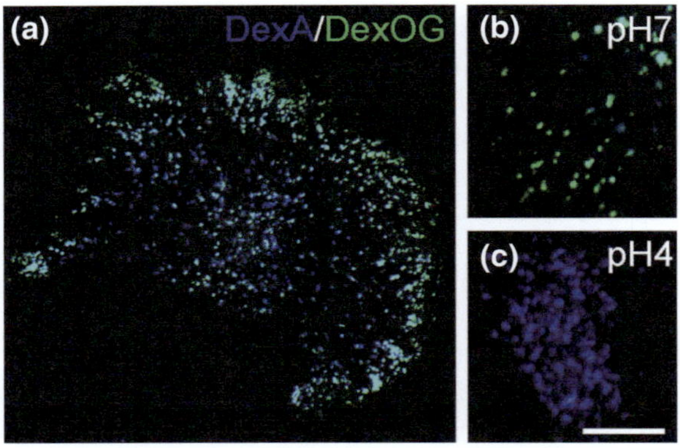

Fig. 6.2 Ratiometric imaging of lysosomal pH in NRK cells. NRK (normal rat kidney) cells were incubated in tissue culture medium containing both 0.5 mg/ml Dextran-Oregon Green 488 as the acid-sensitive fluorochrome and 0.5 mg/ml Dextran-Alexa 647 as the acid-insensitive fluorochrome for 4 h followed by a 20-h chase in conjugate-free medium to load terminal endocytic compartments. Dual-fluorochrome ratiometric confocal microscopy showed that acidic organelles (blue) were more centrally located, whereas less acidic/more neutral organelles (cyan and green) were concentrated towards the cell periphery in a representative cell (**a**). To construct a pH calibration curve, cells were clamped at pH 4–7 in buffer containing 10 μM nigericin and 10 μM monensin for 5 min and representative confocal images are shown from a cell clamped at pH 7 (**b**) or pH 4 (**c**) using identical imaging parameters used to collect the unclamped image. Scale bar = 10 μm

2016). Conversely, if lysosomes were driven towards the microtubule organizing centre/nucleus, there was an increase in acidity. In a separate study, it was concluded that the more acidic lysosomes were in fact endolysosomes, acting as the principal sites of proteolytic degradation, with the less acidic lysosomes being in the latter stages of lysosome re-formation, such that the least acidic form a pool of terminal storage lysosomes (Bright et al. 2016). Although one of these studies concluded that lysosome position within the cell determines acidity and the other that it is the stage of the lysosome fusion/regeneration cycle, which is important, the conclusions are not incompatible. At present, little is known about the role, if any, of Rab7, Arl8b, their effectors and microtubule motor proteins in lysosome re-formation from endolysosomes. The reduced acidity of some lysosomes appears to be due to an increased passive (leak) permeability to protons together with reduced V-ATPase activity (Johnson et al. 2016). This could be due to recycling of the intact V-ATPase, as has been suggested for the generation of neural post-lysosome compartments in *Dictyostelium discoideum* (Carnell et al. 2011), or the separation of V_0 and V_1 sub-complexes of the V-ATPase, as is observed under nutritional control in yeast (Parra and Kane 1998). Interestingly, in yeast, the late endosomal/vacuolar phosphoinositide PI(3,5)P2 has been shown to stabilize the V_0/V_1 interaction in yeast (Li et al. 2014). There may also be regulation of V-ATPase activity as a result of protein binding, one candidate being the Rab7 effector RILP (Rab-interacting lysosomal protein) (De Luca et al. 2014). In this context, it is interesting to note that whereas acidic juxtanuclear lysosomes/endolysosomes are Rab7-positive, less acidic and more peripheral lysosomes are Rab7-negative (Johnson et al. 2016; Bright et al. 2016). Another protein observed to bind V-ATPase and potentially regulate activity is the interferon-induced protein IFITM3 (interferon-inducible transmembrane protein 3), which may be particularly important in regulating lysosomal acidity and function following viral infection (Wee et al. 2012).

6.3 Lysosomal Signalling and Cell Growth Control

The lysosome acts as an important site of nutrient sensing and metabolic regulation, because it is on the limiting membrane of the lysosome that the kinase activity of mTORC1 is regulated in a nutrient-, energy-, stress- and growth factor-regulated manner to regulate the balance between anabolic and catabolic pathways within the cell [reviewed in (Zoncu et al. 2011b; Lim and Zoncu 2016; Perera and Zoncu 2016; Carroll and Dunlop 2017)]. mTORC1 is a rapamycin-sensitive multi-subunit protein complex that contains the PI3K-related serine/threonine protein kinase mTOR. mTOR also exists in an alternative rapamycin-insensitive protein complex called mTORC2 that phosphorylates the serine/threonine kinase AKT (also known as protein kinase B) in response to growth factor stimulation and participates in modulating many cellular functions including cell survival and actin dynamics (Laplante and Sabatini 2012). Whereas mTORC1 activation is associated primarily with the lysosome membrane, mTORC2 function has been localized to the plasma membrane (Ebner et al. 2017). The convergence of several growth factor-initiated signalling pathways on mTORC1 enables it to participate in many developmental and physiological processes, and it is essential for early embryonic development. Some of these signalling pathways (e.g. insulin stimulation or low energy levels), result in the phosphorylation of a large protein complex, TSC (tuberous sclerosis complex), which regulates the nucleotide binding state of the small GTPase RHEB on the cytosolic surface of lysosomes and controls the kinase activation state of mTORC1 (Sengupta et al. 2010). However, independently of TSC, mTORC1 recruitment to the lysosome cytosolic surface is controlled by nutrient levels (most strongly by amino acids), through regulation of the nucleotide binding state and heterodimerization of the Rag family of small GTPases. Thus, it is the lysosome-limiting membrane where diverse signal inputs are sensed and integrated to control both the localization of mTORC1 and subsequent activation of mTOR kinase activity to exert global effects on cell growth and metabolism. A consequence of the activation of mTORC1 is the phosphorylation of a vast array of substrates that control growth and metabolism. Well-studied targets include the ribosomal protein S6 kinase and eIF4E-binding protein 1, which are proteins that control distinct aspects of mRNA translation and thus control the rate of protein synthesis (Sonenberg and Hinnebusch 2009). Because of its ability to sense nutrient levels, mTORC1 controls the activation state of non-selective macroautophagy through phosphorylation of the autophagy-initiating kinase ULK1 (Kim et al. 2011). When nutrient supply is low, mTORC1 is turned off, thus relieving its inhibitory effect on autophagy and this liberates cellular sources of nutrients through bulk degradation of macromolecules. mTORC1 also controls the activity of several transcription factors including some that are implicated in lipid synthesis and mitochondrial metabolism as well as TFEB and TFE3, members of the MiTF/TFE transcription factor family. TFEB and TFE3 are phosphorylated by mTORC1 at the lysosomal surface and regulate lysosome biogenesis, lysosome secretion and autophagy (Raben and Puertollano 2016; Napolitano and Ballabio 2016).

6.3.1 Lysosomal Localization and Activation of mTORC1

Two protein complexes, an obligate heterodimer of Rag GTPases and a multimeric complex called Ragulator, function together to localize mTORC1 to the lysosomal membrane. The Rag heterodimers comprise RagA or B with RagC or D. When RagA or B is bound to GTP, RagC or D is loaded with GDP. Ragulator is composed of five proteins, LAMTOR 1–5 (lysosomal adaptor and mTOR regulator 1–5), and acts both as a scaffold anchoring the Rag GTPases to the lysosomal membrane and as a guanine nucleotide exchange factor for RagA/B (Bar-Peled et al. 2012). Thus, in the presence of amino acids, Ragulator promotes the loading of GTP onto RagA/B, which dimerizes with GDP-loaded Rag C/D. It is likely that it is the GDP-loaded Rag C/D, which directly recruits mTORC1 to the lysosomal membrane (Tsun et al. 2013). It is proposed that recruitment of mTORC1 to the cytosolic surface of the lysosome brings it into close physical proximity to the small GTPase RHEB, which directly activates mTOR kinase activity. The nucleotide loading state of the Rag heterodimers is also regulated at the lysosomal membrane by a protein called GTPase-activating protein towards Rags 1 (GATOR1) (Bar-Peled et al. 2013), which promotes the inactive GDP-bound state of RagA/B. The activity of GATOR1 is under inhibitory control by a second complex called GATOR2, which itself is controlled by additional proteins including Sestrin2 homodimers and the CASTOR complex (dimer of GATSL1/2). Recently, it has been shown that GATOR1 complex recruitment to the lysosome surface is controlled by a complex termed KICSTOR (Peng et al. 2017; Wolfson et al. 2017). The nucleotide binding state of RagC/D has been shown to be regulated by the tumour suppressor protein folliculin, which forms a complex with FNIP1/2 (folliculin-interacting proteins 1/2). Folliculin is proposed to function as a GTPase-activating protein for Rag C/D (Petit et al. 2013; Tsun et al. 2013), although the mechanism of GTPase activation remains to be elucidated.

A long-standing question is what are the precise mechanisms by which amino acids are sensed to control mTORC1 activation on the lysosome surface? Changes in amino acid concentration, in both the lumen of the lysosome and the cytosol, can regulate the activity of mTORC1, although the signalling pathways are complex and far from completely understood (Manifava et al. 2016). The signalling pathway employed by luminal amino acids to stimulate Ragulator GEF activity towards RagA/B on the cytosolic side of the lysosome's limiting membrane requires both the V-ATPase and the sodium-coupled, amino acid-transporting, 11-pass transmembrane protein SLC38A9. The V-ATPase binds Ragulator and the Rag GTPases in an amino acid-dependent manner, and experimental inhibition of the V-ATPase prevents recruitment of mTORC1 to the lysosome in response to amino acids (Zoncu et al. 2011a). SLC38A9 may play a particularly important role in sensing arginine levels within the lysosome lumen (Jung et al. 2015; Rebsamen et al. 2015; Wang et al. 2015). It binds to the V-ATPase via its C-terminal transmembrane region and to Ragulator and the Rag GTPases via its N-terminal cytosolic domain and may act primarily as a transceptor rather than purely as a transporter. Other lysosomal amino acid transporters have also been implicated in mTORC1 activation [reviewed in (Lim and Zoncu 2016)]. SLC38A9 also enables mTORC1 activation by cholesterol, independently of its arginine-sensing function (Castellano et al. 2017).

A rise in cytosolic amino acid concentration results in Rag complex activation, in part through a reduction in RagA/B inhibition by the negative regulator GATOR1. A rise in the concentration of specific amino acids is detected by the cytosolic leucine sensor Sestrin2 and the cytosolic arginine sensor CASTOR, which become inhibited when bound to their respective amino acid ligands (Chantranupong et al. 2016; Wolfson et al. 2016). When not bound to amino acids, both Sestrin2 and CASTOR inhibit GATOR2 (the GATOR1 inhibitory complex described above), resulting in Rag complex inactivation. Thus, when Sestrin2 and CASTOR are bound by amino acids, the net effect is that GATOR2 is able to inhibit the GATOR1-dependent inhibition of RagA/B. The entirety of the lysosome membrane protein machinery responding to amino acids is shown in Fig. 6.3 and includes both integral membrane proteins, the Ragulator/Rag/mTORC1 complexes and associated proteins on the cytosolic surface of the lysosome. It has been suggested that this entire machinery be referred to as the LYNUS (lysosome nutrient sensing) machinery (Settembre et al. 2013b). One interesting consequence of the critical role played by the lysosomal V-ATPase in amino acid signalling to mTORC1 is that it raises the question of whether juxtanuclear and peripheral lysosomes and/or organelles at different stages of the lysosome regeneration cycle play different roles in this signalling, given their difference in V-ATPase activity and luminal pH (Gowrishankar and Ferguson 2016). There is evidence that intracellular lysosome positioning coordinates metabolic responses to nutrient availability with plasma membrane signalling events (Korolchuk et al. 2011).

The regulation of mTORC1 activity by insulin and growth factors is mediated via phosphorylation of the TSC complex as mentioned above. The TSC complex is a heterotrimer of the tuberous sclerosis complex tumour suppressor genes TSC1 and TSC2 with TBC1D7 (Dibble et al. 2012). TSC functions as an inhibitor of mTORC1 by acting as a GAP (GTPase-activating protein) towards the mTOR kinase activator RHEB. Insulin signalling results in release of TSC from the lysosomal surface following its phosphorylation by the protein kinase AKT (Menon et al. 2014), which is dependent on AKT recruitment to the membrane by interaction with phosphoinositides (Ebner et al. 2017). TSC activity can also be enhanced through phosphorylation via the LKB1/AMPK (liver kinase B1/AMP-activated protein kinase) signalling axis in response to energy stress (Inoki et al. 2003), the activity of which has also been localized to the lysosome surface (Zhang et al. 2014).

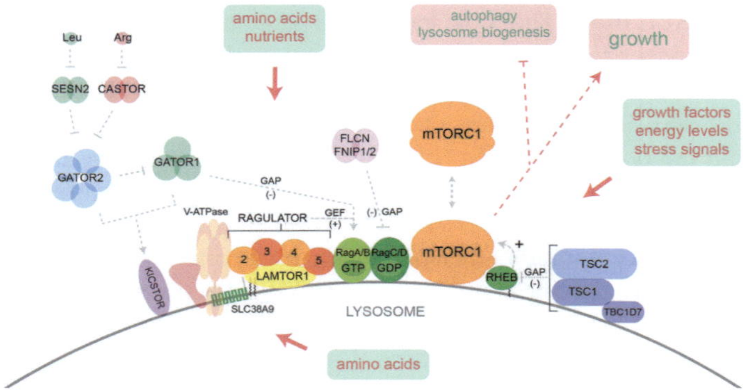

Fig. 6.3 Regulation of mTORC1 activation on the lysosome surface. The mTORC1 complex is recruited to the lysosome surface through direct interaction with the RagA/B(GTP)-RagC/D(GDP) (active state) heterodimer. The Rag heterodimer nucleotide loading state is under complex regulatory control and responds to nutrient levels (especially amino acids). The Ragulator complex acts as a GEF towards RagA/B (thus promoting the active Rag dimer) and functions in concert with SLC38A9 and the V-ATPase in response to amino acid levels in the lysosome lumen. The GATOR1 complex acts as a GAP towards RagA/B (thus promoting the inactive Rag dimer) and is under regulatory control by the GATOR2 complex. GATOR2 inhibition of GATOR1 is regulated by Sestrin2 (SESN2) and CASTOR complexes binding to individual amino acids (Leu and Arg, respectively). The GATOR complexes are recruited to the lysosome surface by the KICSTOR complex. Folliculin (FLCN) and its interacting proteins FNIP1/2 are proposed to act as a GAP towards RagC/D (thus promoting the active Rag dimer). At the lysosome surface, the activation state of the mTOR kinase is regulated by the small GTPase RHEB. The TSC complex (TSC1/2, TBC1D7) acts as a GAP towards RHEB, and is regulated by signals downstream of growth factor receptors, energy levels and cellular stress pathways. Collectively, the integrated regulation of mTORC1 activity on the lysosome surface by nutrients, energy levels, stress and growth signalling pathways exerts global control over the balance between catabolic (e.g. autophagy, lysosome biogenesis) and anabolic cell growth pathways.

Although the best-studied signalling inputs to TSC are those downstream of growth factor receptor activation and energy stress, it has been shown that TSC recruitment to the lysosome surface is also controlled by amino acid availability and may in fact be a general consequence of cellular stress (Demetriades et al. 2014, 2016). Thus, it is spatial control of TSC that contributes to the integration of growth factor signalling, nutrient regulation, and stress signalling and their role in mTORC1 activation on the lysosome surface.

6.3.2 Lysosomal Signalling to the Nucleus: TFEB and the CLEAR Network

In addition to recruiting mTORC1 to the lysosome membrane, active Rag GTPases also promote recruitment of the transcription factor TFEB in an amino acid-dependent manner (Martina and Puertollano 2013). In fully fed cells, TFEB continuously cycles between lysosomes and the cytosol, such that when associated with the lysosome it can be phosphorylated by mTORC1 at several sites, including residue S211 (Settembre et al. 2012; Vega-Rubin-de-Celis et al. 2017). Phosphorylation at S211 promotes interaction with the cytosolic chaperone 14-3-3, resulting in a steady state in which the majority of TFEB is in the cytosol (Roczniak-Ferguson et al. 2012). Nutrient withdrawal or other treatments leading to lysosomal stress, e.g. V-ATPase inhibition, lead to inactivation of mTORC1, since it is released from the lysosomal surface, thus reducing phosphorylation of TFEB. Nutrient withdrawal/lysosomal stress also causes the dephosphorylation of TFEB by the calcium-dependent phosphatase calcineurin, dissociation from 14-3-3 and transport into the nucleus (Fig. 6.4). Activation of calcineurin results from the release of luminal Ca^{2+} from the lysosome via mucolipin1 (Medina et al. 2015). TFEB is a basic helix–loop–helix transcription factor which binds to a palindromic 10-bp (base pair) nucleotide motif, GTCACGTGAC, present (often in multiple copies) in the promoter region of many genes encoding lysosomal enzymes. The palindromic nucleotide motif has been named the CLEAR (coordinated lysosomal expression and regulation) element (Sardiello et al. 2009) and the extensive number of genes affected, the CLEAR network (Palmieri et al. 2011). This network provides a system that regulates the expression, delivery to lysosomes and activity of lysosomal enzymes, which control the degradation of proteins, glycosaminoglycans, sphingolipids and glycogen. The CLEAR network is involved in the regulation of autophagy, exo- and endocytosis, phagocytosis and the immune response, as well as regulating some non-lysosomal enzymes/proteins involved in protein degradation and lipid metabolism (Palmieri et al. 2011; Settembre et al. 2013a). Other members of the MiTF/TFE transcription factor family, in particular TFE3 which also binds CLEAR elements, are regulated in a very similar way to TFEB [reviewed in (Raben and Puertollano 2016; Napolitano and Ballabio 2016)]. TFEB and TFE3 are partially redundant in their ability to induce lysosome biogenesis in response to starvation and both are necessary for a maximal response. However, overall MiTF/TFE transcription factors appear to have limited redundancy and some specific functions. Their ability to heterodimerize with each other has been a complication in studies of their function.

It should be noted that lysosome biogenesis is also affected by mTORC1-independent mechanisms. Thus, protein kinase C couples activation of TFEB with inactivation of the transcriptional repressor ZKSCAN3 via parallel signalling cascades (Li et al. 2016) and an mTORC1-independent pathway mediated via protein kinase RNA-like endoplasmic reticulum kinase (PERK) has been shown to regulate TFEB/TFE3 translocation to the nucleus in response to ER stress (Martina et al. 2016). Recently, it has been demonstrated that AKT modulates TFEB activity by phosphorylation at S467 and that trehalose, an mTOR-independent autophagy enhancer, promotes nuclear translocation of TFEB by inhibiting AKT (Palmieri et al. 2017). These observations are especially interesting because they have suggested that AKT control of TFEB activity may be a useful mTORC1-independent target for pharmacological treatment of neurodegenerative lysosomal storage diseases to stimulate cellular clearance of the storage material.

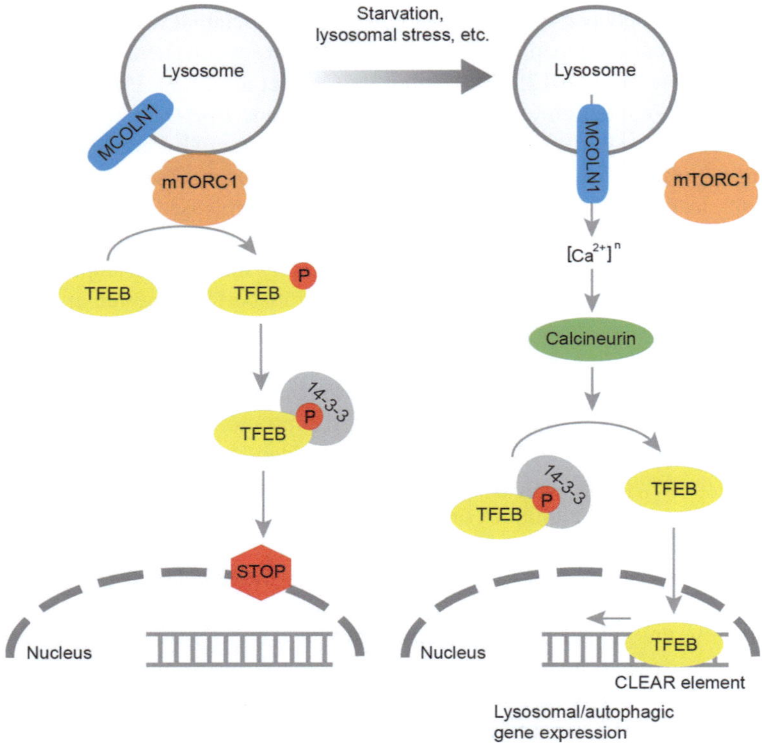

Fig. 6.4 Regulation of TFEB by lysosomal Ca^{2+}. In fully fed cells, TFEB is phosphorylated by mTORC1 on the lysosomal membrane. Binding of phosphorylated TFEB by 14-3-3 in the cytosol prevents its entry into the nucleus (left). Upon starvation or lysosomal stress, mucolipin1 (MCOLN1) releases lysosomal Ca^{2+}, which leads to local calcineurin activation in the cytoplasm and TFEB dephosphorylation. Dephosphorylated TFEB cannot be bound by 14-3-3 proteins and translocates into the nucleus where it binds to CLEAR elements on the DNA and activates the transcription of lysosomal/autophagic genes (right).

6.4 Lysosomal Ca^{2+} Signalling

Mammalian, acidic lysosomes/endolysosomes contain a significant store of intracellular free Ca^{2+}, measured as being ~ 0.5 mM (Christensen et al. 2002; Lloyd-Evans et al. 2008; Ronco et al. 2015). This is within the range of estimates of the steady-state luminal free concentration in the ER and is > 3 orders of magnitude higher than the cytosolic Ca^{2+} concentration. Thus, release of Ca^{2+} through any of the identified lysosomal channels has the potential to affect a range of cytosolic functions. Regulated release of lysosomal Ca^{2+} is implicated in fusion/fission events and the activation of calcineurin to dephosphorylate TFEB and upregulate genes with CLEAR elements as described above. However, the effect of releasing lysosomal Ca^{2+} on cytosolic

Ca^{2+} concentration can be amplified by stimulation of ER Ca^{2+} release, facilitating its involvement in a range of other cellular processes including muscle contraction, neurite extension and differentiation [reviewed in (Morgan et al. 2011; Penny et al. 2015)]. Ca^{2+} release through two-pore channels in the endolysosomal system has also been implicated in metastasis (Nguyen et al. 2017) and in Ebola virus entry into host cells (Sakurai et al. 2015). Defects in lysosomal Ca^{2+} signalling and homeostasis have been suggested to play a role in lysosomal storage disease pathogenesis (Lloyd-Evans and Platt 2011). One of the most significant developments in understanding a role for lysosomes in intracellular signalling came from the discovery that release of Ca^{2+} from acidic LROs in sea urchin eggs is stimulated by the pyridine nucleotide metabolite NAADP (Clapper et al. 1987). Whilst the physiological production and degradation of NAADP are not fully understood, it clearly functions as an intracellular second messenger in mammalian cells (Yamasaki et al. 2005), not just in sea urchin eggs, and a major intracellular target activated by NAADP is the lysosomal two-pore channel TPC2 (Pitt et al. 2010). The regulation of the release of Ca^{2+} through TPC2 is also affected by lysosomal Ca^{2+} concentration and lysosomal pH. As discussed above, the acidic lysosomal pH is generated through the activity of the lysosomal V-ATPase, with charge compensation provided via unspecified cation channels, the lysosomal Cl^-/H^+ antiporter ClC-7/Ostm1 and/or alternative counter-ion pathways (Steinberg et al. 2010). In some cell types, lysosomal pH can be regulated by signalling pathways affecting V-ATPase trafficking or charge compensation, e.g. pathways involving a cell surface G protein-coupled receptor, cyclic AMP and protein kinase A (Lassen et al. 2016; Folts et al. 2016), thus potentially also affecting lysosomal Ca^{2+} release. Re-filling of lysosomal Ca^{2+} stores may also play a role in signalling. The lysosomal Ca^{2+}/H^+ exchanger CAX has been shown to play a role in cell migration during frog development, but does not appear to have an ortholog in placental mammals (Melchionda et al. 2016). In mammalian cells, the ER is the primary source of Ca^{2+} for the lysosome (Garrity et al. 2016) and it has been proposed that selective accumulation of Ca^{2+} released from the ER may allow lysosomes to play a role in shaping cytosolic Ca^{2+} signals caused by release of ER Ca^{2+} (Lopez-Sanjurjo et al. 2013). The functional relationship(s) between lysosomal and ER Ca^{2+} stores are likely affected by the close physical proximity of these organelles and the formation of ER-lysosome contact sites (Penny et al. 2015; Lopez-Sanjurjo et al. 2013; Ronco et al. 2015; Sbano et al. 2017). Membrane contact sites (MCS) between intracellular organelles, especially those involving the ER, are currently the subject of much investigation [reviewed in (Gatta and Levine 2016; Zhang and Hu 2016; Hariri et al. 2016; Raffaello et al. 2016)], since they enable non-vesicular communication, for example for the transfer of cholesterol between endolysosomes and the ER (Du et al. 2011), as well as marking sites of organelle fission of both mitochondria (Friedman et al. 2011) and endosomes (Rowland et al. 2014; Allison et al. 2017) and regulating the final steps of autophagy (Wijdeven et al. 2016). In the context of lysosomal signalling, one especially interesting observation was the induction of NAADP-dependent microdomains of high Ca^{2+} concentration between lysosomes and the sarcoplasmic reticulum in response to beta-adrenoceptor activation in cardiac myocytes (Capel et al. 2015).

6.5 Other Lysosomal Signalling Pathways

6.5.1 Toll-like Receptors

Lysosomes play an important role in the innate immune system, through the function of an intracellular subgroup of the toll-like receptor (TLR) family of pattern recognition receptors, comprising TLR3,7,8 and 9 [reviewed in (Kawai and Akira 2010; Majer et al. 2016)]. These type I integral membrane proteins are widely expressed, require proteolytic cleavage to become functional receptors and bind nucleic acids. This results in the activation of a signalling cascade from their cytosolic domains. TLRs play an important role in initiating and enhancing adaptive immune responses to invading pathogens. TLR9 is one of the best-studied TLRs, including the regulation of its trafficking from the ER to lysosomes, and requires cleavage in an acidic environment by endolysosomal proteases to become functional. TLR9 binds unmethylated CpG motifs in DNA and in addition to its role in innate immunity, responding to bacterial or viral infection, it has also been implicated in regulation of the autophagic pathway, as described above, through recognition of mitochondrial DNA.

6.5.2 Regulation of Lysosome Membrane Permeability

An aspect of lysosomal signalling that merits wider consideration is partial and selective permeabilization of the limiting membrane which can trigger cell death as a consequence of cathepsin release into the cytosol [reviewed in (Serrano-Puebla and Boya 2016)]. One relatively well-studied instance is the post-lactational involution of the mammary gland caused by lysosome-mediated, non-apoptotic, programmed cell death. The key trigger of this process is milk itself, with a Stat3-dependent pathway involving increased phagocytic uptake of milk fat globules by mammary epithelial cells, resulting in degradation of milk triglycerides and the generation of an increased concentration of oleic acid that disrupts the lysosomal membrane enabling cathepsin release (Sargeant et al. 2014). Regulation of lysosomal membrane permeabilization likely plays a wider role in health and disease although the complexities and molecular mechanisms are poorly understood (Serrano-Puebla and Boya 2016). In cancer, it has been suggested that minor lysosomal leakage may not necessarily be lethal and that release of lysosomal cathepsins may be anti-apoptotic (Pislar et al. 2015). In addition, what has been described as lysosome hyperactivity in some cancer cells can result in increased lysosomal membrane vulnerability—a frailty that might be exploited therapeutically by drugs that can induce lysosomal damage preferentially in cancer cells [reviewed in (Hamalisto and Jaattela 2016)].

6.5.3 Lysosomal Signalling, Ageing and Longevity

Lysosomal dysfunction has long been associated with cellular ageing and reduced longevity in animals [reviewed in (Carmona-Gutierrez et al. 2016; Colacurcio and Nixon 2016)]. Whilst much of this is associated with alterations in the degradative and signalling functions discussed above, it has been suggested that an additional signalling pathway between lysosomes and the nucleus may play a role. In *Caenorhabditis elegans,* increased lifespan occurs as a consequence of lysosomal production of the bioactive lipid oleoylethanolamide, which is translocated into the nucleus by a chaperone protein and affects the transcription of genes that regulate longevity. Interestingly, the components of this signalling pathway are conserved in mammals (Folick et al. 2015).

6.6 Conclusions

Whilst the lysosome was for a long time simply regarded as the terminal degradative compartment of the cell's endocytic and autophagic pathways, it is clear that it is in fact a multifunctional signalling hub. The cell's lysosomal compartment is functionally heterogeneous and includes endolysosomes, autolysosomes, storage lysosomes and organelles at different stages of the lysosome fusion/regeneration cycle. It is in constant dynamic exchange with endosomal and autophagosomal compartments, links nutrient status to gene transcription, integrates hormonal and nutrient signalling, signals to other intracellular organelles, ensures plasma membrane integrity and plays a role in regulating cell death and in the ageing and longevity of both individual cells and the metazoan organism. There are many remaining important questions to address concerning the lysosomal compartment and signalling. A non-exhaustive list of such questions, where we currently have at best only partial answers, includes how the various fusion/fission events undertaken by organelles in the lysosomal compartment are regulated and coordinated, how lysosomal acidity is regulated, how calcium accumulation/release in the compartment is regulated, whether there is additional molecular machinery to be discovered in the signalling pathways between the lysosome, the nucleus and other organelles and whether there is physiologically significant heterogeneity in the signalling capacity of endolysosomes/lysosomes based on intracellular positioning and/or stage in the lysosome fusion/regeneration cycle.

Acknowledgements GGH is supported by a Basic Research Fellowship from Parkinson Canada. Recent experimental work from JPL and NAB referred to in this chapter is supported by UK MRC research grant MR/M010007/1. LJD is supported by a BBSRC industrial CASE studentship with GSK Research and Development Ltd. The Cambridge Institute for Medical Research is supported by Wellcome Trust Strategic Award 100140.

References

Allison R, Edgar JR, Pearson G, Rizo T, Newton T, Gunther S, Berner F, Hague J, Connell JW, Winkler J, Lippincott-Schwartz J, Beetz C, Winner B, Reid E (2017) Defects in ER-endosome contacts impact lysosome function in hereditary spastic paraplegia. J Cell Biol 216(5):1337–1355. https://doi.org/10.1083/jcb.201609033

Alonso V, Friedman PA (2013) Minireview: ubiquitination-regulated G protein-coupled receptor signaling and trafficking. Mol Endocrinol 27(4):558–572. https://doi.org/10.1210/me.2012-1404 me.2012-1404 [pii]

Antonioli M, Di Rienzo M, Piacentini M, Fimia GM (2016) Emerging mechanisms in initiating and terminating autophagy. Trends Biochem Sci. https://doi.org/10.1016/j.tibs.2016.09.008

Bar-Peled L, Schweitzer LD, Zoncu R, Sabatini DM (2012) Ragulator is a GEF for the rag GTPases that signal amino acid levels to mTORC1. Cell 150(6):1196–1208. https://doi.org/10.1016/j.cell.2012.07.032

Bar-Peled L, Chantranupong L, Cherniack AD, Chen WW, Ottina KA, Grabiner BC, Spear ED, Carter SL, Meyerson M, Sabatini DM (2013) A Tumor suppressor complex with GAP activity for the Rag GTPases that signal amino acid sufficiency to mTORC1. Science 340(6136):1100–1106. https://doi.org/10.1126/science.1232044

Bento CF, Renna M, Ghislat G, Puri C, Ashkenazi A, Vicinanza M, Menzies FM, Rubinsztein DC (2016) Mammalian autophagy: how does it work? Annu Rev Biochem 85:685–713. https://doi.org/10.1146/annurev-biochem-060815-014556

Bowman SL, Shiwarski DJ, Puthenveedu MA (2016) Distinct G protein-coupled receptor recycling pathways allow spatial control of downstream G protein signaling. J Cell Biol 214(7):797–806. https://doi.org/10.1083/jcb.201512068

Bright NA, Reaves BJ, Mullock BM, Luzio JP (1997) Dense core lysosomes can fuse with late endosomes and are re-formed from the resultant hybrid organelles. J Cell Sci 110(Pt 17):2027–2040

Bright NA, Gratian MJ, Luzio JP (2005) Endocytic delivery to lysosomes mediated by concurrent fusion and kissing events in living cells. Curr Biol 15(4):360–365. https://doi.org/10.1016/j.cub.2005.01.049 S0960982205001053 [pii]

Bright NA, Davis LJ, Luzio JP (2016) Endolysosomes are the principal intracellular sites of acid hydrolase activity. Curr Biol 26(17):2233–2245. https://doi.org/10.1016/j.cub.2016.06.046 S0960-9822(16)30687-X [pii]

Butor C, Griffiths G, Aronson NN Jr, Varki A (1995) Co-localization of hydrolytic enzymes with widely disparate pH optima: implications for the regulation of lysosomal pH. J Cell Sci 108(Pt 6):2213–2219

Cao Q, Zhong XZ, Zou Y, Murrell-Lagnado R, Zhu MX, Dong XP (2015) Calcium release through P2X4 activates calmodulin to promote endolysosomal membrane fusion. J Cell Biol 209(6):879–894. https://doi.org/10.1083/jcb.201409071

Capel RA, Bolton EL, Lin WK, Aston D, Wang Y, Liu W, Wang X, Burton RA, Bloor-Young D, Shade KT, Ruas M, Parrington J, Churchill GC, Lei M, Galione A, Terrar DA (2015) Two-pore channels (TPC2 s) and nicotinic acid adenine dinucleotide phosphate (NAADP) at lysosomal-sarcoplasmic reticular junctions contribute to acute and chronic beta-adrenoceptor signaling in the heart. J Biol Chem 290(50):30087–30098. https://doi.org/10.1074/jbc.M115.684076

Carmona-Gutierrez D, Hughes AL, Madeo F, Ruckenstuhl C (2016) The crucial impact of lysosomes in aging and longevity. Ageing Res Rev 32:2–12. https://doi.org/10.1016/j.arr.2016.04.009

Carnell M, Zech T, Calaminus SD, Ura S, Hagedorn M, Johnston SA, May RC, Soldati T, Machesky LM, Insall RH (2011) Actin polymerization driven by WASH causes V-ATPase retrieval and vesicle neutralization before exocytosis. J Cell Biol 193(5):831–839. https://doi.org/10.1083/jcb.201009119

Carroll B, Dunlop EA (2017) The lysosome: a crucial hub for AMPK and mTORC1 signalling. Biochem J 474(9):1453–1466. https://doi.org/10.1042/BCJ20160780

Castellano BM, Thelen AM, Moldavski O, Feltes M, van der Welle RE, Mydock-McGrane L, Jiang X, van Eijkeren RJ, Davis OB, Louie SM, Perera RM, Covey DF, Nomura DK, Ory DS, Zoncu R

(2017) Lysosomal cholesterol activates mTORC1 via an SLC38A9-Niemann-Pick C1 signaling complex. Science 355(6331):1306–1311. https://doi.org/10.1126/science.aag1417
Castro-Gomes T, Corrotte M, Tam C, Andrews NW (2016) Plasma membrane repair is regulated extracellularly by proteases released from lysosomes. PLoS ONE 11(3):e0152583. https://doi.org/10.1371/journal.pone.0152583
Chantranupong L, Scaria SM, Saxton RA, Gygi MP, Shen K, Wyant GA, Wang T, Harper JW, Gygi SP, Sabatini DM (2016) The CASTOR proteins are arginine sensors for the mTORC1 pathway. Cell 165(1):153–164. https://doi.org/10.1016/j.cell.2016.02.035
Chou HT, Dukovski D, Chambers MG, Reinisch KM, Walz T (2016) CATCHR, HOPS and CORVET tethering complexes share a similar architecture. Nat Struct Mol Biol 23(8):761–763. https://doi.org/10.1038/nsmb.3264 nsmb.3264 [pii]
Christensen KA, Myers JT, Swanson JA (2002) pH-dependent regulation of lysosomal calcium in macrophages. J Cell Sci 115(Pt 3):599–607
Clapper DL, Walseth TF, Dargie PJ, Lee HC (1987) Pyridine nucleotide metabolites stimulate calcium release from sea urchin egg microsomes desensitized to inositol trisphosphate. J Biol Chem 262(20):9561–9568
Colacurcio DJ, Nixon RA (2016) Disorders of lysosomal acidification-The emerging role of v-ATPase in aging and neurodegenerative disease. Ageing Res Rev 32:75–88. https://doi.org/10.1016/j.arr.2016.05.004
Coonrod EM, Graham LA, Carpp LN, Carr TM, Stirrat L, Bowers K, Bryant NJ, Stevens TH (2013) Homotypic vacuole fusion in yeast requires organelle acidification and not the V-ATPase membrane domain. Dev Cell 27(4):462–468. https://doi.org/10.1016/j.devcel.2013.10.014
de Duve C (2005) The lysosome turns fifty. Nat Cell Biol 7(9):847–849. https://doi.org/10.1038/ncb0905-847 ncb0905-847 [pii]
De Leo MG, Staiano L, Vicinanza M, Luciani A, Carissimo A, Mutarelli M, Di Campli A, Polishchuk E, Di Tullio G, Morra V, Levtchenko E, Oltrabella F, Starborg T, Santoro M, di Bernardo D, Devuyst O, Lowe M, Medina DL, Ballabio A, De Matteis MA (2016) Autophagosome-lysosome fusion triggers a lysosomal response mediated by TLR9 and controlled by OCRL. Nat Cell Biol 18(8):839–850. https://doi.org/10.1038/ncb3386
De Luca M, Cogli L, Progida C, Nisi V, Pascolutti R, Sigismund S, Di Fiore PP, Bucci C (2014) RILP regulates vacuolar ATPase through interaction with the V1G1 subunit. J Cell Sci 127(Pt 12):2697–2708. https://doi.org/10.1242/jcs.142604
Demetriades C, Doumpas N, Teleman AA (2014) Regulation of TORC1 in response to amino acid starvation via lysosomal recruitment of TSC2. Cell 156(4):786–799. https://doi.org/10.1016/j.cell.2014.01.024
Demetriades C, Plescher M, Teleman AA (2016) Lysosomal recruitment of TSC2 is a universal response to cellular stress. Nat Commun 7:10662. https://doi.org/10.1038/ncomms10662
Desfougeres Y, Vavassori S, Rompf M, Gerasimaite R, Mayer A (2016) Organelle acidification negatively regulates vacuole membrane fusion in vivo. Sci Rep 6:29045. https://doi.org/10.1038/srep29045
Diao J, Liu R, Rong Y, Zhao M, Zhang J, Lai Y, Zhou Q, Wilz LM, Li J, Vivona S, Pfuetzner RA, Brunger AT, Zhong Q (2015) ATG14 promotes membrane tethering and fusion of autophagosomes to endolysosomes. Nature 520(7548):563–566. https://doi.org/10.1038/nature14147
Dibble CC, Elis W, Menon S, Qin W, Klekota J, Asara JM, Finan PM, Kwiatkowski DJ, Murphy LO, Manning BD (2012) TBC1D7 is a third subunit of the TSC1-TSC2 complex upstream of mTORC1. Mol Cell 47(4):535–546. https://doi.org/10.1016/j.molcel.2012.06.009
Dong XP, Shen D, Wang X, Dawson T, Li X, Zhang Q, Cheng X, Zhang Y, Weisman LS, Delling M, Xu H (2010) PI(3,5)P(2) controls membrane trafficking by direct activation of mucolipin Ca^{2+} release channels in the endolysosome. Nat Commun 1:38. https://doi.org/10.1038/ncomms1037
Dove SK, Dong K, Kobayashi T, Williams FK, Michell RH (2009) Phosphatidylinositol 3,5-bisphosphate and Fab1p/PIKfyve underPPIn endo-lysosome function. Biochem J 419(1):1–13. https://doi.org/10.1042/BJ20081950

Du X, Kumar J, Ferguson C, Schulz TA, Ong YS, Hong W, Prinz WA, Parton RG, Brown AJ, Yang H (2011) A role for oxysterol-binding protein-related protein 5 in endosomal cholesterol trafficking. J Cell Biol 192(1):121–135. https://doi.org/10.1083/jcb.201004142
Du W, Su QP, Chen Y, Zhu Y, Jiang D, Rong Y, Zhang S, Zhang Y, Ren H, Zhang C, Wang X, Gao N, Wang Y, Sun L, Sun Y, Yu L (2016) Kinesin 1 drives autolysosome tubulation. Dev Cell 37(4):326–336. https://doi.org/10.1016/j.devcel.2016.04.014
Ebner M, Lucic I, Leonard TA, Yudushkin I (2017) PI(3,4,5)P3 engagement restricts Akt activity to cellular membranes. Mol Cell 65(3):416–431(e416). https://doi.org/10.1016/j.molcel.2016.12.028
Edgar JR, Eden ER, Futter CE (2014) Hrs- and CD63-dependent competing mechanisms make different sized endosomal intraluminal vesicles. Traffic 15(2):197–211. https://doi.org/10.1111/tra.12139
Eltschinger S, Loewith R (2016) TOR complexes and the maintenance of cellular homeostasis. Trends Cell Biol 26(2):148–159. https://doi.org/10.1016/j.tcb.2015.10.003
Fader CM, Sanchez DG, Mestre MB, Colombo MI (2009) TI-VAMP/VAMP7 and VAMP3/cellubrevin: two v-SNARE proteins involved in specific steps of the autophagy/multivesicular body pathways. Biochim Biophys Acta 1793(12):1901–1916. https://doi.org/10.1016/j.bbamcr.2009.09.011 S0167-4889(09)00236-5 [pii]
Feng Y, He D, Yao Z, Klionsky DJ (2014) The machinery of macroautophagy. Cell Res 24(1):24–41. https://doi.org/10.1038/cr.2013.168 cr2013168 [pii]
Folick A, Oakley HD, Yu Y, Armstrong EH, Kumari M, Sanor L, Moore DD, Ortlund EA, Zechner R, Wang MC (2015) Aging. Lysosomal signaling molecules regulate longevity in *Caenorhabditis elegans*. Science 347(6217):83–86. https://doi.org/10.1126/science.1258857
Folts CJ, Scott-Hewitt N, Proschel C, Mayer-Proschel M, Noble M (2016) Lysosomal re-acidification prevents lysosphingolipid-induced lysosomal impairment and cellular toxicity. PLoS Biol 14(12):e1002583. https://doi.org/10.1371/journal.pbio.1002583
Fraldi A, Annunziata F, Lombardi A, Kaiser HJ, Medina DL, Spampanato C, Fedele AO, Polishchuk R, Sorrentino NC, Simons K, Ballabio A (2010) Lysosomal fusion and SNARE function are impaired by cholesterol accumulation in lysosomal storage disorders. EMBO J 29(21):3607–3620. https://doi.org/10.1038/emboj.2010.237
Friedman JR, Lackner LL, West M, DiBenedetto JR, Nunnari J, Voeltz GK (2011) ER tubules mark sites of mitochondrial division. Science 334(6054):358–362. https://doi.org/10.1126/science.1207385
Fujiwara T, Ye S, Castro-Gomes T, Winchell CG, Andrews NW, Voth DE, Varughese KI, Mackintosh SG, Feng Y, Pavlos N, Nakamura T, Manolagas SC, Zhao H (2016) PLEKHM1/DEF8/RAB7 complex regulates lysosome positioning and bone homeostasis. JCI Insight 1(17):e86330. https://doi.org/10.1172/jci.insight.86330
Garrity AG, Wang W, Collier CM, Levey SA, Gao Q, Xu H (2016) The endoplasmic reticulum, not the pH gradient, drives calcium refilling of lysosomes. Elife 5. https://doi.org/10.7554/elife.15887
Gatta AT, Levine TP (2016) Piecing Together the Patchwork of Contact Sites. Trends Cell Biol. https://doi.org/10.1016/j.tcb.2016.08.010
Goh LK, Sorkin A (2013) Endocytosis of receptor tyrosine kinases. Cold Spring Harb Perspect Biol 5(5):a017459. https://doi.org/10.1101/cshperspect.a017459 a017459, 5/5/a017459 [pii]
Goldman SD, Krise JP (2010) Niemann-Pick C1 functions independently of Niemann-Pick C2 in the initial stage of retrograde transport of membrane-impermeable lysosomal cargo. J Biol Chem 285(7):4983–4994. https://doi.org/10.1074/jbc.M109.037622
Gowrishankar S, Ferguson SM (2016) Lysosomes relax in the cellular suburbs. J Cell Biol 212(6):617–619. https://doi.org/10.1083/jcb.201602082
Grimm C, Holdt LM, Chen CC, Hassan S, Muller C, Jors S, Cuny H, Kissing S, Schroder B, Butz E, Northoff B, Castonguay J, Luber CA, Moser M, Spahn S, Lullmann-Rauch R, Fendel C, Klugbauer N, Griesbeck O, Haas A, Mann M, Bracher F, Teupser D, Saftig P, Biel M, Wahl-

Schott C (2014) High susceptibility to fatty liver disease in two-pore channel 2-deficient mice. Nat Commun 5:4699. https://doi.org/10.1038/ncomms5699

Guardia CM, Farias GG, Jia R, Pu J, Bonifacino JS (2016) BORC functions upstream of kinesins 1 and 3 to coordinate regional movement of lysosomes along different microtubule tracks. Cell Rep 17(8):1950–1961. https://doi.org/10.1016/j.celrep.2016.10.062

Guo B, Liang Q, Li L, Hu Z, Wu F, Zhang P, Ma Y, Zhao B, Kovacs AL, Zhang Z, Feng D, Chen S, Zhang H (2014) O-GlcNAc-modification of SNAP-29 regulates autophagosome maturation. Nat Cell Biol 16(12):1215–1226. https://doi.org/10.1038/ncb3066

Hamalisto S, Jaattela M (2016) Lysosomes in cancer-living on the edge (of the cell). Curr Opin Cell Biol 39:69–76. https://doi.org/10.1016/j.ceb.2016.02.009

Hammond GR, Takasuga S, Sasaki T, Balla T (2015) The ML1Nx2 phosphatidylinositol 3,5-bisphosphate probe shows poor selectivity in cells. PLoS ONE 10(10):e0139957. https://doi.org/10.1371/journal.pone.0139957

Hariri H, Ugrankar R, Liu Y, Henne WM (2016) Inter-organelle ER-endolysosomal contact sites in metabolism and disease across evolution. Commun Integr Biol 9(3):e1156278. https://doi.org/10.1080/19420889.2016.1156278

Hickey CM, Wickner W (2010) HOPS initiates vacuole docking by tethering membranes before trans-SNARE complex assembly. Mol Biol Cell 21(13):2297–2305. https://doi.org/10.1091/mbc.E10-01-0044 E10-01-0044 [pii]

Hirst J, Edgar JR, Esteves T, Darios F, Madeo M, Chang J, Roda RH, Durr A, Anheim M, Gellera C, Li J, Zuchner S, Mariotti C, Stevanin G, Blackstone C, Kruer MC, Robinson MS (2015) Loss of AP-5 results in accumulation of aberrant endolysosomes: defining a new type of lysosomal storage disease. Hum Mol Genet 24(17):4984–4996. https://doi.org/10.1093/hmg/ddv220

Holland P, Torgersen ML, Sandvig K, Simonsen A (2014) LYST affects lysosome size and quantity, but not trafficking or degradation through autophagy or endocytosis. Traffic 15(12):1390–1405. https://doi.org/10.1111/tra.12227

Huotari J, Helenius A (2011) Endosome maturation. EMBO J 30(17):3481–3500. https://doi.org/10.1038/emboj.2011.286 emboj2011286 [pii]

Inoki K, Zhu T, Guan KL (2003) TSC2 mediates cellular energy response to control cell growth and survival. Cell 115(5):577–590

Itakura E, Kishi-Itakura C, Mizushima N (2012) The hairpin-type tail-anchored SNARE syntaxin 17 targets to autophagosomes for fusion with endosomes/lysosomes. Cell 151(6):1256–1269. https://doi.org/10.1016/j.cell.2012.11.001 S0092-8674(12)01336-0 [pii]

Jiang P, Nishimura T, Sakamaki Y, Itakura E, Hatta T, Natsume T, Mizushima N (2014) The HOPS complex mediates autophagosome-lysosome fusion through interaction with syntaxin 17. Mol Biol Cell 25(8):1327–1337. https://doi.org/10.1091/mbc.E13-08-0447

Jin N, Lang MJ, Weisman LS (2016) Phosphatidylinositol 3,5-bisphosphate: regulation of cellular events in space and time. Biochem Soc Trans 44(1):177–184. https://doi.org/10.1042/BST20150174

Johnson DE, Ostrowski P, Jaumouille V, Grinstein S (2016) The position of lysosomes within the cell determines their luminal pH. J Cell Biol 212(6):677–692. https://doi.org/10.1083/jcb.201507112

Jongsma ML, Berlin I, Wijdeven RH, Janssen L, Janssen GM, Garstka MA, Janssen H, Mensink M, van Veelen PA, Spaapen RM, Neefjes J (2016) An ER-associated pathway defines endosomal architecture for controlled cargo transport. Cell 166(1):152–166. https://doi.org/10.1016/j.cell.2016.05.078

Jordens I, Fernandez-Borja M, Marsman M, Dusseljee S, Janssen L, Calafat J, Janssen H, Wubbolts R, Neefjes J (2001) The Rab7 effector protein RILP controls lysosomal transport by inducing the recruitment of dynein-dynactin motors. Curr Biol 11(21):1680–1685

Jung J, Genau HM, Behrends C (2015) Amino acid-dependent mTORC1 regulation by the lysosomal Membrane protein SLC38A9. Mol Cell Biol 35(14):2479–2494. https://doi.org/10.1128/MCB.00125-15

Kawai T, Akira S (2010) The role of pattern-recognition receptors in innate immunity: update on Toll-like receptors. Nat Immunol 11(5):373–384. https://doi.org/10.1038/ni.1863

Khatter D, Raina VB, Dwivedi D, Sindhwani A, Bahl S, Sharma M (2015) The small GTPase Arl8b regulates assembly of the mammalian HOPS complex on lysosomes. J Cell Sci 128(9):1746–1761. https://doi.org/10.1242/jcs.162651 jcs.162651 [pii]

Kim J, Kundu M, Viollet B, Guan KL (2011) AMPK and mTOR regulate autophagy through direct phosphorylation of Ulk1. Nat Cell Biol 13(2):132–141. https://doi.org/10.1038/ncb2152

Kim YM, Jung CH, Seo M, Kim EK, Park JM, Bae SS, Kim DH (2015) mTORC1 phosphorylates UVRAG to negatively regulate autophagosome and endosome maturation. Mol Cell 57(2):207–218. https://doi.org/10.1016/j.molcel.2014.11.013

Korolchuk VI, Saiki S, Lichtenberg M, Siddiqi FH, Roberts EA, Imarisio S, Jahreiss L, Sarkar S, Futter M, Menzies FM, O'Kane CJ, Deretic V, Rubinsztein DC (2011) Lysosomal positioning coordinates cellular nutrient responses. Nat Cell Biol 13(4):453–460. https://doi.org/10.1038/ncb2204

Kruppa AJ, Kendrick-Jones J, Buss F (2016) Myosins, actin and autophagy. Traffic 17(8):878–890. https://doi.org/10.1111/tra.12410

Laplante M, Sabatini DM (2012) mTOR signaling in growth control and disease. Cell 149(2):274–293. https://doi.org/10.1016/j.cell.2012.03.017

Lassen KG, McKenzie CI, Mari M, Murano T, Begun J, Baxt LA, Goel G, Villablanca EJ, Kuo SY, Huang H, Macia L, Bhan AK, Batten M, Daly MJ, Reggiori F, Mackay CR, Xavier RJ (2016) Genetic coding variant in GPR65 alters lysosomal pH and links lysosomal dysfunction with colitis risk. Immunity 44(6):1392–1405. https://doi.org/10.1016/j.immuni.2016.05.007

Li X, Wang X, Zhang X, Zhao M, Tsang WL, Zhang Y, Yau RG, Weisman LS, Xu H (2013) Genetically encoded fluorescent probe to visualize intracellular phosphatidylinositol 3,5-bisphosphate localization and dynamics. Proc Natl Acad Sci U S A 110(52):21165–21170. https://doi.org/10.1073/pnas.1311864110

Li SC, Diakov TT, Xu T, Tarsio M, Zhu W, Couoh-Cardel S, Weisman LS, Kane PM (2014) The signaling lipid PI(3,5)P(2) stabilizes V(1)-V(0) sector interactions and activates the V-ATPase. Mol Biol Cell 25(8):1251–1262. https://doi.org/10.1091/mbc.E13-10-0563

Li Y, Xu M, Ding X, Yan C, Song Z, Chen L, Huang X, Wang X, Jian Y, Tang G, Tang C, Di Y, Mu S, Liu X, Liu K, Li T, Wang Y, Miao L, Guo W, Hao X, Yang C (2016) Protein kinase C controls lysosome biogenesis independently of mTORC1. Nat Cell Biol 18(10):1065–1077. https://doi.org/10.1038/ncb3407

Lim CY, Zoncu R (2016) The lysosome as a command-and-control center for cellular metabolism. J Cell Biol 214(6):653–664. https://doi.org/10.1083/jcb.201607005

Lin X, Yang T, Wang S, Wang Z, Yun Y, Sun L, Zhou Y, Xu X, Akazawa C, Hong W, Wang T (2014) RILP interacts with HOPS complex via VPS41 subunit to regulate endocytic trafficking. Sci Rep 4:7282. https://doi.org/10.1038/srep07282 srep07282 [pii]

Lloyd-Evans E, Platt FM (2011) Lysosomal Ca^{2+} homeostasis: role in pathogenesis of lysosomal storage diseases. Cell Calcium 50(2):200–205. https://doi.org/10.1016/j.ceca.2011.03.010

Lloyd-Evans E, Morgan AJ, He X, Smith DA, Elliot-Smith E, Sillence DJ, Churchill GC, Schuchman EH, Galione A, Platt FM (2008) Niemann-Pick disease type C1 is a sphingosine storage disease that causes deregulation of lysosomal calcium. Nat Med 14(11):1247–1255. https://doi.org/10.1038/nm.1876

Lopez-Sanjurjo CI, Tovey SC, Prole DL, Taylor CW (2013) Lysosomes shape Ins(1,4,5)P3-evoked Ca^{2+} signals by selectively sequestering Ca^{2+} released from the endoplasmic reticulum. J Cell Sci 126(Pt 1):289–300. https://doi.org/10.1242/jcs.116103

Luzio JP, Pryor PR, Bright NA (2007) Lysosomes: fusion and function. Nat Rev Mol Cell Biol 8(8):622–632. https://doi.org/10.1038/nrm2217 nrm2217 [pii]

Luzio JP, Hackmann Y, Dieckmann NM, Griffiths GM (2014) The biogenesis of lysosomes and lysosome-related organelles. Cold Spring Harb Perspect Biol 6(9):a016840. https://doi.org/10.1101/cshperspect.a016840 a016840, 6/9/a016840 [pii]

MacDonald C, Stamnes MA, Katzmann DJ, Piper RC (2015) Tetraspan cargo adaptors usher GPI-anchored proteins into multivesicular bodies. Cell Cycle 14(23):3673–3678. https://doi.org/10.1080/15384101.2015.1100773

Majer O, Liu B, Barton GM (2016) Nucleic acid-sensing TLRs: trafficking and regulation. Curr Opin Immunol 44:26–33. https://doi.org/10.1016/j.coi.2016.10.003

Manifava M, Smith M, Rotondo S, Walker S, Niewczas I, Zoncu R, Clark J, Ktistakis NT (2016) Dynamics of mTORC1 activation in response to amino acids. Elife 5. https://doi.org/10.7554/elife.19960

Martens S, Nakamura S, Yoshimori T (2016) Phospholipids in autophagosome formation and fusion. J Mol Biol. https://doi.org/10.1016/j.jmb.2016.10.029

Martina JA, Puertollano R (2013) Rag GTPases mediate amino acid-dependent recruitment of TFEB and MITF to lysosomes. J Cell Biol 200(4):475–491. https://doi.org/10.1083/jcb.201209135

Martina JA, Diab HI, Brady OA, Puertollano R (2016) TFEB and TFE3 are novel components of the integrated stress response. EMBO J 35(5):479–495. https://doi.org/10.15252/embj.201593428

Marwaha R, Arya SB, Jagga D, Kaur H, Tuli A, Sharma M (2017) The Rab7 effector PLEKHM1 binds Arl8b to promote cargo traffic to lysosomes. J Cell Biol 216(4):1051–1070. https://doi.org/10.1083/jcb.201607085

Mauvezin C, Nagy P, Juhasz G, Neufeld TP (2015) Autophagosome-lysosome fusion is independent of V-ATPase-mediated acidification. Nat Commun 6:7007. https://doi.org/10.1038/ncomms8007

McEwan DG, Dikic I (2015) PLEKHM1: adapting to life at the lysosome. Autophagy 11(4):720–722. https://doi.org/10.1080/15548627.2015.1034419

McEwan DG, Popovic D, Gubas A, Terawaki S, Suzuki H, Stadel D, Coxon FP, Miranda de Stegmann D, Bhogaraju S, Maddi K, Kirchof A, Gatti E, Helfrich MH, Wakatsuki S, Behrends C, Pierre P, Dikic I (2015) PLEKHM1 regulates autophagosome-lysosome fusion through HOPS complex and LC3/GABARAP proteins. Mol Cell 57(1):39–54. https://doi.org/10.1016/j.molcel.2014.11.006

Medina DL, Fraldi A, Bouche V, Annunziata F, Mansueto G, Spampanato C, Puri C, Pignata A, Martina JA, Sardiello M, Palmieri M, Polishchuk R, Puertollano R, Ballabio A (2011) Transcriptional activation of lysosomal exocytosis promotes cellular clearance. Dev Cell 21(3):421–430. https://doi.org/10.1016/j.devcel.2011.07.016

Medina DL, Di Paola S, Peluso I, Armani A, De Stefani D, Venditti R, Montefusco S, Scotto-Rosato A, Prezioso C, Forrester A, Settembre C, Wang W, Gao Q, Xu H, Sandri M, Rizzuto R, De Matteis MA, Ballabio A (2015) Lysosomal calcium signalling regulates autophagy through calcineurin and TFEB. Nat Cell Biol 17(3):288–299. https://doi.org/10.1038/ncb3114

Melchionda M, Pittman JK, Mayor R, Patel S (2016) Ca^{2+}/H^{+} exchange by acidic organelles regulates cell migration in vivo. J Cell Biol 212(7):803–813. https://doi.org/10.1083/jcb.201510019

Menon S, Dibble CC, Talbott G, Hoxhaj G, Valvezan AJ, Takahashi H, Cantley LC, Manning BD (2014) Spatial control of the TSC complex integrates insulin and nutrient regulation of mTORC1 at the lysosome. Cell 156(4):771–785. https://doi.org/10.1016/j.cell.2013.11.049

Metcalf D, Isaacs AM (2010) The role of ESCRT proteins in fusion events involving lysosomes, endosomes and autophagosomes. Biochem Soc Trans 38(6):1469–1473. https://doi.org/10.1042/BST0381469

Miller A, Schafer J, Upchurch C, Spooner E, Huynh J, Hernandez S, McLaughlin B, Oden L, Fares H (2015) Mucolipidosis type IV protein TRPML1-dependent lysosome formation. Traffic 16(3):284–297. https://doi.org/10.1111/tra.12249

Morgan AJ, Platt FM, Lloyd-Evans E, Galione A (2011) Molecular mechanisms of endolysosomal Ca^{2+} signalling in health and disease. Biochem J 439(3):349–374. https://doi.org/10.1042/BJ20110949

Napolitano G, Ballabio A (2016) TFEB at a glance. J Cell Sci 129(13):2475–2481. https://doi.org/10.1242/jcs.146365

Nguyen TN, Padman BS, Usher J, Oorschot V, Ramm G, Lazarou M (2016) Atg8 family LC3/GABARAP proteins are crucial for autophagosome-lysosome fusion but not autophago-

some formation during PINK1/Parkin mitophagy and starvation. J Cell Biol 215(6):857–874. https://doi.org/10.1083/jcb.201607039
Nguyen ON, Grimm C, Schneider LS, Chao YK, Atzberger C, Bartel K, Watermann A, Ulrich M, Mayr D, Wahl-Schott C, Biel M, Vollmar AM (2017) Two-pore channel function is crucial for migration of invasive cancer cells. Cancer Res. https://doi.org/10.1158/0008-5472.CAN-16-0852
Palmieri M, Impey S, Kang H, di Ronza A, Pelz C, Sardiello M, Ballabio A (2011) Characterization of the CLEAR network reveals an integrated control of cellular clearance pathways. Hum Mol Genet 20(19):3852–3866. https://doi.org/10.1093/hmg/ddr306
Palmieri M, Pal R, Nelvagal HR, Lotfi P, Stinnett GR, Seymour ML, Chaudhury A, Bajaj L, Bondar VV, Bremner L, Saleem U, Tse DY, Sanagasetti D, Wu SM, Neilson JR, Pereira FA, Pautler RG, Rodney GG, Cooper JD, Sardiello M (2017) mTORC1-independent TFEB activation via Akt inhibition promotes cellular clearance in neurodegenerative storage diseases. Nat Commun 8:14338. https://doi.org/10.1038/ncomms14338
Parra KJ, Kane PM (1998) Reversible association between the V1 and V0 domains of yeast vacuolar H^+ -ATPase is an unconventional glucose-induced effect. Mol Cell Biol 18(12):7064–7074
Peng M, Yin N, Li MO (2017) SZT2 dictates GATOR control of mTORC1 signalling. Nature. https://doi.org/10.1038/nature21378
Penny CJ, Kilpatrick BS, Eden ER, Patel S (2015) Coupling acidic organelles with the ER through Ca^{2+} microdomains at membrane contact sites. Cell Calcium 58(4):387–396. https://doi.org/10.1016/j.ceca.2015.03.006
Perera RM, Zoncu R (2016) The lysosome as a regulatory hub. Annu Rev Cell Dev Biol 32:223–253. https://doi.org/10.1146/annurev-cellbio-111315-125125
Petit CS, Roczniak-Ferguson A, Ferguson SM (2013) Recruitment of folliculin to lysosomes supports the amino acid-dependent activation of Rag GTPases. J Cell Biol 202(7):1107–1122. https://doi.org/10.1083/jcb.201307084
Pislar A, Perisic Nanut M, Kos J (2015) Lysosomal cysteine peptidases—molecules signaling tumor cell death and survival. Semin Cancer Biol 35:168–179. https://doi.org/10.1016/j.semcancer.2015.08.001
Pitt SJ, Funnell TM, Sitsapesan M, Venturi E, Rietdorf K, Ruas M, Ganesan A, Gosain R, Churchill GC, Zhu MX, Parrington J, Galione A, Sitsapesan R (2010) TPC2 is a novel NAADP-sensitive Ca^{2+} release channel, operating as a dual sensor of luminal pH and Ca^{2+}. J Biol Chem 285(45):35039–35046. https://doi.org/10.1074/jbc.M110.156927
Platt FM, Boland B, van der Spoel AC (2012) The cell biology of disease: lysosomal storage disorders: the cellular impact of lysosomal dysfunction. J Cell Biol 199(5):723–734. https://doi.org/10.1083/jcb.201208152
Platta HW, Stenmark H (2011) Endocytosis and signaling. Curr Opin Cell Biol 23(4):393–403. https://doi.org/10.1016/j.ceb.2011.03.008 S0955-0674(11)00025-1 [pii]
Pols MS, ten Brink C, Gosavi P, Oorschot V, Klumperman J (2013) The HOPS proteins hVps41 and hVps39 are required for homotypic and heterotypic late endosome fusion. Traffic 14(2):219–232. https://doi.org/10.1111/tra.12027
Pryor PR, Mullock BM, Bright NA, Gray SR, Luzio JP (2000) The role of intraorganellar Ca^{2+} in late endosome-lysosome heterotypic fusion and in the reformation of lysosomes from hybrid organelles. J Cell Biol 149(5):1053–1062
Pryor PR, Mullock BM, Bright NA, Lindsay MR, Gray SR, Richardson SC, Stewart A, James DE, Piper RC, Luzio JP (2004) Combinatorial SNARE complexes with VAMP7 or VAMP8 define different late endocytic fusion events. EMBO Rep 5(6):590–595. https://doi.org/10.1038/sj.embor.7400150 7400150 [pii]
Pu J, Schindler C, Jia R, Jarnik M, Backlund P, Bonifacino JS (2015) BORC, a multisubunit complex that regulates lysosome positioning. Dev Cell 33(2):176–188. https://doi.org/10.1016/j.devcel.2015.02.011
Raben N, Puertollano R (2016) TFEB and TFE3: linking lysosomes to cellular adaptation to stress. Annu Rev Cell Dev Biol 32:255–278. https://doi.org/10.1146/annurev-cellbio-111315-125407

Raffaello A, Mammucari C, Gherardi G, Rizzuto R (2016) Calcium at the center of cell signaling: interplay between endoplasmic reticulum, mitochondria, and lysosomes. Trends Biochem Sci 41(12):1035–1049. https://doi.org/10.1016/j.tibs.2016.09.001

Rao SK, Huynh C, Proux-Gillardeaux V, Galli T, Andrews NW (2004) Identification of SNAREs involved in synaptotagmin VII-regulated lysosomal exocytosis. J Biol Chem 279(19):20471–20479. https://doi.org/10.1074/jbc.M400798200

Rebsamen M, Pochini L, Stasyk T, de Araujo ME, Galluccio M, Kandasamy RK, Snijder B, Fauster A, Rudashevskaya EL, Bruckner M, Scorzoni S, Filipek PA, Huber KV, Bigenzahn JW, Heinz LX, Kraft C, Bennett KL, Indiveri C, Huber LA, Superti-Furga G (2015) SLC38A9 is a component of the lysosomal amino acid sensing machinery that controls mTORC1. Nature 519(7544):477–481. https://doi.org/10.1038/nature14107

Roczniak-Ferguson A, Petit CS, Froehlich F, Qian S, Ky J, Angarola B, Walther TC, Ferguson SM (2012) The transcription factor TFEB links mTORC1 signaling to transcriptional control of lysosome homeostasis. Sci Signal 5(228):ra42. https://doi.org/10.1126/scisignal.2002790

Ronco V, Potenza DM, Denti F, Vullo S, Gagliano G, Tognolina M, Guerra G, Pinton P, Genazzani AA, Mapelli L, Lim D, Moccia F (2015) A novel Ca^{2+}-mediated cross-talk between endoplasmic reticulum and acidic organelles: implications for NAADP-dependent Ca^{2+} signalling. Cell Calcium 57(2):89–100. https://doi.org/10.1016/j.ceca.2015.01.001

Rong Y, McPhee CK, Deng S, Huang L, Chen L, Liu M, Tracy K, Baehrecke EH, Yu L, Lenardo MJ (2011) Spinster is required for autophagic lysosome reformation and mTOR reactivation following starvation. Proc Natl Acad Sci U S A 108(19):7826–7831. https://doi.org/10.1073/pnas.1013800108

Rong Y, Liu M, Ma L, Du W, Zhang H, Tian Y, Cao Z, Li Y, Ren H, Zhang C, Li L, Chen S, Xi J, Yu L (2012) Clathrin and phosphatidylinositol-4,5-bisphosphate regulate autophagic lysosome reformation. Nat Cell Biol 14(9):924–934. https://doi.org/10.1038/ncb2557

Rosa-Ferreira C, Munro S (2011) Arl8 and SKIP act together to link lysosomes to kinesin-1. Dev Cell 21(6):1171–1178. https://doi.org/10.1016/j.devcel.2011.10.007

Rowland AA, Chitwood PJ, Phillips MJ, Voeltz GK (2014) ER contact sites define the position and timing of endosome fission. Cell 159(5):1027–1041. https://doi.org/10.1016/j.cell.2014.10.023

Saarikangas J, Zhao H, Lappalainen P (2010) Regulation of the actin cytoskeleton-plasma membrane interplay by phosphoinositides. Physiol Rev 90(1):259–289. https://doi.org/10.1152/physrev.00036.2009

Sakurai Y, Kolokoltsov AA, Chen CC, Tidwell MW, Bauta WE, Klugbauer N, Grimm C, Wahl-Schott C, Biel M, Davey RA (2015) Ebola virus. Two-pore channels control Ebola virus host cell entry and are drug targets for disease treatment. Science 347(6225):995–998. https://doi.org/10.1126/science.1258758

Sancak Y, Peterson TR, Shaul YD, Lindquist RA, Thoreen CC, Bar-Peled L, Sabatini DM (2008) The Rag GTPases bind raptor and mediate amino acid signaling to mTORC1. Science 320(5882):1496–1501. https://doi.org/10.1126/science.1157535 1157535 [pii]

Sardiello M, Palmieri M, di Ronza A, Medina DL, Valenza M, Gennarino VA, Di Malta C, Donaudy F, Embrione V, Polishchuk RS, Banfi S, Parenti G, Cattaneo E, Ballabio A (2009) A gene network regulating lysosomal biogenesis and function. Science 325(5939):473–477. https://doi.org/10.1126/science.1174447 1174447 [pii]

Sargeant TJ, Lloyd-Lewis B, Resemann HK, Ramos-Montoya A, Skepper J, Watson CJ (2014) Stat3 controls cell death during mammary gland involution by regulating uptake of milk fat globules and lysosomal membrane permeabilization. Nat Cell Biol 16(11):1057–1068. https://doi.org/10.1038/ncb3043

Sbano L, Bonora M, Marchi S, Baldassari F, Medina DL, Ballabio A, Giorgi C, Pinton P (2017) TFEB-mediated increase in peripheral lysosomes regulates store-operated calcium entry. Sci Rep 7:40797. https://doi.org/10.1038/srep40797

Schafer IB, Hesketh GG, Bright NA, Gray SR, Pryor PR, Evans PR, Luzio JP, Owen DJ (2012) The binding of Varp to VAMP7 traps VAMP7 in a closed, fusogenically inactive conformation. Nat Struct Mol Biol 19(12):1300–1309. https://doi.org/10.1038/nsmb.2414

Schmid JA, Mach L, Paschke E, Glossl J (1999) Accumulation of sialic acid in endocytic compartments interferes with the formation of mature lysosomes. Impaired proteolytic processing of cathepsin B in fibroblasts of patients with lysosomal sialic acid storage disease. J Biol Chem 274(27):19063–19071

Schulze RJ, Weller SG, Schroeder B, Krueger EW, Chi S, Casey CA, McNiven MA (2013) Lipid droplet breakdown requires dynamin 2 for vesiculation of autolysosomal tubules in hepatocytes. J Cell Biol 203(2):315–326. https://doi.org/10.1083/jcb.201306140

Seaman MN, Ball CL, Robinson MS (1993) Targeting and mistargeting of plasma membrane adaptors in vitro. J Cell Biol 123(5):1093–1105

Sengupta S, Peterson TR, Sabatini DM (2010) Regulation of the mTOR complex 1 pathway by nutrients, growth factors, and stress. Mol Cell 40(2):310–322. https://doi.org/10.1016/j.molcel.2010.09.026

Serrano-Puebla A, Boya P (2016) Lysosomal membrane permeabilization in cell death: new evidence and implications for health and disease. Ann N Y Acad Sci 1371(1):30–44. https://doi.org/10.1111/nyas.12966

Settembre C, Zoncu R, Medina DL, Vetrini F, Erdin S, Huynh T, Ferron M, Karsenty G, Vellard MC, Facchinetti V, Sabatini DM, Ballabio A (2012) A lysosome-to-nucleus signalling mechanism senses and regulates the lysosome via mTOR and TFEB. EMBO J 31(5):1095–1108. https://doi.org/10.1038/emboj.2012.32 emboj201232 [pii]

Settembre C, De Cegli R, Mansueto G, Saha PK, Vetrini F, Visvikis O, Huynh T, Carissimo A, Palmer D, Klisch TJ, Wollenberg AC, Di Bernardo D, Chan L, Irazoqui JE, Ballabio A (2013a) TFEB controls cellular lipid metabolism through a starvation-induced autoregulatory loop. Nat Cell Biol 15(6):647–658. https://doi.org/10.1038/ncb2718

Settembre C, Fraldi A, Medina DL, Ballabio A (2013b) Signals from the lysosome: a control centre for cellular clearance and energy metabolism. Nat Rev Mol Cell Biol 14(5):283–296. https://doi.org/10.1038/nrm3565

Sonenberg N, Hinnebusch AG (2009) Regulation of translation initiation in eukaryotes: mechanisms and biological targets. Cell 136(4):731–745. https://doi.org/10.1016/j.cell.2009.01.042

Sorkin A, Goh LK (2008) Endocytosis and intracellular trafficking of ErbBs. Exp Cell Res 314(17):3093–3106. https://doi.org/10.1016/j.yexcr.2008.08.013

Spampanato C, Feeney E, Li L, Cardone M, Lim JA, Annunziata F, Zare H, Polishchuk R, Puertollano R, Parenti G, Ballabio A, Raben N (2013) Transcription factor EB (TFEB) is a new therapeutic target for Pompe disease. EMBO Mol Med 5(5):691–706. https://doi.org/10.1002/emmm.201202176

Steger M, Tonelli F, Ito G, Davies P, Trost M, Vetter M, Wachter S, Lorentzen E, Duddy G, Wilson S, Baptista MA, Fiske BK, Fell MJ, Morrow JA, Reith AD, Alessi DR, Mann M (2016) Phosphoproteomics reveals that Parkinson's disease kinase LRRK2 regulates a subset of Rab GTPases. Elife 5. https://doi.org/10.7554/elife.12813

Steinberg BE, Huynh KK, Brodovitch A, Jabs S, Stauber T, Jentsch TJ, Grinstein S (2010) A cation counterflux supports lysosomal acidification. J Cell Biol 189(7):1171–1186. https://doi.org/10.1083/jcb.200911083

Tian X, Gala U, Zhang Y, Shang W, Nagarkar Jaiswal S, di Ronza A, Jaiswal M, Yamamoto S, Sandoval H, Duraine L, Sardiello M, Sillitoe RV, Venkatachalam K, Fan H, Bellen HJ, Tong C (2015) A voltage-gated calcium channel regulates lysosomal fusion with endosomes and autophagosomes and is required for neuronal homeostasis. PLoS Biol 13(3):e1002103. https://doi.org/10.1371/journal.pbio.1002103

Toyofuku T, Morimoto K, Sasawatari S, Kumanogoh A (2015) Leucine-Rich repeat kinase 1 regulates autophagy through turning on TBC1D2-dependent Rab7 inactivation. Mol Cell Biol 35(17):3044–3058. https://doi.org/10.1128/MCB.00085-15

Tsun ZY, Bar-Peled L, Chantranupong L, Zoncu R, Wang T, Kim C, Spooner E, Sabatini DM (2013) The folliculin tumor suppressor is a GAP for the RagC/D GTPases that signal amino acid levels to mTORC1. Mol Cell 52(4):495–505. https://doi.org/10.1016/j.molcel.2013.09.016

van der Kant R, Fish A, Janssen L, Janssen H, Krom S, Ho N, Brummelkamp T, Carette J, Rocha N, Neefjes J (2013) Late endosomal transport and tethering are coupled processes controlled by RILP and the cholesterol sensor ORP1L. J Cell Sci 126(Pt 15):3462–3474. https://doi.org/10.1242/jcs.129270

van der Kant R, Jonker CT, Wijdeven RH, Bakker J, Janssen L, Klumperman J, Neefjes J (2015) Characterization of the mammalian CORVET and HOPS complexes and their modular restructuring for endosome specificity. J Biol Chem 290(51):30280–30290. https://doi.org/10.1074/jbc.M115.688440 M115.688440 [pii]

Vega-Rubin-de-Celis S, Pena-Llopis S, Konda M, Brugarolas J (2017) Multistep regulation of TFEB by MTORC1. Autophagy. https://doi.org/10.1080/15548627.2016.1271514

Wang S, Tsun ZY, Wolfson RL, Shen K, Wyant GA, Plovanich ME, Yuan ED, Jones TD, Chantranupong L, Comb W, Wang T, Bar-Peled L, Zoncu R, Straub C, Kim C, Park J, Sabatini BL, Sabatini DM (2015) Metabolism. Lysosomal amino acid transporter SLC38A9 signals arginine sufficiency to mTORC1. Science 347(6218):188–194. https://doi.org/10.1126/science.1257132

Wang Z, Miao G, Xue X, Guo X, Yuan C, Wang Z, Zhang G, Chen Y, Feng D, Hu J, Zhang H (2016) The Vici syndrome protein EPG5 is a Rab7 effector that determines the fusion specificity of autophagosomes with late endosomes/lysosomes. Mol Cell 63(5):781–795. https://doi.org/10.1016/j.molcel.2016.08.021

Wartosch L, Gunesdogan U, Graham SC, Luzio JP (2015) Recruitment of VPS33A to HOPS by VPS16 Is required for lysosome fusion with endosomes and autophagosomes. Traffic 16(7):727–742. https://doi.org/10.1111/tra.12283

Wee YS, Roundy KM, Weis JJ, Weis JH (2012) Interferon-inducible transmembrane proteins of the innate immune response act as membrane organizers by influencing clathrin and v-ATPase localization and function. Innate Immun 18(6):834–845. https://doi.org/10.1177/1753425912443392

White IJ, Bailey LM, Aghakhani MR, Moss SE, Futter CE (2006) EGF stimulates annexin 1-dependent inward vesiculation in a multivesicular endosome subpopulation. EMBO J 25(1):1–12. https://doi.org/10.1038/sj.emboj.7600759

Wijdeven RH, Janssen H, Nahidiazar L, Janssen L, Jalink K, Berlin I, Neefjes J (2016) Cholesterol and ORP1L-mediated ER contact sites control autophagosome transport and fusion with the endocytic pathway. Nat Commun 7:11808. https://doi.org/10.1038/ncomms11808

Wolfson RL, Chantranupong L, Saxton RA, Shen K, Scaria SM, Cantor JR, Sabatini DM (2016) Sestrin2 is a leucine sensor for the mTORC1 pathway. Science 351(6268):43–48. https://doi.org/10.1126/science.aab2674

Wolfson RL, Chantranupong L, Wyant GA, Gu X, Orozco JM, Shen K, Condon KJ, Petri S, Kedir J, Scaria SM, Abu-Remaileh M, Frankel WN, Sabatini DM (2017) KICSTOR recruits GATOR1 to the lysosome and is necessary for nutrients to regulate mTORC1. Nature. https://doi.org/10.1038/nature21423

Woodman PG, Futter CE (2008) Multivesicular bodies: co-ordinated progression to maturity. Curr Opin Cell Biol 20(4):408–414. https://doi.org/10.1016/j.ceb.2008.04.001 S0955-0674(08)00072-0 [pii]

Yamasaki M, Thomas JM, Churchill GC, Garnham C, Lewis AM, Cancela JM, Patel S, Galione A (2005) Role of NAADP and cADPR in the induction and maintenance of agonist-evoked Ca^{2+} spiking in mouse pancreatic acinar cells. Curr Biol 15(9):874–878. https://doi.org/10.1016/j.cub.2005.04.033

Yamashiro DJ, Maxfield FR (1987) Acidification of morphologically distinct endosomes in mutant and wild-type Chinese hamster ovary cells. J Cell Biol 105(6 Pt 1):2723–2733

Yu L, McPhee CK, Zheng L, Mardones GA, Rong Y, Peng J, Mi N, Zhao Y, Liu Z, Wan F, Hailey DW, Oorschot V, Klumperman J, Baehrecke EH, Lenardo MJ (2010) Termination of autophagy and reformation of lysosomes regulated by mTOR. Nature 465(7300):942–946. https://doi.org/10.1038/nature09076

Zhang H, Hu J (2016) Shaping the Endoplasmic Reticulum into a Social Network. Trends Cell Biol 26(12):934–943. https://doi.org/10.1016/j.tcb.2016.06.002

Zhang CS, Jiang B, Li M, Zhu M, Peng Y, Zhang YL, Wu YQ, Li TY, Liang Y, Lu Z, Lian G, Liu Q, Guo H, Yin Z, Ye Z, Han J, Wu JW, Yin H, Lin SY, Lin SC (2014) The lysosomal v-ATPase-Ragulator complex is a common activator for AMPK and mTORC1, acting as a switch between catabolism and anabolism. Cell Metab 20(3):526–540. https://doi.org/10.1016/j.cmet.2014.06.014

Zoncu R, Bar-Peled L, Efeyan A, Wang S, Sancak Y, Sabatini DM (2011a) mTORC1 senses lysosomal amino acids through an inside-out mechanism that requires the vacuolar H^+-ATPase. Science 334(6056):678–683. https://doi.org/10.1126/science.1207056

Zoncu R, Efeyan A, Sabatini DM (2011b) mTOR: from growth signal integration to cancer, diabetes and ageing. Nat Rev Mol Cell Biol 12(1):21–35. https://doi.org/10.1038/nrm3025

Zou J, Hu B, Arpag S, Yan Q, Hamilton A, Zeng YS, Vanoye CG, Li J (2015) Reactivation of lysosomal $Ca2^+$ efflux rescues abnormal lysosomal storage in FIG4-deficient cells. J Neurosci 35(17):6801–6812. https://doi.org/10.1523/JNEUROSCI.4442-14.2015

Chapter 7
Interplay of Endocytosis and Growth Factor Receptor Signalling

Rachel Moore, Marta Giralt Pujol, Zhou Zhu and Elizabeth Smythe

Contents

Abstract Growth factor receptors play a variety of roles during embryonic development and in adult homeostasis. These receptors are activated repeatedly in different cellular contexts and with different cellular outcomes. This begs the question as to how cells in a particular developmental, spatial and temporal context, or in adult tissue, interpret signalling by growth factor receptors in order to deliver qualitatively

R. Moore · M. G. Pujol · Z. Zhu · E. Smythe (✉)
Department of Biomedical Science, Centre for Membrane Interactions and Dynamics, University of Sheffield, Sheffield S10 2TN, UK
e-mail: e.smythe@sheffield.ac.uk

R. Moore
e-mail: Rkmoore1@sheffield.ac.uk

M. G. Pujol
e-mail: mdp12mg@sheffield.ac.uk

Z. Zhu
e-mail: mdp12zz@sheffield.ac.uk

C. Lamaze and I. Prior (eds.), *Endocytosis and Signaling*, Progress in Molecular and Subcellular Biology 57, https://doi.org/10.1007/978-3-319-96704-2_7

different signalling outputs. One mechanism by which this could occur is via endocytic regulation. The original paradigm for the role of endocytosis in growth factor receptor signalling was that receptor uptake has a quantitative role in signalling by reducing the number of cell surface receptors available for activation and targeting activated receptors for degradation. However, a range of studies over the last several years, in many different experimental systems, has demonstrated an additional qualitative role for endocytic trafficking in receptor signalling, with specific outcomes depending on the location of the signalling complex. Confinement of receptors within endosomes can spatially regulate signalling, facilitating specific protein interactions or post-translational modifications that alter throughout the trafficking process. Therefore, endocytosis does not simply regulate cell surface expression, but tightly controls protein interactions and function to produce distinct outcomes.

7.1 Introduction

Throughout development and in the adult, cells respond to a variety of growth factors by binding to high-affinity cell surface transmembrane receptors (Davidson 1993). This review will specifically focus on growth factor receptors that contain a tyrosine or serine/threonine kinase domain within their cytoplasmic tail. Engagement of growth factors, such as epidermal growth factor (EGF), vascular endothelial growth factor (VEGF), transforming growth factor alpha and beta (TGF-α and TGF-β), fibroblast growth factor (FGF), hepatocyte growth factor (HGF) and angiopoietin (Ang1) with their cognate receptors results in receptor oligomerisation and transphosphorylation. This is followed by activation of signalling cascades that result in many different cellular outcomes including cell proliferation, migration, differentiation and survival (Blume-Jensen and Hunter 2001; Lemmon and Schlessinger 2010). The precise outcome of growth factor receptor signalling is context-specific, depending on the environment in which a cell finds itself. Additionally, growth factor receptors do not act in isolation. Rather they are part of complex signalling networks, which allow flexibility and diversity of signalling and consequent cell behaviours. Context-dependent signalling thus requires very tight regulation in order to ensure precise signalling outcomes. Endocytosis is increasingly recognised as a key mechanism to allow subtle and sophisticated nuancing of signal interpretation.

The endocytic pathway has the potential to act as a regulator of signalling both quantitatively and qualitatively. Receptor uptake can have a quantitative role on signalling by reducing the number of cell surface receptors available for activation. Qualitative regulation of signalling by endocytosis can operate in multiple ways. These include individual signalling outputs arising from different locations on the endocytic pathway, with activated receptors at the cell surface resulting in a particular cell behaviour, and receptors within endosomes causing a different signalling outcome. Alternatively, receptor signalling could be modulated depending on the endocytic portal used by growth factor receptors for their uptake or they may require

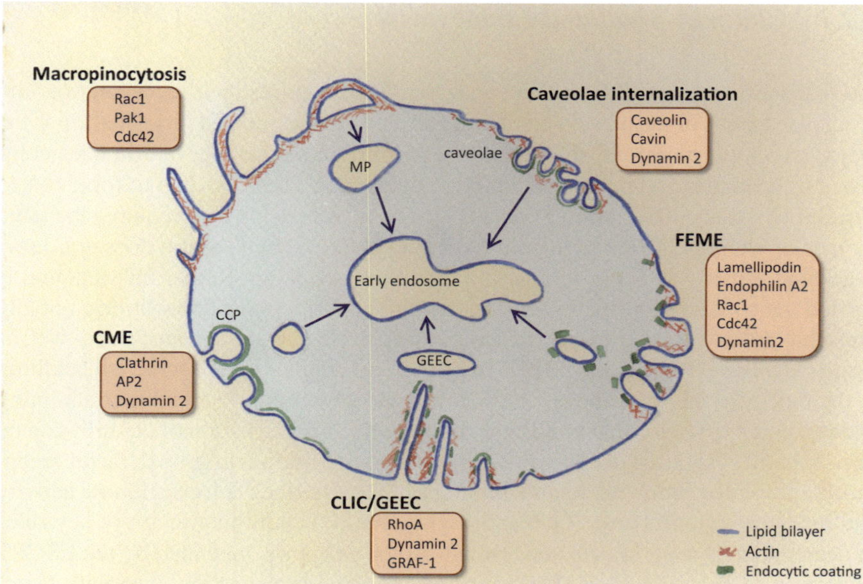

Fig. 7.1 *Multiple internalisation pathways allow growth factor receptor uptake* Activated growth factor receptors can be endocytosed by different routes depending on the context in which a cell finds itself. The main components of the machineries that are required for uptake are indicated in the figure. (CCP: clathrin coated pit; MP: macropinosome; GEEC: GPI-AP enriched early endosome compartments; FEME: Fast endophilin-mediated endocytosis)

delivery to a particular subcellular compartment for effective signalling. There is evidence to support all of the scenarios outlined above, and some examples will be discussed further and highlighted in the course of this review.

7.2 Overview of Endocytic Pathways

Cells use a variety of pathways to internalise growth factor receptors after they have been activated by their cognate ligands (Fig. 7.1). These include both clathrin-mediated endocytosis (CME) and clathrin-independent endocytosis (CIE) pathways. These pathways mostly require selection of material into a specialised area of the cell surface, a membrane microdomain, which then invaginates and eventually pinches off to form an endocytic vesicle. The molecular machinery required to internalise cargo, such as activated growth factor receptors, is increasingly well defined (Mayor et al. 2014; Schmid et al. 2014) although many questions about the regulation of these pathways remain, with direct implications for the interplay of endocytosis and signalling.

7.2.1 Clathrin-Mediated Endocytosis

Internalisation of growth factor receptors via CME requires their incorporation into specialised areas of the plasma membrane termed clathrin-coated pits. Clathrin is the major structural protein of the pit and clathrin trimers assemble into lattices composed of hexagons and pentagons to form coated pits. These can bud to form coated vesicles (Brodsky 2012; Maib et al. 2017). The final scission step requires the large molecular weight GTPase, dynamin (Antonny et al. 2016). Clathrin does not, however, bind directly to either the plasma membrane or cargo. Rather this function is carried out by adaptor proteins that can bind to clathrin, phosphoinositides, notably phosphatidylinositol, 4,5-bisphosphate, [PtdIns(4,5)P_2], at the cell surface, as well as cargo (Traub and Bonifacino 2013). The canonical adaptor is AP2 which, in addition to the functions outlined above, can also interact with multiple membrane sculpting and regulatory proteins that contribute to the successful formation of clathrin-coated pits (Schmid et al. 2014; Robinson 2015). The internalisation of growth factor receptors is stimulated following ligand binding and requires their intrinsic kinase activity and receptor ubiquitination (Goh and Sorkin 2013). Ubiquitination plays key roles in degradative sorting of activated growth factor receptors, mediated by the ESCRT proteins (see Sect. 7.2.3). It also contributes to the internalisation of growth factor receptors. EGF receptor (EGFR), for example, has multiple, redundant internalisation motifs including AP2 binding and ubiquitination sites that operate to promote efficient internalisation, most likely in a cell-type-specific manner (Goh et al. 2010; Fortian et al. 2015). Ubiquitinated growth factor receptors may be recognised by cargo-specific clathrin-associated sorting proteins (so-called CLASPs) such as epsin and eps15 which contain ubiquitin-binding domains (Traub and Bonifacino 2013).

7.2.2 Clathrin-Independent Endocytosis

CIE internalisation pathways (Fig. 7.1) are defined by the molecular machinery used to uptake cargo. Caveolae are cup-shaped structures, enriched in cholesterol and characterised by the presence of caveolin and cavins. These proteins have been implicated in endocytosis, transcytosis in endothelial cells as well as in mechanosensing (Mayor et al. 2014). The fast endophilin-mediated endocytosis (FEME) pathway requires RhoA, endophilin and dynamin and appears to be important for rapid uptake of various growth factor, cytokine and G protein-coupled receptors, often at the leading edge of cells (Boucrot et al. 2015). The CLIC (clathrin-independent carriers)/GEEC (GPI-enriched protein endosomal compartments) pathway requires the small GTPases, Arf1 and Cdc42, cholesterol and actin but is independent of clathrin and dynamin (Mayor and Riezman 2004). Arf6 defines a further CIE pathway important for the uptake of MHC class 1 molecules (Donaldson et al. 2016).

When portions of the plasma membrane ruffle and protruding edges of a ruffle fuse, macropinocytosis ensues, resulting in the non-selective uptake of extracellular material. Interestingly, this is a process that is often stimulated by high concentrations of growth factors and may represent a cellular response to high extracellular ligand concentrations (Buckley and King 2017).

7.2.3 Delivery to Endosomes and Lysosomes

Following internalisation, endocytic vesicles are delivered to the early endosome from where material can either be targeted for degradation via late endosomes, recycled to the cell surface or trafficked to further intracellular organelles such as the trans-Golgi network. Movement through the endocytic pathway is regulated by the rab family of small GTPases. Rabs are molecular switches that cycle between a GDP-bound inactive form in the cytoplasm and a GTP-bound active form when membrane-associated. Switching is brought about by guanine nucleotide exchange factors (GEFs) that mediate the exchange of GDP for GTP and GTPase-activating proteins (GAPs) that stimulate GTP hydrolysis. In their active form, rabs mediate cargo selection, vesicle movement, interactions with the cytoskeleton, fidelity of vesicle fusion and signalling, by interacting with a wide range of cytoplasmic effector proteins (Christoforidis et al. 1999; Stenmark 2009; Wandinger-Ness and Zerial 2014). Rab5 is a major regulator of the early endocytic pathway, modulating traffic between the plasma membrane and early endosomes. Rab4 and Rab11 regulate recycling pathways while Rab7 mediates the transition from early to late endosomes.

Rab proteins are important for organelle identity. One way in which they contribute to this is by selective recruitment of combinations of lipid kinases and/or phosphatases that ensure organelles are enriched in specific phosphoinositides (Shin et al. 2005; Di Paolo and De Camilli 2006). These phosphoinositides are key for organelle identity (Behnia and Munro 2005) and cooperate with rab proteins to ensure the so-called coincidence detection, where bivalent interactions of effector molecules are required for membrane association, thus ensuring location specificity (see Sect. 7.3.2). Maturation of endosomes is marked by changes in the complement of rab proteins that localise to the cytoplasmic surface of the endosome through a process termed rab conversion, where a cascade of rab activity occurs with active Rab5 recruiting a GEF to activate Rab7, which in turn recruits a GAP to switch off Rab5 (Rink et al. 2005; Del Conte-Zerial et al. 2008). Similarly rabaptin5, which binds both Rab5GTP and Rab4GTP (De Renzis et al. 2002), is recruited to early endosomes in complex with the Rab5GEF, Rabex5 (Horiuchi et al. 1997). Recent studies indicate that Rab4GTP together with ubiquitinated cargo, such as activated growth factor receptors, is primarily responsible for recruitment of rabaptin5 to endosomes which then allows for activation of Rab5 by Rabex5 (Kalin et al. 2015). Rab conversion thus ensures directionality of cargo flux along the endocytic pathway (Pfeffer 2017). Changes in phosphoinositide composition, along the endocytic pathway, also contribute to rab conversion (Cauvin et al. 2016; Liu et al. 2017).

Cargo destined for degradation is sorted into areas of the endosome membrane that bud inwards to form intraluminal vesicles (ILVs) within late endosomes resulting

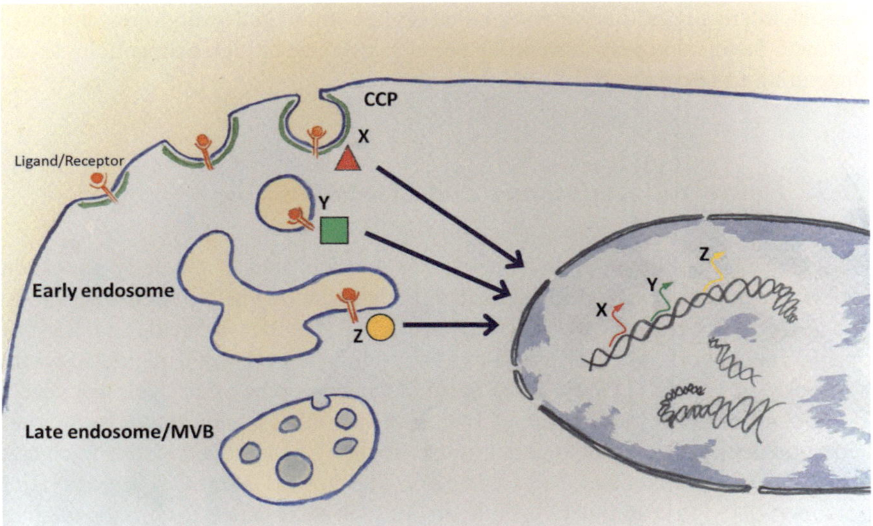

Fig. 7.2 *Signalosome formation on the endocytic pathway* Following engagement with ligand, activated growth factor receptors continue to signal as they traverse the endocytic pathway. This allows entry into membrane microdomains that facilitates specific downstream signaling outputs, which may include transcriptional programs. For example, activated receptors may signal from a microenvironment at the cell surface (triangle) which activates a complement of genes X, at the early endosome (square) which activates a complement of genes Y, or at late endosomes (circle) which activates a complement of genes, Z

in the formation of multivesicular bodies (MVBs) (Fig. 7.2). This process sorts cargo such as growth factor receptors for degradation and attenuates signalling by preventing the cytoplasmic tails of receptors from interacting with kinases or phosphatases. ILV formation is the topological inverse of budding from the cell surface and is mediated by the endosomal sorting complex required for transport (ESCRT), a family of four protein complexes that mediate cargo sorting and inclusion into ILVs. Growth factor receptors that are ubiquitinated are recognised by ESCRT-0. ESCRT-0 is a heterodimer comprising HGF-regulated tyrosine kinase substrate (Hrs) and signal transducing adaptor molecule (STAM), which is crucial in initiating ILV production by binding ubiquitin-tagged proteins. Hrs localises to endosomal membranes through a FYVE domain that binds phosphatidylinositol-3-phosphate (PtdIns3P). A distinct domain of ESCRT-I then binds to ESCRT-II, which can associate with the endosomal membrane, cargo and ESCRT-III. Transient assembly of ESCRT-III results in membrane scission and recruitment of Vps4, which results in the termination of MVB formation and recycling of ESCRT complexes. MVBs then fuse with lysosomes, ensuring cargo is inactivated and degraded (Schmidt and Teis 2012; Hurley and Emr 2006; Schoneberg et al. 2017).

7.3 How Does Endocytosis Regulate Signalling?

7.3.1 Internalisation Quantitatively Affects Signalling

Early studies on growth factor receptor signalling were based on the assumption that all signalling occur at the cell surface and that internalisation largely plays a housekeeping role in removing available receptors. The level of signalling can be regulated by whether activated receptors are targeted for degradation to the lysosomes or whether receptors are recycled back to the cell surface for further rounds of uptake. Recent studies support such a role for endocytosis in this context. In studies using genome-edited cells expressing mVenus-HRas at endogenous levels, it was shown that a small pool of activated EGF receptor (EGFR) maintains active HRas predominantly at the plasma membrane and, surprisingly, that this is sufficient to ensure sustained MAPK signalling even when the bulk of the active EGFR complexes have been delivered to endosomes (Pinilla-Macua et al. 2016). It is unclear whether this pool of active EGFR is maintained by very rapid recycling [e.g. (Montagnac et al. 2011)] or by being tethered in a plasma membrane microdomain (Grecco et al. 2011). Whatever the mechanism, in this case, endocytosis appears to modulate the signal by separation of the bulk of the active EGFR complexes from active HRas.

7.3.2 Endocytic Portals Determine Signalling Outcome

TGF-β receptor signalling regulates cell growth, proliferation and apoptosis in diverse cell types. Internalisation of activated TGF-β receptors provides a good example for how routes of entry and particular endosomal environments can influence signalling outcome. The intracellular effectors of TGF-β receptors, the R-SMAD transcription factors, translocate into the nucleus following activation, where they regulate transcriptional output. Interaction with inhibitory I-SMADs increases degradation of the activated receptor through recruitment of a ubiquitin ligase, SMURF7 (Schmierer and Hill 2007). TGF-β receptors can be internalised by both CME and CIE. Entry via CME results in delivery to endosomes that are positive for the early endosome marker, EEA1, which associates with early endosomes because it binds Rab5GTP and also has a FYVE domain that binds PtdIns3P (Simonsen et al. 1998). The enrichment of PtdIns3P in the early endosome also facilitates the recruitment and accumulation of another FYVE domain protein, SARA, which provides a bridge between the activated receptor and the R-SMAD, resulting in signal propagation (Tsukazaki et al. 1998). By contrast in the presence of higher ligand concentrations, the activated receptor is taken up via a caveolin-dependent CIE pathway. Endocytosed receptor/ligand complexes avoid delivery to the early endosome and instead target the excess ligand for degradation (Di Guglielmo et al. 2003). Targeting to the degradative pathway is largely due to interactions of the activated receptor with I-SMADs and recruitment of SMURF7. The choice of alternative endocytic pathways is proposed as a mechanism by which cells may limit their signalling in the context of high external ligand concentrations as well as providing evidence that endosomal localisation is essential for signalling. However, the system may be even further nuanced, as SARA can associate with activated TGF-β receptors at the cell surface (Runyan et al. 2005). Thus, one hypothesis is that its accumulation at endosomes contributes to the release of R-SMADs, allowing them to translocate to the nucleus (Corallino et al. 2015). These data together support a role for endocytic flux in the modulation of TGF-β signalling. This is further supported by the large number of endocytic proteins identified in a recent analysis of changes in the proteome and phosphoproteome that occur following TGF-β mediated epithelial to mesenchymal transition (D'Souza et al. 2014).

EGFR endocytosis shows a similar sensitivity to ligand concentration: following activation at low EGF concentrations (1–5 ng/ml), EGFR enters cells via CME, switching to a CIE pathway at increasing EGF concentrations and, at > 100 ng/ml, using the non-selective macropinocytic pathway. Similar to TGF-β, the switch between endocytic pathways is thought to allow cells to manage high concentrations of ligand, as EGFR uptake via CIE removes excess ligand and targets it for degradation (Sigismund et al. 2005). EGFR internalisation via CME, like that of TGF-β-R, is also required for sustained EGF signalling (Sigismund et al. 2008).

This raises the question as to how excess ligand is interpreted by cells. As discussed above, cells respond to threshold levels of EGF, which influences which internalisation pathway is followed. While EGFR receptor shows a gradual increase in receptor phosphorylation following binding of EGF, receptor ubiquitination exhibits a threshold response, which is regulated by recruitment of the Cbl ubiquitin ligase via the EGFR binding partner, Grb2 (Goh and Sorkin 2013). The dose-response of receptor ubiquitination was shown to mirror the dose-response of switching to the cholesterol-dependent CIE pathway (Sigismund et al. 2013). Uptake via this latter pathway has been shown to be dependent on membrane contact sites between the endoplasmic reticulum and the plasma membrane (Caldieri et al. 2017).

In contrast to a requirement for CME for productive EGF and TGF-β signalling, other activated growth factor receptors utilise alternative pathways for effective signalling depending on cellular contexts. In endothelial cells, FGF signalling to MAPK occurs via macropinocytosis of activated receptors (Elfenbein et al. 2012). Similarly, in the case of VEGFR, CME is responsible for the constitutive uptake of receptor in the absence of ligand. Engagement of the receptor with VEGFA stimulates macropinocytosis, and its uptake via this pathway is key for signalling outcomes such as cell survival and migration in in vitro angiogenesis assays (Basagiannis et al. 2016).

7.3.3 Receptor Recycling

Recycling of receptors from the early endosome via a Rab4- or Rab11-dependent pathway will potentiate signalling by maintaining a surface pool of receptors. The proportion of receptors that recycle depends on ligand concentration as well as level of receptor expression since the degradative pathway appears to be saturable (French et al. 1994). Because sorting into intraluminal vesicles is dependent on ubiquitination via association of ubiquitin ligases such as Cbl that are recruited to activated receptors, recycling is more likely for those receptor-ligand complexes where the ligand dissociates in early endosomes. Both EGF and TGF-α bind to EGFR but elicit very different cellular outcomes (Francavilla et al. 2016). Binding of TGF-α is more sensitive than EGF to the acidic conditions of the early endosome, and hence, it dissociates from EGFR, promoting its recycling and continued signalling. The recycling pathway has traditionally been considered a default pathway, transporting cargo that lacks specific signals for delivery to lysosomes. Emerging evidence, however, suggests that selection for recycling also requires sorting determinants and that, as for G protein-coupled receptors (Bahouth and Nooh 2017), growth factor receptors may also have signals targeting them to the recycling pathway (Parachoniak et al. 2011). The requirement for signal-mediated recycling makes sense given that rapid recycling allows cells to become sensitised to low levels of ligand.

To address the molecular mechanism underpinning the difference in receptor behaviour following engagement of EGF or TGF-α, Olsen and colleagues used quantitative mass spectrometry to measure changes in the receptor interactome as well as phosphorylation and ubiquitination following stimulation with either ligand (Francavilla et al. 2016). There were significant changes in many (>65) proteins. Specifically, Rab7 phosphorylation was identified as essential for EGFR degradation while the interaction of EGFR with the Rab11 effector, rab coupling protein (RCP), regulated receptor recycling, resulting in sustained signalling, cell proliferation and migration. Strikingly, modulation of either Rab7 phosphorylation or interaction with RCP allowed conversion of an EGF-like response to a TGF-α response (Francavilla et al. 2016). These data are supported by experiments in physiological contexts, e.g. TNF-α-mediated EGFR recycling prolongs EGFR signalling and thus facilitates more effective wound healing in the cornea (McClintock and Ceresa 2010).

7.4 Endocytosis Qualitatively Regulates Growth Factor Signalling

The first indication that receptor signalling was not simply a cell surface event came from studies in the Bergeron laboratory using plasma membrane and endosomal fractions from rat liver. These experiments demonstrated that EGFR could signal more effectively from endosomes than from the plasma membrane. This contrasted with the insulin growth factor receptor, which appeared to signal solely at the cell surface

(Di Guglielmo et al. 1994). It was shown, using dominant negative dynamin as an inhibitor of endocytosis, that EGF-dependent signalling to MAPK was inhibited if the receptor was trapped at the cell surface although other EGF-dependent events were not affected (Vieira et al. 1996). Subsequent studies largely supported the idea of qualitative regulation of signalling by the endocytic pathway (Sorkin and von Zastrow 2009) although there have been dissenting views. For example, it was shown that, in HeLa cells, blocking activated EGFR at the cell surface results in enhanced signalling, similar to that observed in the presence of high levels of EGFR expression and, moreover, that there were no significant qualitative differences in transcriptional output, arguing against a need for delivery to a particular domain for specialised signalling (Brankatschk et al. 2012). However, cytokine receptor signalling was much more sensitive to an endocytic block in this study, suggesting a global role for endocytic regulation of intracellular signalling. Furthermore, the availability of dynamin knockout mice allowed analysis of EGFR signalling in fibroblasts where endocytosis was inhibited. These studies indicated that, in these cells, MAP kinase signalling occurs primarily at the plasma membrane (Sousa et al. 2012). However in the latter study, there were differences in the kinetics of MAPK activation at low EGF concentrations and more sustained Akt signalling over a range of EGF concentrations. This suggests that although signalling can still occur when endocytosis is inhibited, there are likely to be qualitatively different signalling outputs resulting from disruptions in signal magnitude and timing (Villasenor et al. 2016), which reinforces the importance of endocytic flux in regulation of signalling.

7.4.1 Endosomes as Signalosomes

If endocytosis is responsible for qualitative differences in growth factor signalling, it follows that signalling receptors must be located within a particular membrane environment which is competent for a subset of signalling, a so-called signalosome. Signalosome formation is a very effective way to ensure signal specificity and to regulate signal strength (Kholodenko 2003). Membrane environments on the endocytic pathway could be established either by having different subpopulations of endosomes or membrane microdomains within the same limiting membrane, both of which would allow for the recruitment of specific scaffolds, adaptors, kinases and phosphatases, in response to growth factor activation (Fig. 7.3a). There is evidence to support a role for membrane microdomains in targeting growth factor receptors to the degradative pathway. Ubiquitinated receptors, destined for lysosomes, are recognised by the Hrs component of the ESCRT-0 complex and accumulate in domains enriched in Hrs that are separate from EEA1-positive microdomains. This segregation is dependent on clathrin, which forms flat lattices on endosomes and interacts directly with Hrs. This sorting mechanism is thought to be a prerequisite for recognition and subsequent incorporation in intraluminal vesicles in MVBs (Raiborg et al. 2002). By taking advantage of the large endosomes in the coelomocytes of *C.elegans*, it was demonstrated that within the same limiting endosomal

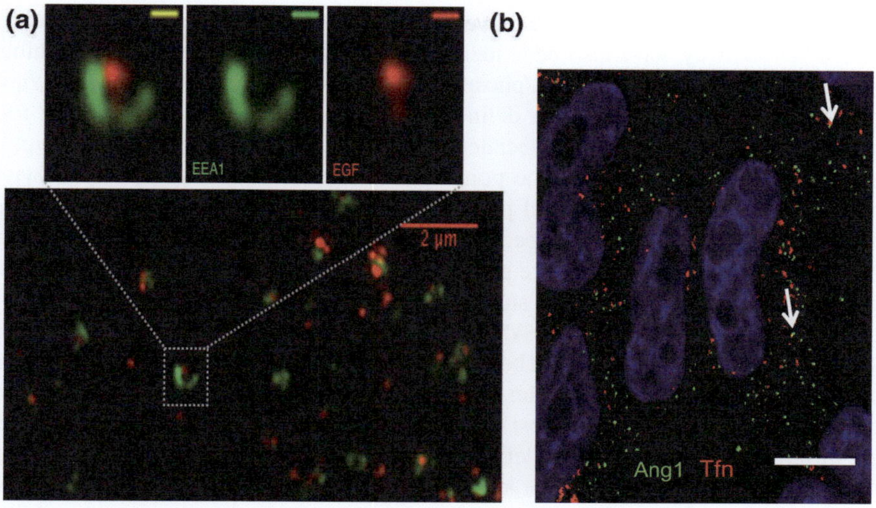

Fig. 7.3 *Multiple endosomal subdomains and compartments exist within cells* **a** Membrane microdomains exist within endosomal compartments. HeLa cells were incubated Alexa555-EGF (red) for 20 min at 37°C and then fixed and co-stained with rabbit anti-EEA1 antibodies followed by Alexa488- labelled (green) goat anti-rabbit antibodies. The insert shows Alexa555-EGF clustered in a microdomain of an early endosome labelled by EEA1. Images were collected on a DeltaVision/ GE OMX optical microscope for structured illumination (3D-SIM). Scale bar on insert is 0.25 μm. **b** HeLa cells, expressing the Tie2 receptor tyrosine kinase at endogenous levels, were incubated with fluorescently labeled Alexa 488-Ang1 (the ligand for Tie2) and fluorescently labeled transferrin (Alexa568-Tfn) for 5 minutes at 37°C. Images were captured on a Nikon Inverted Ti eclipse Dual camera system and deconvolved with the Nikon NIS elements software. The ROI is from a single Z-stack and the scale bar equals 10 μm. There is significant segregation of the two cargoes in different endosomal populations after this time of incubation. Arrows indicate rare examples of overlap

membrane, microdomains enriched in recycling cargo destined for the trans-Golgi network and driven by SNX1 and Rme-8 were distinct from Hrs-positive degradative microdomains. Furthermore, competition between these two domains regulates the balance of cargo flux through the recycling and degradative compartments (Norris et al. 2017). At the cell surface, there is evidence to suggest that some growth factor receptors are held within cholesterol-rich microdomains although this appears to be a prelude to their release for internalisation via CME following activation (Mineo et al. 1999; Foti et al. 2004; Hommelgaard et al. 2004). The relationship of receptor localisation and productive signalling in microdomains in the plasma membrane has been very elegantly illustrated for the interferon-gamma receptor (Blouin et al. 2016).

There is also evidence that endosomal subpopulations contribute to qualitative regulation of growth factor signalling. The canonical early endosome compartment is marked by the presence of EEA1 and is usually identified by the delivery of transferrin after 5-15 min of incubation. A number of other endosomal compartments, which play important roles in the movement of growth factor receptors through the endocytic pathway, have also been identified. APPL1 (adaptor proteins containing the pleckstrin homology domain, phosphotyrosine binding domain and leucine zipper motif) is a Rab5 effector that defines a population of endosomes important for signalling by EGFR. APPL1-associated endosomes are located in the cell periphery and are reached at early time points following EGFR stimulation (Miaczynska et al. 2004). This endosomal population shows relatively little labelling with transferrin suggesting a cargo-specific role. Live cell imaging has suggested that APPL1-associated endosomes are precursors of EEA1-positive endosomes (Zoncu et al. 2009) although more recent studies have defined endosomes that were APPL1-positive, EEA1-positive and both APPL1- and EEA1-positive, all of which behave as stable structures in that cargo can be reversibly delivered between them (Kalaidzidis et al. 2015). Furthermore, they can be reached by cargo which has entered cells by both CME and CIE. Perhaps, most important has been the demonstration that there are differences in flux of two model cargoes (transferrin as a nutrient, housekeeping, cargo and EGF as a signalling cargo) through these compartments. This has important implications for signalling because it suggests that there are components of the sorting and trafficking machinery which are sensitive to different cargoes and can thus adjust the rate at which a specific cargo fluxes through the pathway. Interestingly following a pulse of EGF, APPL1 endosomes appear to become refractory to the delivery of EGF (Miaczynska et al. 2004; Villasenor et al. 2016), suggesting a further potential mechanism of signal regulation.

Activated EGFR is reported to transit through another APPL1-independent endosome en route to EEA1-positive endosomes via an endosomal compartment labelled by SNX15 (Danson et al. 2013), a member of the sorting nexin family, which regulates membrane traffic often by binding/tubulating membranes (Cullen and Korswagen 2011). Moreover, there is also growing evidence to suggest that endosomes are very plastic with a recent report demonstrating that ESCRT-0 and ESCRT-I, normally considered as markers of the transition to late endosomes, are required for delivery of internalised EGF to EEA1-positive endosomes (Flores-Rodriguez et al. 2015).

EGFR can undergo ligand-independent internalisation in response to stresses such as UV irradiation. This is mediated by p38 and results in delivery to a subset of MVBs that are distinct from those that sequester ligand-activated receptors and target them for degradation. Sorting to this class of MVBs is mediated by actin regulators at the early endosome. It requires the ESCRT proteins, although it is independent of ubiquitination, and is necessary for ligand-independent signalling and cell survival. For this reason, this pathway may contribute to resistance to chemotherapy (Tomas et al. 2015).

7.4.2 Endosomal Microdomains as Signalosomes

An early example of how endosomal compartments can provide specific environments for optimal, context-specific, signalling is the localisation of the extracellular signal regulated kinase (ERK) scaffold LAMTOR3/MP1 (MEK1 partner) to late endosomes via LAMTOR2/p14. This localisation was shown to be essential for effective ERK1 signalling (Teis et al. 2002). The importance of the LAMTOR complex is reinforced by mice studies showing that knockout of LAMTOR1/p18 is embryonic lethal (Nada et al. 2009) while targeted gene disruption of LAMTOR2/p14 severely disrupts tissue homeostasis (Teis et al. 2006). Subsequent studies have revealed that another function of the LAMTOR2/3 complex is in regulating cell migration where the late endosome population containing the LAMTOR2/3 complex is delivered to mature focal adhesions (FAs). Delivery is essential for FA turnover and cell migration although the molecular mechanisms have yet to be elucidated (Schiefermeier et al. 2014). Recent morphological studies comparing mouse embryonic fibroblasts from wild-type and LAMTOR2 knockout mice suggest roles for the complex in the maintenance of recycling tubules from MVBs. This again highlights the complex interplay between signalling and endocytic machineries (Vogel et al. 2015).

Another good example of how different endosomal compartments can nuance a downstream signal is the activation of the transcription factor STAT3 and its translocation into the nucleus in response to either oncostatin M or c-Met activation. Oncostatin M activation of its cognate receptor elicits strong and rapid phosphorylation of STAT3 and its nuclear accumulation is independent of endocytosis. In contrast, HGF engagement of the c-Met receptor induces lower levels of phosphorylation over a longer time course. In this case, nuclear translocation of activated STAT3 requires microtubule-dependent delivery to a perinuclear endosome, enriched in Rab7. It is proposed that the kinase/phosphatase equilibrium that maintains the threshold level of STAT3 phosphorylation required for nuclear translocation is made possible by delivery of the activated receptor to the endosomal population that is in close proximity to the nucleus (Kermorgant and Parker 2008). More recently, delivery of activated c-Met to a perinuclear endosomal population was shown to be required to sustain Rac1 activation and promote cell migration in an invasive breast cancer cell model. Although c-Met was activated to the same extent in an earlier endosome compartment, this was not sufficient for optimal Rac1 activation, further supporting a role for regulated spatiotemporal flux through particular endosomal populations in specifying signalling outcomes (Menard et al. 2014).

All of the above illustrates the heterogeneity and dynamic organisation of the early endosomal network and is consistent with observations that internalised cargoes are often spatially segregated (Kalaidzidis et al. 2015) (Fig. 7.3b). The existence of multiple plastic compartments could be easily utilised by cells in a context-specific manner to modulate signalling.

7.4.3 Endocytic Flux: A Key Element of Signal Regulation

Both qualitative and quantitative regulations of growth factor receptor signalling will depend on the overall rate of endocytic flux, i.e. the rate at which signalling receptors are delivered along the endocytic pathway or to a particular endosomal environment. Therefore, it is not surprising that regulators of endocytic kinetics, such as Rab5GEFs and ESCRT complexes, regulate signalling. In cell culture models, overexpression or knockdown of Rab5GEFs and GAPs (Haas et al. 2005; Kong et al. 2007; Tall et al. 2001; Su et al. 2007; Balaji and Colicelli 2013) or ESCRT proteins (Doyotte et al. 2005; Malerod et al. 2007; Doyotte et al. 2008) prevents degradation of growth factors such as EGFR and accumulation of activated receptors on endosomes. However, the downstream effect on tumourigenesis, for example, can be much more complex (Mattissek and Teis 2014), indicating the need to investigate more thoroughly the effects of such changes in receptor degradation on transcriptional networks (see Future Directions).

7.4.4 Evidence for Endocytic Regulation of Signalling In Vivo

There have been several compelling studies which have used knockdown or over-expression of dominant negative endocytic proteins such as rabs and ESCRT components to demonstrate the importance of regulated endocytic flux in growth factor signalling in vivo (Lloyd et al. 2002; Vaccari and Bilder 2005; Lu and Bilder 2005). While these have been key in establishing regulation of endocytic flux as a paradigm, there is always concern as to whether there is a direct causal relationship between endocytosis and signalling in whole organisms. Inhibition of a process as fundamental as endocytosis is likely to have quite pleiotropic effects. It is also likely that endocytosis in vivo will be much more nuanced and that regulators of key proteins such as rabs will play major roles in regulating growth factor signalling output.

Rabex5 is the canonical Rab5GEF and also has a ubiquitin ligase domain which can ubiquitinate ras. Rabex5-mediated ubiquitination has been implicated in ras signalling in Drosophila, controlling size, wing vein development and cell fate decisions (Yan et al. 2010). While the functions of Rabex-5 in regulation of growth factor signalling in flies appears to be distinct, it is notable that other Rab5GEFs have either rasGAP (Rme-6/GAPex5) or ras-association (Rin1, 2 and 3) domains and the trafficking and signalling functions of the molecules appear to be linked for Rin1 (Tall et al. 2001) and for Rme-6 (Zhu, Ferreira and Smythe, unpublished results). Having trafficking and signalling modules within the same proteins can provide very sensitive integration of trafficking and signalling of growth factor receptors.

One way by which rab proteins are thought to effect biological processes is by formation of membrane microdomains through recruitment of subsets of effector molecules such as lipid kinases, which generate membrane microdomains conducive to the recruitment of further effector proteins. An example of positive feedback occurs at the early endosome where the recruitment by active Rab5 of its effector, rabaptin5 in complex with rabex5, is thought to increase local concentrations of active Rab5 (Horiuchi et al. 1997). This in turn promotes recruitment of Vps34, a PI3 kinase, which generates PtdIns3P. EEA1 then binds to the early endosome membrane by coincidence detection of Rab5GTP and PtdIns3P and generates a fusion-competent microdomain (Simonsen et al. 1998). Another RabGEF, Rme-6, is involved in the spatial activation of a functional pool of Rab5 during the process of clathrin-coated vesicle uncoating (Semerdjieva et al. 2008). This model of GEFs driving the spatial and temporal activation of rabs is further supported by the demonstration that GEFs are the major determinants for membrane targeting of rab proteins (Blumer et al. 2013).

RabGEFs and their effectors are thus excellent candidates to drive the assembly of membrane-associated signalosomes. A seminal study in zebrafish demonstrated that the Rab5 effectors, APPL1 and APPL2, are necessary to establish an endosomal subdomain in vivo that is specific for a subset of Akt signalling. Morpholino knockdown of the APPL proteins resulted in increased apoptosis in zebra fish. This was because Akt signalling for GSK3-β-mediated cell survival was compromised, while Akt-dependent activation of Tsc2, which regulates growth control, was unaffected. Endosomal localisation of APPL1 was shown to be critical for its role in cell survival (Schenck et al. 2008). This was a very elegant demonstration of how rab effectors can nuance signalling pathways and begins to explain how one downstream signalling output might be selected over another in vivo.

7.4.5 Growth Factor Receptors Regulate the Endocytic Machinery

The crosstalk between endocytosis and signalling can be bidirectional with growth factor receptors directly modulating the endocytic machinery. As mentioned above, EGFR activation of ras results in activation of the Rab5GEF activity of Rin1 thus promoting EGFR internalisation (Tall et al. 2001). Akt-dependent phosphorylation of PIP fyve which generates PtdIns(3,5)P_2, a prerequisite for ILV formation, is required for targeting EGFR to the degradative pathway (Er et al. 2013). Akt-dependent GSK3-β phosphorylation also activates the neuronal form of dynamin in non-small cell lung carcinoma cells to stimulate EGFR internalisation (Reis et al. 2015), and the upregulation of this pathway correlates with increased metastasis (Chen et al. 2017).

A recent study proposed that cells respond to EGF by an ‘analogue-to-digital’ conversion of the signal. This study observed that similar ‘quanta’ of activated EGFR were present in individual endosomes and proposed that cells respond to different extracellular ligand concentrations by altering the number of endosomes rather than the amount of ligand per endosome. Cells exposed to nerve growth factor (NGF) show more sustained activation of ERK compared to treatment with EGF. By comparing the numbers and sizes of endosomes labelled by different growth factors, the authors were able to show that exposing cells to NGF resulted in an increased number of small endosomes which correlated with sustained ERK activation. Strikingly, it was possible to prolong EGF activation of ERK signalling, converting it to an NGF-like outcome, by inhibiting homotypic fusion of early endosomes by siRNA knockdown of the fusion machinery (Villasenor et al. 2015). This provides strong support that individual growth factors can regulate their own flux through the pathway by modulation of the fusion machinery, as had been suggested by previous studies (Cavalli et al. 2001; Mace et al. 2005).

7.5 Future Directions

While a large body of evidence supports a role for endocytic trafficking in both qualitative and quantitative regulation of growth factor receptor signalling, the mechanisms by which these are achieved are functionally diverse. Are the broad mechanisms outlined above, endocytic flux, signal determination via ports of entry, signalosome formation, conversion of packets of signals to a digital output, sufficient to account for regulatory mechanisms which may differ in molecular detail? This raises the question as to whether there are fundamental principles governing signal regulation by endocytosis. Resolution of this issue will require a deeper understanding of the underlying molecular mechanisms.

Mathematical modelling has been used for many years to understand signalling networks (Kholodenko and Birtwistle 2009). Increasingly, proteomic approaches reveal how interactomes and post-translational modifications alter in response to growth factors and together these data can be combined with greater understanding of the spatiotemporal regulation of signalling networks to develop models that are directly testable. Advanced light microscopy, including super resolution (Li et al. 2015) and lattice sheet microscopy (Aguet et al. 2016) will allow us to understand the dynamics of the assembly and disassembly of signalosomes and to visualise the membrane environments of signalosomes, at a level of detail, unimaginable five years ago. One of the greatest challenges for the future is to understand how endocytic regulation of signalling occurs in vivo. The development of more relevant in vitro models to measure signalling and trafficking are likely to include the use of primary cells in 3D cultures (Zeigerer et al. 2017) and 'organ-on-a-chip' (Zhang et al. 2017). These systems will allow us to validate and extend studies from 2D systems in more physiological contexts, thus paving the way for in vivo studies. Together these approaches will progress towards a full understanding of how growth factors induce context-dependent signalling in vivo with profound importance in health and disease.

References

Aguet F, Upadhyayula S, Gaudin R, Chou YY, Cocucci E, He K, Chen BC, Mosaliganti K, Pasham M, Skillern W, Legant WR, Liu TL, Findlay G, Marino E, Danuser G, Megason S, Betzig E, Kirchhausen T (2016) Membrane dynamics of dividing cells imaged by lattice light-sheet microscopy. Mol Biol Cell 27:3418–3435

Antonny B, Burd C, De Camilli P, Chen E, Daumke O, Faelber K, Ford M, Frolov VA, Frost A, Hinshaw JE, Kirchhausen T, Kozlov MM, Lenz M, Low HH, Mcmahon H, Merrifield C, Pollard

TD, Robinson PJ, Roux A, Schmid S (2016) Membrane fission by dynamin: what we know and what we need to know. EMBO J 35:2270–2284
Bahouth SW, Nooh MM (2017) Barcoding of GPCR trafficking and signaling through the various trafficking roadmaps by compartmentalized signaling networks. Cell Signal 36:42–55
Balaji K, Colicelli J (2013) RIN1 regulates cell migration through Rab5 GTPases and ABL tyrosine kinases. Commun Integr Biol 6:e25421
Basagiannis D, Zografou S, Murphy C, Fotsis T, Morbidelli L, Ziche M, Bleck C, Mercer J, Christoforidis S (2016) VEGF induces signalling and angiogenesis by directing VEGFR2 internalisation through macropinocytosis. J Cell Sci 129:4091–4104
Behnia R, Munro S (2005) Organelle identity and the signposts for membrane traffic. Nature 438:597–604
Blouin CM, Hamon Y, Gonnord P, Boularan C, Kagan J, Viaris de Lesegno C, Ruez R, Mailfert S, Bertaux N, Loew D, Wunder C, Johannes L, Vogt G, Contreras FX, Marguet D, Casanova JL, Gales C, He HT, Lamaze C (2016) Glycosylation-dependent IFN-gammaR partitioning in lipid and actin nanodomains is critical for JAK activation. Cell 166:920–934
Blume-Jensen P, Hunter T (2001) Oncogenic kinase signalling. Nature 411:355–365
Blumer J, Rey J, Dehmelt L, Mazel T, Wu YW, Bastiaens P, Goody RS, Itzen A (2013) RabGEFs are a major determinant for specific Rab membrane targeting. J Cell Biol 200:287–300
Boucrot E, Ferreira AP, Almeida-Souza L, Debard S, Vallis Y, Howard G, Bertot L, Sauvonnet N, Mcmahon HT (2015) Endophilin marks and controls a clathrin-independent endocytic pathway. Nature 517:460–465
Brankatschk B, Wichert SP, Johnson SD, Schaad O, Rossner MJ, Gruenberg J (2012) Regulation of the EGF transcriptional response by endocytic sorting. Sci Signal 5:ra21
Brodsky FM (2012) Diversity of clathrin function: new tricks for an old protein. Annu Rev Cell Dev Biol 28:309–336
Buckley CM, King JS (2017) Drinking problems: Mechanisms of macropinosome formation and maturation. FEBS J
Caldieri G, Barbieri E, Nappo G, Raimondi A, Bonora M, Conte A, Verhoef L, Confalonieri S, Malabarba MG, Bianchi F, Cuomo A, Bonaldi T, Martini E, Mazza D, Pinton P, Tacchetti C, Polo S, Di Fiore PP, Sigismund S (2017) Reticulon 3-dependent ER-PM contact sites control EGFR nonclathrin endocytosis. Science 356:617–624
Cauvin C, Rosendale M, Gupta-Rossi N, Rocancourt M, Larraufie P, Salomon R, Perrais D, Echard A (2016) Rab35 GTPase triggers switch-like recruitment of the lowe syndrome lipid phosphatase OCRL on newborn endosomes. Curr Biol 26:120–128
Cavalli V, Vilbois F, Corti M, Marcote MJ, Tamura K, Karin M, Arkinstall S, Gruenberg J (2001) The stress-induced MAP kinase p38 regulates endocytic trafficking via the GDI:Rab5 complex. Mol Cell 7:421–432
Chen PH, Bendris N, Hsiao YJ, Reis CR, Mettlen M, Chen HY, Yu SL, Schmid SL (2017) Crosstalk between CLCb/Dyn1-mediated adaptive Clathrin-mediated endocytosis and epidermal growth factor receptor signaling increases metastasis. Dev Cell 40(278–288):e5
Christoforidis S, Mcbride HM, Burgoyne RD, Zerial M (1999) The Rab5 effector EEA1 is a core component of endosome docking. Nature 397:621–625
Corallino S, Malabarba MG, Zobel M, Di Fiore PP, Scita G (2015) Epithelial-to-Mesenchymal plasticity harnesses endocytic circuitries. Front Oncol 5:45
Cullen PJ, Korswagen HC (2011) Sorting nexins provide diversity for retromer-dependent trafficking events. Nat Cell Biol 14:29–37
D'Souza RC, Knittle AM, Nagaraj N, Van Dinther M, Choudhary C, Ten Dijke P, Mann M, Sharma K (2014) Time-resolved dissection of early phosphoproteome and ensuing proteome changes in response to TGF-beta. Sci Signal 7:rs5
Danson C, Brown E, Hemmings OJ, Mcgough IJ, Yarwood S, Heesom KJ, Carlton JG, Martin-Serrano J, May MT, Verkade P, Cullen PJ (2013) SNX15 links clathrin endocytosis to the PtdIns3P early endosome independently of the APPL1 endosome. J Cell Sci 126:4885–4899

Davidson EH (1993) Later embryogenesis: regulatory circuitry in morphogenetic fields. Development 118:665–690
De Renzis S, Sonnichsen B, Zerial M (2002) Divalent Rab effectors regulate the sub-compartmental organization and sorting of early endosomes. Nat Cell Biol 14:14
Del Conte-Zerial P, Brusch L, Rink JC, Collinet C, Kalaidzidis Y, Zerial M, Deutsch A (2008) Membrane identity and GTPase cascades regulated by toggle and cut-out switches. Mol Syst Biol 4:206
Di Guglielmo GM, Baass PC, Ou WJ, Posner BI, Bergeron JJ (1994) Compartmentalization of SHC, GRB2 and mSOS, and hyperphosphorylation of Raf-1 by EGF but not insulin in liver parenchyma. EMBO J 13:4269–4277
Di Guglielmo GM, Le Roy C, Goodfellow AF, Wrana JL (2003) Distinct endocytic pathways regulate TGF-beta receptor signalling and turnover. Nat Cell Biol 5:410–421
Di Paolo G, De Camilli P (2006) Phosphoinositides in cell regulation and membrane dynamics. Nature 443:651–657
Donaldson JG, Johnson DL, Dutta D (2016) Rab and Arf G proteins in endosomal trafficking and cell surface homeostasis. Small GTPases 7:247–251
Doyotte A, Russell MR, Hopkins CR, Woodman PG (2005) Depletion of TSG101 forms a mammalian "Class E" compartment: a multicisternal early endosome with multiple sorting defects. J Cell Sci 118:3003–3017
Doyotte A, Mironov A, Mckenzie E, Woodman P (2008) The Bro1-related protein HD-PTP/PTPN23 is required for endosomal cargo sorting and multivesicular body morphogenesis. Proc Natl Acad Sci U S A 105:6308–6313
Elfenbein A, Lanahan A, Zhou TX, Yamasaki A, Tkachenko E, Matsuda M, Simons M (2012) Syndecan 4 regulates FGFR1 signaling in endothelial cells by directing macropinocytosis. Sci Signal 5:ra36
Er EE, Mendoza MC, Mackey AM, Rameh LE, Blenis J (2013) AKT facilitates EGFR trafficking and degradation by phosphorylating and activating PIKfyve. Sci Signal 6:ra45
Flores-Rodriguez N, Kenwright DA, Chung PH, Harrison AW, Stefani F, Waigh TA, Allan VJ, Woodman PG (2015) ESCRT-0 marks an APPL1-independent transit route for EGFR between the cell surface and the EEA1-positive early endosome. J Cell Sci 128:755–767
Fortian A, Dionne LK, Hong SH, Kim W, Gygi SP, Watkins SC, Sorkin A (2015) Endocytosis of ubiquitylation-deficient EGFR mutants via Clathrin-coated pits is mediated by ubiquitylation. Traffic 16:1137–1154
Foti M, Moukil MA, Dudognon P, Carpentier JL (2004) Insulin and IGF-1 receptor trafficking and signalling. Novartis Found Symp 262:125–141; discussion 141–147, 265–268
Francavilla C, Papetti M, Rigbolt KT, Pedersen AK, Sigurdsson JO, Cazzamali G, Karemore G, Blagoev B, Olsen JV (2016) Multilayered proteomics reveals molecular switches dictating ligand-dependent EGFR trafficking. Nat Struct Mol Biol 23:608–618
French AR, Sudlow GP, Wiley HS, Lauffenburger DA (1994) Postendocytic trafficking of epidermal growth factor-receptor complexes is mediated through saturable and specific endosomal interactions. J Biol Chem 269:15749–15755
Goh LK, Sorkin A (2013) Endocytosis of receptor tyrosine kinases. Cold Spring Harb Perspect Biol 5:a017459
Goh LK, Huang F, Kim W, Gygi S, Sorkin A (2010) Multiple mechanisms collectively regulate clathrin-mediated endocytosis of the epidermal growth factor receptor. J Cell Biol 189:871–883
Grecco HE, Schmick M, Bastiaens PI (2011) Signaling from the living plasma membrane. Cell 144:897–909
Haas AK, Fuchs E, Kopajtich R, Barr FA (2005) A GTPase-activating protein controls Rab5 function in endocytic trafficking. Nat Cell Biol 7:887–893
Hommelgaard AM, Lerdrup M, Van Deurs B (2004) Association with membrane protrusions makes ErbB2 an internalization-resistant receptor. Mol Biol Cell 15:1557–1567

Horiuchi H, Lippe R, Mcbride HM, Rubino M, Woodman P, Stenmark H, Rybin V, Wilm M, Ashman K, Mann M, Zerial M (1997) A novel Rab5 GDP/GTP exchange factor complexed to Rabaptin-5 links nucleotide exchange to effector recruitment and function. Cell 90:1149–1159

Hurley JH, Emr SD (2006) The ESCRT complexes: structure and mechanism of a membrane-trafficking network. Annu Rev Biophys Biomol Struct 35:277–298

Kalaidzidis I, Miaczynska M, Brewinska-Olchowik M, Hupalowska A, Ferguson C, Parton RG, Kalaidzidis Y, Zerial M (2015) APPL endosomes are not obligatory endocytic intermediates but act as stable cargo-sorting compartments. J Cell Biol 211:123–144

Kalin S, Hirschmann DT, Buser DP, Spiess M (2015) Rabaptin5 is recruited to endosomes by Rab4 and Rabex5 to regulate endosome maturation. J Cell Sci 128:4126–4137

Kermorgant S, Parker PJ (2008) Receptor trafficking controls weak signal delivery: a strategy used by c-Met for STAT3 nuclear accumulation. J Cell Biol 182:855–863

Kholodenko BN (2003) Four-dimensional organization of protein kinase signaling cascades: the roles of diffusion, endocytosis and molecular motors. J Exp Biol 206:2073–2082

Kholodenko BN, Birtwistle MR (2009) Four-dimensional dynamics of MAPK information processing systems. Wiley Interdiscip Rev Syst Biol Med 1:28–44

Kong C, Su X, Chen PI, Stahl PD (2007) Rin1 interacts with signal-transducing adaptor molecule (STAM) and mediates epidermal growth factor receptor trafficking and degradation. J Biol Chem 282:15294–15301

Lemmon MA, Schlessinger J (2010) Cell signaling by receptor tyrosine kinases. Cell 141:1117–1134

Li D, Shao L, Chen BC, Zhang X, Zhang M, Moses B, Milkie DE, Beach JR, Hammer JA, Pasham M, Kirchhausen T, Baird MA, Davidson MW, Xu P, Betzig E (2015) Advanced imaging. Extended-resolution structured illumination imaging of endocytic and cytoskeletal dynamics. Science 349:aab3500

Liu K, Xing R, Jian Y, Gao Z, Ma X, Sun X, Li Y, Xu M, Wang X, Jing Y, Guo W, Yang C (2017) WDR91 is a Rab7 effector required for neuronal development. J Cell Biol 216:3307–3321

Lloyd TE, Atkinson R, Wu MN, Zhou Y, Pennetta G, Bellen HJ (2002) Hrs regulates endosome membrane invagination and tyrosine kinase receptor signaling in Drosophila. Cell 108:261–269

Lu H, Bilder D (2005) Endocytic control of epithelial polarity and proliferation in Drosophila. Nat Cell Biol 7:1232–1239

Mace G, Miaczynska M, Zerial M, Nebreda AR (2005) Phosphorylation of EEA1 by p38 MAP kinase regulates mu opioid receptor endocytosis. EMBO J 24:3235–3246

Maib H, Smythe E, Ayscough K (2017) Forty years on: clathrin-coated pits continue to fascinate. Mol Biol Cell 28:843–847

Malerod L, Stuffers S, Brech A, Stenmark H (2007) Vps22/EAP30 in ESCRT-II mediates endosomal sorting of growth factor and chemokine receptors destined for lysosomal degradation. Traffic 8:1617–1629

Mattissek C, Teis D (2014) The role of the endosomal sorting complexes required for transport (ESCRT) in tumorigenesis. Mol Membr Biol 31:111–119

Mayor S, Riezman H (2004) Sorting GPI-anchored proteins. Nat Rev Mol Cell Biol 5:110–120

Mayor S, Parton RG, Donaldson JG (2014) Clathrin-independent pathways of endocytosis. Cold Spring Harb Perspect Biol 6

Mcclintock JL, Ceresa BP (2010) Transforming growth factor-{alpha} enhances corneal epithelial cell migration by promoting EGFR recycling. Invest Ophthalmol Vis Sci 51:3455–3461

Menard L, Parker PJ, Kermorgant S (2014) Receptor tyrosine kinase c-Met controls the cytoskeleton from different endosomes via different pathways. Nat Commun 5:3907

Miaczynska M, Christoforidis S, Giner A, Shevchenko A, Uttenweiler-Joseph S, Habermann B, Wilm M, Parton RG, Zerial M (2004) APPL proteins link Rab5 to nuclear signal transduction via an endosomal compartment. Cell 116:445–456

Mineo C, Gill GN, Anderson RG (1999) Regulated migration of epidermal growth factor receptor from caveolae. J Biol Chem 274:30636–30643

Montagnac G, De Forges H, Smythe E, Gueudry C, Romao M, Salamero J, Chavrier P (2011) Decoupling of activation and effector binding underlies ARF6 priming of fast endocytic recycling. Curr Biol 21:574–579

Nada S, Hondo A, Kasai A, Koike M, Saito K, Uchiyama Y, Okada M (2009) The novel lipid raft adaptor p18 controls endosome dynamics by anchoring the MEK-ERK pathway to late endosomes. EMBO J 28:477–489

Norris A, Tammineni P, Wang S, Gerdes J, Murr A, Kwan KY, Cai Q, Grant BD (2017) SNX-1 and RME-8 oppose the assembly of HGRS-1/ESCRT-0 degradative microdomains on endosomes. Proc Natl Acad Sci U S A 114:E307–E316

Parachoniak CA, Luo Y, Abella JV, Keen JH, Park M (2011) GGA3 functions as a switch to promote Met receptor recycling, essential for sustained ERK and cell migration. Dev Cell 20:751–763

Pfeffer SR (2017) Rab GTPases: master regulators that establish the secretory and endocytic pathways. Mol Biol Cell 28:712–715

Pinilla-Macua I, Watkins SC, Sorkin A (2016) Endocytosis separates EGF receptors from endogenous fluorescently labeled HRas and diminishes receptor signaling to MAP kinases in endosomes. Proc Natl Acad Sci U S A 113:2122–2127

Raiborg C, Bache KG, Gillooly DJ, Madshus IH, Stang E, Stenmark H (2002) Hrs sorts ubiquitinated proteins into clathrin-coated microdomains of early endosomes. Nat Cell Biol 4:394–398

Reis CR, Chen PH, Srinivasan S, Aguet F, Mettlen M, Schmid SL (2015) Crosstalk between Akt/GSK3beta signaling and dynamin-1 regulates clathrin-mediated endocytosis. EMBO J 34:2132–2146

Rink J, Ghigo E, Kalaidzidis Y, Zerial M (2005) Rab conversion as a mechanism of progression from early to late endosomes. Cell 122:735–749

Robinson MS (2015) Forty Years of Clathrin-coated Vesicles. Traffic 16:1210–1238

Runyan CE, Schnaper HW, Poncelet AC (2005) The role of internalization in transforming growth factor beta1-induced Smad2 association with Smad anchor for receptor activation (SARA) and Smad2-dependent signaling in human mesangial cells. J Biol Chem 280:8300–8308

Schenck A, Goto-Silva L, Collinet C, Rhinn M, Giner A, Habermann B, Brand M, Zerial M (2008) The endosomal protein Appl1 mediates Akt substrate specificity and cell survival in vertebrate development. Cell 133:486–497

Schiefermeier N, Scheffler JM, De Araujo ME, Stasyk T, Yordanov T, Ebner HL, Offterdinger M, Munck S, Hess MW, Wickstrom SA, Lange A, Wunderlich W, Fassler R, Teis D, Huber LA (2014) The late endosomal p14-MP1 (LAMTOR2/3) complex regulates focal adhesion dynamics during cell migration. J Cell Biol 205:525–540

Schmid SL, Sorkin A, Zerial M (2014) Endocytosis: past, present, and future. Cold Spring Harb Perspect Biol 6:a022509

Schmidt O, Teis D (2012) The ESCRT machinery. Curr Biol 22:R116–R120

Schmierer B, Hill CS (2007) TGFbeta-SMAD signal transduction: molecular specificity and functional flexibility. Nat Rev Mol Cell Biol 8:970–982

Schoneberg J, Lee IH, Iwasa JH, Hurley JH (2017) Reverse-topology membrane scission by the ESCRT proteins. Nat Rev Mol Cell Biol 18:5–17

Semerdjieva S, Shortt B, Maxwell E, Singh S, Fonarev P, Hansen J, Schiavo G, Grant BD, Smythe E (2008) Co-ordinated regulation of AP2 uncoating from clathrin coated vesicles by Rab5 and hRME-6. J Cell Biol 183:499–511

Shin HW, Hayashi M, Christoforidis S, Lacas-Gervais S, Hoepfner S, Wenk MR, Modregger J, Uttenweiler-Joseph S, Wilm M, Nystuen A, Frankel WN, Solimena M, De Camilli P, Zerial M (2005) An enzymatic cascade of Rab5 effectors regulates phosphoinositide turnover in the endocytic pathway. J Cell Biol 170:607–618

Sigismund S, Woelk T, Puri C, Maspero E, Tacchetti C, Transidico P, Di Fiore PP, Polo S (2005) Clathrin-independent endocytosis of ubiquitinated cargos. Proc Natl Acad Sci U S A 102:2760–2765

Sigismund S, Argenzio E, Tosoni D, Cavallaro E, Polo S, Di Fiore PP (2008) Clathrin-mediated internalization is essential for sustained EGFR signaling but dispensable for degradation. Dev Cell 15:209–219

Sigismund S, Algisi V, Nappo G, Conte A, Pascolutti R, Cuomo A, Bonaldi T, Argenzio E, Verhoef LG, Maspero E, Bianchi F, Capuani F, Ciliberto A, Polo S, Di Fiore PP (2013) Threshold-controlled ubiquitination of the EGFR directs receptor fate. EMBO J 32:2140–2157

Simonsen A, Lippe R, Christoforidis S, Gaullier JM, Brech A, Callaghan J, Toh BH, Murphy C, Zerial M, Stenmark H (1998) EEA1 links PI(3)K function to Rab5 regulation of endosome fusion. Nature 394:494–498

Sorkin A, von Zastrow M (2009) Endocytosis and signalling: intertwining molecular networks. Nat Rev Mol Cell Biol 10:609–622

Sousa LP, Lax I, Shen H, Ferguson SM, De Camilli P, Schlessinger J (2012) Suppression of EGFR endocytosis by dynamin depletion reveals that EGFR signaling occurs primarily at the plasma membrane. Proc Natl Acad Sci U S A 109:4419–4424

Stenmark H (2009) Rab GTPases as coordinators of vesicle traffic. Nat Rev Mol Cell Biol 10:513–525

Su X, Kong C, Stahl PD (2007) GAPex-5 mediates ubiquitination, trafficking, and degradation of epidermal growth factor receptor. J Biol Chem 282:21278–21284

Tall GG, Barbieri MA, Stahl PD, Horazdovsky BF (2001) Ras-activated endocytosis is mediated by the Rab5 guanine nucleotide exchange activity of RIN1. Dev Cell 1:73–82

Teis D, Wunderlich W, Huber LA (2002) Localization of the MP1-MAPK scaffold complex to endosomes is mediated by p14 and required for signal transduction. Dev Cell 3:803–814

Teis D, Taub N, Kurzbauer R, Hilber D, De Araujo ME, Erlacher M, Offterdinger M, Villunger A, Geley S, Bohn G, Klein C, Hess MW, Huber LA (2006) p14-MP1-MEK1 signaling regulates endosomal traffic and cellular proliferation during tissue homeostasis. J Cell Biol 175:861–868

Tomas A, Vaughan SO, Burgoyne T, Sorkin A, Hartley JA, Hochhauser D, Futter CE (2015) WASH and Tsg101/ALIX-dependent diversion of stress-internalized EGFR from the canonical endocytic pathway. Nat Commun 6:7324

Traub LM, Bonifacino JS (2013) Cargo Recognition in Clathrin-Mediated Endocytosis. Cold Spring Harb Perspect Biol 5

Tsukazaki T, Chiang TA, Davison AF, Attisano L, Wrana JL (1998) SARA, a FYVE domain protein that recruits Smad2 to the TGFbeta receptor. Cell 95:779–791

Vaccari T, Bilder D (2005) The Drosophila tumor suppressor vps25 prevents nonautonomous over-proliferation by regulating notch trafficking. Dev Cell 9:687–698

Vieira AV, Lamaze C, Schmid SL (1996) Control of egf receptor signaling by clathrin-mediated endocytosis. Science 274:2086–2089

Villasenor R, Nonaka H, Del Conte-Zerial P, Kalaidzidis Y, Zerial M (2015) Regulation of EGFR signal transduction by analogue-to-digital conversion in endosomes. Elife 4

Villasenor R, Kalaidzidis Y, Zerial M (2016) Signal processing by the endosomal system. Curr Opin Cell Biol 39:53–60

Vogel GF, Ebner HL, De Araujo ME, Schmiedinger T, Eiter O, Pircher H, Gutleben K, Witting B, Teis D, Huber LA, Hess MW (2015) Ultrastructural morphometry points to a new role for LAMTOR2 in regulating the Endo/Lysosomal system. Traffic 16:617–634

Wandinger-Ness A, Zerial M (2014) Rab proteins and the compartmentalization of the endosomal system. Cold Spring Harb Perspect Biol 6:a022616

Yan H, Jahanshahi M, Horvath EA, Liu HY, Pfleger CM (2010) Rabex-5 ubiquitin ligase activity restricts Ras signaling to establish pathway homeostasis in Drosophila. Curr Biol 20:1378–1382

Zeigerer A, Wuttke A, Marsico G, Seifert S, Kalaidzidis Y, Zerial M (2017) Functional properties of hepatocytes in vitro are correlated with cell polarity maintenance. Exp Cell Res 350:242–252

Zhang YS, Aleman J, Shin SR, Kilic T, Kim D, Mousavi Shaegh SA, Massa S, Riahi R, Chae S, Hu N, Avci H, Zhang W, Silvestri A, Sanati Nezhad A, Manbohi A, De Ferrari F, Polini A, Calzone G, Shaikh N, Alerasool P, Budina E, Kang J, Bhise N, Ribas J, Pourmand A, Skardal A, Shupe T, Bishop CE, Dokmeci MR, Atala A, Khademhosseini A (2017) Multisensor-integrated

organs-on-chips platform for automated and continual in situ monitoring of organoid behaviors. Proc Natl Acad Sci U S A 114:E2293–E2302
Zoncu R, Perera RM, Balkin DM, Pirruccello M, Toomre D, De Camilli P (2009) A phosphoinositide switch controls the maturation and signaling properties of APPL endosomes. Cell 136:1110–1121

Chapter 8
Role of the Endocytosis of Caveolae in Intracellular Signaling and Metabolism

Olivia Muriel, Miguel Sánchez-Álvarez, Raffaele Strippoli and Miguel Angel del Pozo

Contents

Abstract Caveolae are 60–80 nm invaginated plasma membrane (PM) nanodomains, with a specific lipid and protein composition, which assist and regulate multiple processes in the plasma membrane—ranging from the organization of signalling complexes to the mechanical adaptation to changes in PM tension. However, since their initial descriptions, these structures have additionally been found tightly linked to internalization processes, mechanoadaptation, to the regulation of

Muriel, Sánchez-Álvarez, Strippoli—These authors contributed equally to this work.

O. Muriel · M. Sánchez-Álvarez · R. Strippoli · M. A. del Pozo (✉)
Mechanoadaptation and Caveolae Biology Lab, Cell and Developmental Biology Area, Centro Nacional de Investigaciones Cardiovasculares, 28029 Madrid, Spain
e-mail: madelpozo@cnic.es

O. Muriel
Department of Fundamental Microbiology, University of Lausanne, Quartier UNIL-Sorge. Bâtiment Biophore, 1015 Lausanne, Switzerland

R. Strippoli
Istituto Pasteur Italia-Fondazione Cenci Bolognetti, Department of Cellular Biotechnologies and Hematology, Sapienza University of Rome, Rome, Italy

C. Lamaze and I. Prior (eds.), *Endocytosis and Signaling*, Progress in Molecular and Subcellular Biology 57, https://doi.org/10.1007/978-3-319-96704-2_8

signalling events and of endosomal trafficking. Here, we review caveolae biology from this perspective, and its implications for cell physiology and disease.

8.1 Introduction

Caveolae are 60–80 nm invaginated plasma membrane (PM) nanodomains with a specific lipid and protein composition. Although they were initially described as flask-shaped with a constricted neck (Palade 1953; Yamada 1955), later cryo-EM studies support a wider range of caveolar opening sizes (Richter et al. 2008; Schlormann et al. 2010). Recent structural studies have shown that caveolae may have a polygonal shape (Ludwig et al. 2016; Stoeber et al. 2016; Walser et al. 2012).

They are present in most vertebrate cell types, and their density varies with (i) cell type—tissues subjected to large variations in mechanical challenges being those with a larger caveolar density (adipose tissue, vascular endothelium and muscle) (Parton and del Pozo 2013; Thorn et al. 2003); (ii) external cues—such as shear stress in endothelial cells (Rizzo et al. 2003), and (iii) subcellular location—for example, *caveolae* are concentrated at the basolateral surface of epithelial cells and at the rear of migrating cells (Parat et al. 2003; Parton and del Pozo 2013; Scheiffele et al. 1998).

Caveolae functions are not fully understood, being important for signaling organization, lipid composition control, and mechanoreception and compliance to tensile stress.

Here, we review current knowledge and standing questions regarding the biology of *caveolae* and their components, with a focus on the links between regulation of endocytosis, cell signaling, and metabolism.

8.2 Caveolar Components

8.2.1 Caveolins

Evolutionarily restricted to metazoans (Kirkham et al. 2008), three paralogs have been identified in vertebrates: caveolin 1 (CAV1), caveolin 2 (CAV2), and caveolin 3 (CAV3). (Williams et al. 2004). While CAV1 and CAV2 are simultaneously expressed in all tissues except skeletal muscle (Scherer et al. 1996, 1997), CAV3 is exclusive to striated muscle cells (Song et al. 1996; Tang et al. 1996). In a few specific cases, such as smooth muscle cells, cardiomyocytes or zebrafish notochord cells, all three caveolins are simultaneously expressed (Head et al. 2006; Nixon et al. 2007; Patel et al. 2007; Robenek et al. 2008). By virtue of their predicted "hairpin-like" structure, they are inserted across their inner domains through the cytoplasmic leaflet of cellular membranes, leaving accessible N- and C-terminal regions for interaction with other cellular components (Dupree et al. 1993; Monier et al. 1995).

CAV1 is the most studied caveolin and the most pervasively required (with the exception of skeletal muscle) for the assembly of *caveolae,* whereas EHD2 (Eps-15 homology domain-containing protein 2) and CAVINS are required for stabilization and morphology, respectively. Its caveolin-scaffolding domain (CSD), encompassing residues 82-101, critically contributes to target caveolin to membranes (Schlegel et al. 1999). Despite being previously conceived as a key element for the signaling functions of CAV1 through enabling its interaction with signaling proteins and cholesterol (Bernatchez et al. 2005), this view has been recently challenged (Byrne et al. 2012; Collins et al. 2012).

Posttranscriptional modifications have a relevant role in the regulation of CAV1 biology. Palmitoylation seems to be important to stabilize the tertiary structure of the protein (Dietzen et al. 1995; Monier et al. 1996) and favors the interaction between CAV1 and cholesterol (Uittenbogaard and Smart 2000). Ser80 phosphorylation regulates the localization to endoplasmic reticulum membranes and the access to the secretory route (Schlegel et al. 2001). Tyr14 phosphorylation is crucial for the regulatory influence of CAV1 on several signaling pathways (Cao et al. 2002; Labrecque et al. 2004) by modulating the interaction of CAV1 with different signaling mediators and by determining potentially associated processes (i.e., internalization of *caveolae* (del Pozo et al. 2005), persistent/directional migration (Grande-Garcia et al. 2007), and cellular 3D microenvironment remodeling (Goetz et al. 2011). Together with CSD functional relevance, Tyr14 phosphorylation is further discussed later in this chapter.

8.2.2 *Cavins*

Cavins are cytoplasmic proteins exclusively found in vertebrates, proposed to play adaptor/scaffolding roles in *caveolae* by modulating the organization and functions of caveolins (Chidlow and Sessa 2010). Four paralogs have been described: Cavin-1, Cavin-2, Cavin-3, and Cavin-4 (Bastiani et al. 2009; Hill et al. 2008; McMahon et al. 2009; Vinten et al. 2005). All cavins form large heteromeric complexes through their N-terminal coiled-coil domains, subsequently recruited to *caveolae* in cells expressing caveolins. Another common characteristic of cavins is their ability to bind phosphatidylserine directly and to undergo profuse post-translational modifications, such as phosphorylation (Rahman and Sward 2009).

Cavin-1, also called polimerase I and transcript release factor (PTRF), was originally identified as a nuclear protein with transcriptional regulation activity in the nucleolus (Jansa et al. 1998). Cavin-1 complexes with CAV1 and CAV2 and is essential for the formation of *caveolae* (Aboulaich et al. 2004; Hill et al. 2008; Liu et al. 2008; Liu and Pilch 2008; Ludwig et al. 2013) but may be indirect (Liu and Pilch 2008). Downregulation of Cavin-1 leads to disassembly of *caveolae*, increased lateral diffusion and subsequent release and lysosomal degradation of CAV1, because Cavin-1 stabilizes CAV1 oligomers at the PM (Hansen et al. 2013; Hill et al. 2008; Liu and Pilch 2008). Since Cavin-1 traffics to the PM independently from CAV1 (Hayer et al. 2010a) and can be released from *caveolae* upon osmotic swelling (Sinha et al. 2011), non-caveolar roles of Cavin-1 have been also proposed. Recently, Cavin1 has been shown to regulate ribosomal RNA synthesis in response to metabolic challenges in mature adipocytes. This regulation of RNA polymerase I transcriptional activity by Cavin1 may contribute to lipodystrophy phenotypes observed in Cavin1-deficient mice and humans, and be integral to Cavin 1-dependent lipid regulation (Liu and Pilch 2016).

Cavin-2 or serum deprivation protein response (SDPR) is also necessary for the formation *caveolae* in lung endothelium and adipose tissue, presumably by promoting membrane curvature (Hansen et al. 2009; Nabi 2009). Cavin-2 interacts directly with Cavin-1, and it is necessary to maintain stable expression levels of Cavin-1 and CAV1. Cavin-3 or Sdr-related gene product that binds to c-kinase (SRBC) participates in the formation of caveolar endocytic carriers (McMahon et al. 2009). However, Cavin-3 is not involved in the morphogenesis of *caveolae* (Hansen et al. 2013), although it regulates its dynamics (Mohan et al. 2015). Finally, Cavin-4 or muscle-restricted, coiled-coil protein (MURC) is highly specific for striated muscle, where it colocalizes with CAV3 (Bastiani et al. 2009; Ogata et al. 2008; Tagawa et al. 2008).

In non-muscle cells CAV1, CAV2, Cavin-1, Cavin-2, and Cavin-3 presumably conform the core scaffold determining the characteristic shape of *caveolae*. Stoichiometry is essential and is kept constant—hence the pervasive cross-regulation at expression and turnover levels across most components. For example, Cavin-1 is essential for the expression and stability of all the other cavins and of caveolins, while CAV1 is also required for the expression and stability of all cavins except for Cavin-4 (Hansen et al. 2013). In HeLa cells, CAV1 and CAV2 are the most abundant proteins in the complex—Cavin-1 is fourfold less represented, and Cavin-2 and Cavin-3 are the least abundant and compete with each other for binding the complex (Ludwig et al. 2013).

8.2.3 Other Regulators

Apart from caveolins and cavins, the inherent components of *caveolae*, EHD2, dynamin-related ATPase protein family and PACSIN2 (PKC and casein kinase substrate in neurons 2), two proteins that interact with each other have been identified as additional components of *caveolae* (Parton and del Pozo 2013). EHD1 and EHD4 also associate with *caveolae*, but EHD2 has been characterized in greater detail. EHD2 is not required for *caveolae* formation, but it is involved in the formation of clusters of *caveolae* conforming membrane domains with stretch-buffering capacity (Yeow et al. 2017). EHD2 is a negative regulator of the internalization and dynamics of *caveolae* (Moren et al. 2012; Stoeber et al. 2012; Hoernke et al. 2017). It binds to phosphatidylinositol-4,5-bisphosphate (PtdIns(4,5)P_2) (Daumke et al. 2007)—which is enriched in the caveolar opening (Fujita et al. 2009). EDH family shares important properties with proteins belonging to dynamin family: Both oligomerize into rings, undergo lipid-stimulated nucleotide hydrolysis, and tubulate membranes (Daumke et al. 2007). PACSIN2 is a bin/amphiphysin/Rvs (BAR) domain protein that can sense and modulate membrane curvature involved in clathrin-mediated endocytosis. It binds CAV1 (Senju et al. 2011) and is essential for the maintenance of *caveolae* structure (Hansen et al. 2011; Koch et al. 2012). Recently, PACSIN3, also called syndapin III, has been shown to be crucial for caveolar invagination and membrane tension (Seemann et al. 2017).

Other proteins, such as integrins, are not required for the shaping of *caveolae*, but play a major role in the localization of CAV1 at the PM level (del Pozo et al. 2005; Echarri et al. 2007). CAV1, on the other hand, may control integrin endocytosis (Bass et al. 2011; Shi and Sottile 2008). The role of integrins in the internalization of *caveolae* will be discussed in detail later in this chapter.

8.3 Functions of *Caveolae*

Due to its characteristic shape, caveolar invaginations appeared in early studies as good candidates to embody endocytic routes alternative to clathrin (Kurzchalia and Parton 1999; Williams and Lisanti 2005). Their functional relevance to multiple cellular processes was revealed soon after. *Caveolae* contribute to the regulation and spatial compartmentalization of several signaling cascades serving as platforms to integrate the activity and/or localization of such molecules with their effectors (Anderson 1998; Drab et al. 2001). Through the modulation of several signaling pathways, *caveolae* contribute to the regulation of cell polarization, cell migration, cell cycle, apoptosis, and extracellular matrix remodeling (Del Pozo and Schwartz 2007; Goetz et al. 2011; Grande-Garcia et al. 2007; Guan 2004; Le Roy and Wrana 2005; Simons and Toomre 2000). Caveolar components emerge as pivotal regulators of cholesterol homeostasis, lipid sorting and metabolism, and intermembrane transactions (Fu et al. 2004; Razani et al. 2002).

8.3.1 Internalization of Caveolae

Internalization of *caveolae* was first reported by Parton and Simons in 1994 as phosphorylation- and actin cytoskeleton-dependent (Parton et al. 1994). Internalized CAV1-containing vesicles undergo long-range microtubule-dependent trafficking (Conrad et al. 1995; McMahon et al. 2009; Mundy et al. 2002) or can engage in a particular, transient contact with and separation from the PM, termed "kiss-and-run" (Pelkmans and Zerial 2005).

Once internalized, *caveolae* fuse with early endosomes from where components of *caveolae* can recycle back to the PM, equilibrium believed to be essential to maintain a constant caveolar density (Boucrot et al. 2011; Parton and del Pozo 2013). Indeed, loss of cavins or perturbation of cholesterol levels leads to internalization and regulated degradation of non-caveolar caveolin (Hayer et al. 2010b; Hill et al. 2008). Moreover, in the absence of cavins (and hence, of caveolae), CAV1 is endocytosed preferentially through the CLIC/GEEC pathway (Chaudhary et al. 2014). Oligomers of internalized non-caveolar caveolins are mono-ubiquitylated and recognized by VCP (AAA-ATPase vasolin-containing protein), which binds to them to form a complex with UBXD1 (UBX domain-containing protein 1). This complex is then incorporated into intraluminal vesicle (ILVs) of multivesicular bodies (MVBs) (Ritz et al. 2011). Intriguingly, CAV1 also modulates the substrate specificity of VCP/p97 through as yet poorly understood mechanisms (Chen et al. 2013), and might thus act as a co-regulator of cargo sorting toward MVBs and lysosomes.

Despite these lines of evidence, a specific active role for caveolar internalization as an endocytic pathway itself is still subject to debate, largely because specific candidate cargoes have not been identified to date. SV40 and cholera toxin, both ligands of GM1 (highly enriched in *caveolae*), were originally believed to be specific substrates for *caveolae*-dependent endocytosis (Pelkmans et al. 2001), and their use became a standard in the field (Echarri and del Pozo 2015). However, their caveolin-independent internalization is currently well established (Damm et al. 2005; Ewers et al. 2010; Parton and del Pozo 2013). Experimental settings relying on the overexpression of caveolar components have further added to challenges in interpretation across studies (Parton and del Pozo 2013). Furthermore, the basal rate of internalization of *caveolae* is low (~5% of the whole population), and is reflected by a morphological continuum (Bitsikas et al. 2014; Kirkham et al. 2005)—thus distinguishing between endocytosed caveolar vesicles and membrane folds including *caveolae* still connected with external cell surface can be remarkably challenging (Parton et al. 2002).

Still, regardless of these controversies, *caveolae* internalization does occur and is critical for the regulation of multiple signal transduction pathways (Anderson 1998; Parton and Simons 2007). Table 8.1 summarizes major known stimuli of caveolar internalization and associated regulators. However, the molecular underpinnings of such regulation are not clearly established yet.

The physio-pathological relevance of internalization of *caveolae* is evidenced by (i) its function as an entry gate for a number of relevant intracellular pathogens, such as Streptococcus (Almeida et al. 2010; Benga et al. 2004), SV40 and polioma virus (Pelkmans and Helenius 2002; Pelkmans et al. 2001; Shin et al. 2000), (ii) its role in mechanoreception and mechanotransduction (Sinha et al. 2011), and (iii) its importance in the regulation of lipid trafficking, cholesterol homeostasis, albumin transcytosis, and signaling receptor internalization (Echarri et al. 2007; Minshall et al. 2000; Nichols 2003).

Moreover, multiple levels of crosstalk between caveolar and non-caveolar clathrin-independent endocytosis have been proposed. CAV1, CAV3, Cavin-1, and Cavin-3 are potent inhibitors of the clathrin-independent carriers/GPI-AP enriched early endosomal compartment (CLIC/GEEC) endocytic pathway, in a process independent of caveola formation (Chaudhary et al. 2014).

Table 8.1 Molecular mechanisms reported to drive the internalization of caveolar domains

Regulator	Experimental evidence of *caveolae* endocytosis	References
Actin cytoskeleton	SV40 infection is inhibited by latrunculin A	Pelkmans and Helenius (2002)
	GPI-anchored proteins endocytosis is inhibited by cytochalasin D	Parton et al. (1994)
	CAV1 endocytosis is stimulated by cytochalasin D	Kang et al. (2000)
Microtubules	CAV1 internalization is inhibited by nocodazole	Conrad et al. (1995), Kang et al. (2000), Mundy et al. (2002)
Changes in the lipid composition of the membrane	Albumin endocytosis is stimulated by the addition of external glycosphingolipids and by the increase in cholesterol and GM1	Sharma et al. (2010)
	SV40 infection is inhibited by nistatin and by Nys-Prog (cholesterol synthesis inhibitor)	Pelkmans and Helenius (2002)
Kinase activity	GPI-anchored protein endocytosis is stimulated by okadaic acid (general inhibitor of phosphatases), but it is inhibited by staurosporine	Parton et al. (1994)
	SV40 infection is inhibited by staurosporine or genistein (Tyr-kinase inhibitor) while it is stimulated by okadaic acid and vanadate (Tyr-kinase inhibitor)	Pelkmans and Helenius (2002)
	SV40 infection is regulated by at least 34 kinases	Pelkmans and Zerial (2005)
CAV1 phosphorylation at Tyr14	*Caveolae* endocytosis does not occur in $CAV1^{-/-}$ cells expressing a Tyr14 non-phosphorylatable CAV1 mutant	del Pozo et al. (2005)
Oxidative stress	*Caveolae* endocytosis is stimulated by hydrogen peroxide treatment	Kang et al. (2000)
Hyperosmotic stress	*Caveolae* endocytosis is stimulated by 6 mM sorbitol (hyperosmotic shock)	Kang et al. (2000)
Thermic shock	CAV1 endocytosis is stimulated by thermic shock a 43 °C	Kang et al. (2000)
Loss of ECM adhesion	CAV1 and ChTx endocytosis is stimulated by loss of cell adhesion	del Pozo et al. (2005)
Dynamin2	ChTx internalization is inhibited upon blockade of Dyn2 activity with specific antibodies	Henley et al. (1998)
	SV40 infection is inhibited after the expression of the negative dominant mutant of Dyn2 (K44A)	Pelkmans and Helenius (2002)

(continued)

Table 8.1 (continued)

Regulator	Experimental evidence of *caveolae* endocytosis	References
	CAV1 endocytosis upon loss of cell adhesion is inhibited by expression of Dyn2K44A	del Pozo et al. (2005)
	Albumin endocytosis is inhibited by the expression of a non-phosphorylatable Dyn mutant (Dyn2Y231F/Y597F)	Shajahan et al. (2004)
RhoG	*Caveolae* internalization is altered by a constitutively active form of RhoG (RhoGQ61L)	Prieto-Sanchez et al. (2006)
Src	Albumin endocytosis is inhibited by Src inhibitors (PP2 and herbimicin)	Sharma et al. (2010)
	Gp60 endocytosis is inhibited by the expression of a Src dominant negative mutant (Y527F and K295M)	Minshall et al. (2000)
PKCα	*Caveolae* endocytosis is inhibited by peptidic inhibitors specific for PKCα treatment	Smart et al. (1995)
	GPI-anchored protein endocytosis is stimulated by okadaic acid, but not by PKCα activation.	Parton et al. (1994)
Intersectin	Intersectin, Cdc42 GEF, is located in *caveolae* neck interacting with Dyn2, suggesting its implication in *caveolae* fission	Predescu et al. (2003)
Mitosis	*Caveolae* endocytosis is enhanced during mitosis	Boucrot et al. (2011)
FilaminA	Filamin is necessary for CAV1 internalization and GM1 uptake	Muriel et al. (2011)
	Albumin transcytosis is FilaminA dependent	Sverdlov et al. (2009)
Abl tyrosine kinases	c-Abl and Arg are required for CAV1 internalization and GM1 uptake	Echarri et al. (2012)

8.3.1.1 Internalization of *Caveolae* Induced by Loss of Adhesion

Loss of cell adhesion is one of the most powerful stimuli to induce *caveolae* internalization (del Pozo et al. 2005). Upon loss of cell adhesion, CAV1 accumulates at Rab11-positive recycling endosomes, and upon readhesion, it is recycled back to the PM in an integrin-dependent manner (del Pozo et al. 2005; Lapierre et al. 2012; Muriel et al. 2011). Other key regulators of this process are integrin-like kinase (ILK), mDia1, IQGAP (Wickstrom et al. 2010), the exocyst component Exo70, actin and microtubules (Hertzog et al. 2012). According to the prevailing model (Echarri et al. 2007), steady-state integrin-dependent signals in adherent cells prevent the internalization of *caveolae*. In these conditions, signaling complexes in the PM are stabilized and engaged in normal dynamics of activation and deactivation. Cell detachment leads to the disengagement of integrin-dependent signaling platforms, and the internalization of *caveolae* is enabled—this dynamical change potentially entails extensive changes in the functioning of different signaling effectors associated to the PM. The importance of cell adhesion in regulating caveolae endocytosis has recently been evidenced in vivo (Cota and Davidson 2015).

The physiological relevance of detachment-triggered CAV1 dynamics is highlighted by its influence in Rac1 asymmetrical distribution in the migrating cell, where CAV1 internalization is maximal at the rear of the cell (Grande-Garcia et al. 2007). Interestingly, downregulating CAV1 or intervening CAV1 internalization dynamics elicit a number of pivotal properties, such as adhesion-independent cell growth and proliferation, thus potentially favoring survival to anoikis and enable spreading through the bloodstream for tumor cells (Cerezo et al. 2009). Indeed, some of the signaling pathways regulated by integrin-mediated adhesions are profusely altered in and pivotal for the biology of tumor cells, such as Ras-ERK1/2 and PI3K-AKT. Importantly, CAV1-null cells exhibit abnormally sustained activation of these pathways in suspension, as opposed to wild-type normal cells (del Pozo et al. 2005) and have therefore been hypothesized as a mechanism to bypass integrin-dependent growth control (Cerezo et al. 2009), although the mechanisms by which tumor cells avoid this regulation are likely more complex and not completely understood. Therefore, the mechanistic underpinning of integrin-dependent regulation of CAV1 internalization could well offer new opportunities for oncological therapeutic intervention.

8.3.1.2 Control of Organization and Trafficking of *Caveolae* by Stress Fibers

Caveolae distribute along stress fibers in a RhoA- and filaminA (FLNa)-dependent manner (Echarri et al. 2012; Muriel et al. 2011; Rothberg et al. 1992). Abl tyrosine kinases and the formin mDia1 can also organize pools of CAV1 linked to the actin fiber structures they promote. Importantly, FLNa, Abl, and mDia1 are additionally required for proper internalization of *caveolae* since in their absence of CAV1 clusters at the PM and is unable to be endocytosed (Echarri et al. 2012; Muriel et al. 2011; Sverdlov et al. 2009). However, the major alternative actin polymerization pathway, mediated by Arp2/3, is not required for organization and internalization of *caveolae* (Echarri et al. 2012).

Depletion of FLNa increases caveolar dynamics and disrupts stable anchoring of *caveolae*. During mitosis or upon cell detachment, increased PKCα-mediated phosphorylation of FLNa at Ser2152 allows internalization of *caveolae*, thus controlling Rac1 targeting to the PM and cell cycle progression (Muriel et al. 2011). In smooth muscle, the link between actin stress fibers and *caveolae* is maintained primarily by interactions of caveolin with β-dystroglycan (Sharma et al. 2010).

The tight control exerted by stress fibers on caveolar plasticity, dynamics, and trafficking in response to changes in adhesion suggests that they might induce a mechanosensory response in *caveolae* (Parton and del Pozo 2013; Echarri and del Pozo 2015).

8.3.2 Caveolae *as Mechanosensors and Mediators of Crosstalk Between ECM and Cells*

The exposure to large variations in membrane tension or mechanical stress is a shared key feature of cells with abundant *caveolae*, such as endothelial cells, adipocytes, and muscle cells. *Caveolae* contain a considerable amount of membrane that can be released to provide physical protection against potential damage by changing the volume-to-surface ratio of the cell. Flattening and disassembly of *caveolae* buffer membrane tension during mechanical stress, thus regulating membrane tension (Sinha et al. 2011). In addition, *caveolae* could respond to mechanical cues by activating specific downstream signaling pathways as has been demonstrated for shear stress in endothelial cells and stretching in smooth muscle (Czarny and Schnitzer 2004; Kozera et al. 2009; Rizzo et al. 1998; Sedding et al. 2005; Yu et al. 2006). Upon flattening of *caveolae*, CAV1 diffuses toward the adjacent membrane, while PTRF is released to the cytoplasm (Sinha et al. 2011). These released pools thus constitute mechanical information and likely have intrinsic roles in *caveolae*-mediated mechanosensing. Additionally, CAV1 is phosphorylated at Tyr14 in response to several mechanical stimuli (Joshi et al. 2012; Zhang et al. 2007)—it is expected that these events would thus couple mechanical cues to signaling regulation and caveolar internalization. Free PTRF may also help to propagate a signal through interactions with target effectors in the cytosol or in the nucleus with the final regulation of gene expression, and ribosomal biosynthesis (Bai et al. 2011; Hasegawa et al. 2000; Jansa et al. 1998; Liu and Pilch 2016). Because *caveolae* are frequently associated to stress fibers, a major regulator of membrane tension and cell shape, they might couple mechanotransduction pathways to actin-controlled changes in tension through their association with stress fibers (Echarri and del Pozo 2015).

Cells need to interact with the extracellular matrix (ECM) for developmental morphogenesis and normal tissue architecture. Moreover, tumor growth and progression also requires cell-ECM interactions. Tumor-associated fibroblasts (TAFs) express high levels of CAV1, which regulates the mechanical properties of the tissue microenvironment. *Caveolae* may also mediate mechanotransduction by responding to changes in the ECM via integrin signaling. *Caveolae* formation and function are closely linked to integrins, which are key sensors of the ECM (Parton and del Pozo 2013). CAV1 promotes Rho- and force-dependent actomyosin contraction, which induces ECM fibrillogenesis and stiffens the microenvironment (Goetz et al. 2011). This, in turn, promotes elongation, integrin-mediated adhesiveness and directional migration of cancer cells embedded in this matrix, thus enabling local invasion and distant metastasis. CAV1 also contributes to ECM remodeling in physiological responses such as wound healing and the scarring process (Grande-Garcia et al.

2007). Specific interactions with ECM may control CAV1 endocytosis: interaction of syndecan-4, a PM proteoglycan, with ECM components, may induce integrin internalization via caveolae (Bass et al. 2011). Moreover, CAV1 may directly regulate ECM turnover by controlling endocytosis of FN-bound integrins (Osmani et al. 2017; Shi and Sottile 2008).

8.3.3 Role of CAV1 in Endocytosis-Mediated Metabolic Regulation and Organelle Homeostasis

One of the most prominent, but as yet poorly understood, aspects of caveolin function is that related to metabolic control. The control of the anabolism, storage and intracellular fluxes of different lipid species, including cholesterol and energy management, are cell functions affected profoundly by CAV1 regulation at multiple levels. Because *caveolae* stabilize pools of very defined lipid species, with a marked enrichment in cholesterol, sphingosine and saturated fatty acids, these structures likely contribute to define lipidome fluxes in the cell, including those related to inter-organelle signaling (Pilch and Liu 2011). Accordingly, most syndromes or phenotypes associated with caveolin deficiency display severe lipodystrophy and metabolic dysfunction (Bruno et al. 1993).

Although hereon we summarize current mechanistic knowledge of these aspects separately, pervasive interdependence necessarily exists, and caveolin itself may be proposed as a key integrative regulator among them.

8.3.3.1 Caveolin and Lipid Management

The primary aspect that arises when studying caveolin-dependent metabolism regulation is lipid management, mobilization, and usage. CAV1 can bind newly synthesized cholesterol and traverse endoplasmic reticulum (ER) membranes toward the Golgi-endosomal apparatus (Bosch et al. 2011a; Monier et al. 1996; Schlegel et al. 2001; Smart et al. 1996). CAV1 can also be found in lipid droplets (Pol et al. 2004). Caveolin deficiency leads to impaired fatty acid storage and lipid droplet formation, and genetic ablation of caveolin leads to severe lipodystrophy and a resistance to develop obesity upon exposure to high fat diets (Kim et al. 2008; Mercier et al. 2009). Caveolin knockout (KO) mice also exhibit altered levels of circulating triglycerides, free fatty acids and cholesterol derivatives (Kim et al. 2008) underlying a potential impact on whole organismal regulation.

Caveolin deficiency is associated with an accumulation of cholesterol across intracellular membrane-bound compartments (Bosch et al. 2011a, b), and because of this primary alteration of cholesterol distribution within the cell, the metabolism of several other lipid species can be affected, including potentially those determining endosomal subcompartment trafficking. This cholesterol-dependent influence of CAV1 on the endosomal system may affect their dynamics and fate, as well as the functioning of associated signaling units (see Sect. 8.3.4.7 below). This is well exemplified by the extreme case embodied by lysosomal disorders such as Niemann–Pick disease, whereby severe cholesterol accumulation within late endosomes significantly affects the performance of lysosomal vesicles and partially abrogates autophagic function (Carstea et al. 1997; Sarkar et al. 2013). It is currently unclear whether CAV1 dysregulation can impact to the same extent lysosomal function. However, cholesterol accumulation upon loss of CAV1 does affect functionally other organelles, such as the mitochondria, ER and their interface domains (Bosch et al. 2011b; Sala-Vila et al. 2016). These perturbations might be derived both from broad alterations of membrane properties or more specific effects on particular complexes, as is the case for the mitochondrial 2-oxoglutarate/glutathione antiporter (Mari et al. 2009). Intriguingly, recent lines of research hint at relevant reciprocal links between endosomal regulation and mitochondrial homeostasis and quality control (Lai et al. 2015; Lang et al. 2015; McLelland et al. 2014). It will be of outstanding interest to dissect the role of CAV1 on the underlying molecular mechanisms of this communication.

8.3.3.2 Metabolic Roles of CAV1-Associated Endocytosis

The role of caveolae-mediated endocytosis in the physiopathology of lipid metabolism is proven by multiple experimental evidences. The uptake of dietary triglycerides is moderately reduced in intestinal epithelial cells of CAV1 KO mice, although how direct this influence is, is at present unclear (Siddiqi et al. 2013). Also, because genetic ablation of CAV1 reduces by ~70% the incidence and progression of atherosclerotic lesions in classical genetic hypercholesterolemia models (Frank et al. 2004), the potential role of CAV1 in endothelial uptake and/or trafficking of circulating LDL-cholesterol has received considerable attention. Some studies depict LDL receptor (LDLr) as being spatially and temporally organized by *caveolae*. Indeed, CAV1/*caveolae* have been proven to associate to at least one key co-regulator of LDLr, Lrp6 (Yamamoto et al. 2006; Zilberberg et al. 2004). However, classical models of LDL-cholesterol receptor-mediated endocytosis in endothelial cells are still debated because, although some key regulators of LDLr dynamics, such as PCSK9, are important targets for pharmacological intervention (Maxwell and Breslow 2005; Maxwell et al. 2005), genetic deficiency of LDL receptors per se dramatically increases incidence and severity of atherosclerotic disease, as a result of marked hypercholesterolemia (Ishibashi et al. 1993). Recently, a role for caveolae regulating the amount of lipids at the plasma membrane has been proposed (Shvets et al. 2015).

Alternative specialized internalization processes termed *transcytosis* have been proposed, and experimental evidence suggests that CAV1 is present at transcytotic vesicles (Frank et al. 2009), but the exact functional relevance of this CAV1 pool is at present unknown. Because PCSK9 requires COPII trafficking (presumably for its coordination with cholesterol efflux) (Chen et al. 2013) and CAV1 determines cholesterol efflux through this compartment (Hayer et al. 2010a), this molecule might hold the key to part of these elusive links of CAV1 with atherosclerosis disease.

Besides controlling various aspects of lipid homeostasis, *caveolae* have a role in regulating the trafficking of key membrane metabolic regulators such as glucose transporters [GLUT1, 2 and 4; (Rauch et al. 2006; Scherer et al. 1994)] or calcium channel systems (Yeh and Parekh 2015). Internalization of *caveolae* is indeed a means to regulate the fluxes operated by these entry channels. For example, insulin stimulation selectively primes internalization of *caveolae*-associated GLUT4 receptors in adipocytes (Yuan et al. 2007). The current challenge lies on rationalizing these dynamics with those of other *caveolae*-dependent elements. An unproven, but very appealing model, emerges if we consider *caveolae* and their components as nodes where pivotal, but poorly understood functional crosstalk takes place—such as for example integrating mechanoadaptation and metabolic remodeling.

8.3.4 Regulation of Signaling Pathways by Caveolae

8.3.4.1 *Caveolae* and Cell Signaling

Substantial evidence has been gathered over the years supporting a role for *caveolae* and caveolar components, especially caveolins and cavins, in signal transduction. Implication of *caveolae* in downstream signaling events is strongly suggested by the early observation that *caveolae* are enriched in membrane receptors such as epidermal growth factor receptor (EGFR) transforming growth factor β receptor (TGFβR) and insulin receptor (IR) (Couet et al. 1997; Razani et al. 2001; Yamamoto et al. 1999). Endothelial VEGFR-2 seems to also be a highly *caveolae*-dependent RTK (Labrecque et al. 2004).

Early models proposed *caveolae* form a hub where membrane receptors interact, and engage in fine-tuning with adaptors, kinases and phosphatases (Okamoto et al. 1998; Shaul and Anderson 1998). In this model, the physical interaction between membrane receptors and other signaling molecules within *caveolae* strongly affects the propagation of signals elicited at the PM. *Caveolae* might also provide a means to attain "digital coding" at the PM, by defining discrete "signaling units", perhaps simultaneously and coordinately allocating specific numbers of membrane receptors and regulators (Villasenor et al. 2016). Furthermore, regulated caveolar internalization may constitute an additional layer to control the amount of specific receptors at the PM and the activation of their downstream signaling pathways.

Despite the initial proposition of specific cargoes as markers of caveolar internalization, the endocytosis of *caveolae* and its functional significance have long been debated (Parton and Howes 2010; Pelkmans et al. 2001) as discussed above. However, endocytosis mediated by *caveolae* does have a relevant role in mediating the activation/deactivation of different signaling pathways.

8.3.4.2 Intrinsic Positive and Negative Regulatory Activities of CAV1

Apart from its structural role in caveolar architecture, evidence has accumulated on a specific role of CAV1 in directly modulating the activity of signal transduction proteins. Among the best described proteins associated to CAV1 are H-Ras (Li et al. 1996), members of the family of Src tyrosine kinases (Li et al. 1996; Song et al. 1997), the α subunit of heterotrimeric G proteins (Li et al. 1995; Song et al. 1997), PKA (Razani and Lisanti 2001) and eNOS (Feron et al. 1996; Garcia-Cardena et al. 1996). As will be discussed later, most of these signaling molecules present a caveolin-binding motif (CBM). CAV1 can also mediate the downregulation of MEK1 activity through direct binding to the endogenous Ras/MAPK inhibitor Dok1. Binding with Dok1 results in MEK1 inactivation and in nuclear translocation of peroxisome proliferator-activated receptor γ (PPARγ) (Burgermeister et al. 2011).

A number of studies might propose an intrinsic, predominantly inhibitory activity for CAV1. Experiments based on knockdown (KD) of CAV1 expression using siRNAs suggest that CAV1 often exerts a curbing inhibitory role on a number of membrane-elicited signaling pathways, such as Ras-MEK-ERK1/2 pathway (Galbiati et al. 1998; Murata et al. 2007; Strippoli et al. 2015). CAV1 deficiency has also been reported to lead to upregulation of p38 and JNK activities (Wang et al. 2006). Moreover, expression of CAV1 abolishes the increase of NO activity observed in cells from CAV1$^{-/-}$ mice (Murata et al. 2007). Some observations also suggest a capability for CAV1 to dampen PI3K- and mTOR-dependent cues (Feng et al. 2010; Mercier et al. 2012). However, most influences from CAV1 on cell signaling state have subsequently been found to be highly contextual: for example, depending on the cellular model, CAV1 may inhibit or induce ERK1/2 or PI3K-dependent signals (Moreno-Caceres et al. 2014; Shack et al. 2003; Tang et al. 2015). Further supporting a highly conditional nature for Cav1-dependent regulation of cell signaling, recent reports suggest cell type-specific interactions between caveolar components and regulatory adaptors as key for defining the output of these signaling systems. A recent report exemplifies this in ROR1, a poorly characterized RTK which functions in lung adenocarcinoma as a caveolae-associated scaffold essential for sustaining prosurvival growth factor signaling (Yamaguchi et al. 2016).

Apart from direct associations, by controlling associations with liquid-ordered PM domains, CAV1 may determine the subcellular localization, and thus activity, of several signaling proteins, such as those belonging to the Ras GTPases family and to the Src family. In particular, cell detachment leads to decrease of ERK1/2, PI3K and Rho-GTPase activity, with concomitant increase of cAMP. All of these events were dependent on the expression levels of CAV1 (del Pozo et al. 2005; Norambuena and Schwartz 2011). Interestingly, elevated cAMP is also implicated inhibition of ERK1/2 activity and blockage of cell cycle after detachment (Howe and Juliano 2000). CAV1 can also modulate MEK-ERK1/2 signaling pathway through organization of K-Ras and H-Ras nanoclusters (Ariotti et al. 2014).

8.3.4.3 Specific Roles of the CSD

As mentioned in the introduction, CAV1 contains a specific domain composed of the residues 82-101 designated the caveolin-scaffolding domain (CSD), which contributes to its localization at the PM (Schlegel et al. 1999). CAV1 has been demonstrated to interact through its CSD with several signaling proteins (serine and tyrosine kinases such as Src, AKT; and other signaling relays and mediators such as H-Ras and eNOS), as well as cholesterol, contributing to its organization in membrane nanodomains. A number of these interactions seem to contribute to cell cycle and gene expression regulation (Bernatchez et al. 2005; Couet et al. 1997; Murata et al. 2007).

Among these signaling mediators, eNOS has been particularly intensely studied because of its broad impact on organismal physiology and physiopathology. CAV1 directly interacts with eNOS bound to PM via the CSD, preventing NO production. NO production leads in turn to Src-mediated CAV1 tyrosine phosphorylation, which further promotes binding to and inhibition of eNOS in an efficient, classical negative feedback loop (Cheng and Nichols 2016; Li et al. 1996; Michel et al. 1997). Of note, treatment with a membrane-permeable caveolin-1 CSD peptide leads to a similar reduction of eNOS activity in vivo (Bucci et al. 2000; Tourkina et al. 2008). This approach has also been recently used to outcompete Src activity (Zimnicka et al. 2016).

Inconclusive evidence of direct interaction between the CSD and (CBM) in CAV1-interacting proteins has challenged the CSD/CBM hypothesis. CBM may be not exposed at the surface of many signaling proteins, and the CSD may be either in close proximity to the phospholipid bilayer or completely buried by it, hampering the interaction between both domains (Ariotti et al. 2015, Byrne et al. 2012). Thus, the role of CSD in mediating CAV1 functions is still a subject of intense debate and research.

8.3.4.4 Role of Tyr14 Phosphorylation

Tyr14 phosphorylation is probably the most studied CAV1 post-translational modification, due to its impact on many signaling pathways (Cao et al. 2002; Labrecque et al. 2004). Tyr14 of CAV1 is targeted by several tyrosine kinases, like c-Abl (Sanguinetti and Mastick 2003), Src (Glenney 1989) and Fyn (Sanguinetti et al. 2003), and its dephosphorylation is mediated by PTP1b phosphatase (Lee et al. 2006). Tyr14 phosphorylation regulates the binding with tumor necrosis factor-α-receptor associates factor 2 (TRAF2) and Growth Factor Receptor Bound Protein 7 (Grb7) (Cao et al. 2002; Lee et al. 2000; Parton and del Pozo 2013). Moreover, Tyr14 phosphorylation reinforces the interaction of CAV1 with Src kinase via the binding of Src SH2 domain to phospho-CAV1 and activated Src accumulation in focal adhesions (Gottlieb-Abraham et al. 2013). CAV1 phosphorylated at Tyr14 also interacts with CSK (C-terminal SRC kinase), which is a SRC kinase—inhibitor that mediates actin-reorganization (Cao et al. 2002).

Phosphorylated CAV1 regulates internalization of *caveolae* (del Pozo et al. 2005), persistent/directional migration [through RHO-dependent actomyosin contraction by altering the localization and activity of its endogenous inhibitor, p190RHOGAP) (Grande-Garcia et al. 2007) and cellular 3D microenvironment remodeling (Goetz et al. 2011)]. A recent report indicated that CAV1 phosphorylation at Tyr14 can mediate IGF-/PI3K/AKT/MAPK signaling an inhibit anoikis, a programmed cell death pathway (Tang et al. 2015).

Interestingly, CAV1 phosphorylation at Tyr14 during exposure to cyclic stretch leads to the transcriptional regulation of both CAV1 and Cavin-1 via inhibition the transcription factor early growth response protein 1 (EGR1) in breast carcinoma cells (Joshi et al. 2012), and favors *caveolae* formation in human pancreatic (PANC-1) carcinoma cells (Orlichenko et al. 2006)—hinting at positive self-regulatory loops involving CAV1 and operated through Tyr14 phosphorylation. Recently, it has been proposed that when phosphorylated on this residue, caveolae swell due to the intrinsic properties of phosphorylated tyrosine within oligomeric caveolins (Zimnicka et al. 2016). This indicates that in addition to signaling capacity of this residue, it may also play a role in organizing CAV1 within the plasma membrane.

8.3.4.5 Direct Interactions with Membrane Receptors Present in *Caveolae*

PM receptors are enriched in caveolae. CAV1 was demonstrated to directly associate to EGFR via its CSD (Aboulaich et al. 2004) inhibiting its CAV1tyrosine kinase activity (Couet et al. 1997). EGFR phosphorylation and activity were both linked to exit from *caveolae* (Abulrob et al. 2004).

More recently, it was demonstrated that CAV1 KD enhances EGF-dependent ERK1/2 phosphorylation and nuclear translocation due to an effect on EGFR lateral mobility, which was reduced upon CAV1 ectopic expression (Lajoie et al. 2007). On the other hand, EGFR stimulation was able to downregulate CAV1 expression leading to enhanced snail expression and EMT (Lu et al. 2003).

Similarly to EGFR, CAV1 was found to directly interact with PDGF through interaction with CSD within *caveolae*, leading to inhibition of PDGF activity (Yamamoto et al. 1999).

Differently from other receptors, direct binding of the beta subunit of insulin receptor (IR) to CAV1 was demonstrated to enhance insulin signaling (Nystrom et al. 1999; Yamamoto et al. 1998). IR catalyzes CAV1 tyrosine phosphorylation, which correlates with AKT phosphorylation. At the same time, the expression of glucose transporter Glut4 increased in *caveolae*, leading to increased glucose uptake. CAV1 may also enhance AKT activity promoting IGF-IR expression in liver (Tang et al. 2015).

TGFBRI (ALK1 and ALK5) and TGFBRII are also enriched in *caveolae* (Schwartz et al. 2005). CAV1 directly interacts with TGFBRI within *caveolae* and inhibits TGFβ signaling (Razani et al. 2001). In a subsequent study, CAV1 was reported to inhibit ALK5-SMAD3 activity, while enhancing ALK1-SMAD1 signaling (Santibanez et al. 2008). Interestingly, TGFβ is a general inhibitor of CAV1 expression (Strippoli et al. 2014; Wang et al. 2006)

The role of CAV1 in TGFβ receptor internalization may explain the function of CAV1 in limiting fibrosis in different organs (Del Galdo et al. 2008). Internalization of TGFβ receptors via *caveolae* targets them for proteasomal degradation, whereas internalization via the clathrin pathway elicits recycling of receptors and maintenance of TGFβ signaling (Di Guglielmo et al. 2003). Accordingly, a recent study demonstrated a convergence of clathrin- and *caveolae*-mediated endocytic pathways during TGF-β receptor endocytic trafficking (He et al. 2015).

8.3.4.6 Specific Roles of Cavins

Cavins are essential for the organization of *caveolae* and for the maintenance of *caveolae* stability. In the absence of Cavin-1, the size of caveolin1 oligomers in the PM decreases and their lateral mobility increases, implying a role for Cavin-1 in formation of *caveolae* and sequestration of mobile caveolin into immobile *caveolae* (Hill et al. 2008). Only a limited number of studies dealt so far on regulation of signaling pathways by cavins. Both Cavin-2 and Cavin-3 were originally identified as protein kinase C (PKC) substrates and have been suggested to target PKC to *caveolae* (Izumi et al. 1997; Parton and del Pozo 2013; Xu et al. 2001). However, the effect of cavins on PKC regulation is unclear (Mineo et al. 1998; Xu et al. 2001). More recently, Cavin-3 was demonstrated to balance ERK1/2 versus AKT activities, anchoring the ERK1/2 activation module of *caveolae* (Hernandez et al. 2013).

Cavin-4 facilitates the recruitment and activation of ERK1/2 in response to α1adrenergic receptor stimulation in cardiomyocytes (Ogata et al. 2014). Cavin-4 associates with CAV1 and with p115RhoGEF and RhoA inducing Rho/RHOK signaling in smooth muscle cells, playing a role in the development of pulmonary hypertension (Nakanishi et al. 2016).

8.3.4.7 Caveolin as a Mechanism Integrating Cholesterol Homeostasis with the Endosomal System—Postendocytic Regulation of Cell Signaling

An additional, largely unexplored angle to understand the impact of CAV1 in systems-level signaling through endocytosis is derived from the fact that cholesterol content (and overall relative lipid composition) of intracellular membranes can determine the targeting and relationships of a given subcompartment, as well as modulating the functionality of its associated proteins and receptors (Chevallier et al. 2008; Musiol et al. 2013; Scott et al. 2015; Zhang et al. 2001). Cholesterol fluxes within the endosomal system have been shown to regulate signaling of endocytosed tyrosine kinase receptors (RTKs) such as EGFR, because cholesterol content profoundly affects their interaction with ER domains rich in the deactivating PTP1B phosphatase (Eden et al. 2016). Because CAV1 is an essential regulator of cholesterol efflux and its levels determine the relative retention of this lipid within endomembranes (Bosch et al. 2011a), the regulation of CAV1 levels and subcellular distribution might constitute a means by which cells couple and coordinate cholesterol metabolism, organelle communication, and specific signaling outputs at the endosomal compartment.

8.4 Pending Questions and Future Directions

Despite the enormous amount of studies that have accumulated throughout the years from the first discovery of caveolae in the second-half of the twentieth century, and especially the discovery of caveolin-1 as their first molecular marker in early 1990s, our knowledge regarding the functions of *caveolae* remains limited. One likely reason might well reside on the sheer complexity and contextuality of the interactions among caveolins, cavins, and different PM receptors (see previous section) whose number and activity may vary in different cell types or the same cell type at different conditions—eliciting varied responses to biochemical/biophysical extracellular stimuli (including ECM molecules and cytokines) and widening their coding range. A good example for this concept is the role of CAV1 in tumors; CAV1 expression may curb the progression of some primary tumors, while positively regulating tumor invasiveness at later stages (Lamaze and Torrino 2015).

In more general terms, many questions remain unanswered regarding the structure, dynamics, and functional relevance of *caveolae*, as well as their interplay with endocytosis. The dynamical assembly of lipids, caveolins, and cavins and their structural arrangement is not fully understood. The mechanistics as to how *caveolae* encode and transduce information to downstream events (signaling networks and gene expression programs) remain uncharacterized. In this sense, we dwell in exciting times because of the revolution recently started by the development of sub-diffraction optical super-resolution microscopy, and it is likely that major advancements will be soon reported upon enabling nanoscale live cell imaging, for example.

Furthermore, the pervasive relationships of *caveolae* and their components with other constituents of the cell makes them appealing candidates to function as "coordination hubs" for multiple processes—including their potential interplay with other regulators of endocytic fluxes (Chaudhary et al. 2014; He et al. 2015; Lamaze et al. 2017). For example, mechanically induced disassembly of *caveolae* could encode critical information for the coordinated adaptation of metabolism and associated cell signaling (Nassoy and Lamaze 2012). Integrative approaches exploiting novel, unbiased technologies such as high-content microscopy and functional genomics will likely shed light on these questions.

Finally, beyond these basic queries remains their framing into relevant physiopathological processes associated with dysregulated caveolar functioning. Caveolinopathies are associated with broad metabolic imbalances and inflexibility, including dyslipidemia, altered vascular function, aberrant muscle physiology, and inflammatory/fibrotic diseases. Caveolin-1 and *caveolae* also have a relevant impact in tumor biology, if highly contextual (see above). Thus, exciting discoveries lay ahead

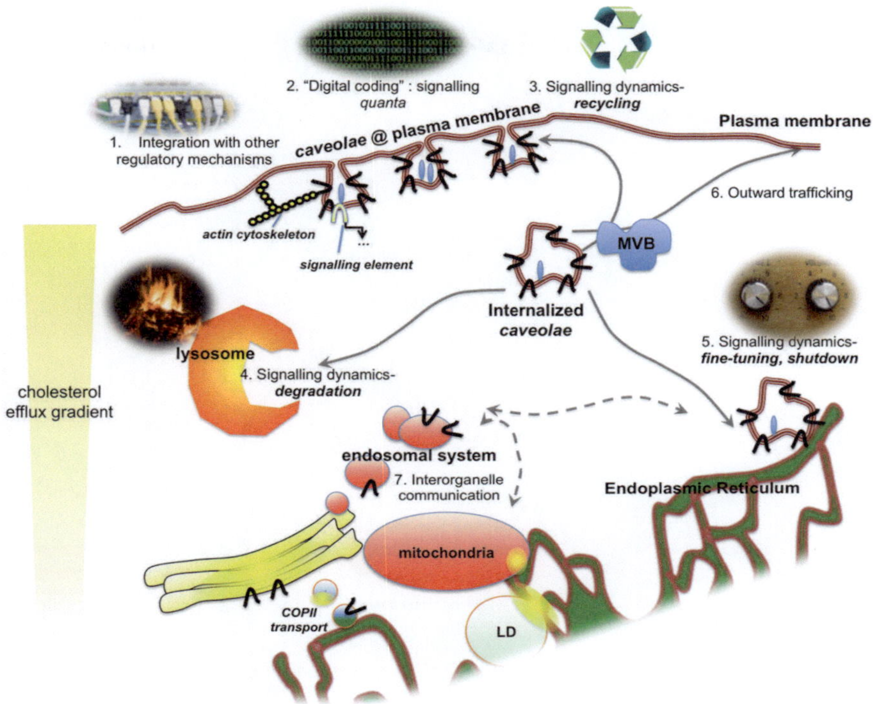

Fig. 8.1 Multiple functions and "fates" of plasma membrane-located and internalized *caveolae*, and their associated components. *Caveolae* at the plasma membrane provide unique signaling platforms where ligand-mediated signaling, mechanotransduction, and membrane/lipid homeostasis converge (1). Importantly, because of their distinct architecture and potential coalescence in larger-order domains (*rosettes*) caveolae might contribute to specify "digital codes" for different signaling pathways, by allocating defined stoichiometries of receptors and coregulators (proposed as signaling *quanta* by Pelkmans, and Zerial 2005) (2). Caveolae dynamics constitute dynamic structures tweaking cholesterol efflux to the plasma membrane and stabilizing local nanodomains with specific biophysical properties, as well as contributing to trafficking from the endosomal system and recycling (3). Internalization of *caveolae* may target their components (including signaling units) for lysosomal degradation (4); specific channels of "signal fine-tuning", such as ER-endosomal contacting (5, see Eden et al. 2016); outward rerouting such as exosomal trafficking (6); or endomembrane transactions such as those established between metabolically relevant organelles (7)

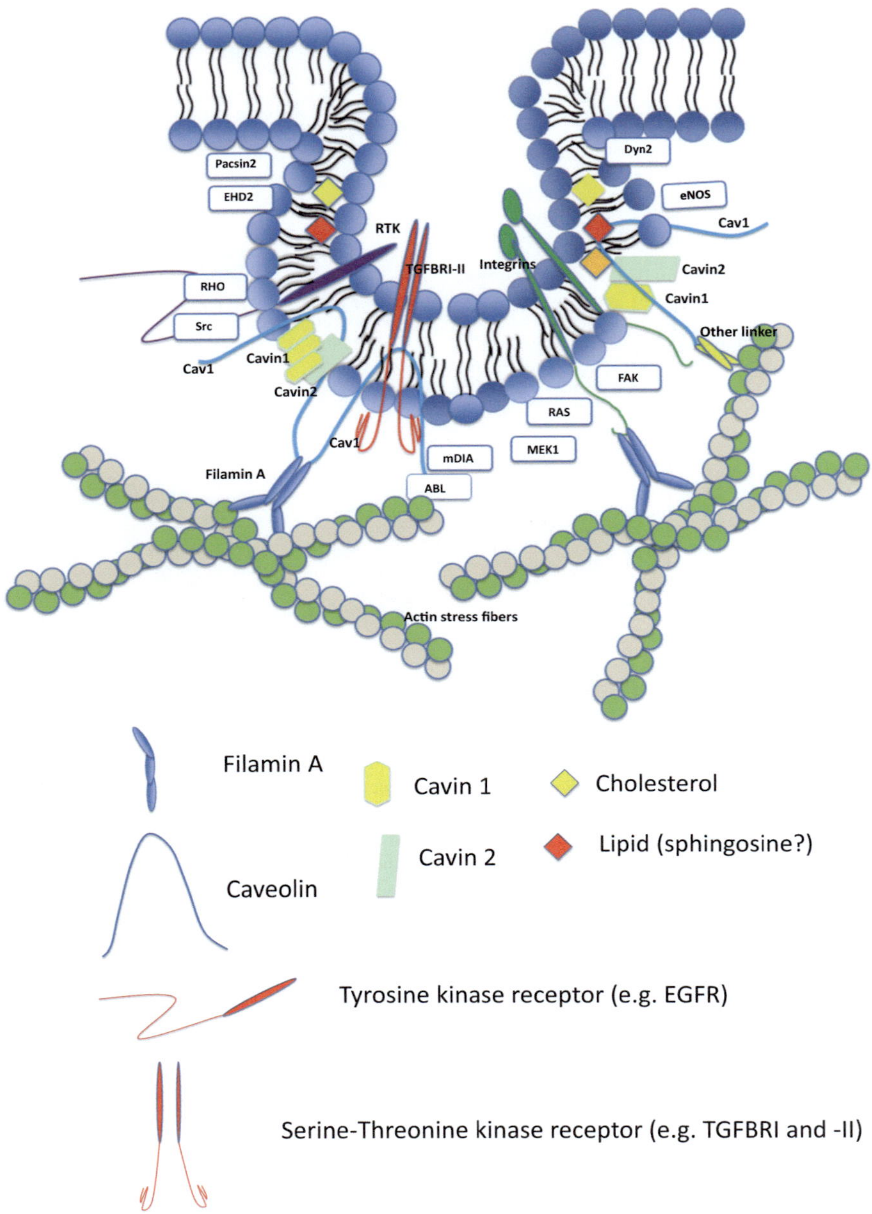
Pacsin2
EHD2
RTK
TGFBRI-II
Integrins
Dyn2
eNOS
Cav1
Cavin2
Cavin1
Other linker
RHO
Src
Cav1
Cavin1
Cavin2
FAK
RAS
MEK1
Cav1
mDIA
ABL
Filamin A
Actin stress fibers
Filamin A
Cavin 1
Cholesterol
Cavin 2
Lipid (sphingosine?)
Caveolin
Tyrosine kinase receptor (e.g. EGFR)
Serine-Threonine kinase receptor (e.g. TGFBRI and -II)

◂**Fig. 8.2** *Caveolae* (shaped by caveolins and cavins and with the presence of other proteins such as Pacsin, EHD2 and dynamin2) may be depicted as hubs where membrane receptors [including tyrosine kinase receptors such as EGFR (purple) or serine-threonine kinase receptors such as TGFBRI-II (red)]; lipids such as cholesterol and sphingomyelin (red and yellow squares); and signaling adaptors and scaffolds are selectively recruited. Integrins (green rods) take part of caveolar dynamics. Non-receptor tyrosine kinases (such as Src and FAK), eNOS, Rho and Ras GTPases and elements of MAPK pathway are also transient components of *caveolae*. Filamin A and other unknown interactors mediate binding with the actin cytoskeleton. Actin-mediated endocytosis of *caveolae* contributes to the regulation of several signaling cascades

in the field of caveolar dynamics and endocytosis, which might lead to a better understanding of these complex diseases and their management (Figs. 8.1 and 8.2).

Acknowledgements This work was supported by grants SAF2011-25047 and CSD2009-0016 from Spanish Ministry of Science and Innovation (MICINN), SAF2014-51876-R from Spanish Ministry of Economy and Competitiveness (MINECO) and co-funded by FEDER funds, 674/C/2013 from Fundació La Marató de TV3, and AICR 15-0404 from the Worldwide Cancer Research Foundation (to M.A.dP.). The CNIC is supported by the Ministry of Economy, Industry and Competitiveness (MEIC), and the Pro CNIC Foundation, and is a Severo Ochoa Center of Excellence (SEV-2015-0505).

Competing Financial Interests
The authors declare no competing financial interests exist.

References

Aboulaich N, Vainonen JP, Stralfors P, Vener AV (2004) Vectorial proteomics reveal targeting, phosphorylation and specific fragmentation of polymerase I and transcript release factor (PTRF) at the surface of caveolae in human adipocytes. Biochem J 383:237–248

Abulrob A, Giuseppin S, Andrade MF, McDermid A, Moreno M, Stanimirovic D (2004) Interactions of EGFR and caveolin-1 in human glioblastoma cells: evidence that tyrosine phosphorylation regulates EGFR association with caveolae. Oncogene 23:6967–6979

Almeida RA, Dunlap JR, Oliver SP (2010) Binding of host factors influences internalization and intracellular trafficking of streptococcus uberis in bovine mammary epithelial cells. Vet Med Int 2010:319192

Anderson RG (1998) The caveolae membrane system. Annu Rev Biochem 67:199–225

Ariotti N, Fernandez-Rojo MA, Zhou Y, Hill MM, Rodkey TL, Inder KL, Tanner LB, Wenk MR, Hancock JF, Parton RG (2014) Caveolae regulate the nanoscale organization of the plasma membrane to remotely control Ras signaling. J Cell Biol 204:777–792

Ariotti N, Rae J, Leneva N, Ferguson C, Loo D, Okano S, Hill MM, Walser P, Collins BM, Parton RG (2015) Molecular characterization of caveolin-induced membrane curvature. J Biol Chem 290:24875–24890

Bai L, Deng X, Li J, Wang M, Li Q, An WAD, Cong YS (2011) Regulation of cellular senescence by the essential caveolar component PTRF/Cavin-1. Cell Res 21:1088–1101

Bass MD, Williamson RC, Nunan RD, Humphries JD, Byron A, Morgan MR, Martin P, Humphries MJ (2011) A syndecan-4 hair trigger initiates wound healing through caveolin- and RhoG-regulated integrin endocytosis. Dev Cell 21:681–693

Bastiani M, Liu L, Hill MM, Jedrychowski MP, Nixon SJ, Lo HP, Abankwa D, Luetterforst R, Fernandez-Rojo M, Breen MR, Gygi SP, Vinten J, Walser PJ, North KN, Hancock JF, Pilch

PF, Parton RG (2009) MURC/Cavin-4 and cavin family members form tissue-specific caveolar complexes. J Cell Biol 185:1259–1273

Benga L, Goethe R, Rohde M, Valentin-Weigand P (2004) Non-encapsulated strains reveal novel insights in invasion and survival of Streptococcus suis in epithelial cells. Cell Microbiol 6:867–881

Bernatchez PN, Bauer PM, Yu J, Prendergast JS, He P, Sessa WC (2005) Dissecting the molecular control of endothelial NO synthase by caveolin-1 using cell-permeable peptides. Proc Natl Acad Sci USA 102:761–766

Bitsikas V, Correa IR, Jr., Nichols BJ (2014) Clathrin-independent pathways do not contribute significantly to endocytic flux. eLife 3:e03970

Bosch M, Mari M, Gross SP, Fernandez-Checa JC, Pol A (2011a) Mitochondrial cholesterol: a connection between caveolin, metabolism, and disease. Traffic 12:1483–1489

Bosch M, Mari M, Herms A, Fernandez A, Fajardo A, Kassan A, Giralt A, Colell A, Balgoma D, Barbero E, Gonzalez-Moreno E, Matias N, Tebar F, Balsinde J, Camps M, Enrich C, Gross SP, Garcia-Ruiz C, Perez-Navarro E, Fernandez-Checa JC et al (2011b) Caveolin-1 deficiency causes cholesterol-dependent mitochondrial dysfunction and apoptotic susceptibility. Current Biol 21:681–686

Boucrot E, Howes MT, Kirchhausen T, Parton RG (2011) Redistribution of caveolae during mitosis. J Cell Sci 124:1965–1972

Bruno C, Sotgia F, Gazzerro M, Minetti C, Lisanti MP (1993) Caveolinopathies. Synonym: Caveolin-3 Deficiency. GeneReviews- University of Washington, Seattle. PMID: 20301559

Bucci M, Gratton JP, Rudic RD, Acevedo L, Roviezzo F, Cirino G, Sessa WC (2000) In vivo delivery of the caveolin-1 scaffolding domain inhibits nitric oxide synthesis and reduces inflammation. Nat Med 6:1362–1367

Burgermeister E, Friedrich T, Hitkova I, Regel I, Einwachter H, Zimmermann W, Rocken C, Perren A, Wright MB, Schmid RM, Seger R, Ebert MP (2011) The Ras inhibitors caveolin-1 and docking protein 1 activate peroxisome proliferator-activated receptor gamma through spatial relocalization at helix 7 of its ligand-binding domain. Mol Cell Biol 31:3497–3510

Byrne DP, Dart C, Rigden DJ (2012) Evaluating caveolin interactions: do proteins interact with the caveolin scaffolding domain through a widespread aromatic residue-rich motif? PLoS ONE 7:e44879

Cao H, Courchesne WE, Mastick CC (2002) A phosphotyrosine-dependent protein interaction screen reveals a role for phosphorylation of caveolin-1 on tyrosine 14: Recruitment of C- terminal Src kinase. J Biol Chem 22:22

Carstea ED, Morris JA, Coleman KG, Loftus SK, Zhang D, Cummings C, Gu J, Rosenfeld MA, Pavan WJ, Krizman DB, Nagle J, Polymeropoulos MH, Sturley SL, Ioannou YA, Higgins ME, Comly M, Cooney A, Brown A, Kaneski CR, Blanchette-Mackie EJ et al (1997) Niemann-Pick C1 disease gene: homology to mediators of cholesterol homeostasis. Science 277:228–231

Cerezo A, Guadamillas MC, Goetz JG, Sanchez-Perales S, Klein E, Assoian RK, del Pozo MA (2009) The absence of caveolin-1 increases proliferation and anchorage- independent growth by a Rac-dependent, Erk-independent mechanism. Mol Cell Biol 29:5046–5059

Collins BM, Davis MJ, Hancock JF, Parton RG (2012) Structure-based reassessment of the caveolin signaling model: do caveolae regulate signaling through caveolin-protein interactions? Dev Cell 23:11–20

Conrad PA, Smart EJ, Ying YS, Anderson RG, Bloom GS (1995) Caveolin cycles between plasma membrane caveolae and the Golgi complex by microtubule-dependent and microtubule-independent steps. J Cell Biol 131:1421–1433

Cota CD, Davidson B (2015) Mitotic membrane turnover coordinates differential induction of the heart progenitor lineage. Dev Cell 34:505–519

Couet J, Sargiacomo M, Lisanti MP (1997) Interaction of a receptor tyrosine kinase, EGF-R, with caveolins. Caveolin binding negatively regulates tyrosine and serine/threonine kinase activities. J Biol Chem 272:30429–30438

Czarny M, Schnitzer JE (2004) Neutral sphingomyelinase inhibitor scyphostatin prevents and ceramide mimics mechanotransduction in vascular endothelium. Am J Physiol Heart Circ Physiol 287:H1344–H1352

Chaudhary N, Gomez GA, Howes MT, Lo HP, McMahon KA, Rae JA, Schieber NL, Hill MM, Gaus K, Yap AS, Parton RG (2014) Endocytic crosstalk: cavins, caveolins, and caveolae regulate clathrin-independent endocytosis. PLoS Biol 12:e1001832

Chen XW, Wang H, Bajaj K, Zhang P, Meng ZX, Ma D, Bai Y, Liu HH, Adams E, Baines A, Yu G, Sartor MA, Zhang B, Yi Z, Lin J, Young SG, Schekman R, Ginsburg D (2013) SEC24A deficiency lowers plasma cholesterol through reduced PCSK9 secretion. eLife 2:e00444

Cheng JP, Nichols BJ (2016) Caveolae: one function or many? Trends Cell Biol 26:177–189

Chevallier J, Chamoun Z, Jiang G, Prestwich G, Sakai N, Matile S, Parton RG, Gruenberg J (2008) Lysobisphosphatidic acid controls endosomal cholesterol levels. J Biol Chem 283:27871–27880

Chidlow JH Jr, Sessa WC (2010) Caveolae, caveolins, and cavins: complex control of cellular signalling and inflammation. Cardiovasc Res 86:219–225

Damm EM, Pelkmans L, Kartenbeck J, Mezzacasa A, Kurzchalia T, Helenius A (2005) Clathrin- and caveolin-1-independent endocytosis: entry of simian virus 40 into cells devoid of caveolae. J Cell Biol 168:477–488

Daumke O, Lundmark R, Vallis Y, Martens S, Butler PJ, McMahon HT (2007) Architectural and mechanistic insights into an EHD ATPase involved in membrane remodelling. Nature 449:923–927

Del Galdo F, Lisanti MP, Jimenez SA (2008) Caveolin-1, transforming growth factor-beta receptor internalization, and the pathogenesis of systemic sclerosis. Curr Opin Rheumatol 20:713–719

del Pozo MA, Balasubramanian N, Alderson NB, Kiosses WB, Grande-Garcia A, Anderson RG, Schwartz MA (2005) Phospho-caveolin-1 mediates integrin-regulated membrane domain internalization. Nat Cell Biol 7:901–908

Del Pozo MA, Schwartz MA (2007) Rac, membrane heterogeneity, caveolin and regulation of growth by integrins. Trends Cell Biol 17:246–250

Di Guglielmo GM, Le Roy C, Goodfellow AF, Wrana JL (2003) Distinct endocytic pathways regulate TGF-beta receptor signalling and turnover. Nat Cell Biol 5:410–421

Dietzen DJ, Hastings WR, Lublin DM (1995) Caveolin is palmitoylated on multiple cysteine residues. Palmitoylation is not necessary for localization of caveolin to caveolae. J Biol Chem 270:6838–6842

Drab M, Verkade P, Elger M, Kasper M, Lohn M, Lauterbach B, Menne J, Lindschau C, Mende F, Luft FC, Schedl A, Haller H, Kurzchalia TV (2001) Loss of caveolae, vascular dysfunction, and pulmonary defects in caveolin-1 gene-disrupted mice. Science 293:2449–2452

Dupree P, Parton RG, Raposo G, Kurzchalia TV, Simons K (1993) Caveolae and sorting in the trans-Golgi network of epithelial cells. EMBO J 12:1597–1605

Echarri A, Del Pozo MA (2015) Caveolae - mechanosensitive membrane invaginations linked to actin filaments. J C Sci

Echarri A, Muriel O, Del Pozo MA (2007) Intracellular trafficking of raft/caveolae domains: insights from integrin signaling. Semin Cell Dev Biol 18:627–637

Echarri A, Muriel O, Pavon DM, Azegrouz H, Escolar F, Terron MC, Sanchez-Cabo F, Martinez F, Montoya MC, Llorca O, Del Pozo MA (2012) Caveolar domain organization and trafficking is regulated by Abl kinases and mDia1. J Cell Sci 125:3097–3113

Eden ER, Sanchez-Heras E, Tsapara A, Sobota A, Levine TP, Futter CE (2016) Annexin A1 tethers membrane contact sites that mediate ER to endosome cholesterol transport. Dev Cell 37:473–483

Ewers H, Romer W, Smith AE, Bacia K, Dmitrieff S, Chai W, Mancini R, Kartenbeck J, Chambon V, Berland L, Oppenheim A, Schwarzmann G, Feizi T, Schwille P, Sens P, Helenius A, Johannes L (2010) GM1 structure determines SV40-induced membrane invagination and infection. Nat Cell Biol 12(Supplementary 11–18):1–12

Feng S, Wang Y, Wang X, Wang Z, Cui Y, Liu J, Zhao C, Jin M, Zou W (2010) Caveolin-1 gene silencing promotes the activation of PI3K/AKT dependent on Erα36 and the transformation of MCF10ACE. Sci China Life Sci 53:598–605

Feron O, Belhassen L, Kobzik L, Smith TW, Kelly RA, Michel T (1996) Endothelial nitric oxide synthase targeting to caveolae. Specific interactions with caveolin isoforms in cardiac myocytes and endothelial cells. J Biol Chem 271:22810–22814

Frank PG, Lee H, Park DS, Tandon NN, Scherer PE, Lisanti MP (2004) Genetic ablation of caveolin-1 confers protection against atherosclerosis. Arterioscler Thromb Vasc Biol 24:98–105

Frank PG, Pavlides S, Lisanti MP (2009) Caveolae and transcytosis in endothelial cells: role in atherosclerosis. Cell Tissue Res 335:41–47

Fu Y, Hoang A, Escher G, Parton RG, Krozowski Z, Sviridov D (2004) Expression of caveolin-1 enhances cholesterol efflux in hepatic cells. J Biol Chem 279:14140–14146

Fujita A, Cheng J, Tauchi-Sato K, Takenawa T, Fujimoto T (2009) A distinct pool of phosphatidylinositol 4,5-bisphosphate in caveolae revealed by a nanoscale labeling technique. Proc Natl Acad Sci USA 106:9256–9561

Galbiati F, Volonte D, Engelman JA, Watanabe G, Burk R, Pestell RG, Lisanti MP (1998) Targeted downregulation of caveolin-1 is sufficient to drive cell transformation and hyperactivate the p42/44 MAP kinase cascade. EMBO J 17:6633–6648

Garcia-Cardena G, Oh P, Liu J, Schnitzer JE, Sessa WC (1996) Targeting of nitric oxide synthase to endothelial cell caveolae via palmitoylation: implications for nitric oxide signaling. Proc Natl Acad Sci USA 93:6448–6453

Glenney JR Jr (1989) Tyrosine phosphorylation of a 22-kDa protein is correlated with transformation by Rous sarcoma virus. J Biol Chem 264:20163–20166

Goetz JG, Minguet S, Navarro-Lerida I, Lazcano JJ, Samaniego R, Calvo E, Tello M, Osteso-Ibanez T, Pellinen T, Echarri A, Cerezo A, Klein-Szanto AJ, Garcia R, Keely PJ, Sanchez-Mateos P, Cukierman E, Del Pozo MA (2011) Biomechanical remodeling of the microenvironment by stromal caveolin-1 favors tumor invasion and metastasis. Cell 146:148–163

Gottlieb-Abraham E, Shvartsman DE, Donaldson JC, Ehrlich M, Gutman O, Martin GS, Henis YI (2013) Src-mediated caveolin-1 phosphorylation affects the targeting of active Src to specific membrane sites. Mol Biol Cell 24:3881–3895

Grande-Garcia A, Echarri A, de Rooij J, Alderson NB, Waterman-Storer CM, Valdivielso JM, del Pozo MA (2007) Caveolin-1 regulates cell polarization and directional migration through Src kinase and Rho GTPases. J Cell Biol 177:683–694

Guan JL (2004) Cell biology. Integrins, rafts, Rac, and Rho. Science 303:773–774

Hansen CG, Bright NA, Howard G, Nichols BJ (2009) SDPR induces membrane curvature and functions in the formation of caveolae. Nat Cell Biol 11:807–814

Hansen CG, Howard G, Nichols BJ (2011) Pacsin 2 is recruited to caveolae and functions in caveolar biogenesis. J Cell Sci 124:2777–2785

Hansen CG, Shvets E, Howard G, Riento K, Nichols BJ (2013) Deletion of cavin genes reveals tissue-specific mechanisms for morphogenesis of endothelial caveolae. Nat Communications 4:1831

Hasegawa T, Takeuchi A, Miyaishi O, Xiao H, Mao J, Isobe K (2000) PTRF (polymerase I and transcript-release factor) is tissue-specific and interacts with the BFCOL1 (binding factor of a type-I collagen promoter) zinc-finger transcription factor which binds to the two mouse type-I collagen gene promoters. Biochem J 347(Pt 1):55–59

Hayer A, Stoeber M, Bissig C, Helenius A (2010a) Biogenesis of caveolae: stepwise assembly of large caveolin and cavin complexes. Traffic 11:361–382

Hayer A, Stoeber M, Ritz D, Engel S, Meyer HH, Helenius A (2010b) Caveolin-1 is ubiquitinated and targeted to intralumenal vesicles in endolysosomes for degradation. J Cell Biol 191:615–629

He K, Yan X, Li N, Dang S, Xu L, Zhao B, Li Z, Lv Z, Fang X, Zhang Y, Chen YG (2015) Internalization of the TGF-beta type I receptor into caveolin-1 and EEA1 double-positive early endosomes. Cell Res 25:738–752

Head BP, Patel HH, Roth DM, Murray F, Swaney JS, Niesman IR, Farquhar MG, Insel PA (2006) Microtubules and actin microfilaments regulate lipid raft/caveolae localization of adenylyl cyclase signaling components. J Biol Chem 281:26391–26399

Henley JR, Krueger EW, Oswald BJ, McNiven MA (1998) Dynamin-mediated internalization of caveolae. J Cell Biol 141:85–99
Hernandez VJ, Weng J, Ly P, Pompey S, Dong H, Mishra L, Schwarz M, Anderson RG, Michaely P (2013) Cavin-3 dictates the balance between ERK and Akt signaling. eLife 2:e00905
Hertzog M, Monteiro P, Le Dez G, Chavrier P (2012) Exo70 subunit of the exocyst complex is involved in adhesion-dependent trafficking of caveolin-1. PLoS ONE 7:e52627
Hill MM, Bastiani M, Luetterforst R, Kirkham M, Kirkham A, Nixon SJ, Walser P, Abankwa D, Oorschot VM, Martin S, Hancock JF, Parton RG (2008) PTRF-Cavin, a conserved cytoplasmic protein required for caveola formation and function. Cell 132:113–124
Hoernke M, Mohan J, Larsson E, Blomberg J, Kahra D, Westenhoff S, Schwieger C, Lundmark R (2017) EHD2 restrains dynamics of caveolae by an ATP-dependent, membrane-bound, open conformation. Proc Natl Acad Sci USA 114:E4360–E4369
Howe AK, Juliano RL (2000) Regulation of anchorage-dependent signal transduction by protein kinase A and p21-activated kinase. Nat Cell Biol 2:593–600
Ishibashi S, Brown MS, Goldstein JL, Gerard RD, Hammer RE, Herz J (1993) Hypercholesterolemia in low density lipoprotein receptor knockout mice and its reversal by adenovirus-mediated gene delivery. J Clin Invest 92:883–893
Izumi Y, Hirai S, Tamai Y, Fujise-Matsuoka A, Nishimura Y, Ohno S (1997) A protein kinase Cδ-binding protein SRBC whose expression is induced by serum starvation. J Biol Chem 272:7381–7389
Jansa P, Mason SW, Hoffmann-Rohrer U, Grummt I (1998) Cloning and functional characterization of PTRF, a novel protein which induces dissociation of paused ternary transcription complexes. EMBO J 17:2855–2864
Joshi B, Bastiani M, Strugnell SS, Boscher C, Parton RG, Nabi IR (2012) Phosphocaveolin-1 is a mechanotransducer that induces caveola biogenesis via Egr1 transcriptional regulation. J Cell Biol 199:425–435
Kang YS, Ko YG, Seo JS (2000) Caveolin internalization by heat shock or hyperosmotic shock. Exp Cell Res 255:221–228
Kim CA, Delepine M, Boutet E, El Mourabit H, Le Lay S, Meier M, Nemani M, Bridel E, Leite CC, Bertola DR, Semple RK, O'Rahilly S, Dugail I, Capeau J, Lathrop M, Magre J (2008) Association of a homozygous nonsense caveolin-1 mutation with Berardinelli-Seip congenital lipodystrophy. J Clin Endocrinol Metab 93:1129–1134
Kirkham M, Fujita A, Chadda R, Nixon SJ, Kurzchalia TV, Sharma DK, Pagano RE, Hancock JF, Mayor S, Parton RG (2005) Ultrastructural identification of uncoated caveolin-independent early endocytic vehicles. J Cell Biol 168:465–476
Kirkham M, Nixon SJ, Howes MT, Abi-Rached L, Wakeham DE, Hanzal-Bayer M, Ferguson C, Hill MM, Fernandez-Rojo M, Brown DA, Hancock JF, Brodsky FM, Parton RG (2008) Evolutionary analysis and molecular dissection of caveola biogenesis. J Cell Sci 121:2075–2086
Koch D, Westermann M, Kessels MM, Qualmann B (2012) Ultrastructural freeze-fracture immunolabeling identifies plasma membrane-localized syndapin II as a crucial factor in shaping caveolae. Histochem Cell Biol 138:215–230
Kozera L, White E, Calaghan S (2009) Caveolae act as membrane reserves which limit mechanosensitive I(Cl, swell) channel activation during swelling in the rat ventricular myocyte. PLoS ONE 4:e8312
Kurzchalia TV, Parton RG (1999) Membrane microdomains and caveolae. Curr Opin Cell Biol 11:424–431
Labrecque L, Nyalendo C, Langlois S, Durocher Y, Roghi C, Murphy G, Gingras D, Beliveau R (2004) Src-mediated tyrosine phosphorylation of caveolin-1 induces its association with membrane type 1 matrix metalloproteinase. J Biol Chem 279:52132–52140
Lai YC, Kondapalli C, Lehneck R, Procter JB, Dill BD, Woodroof HI, Gourlay R, Peggie M, Macartney TJ, Corti O, Corvol JC, Campbell DG, Itzen A, Trost M, Muqit MM (2015) Phosphoproteomic screening identifies Rab GTPases as novel downstream targets of PINK1. EMBO J 34:2840–2861

Lajoie P, Partridge EA, Guay G, Goetz JG, Pawling J, Lagana A, Joshi B, Dennis JW, Nabi IR (2007) Plasma membrane domain organization regulates EGFR signaling in tumor cells. J Cell Biology 179:341–356
Lamaze C, Tardif N, Dewulf M, Vassilopoulos S, Blouin CM (2017) The caveolae dress code: structure and signaling. Curr Opin Cell Biol 47:117–125
Lamaze C, Torrino S (2015) Caveolae and cancer: a new mechanical perspective. Biomed J 38:367–379
Lang AB, John Peter AT, Walter P, Kornmann B (2015) ER-mitochondrial junctions can be bypassed by dominant mutations in the endosomal protein Vps13. J Cell Biol 210:883–890
Lapierre LA, Ducharme NA, Drake KR, Goldenring JR, Kenworthy AK (2012) Coordinated regulation of caveolin-1 and Rab11a in apical recycling compartments of polarized epithelial cells. Exp Cell Res 318:103–113
Le Roy C, Wrana JL (2005) Clathrin- and non-clathrin-mediated endocytic regulation of cell signalling. Nat Rev Mol Cell Biol 6:112–126
Lee H, Volonte D, Galbiati F, Iyengar P, Lublin DM, Bregman DB, Wilson MT, Campos-Gonzalez R, Bouzahzah B, Pestell RG, Scherer PE, Lisanti MP (2000) Constitutive and growth factor-regulated phosphorylation of caveolin-1 occurs at the same site (Tyr-14) in vivo: identification of a c-Src/Cav-1/Grb7 signaling cassette. Mol Endocrinol 14:1750–1775
Lee H, Xie L, Luo Y, Lee SY, Lawrence DS, Wang XB, Sotgia F, Lisanti MP, Zhang ZY (2006) Identification of phosphocaveolin-1 as a novel protein tyrosine phosphatase 1B substrate. Biochemistry 45:234–240
Li G, D'Souza-Schorey C, Barbieri MA, Roberts RL, Klippel A, Williams LT, Stahl PD (1995) Evidence for phosphatidylinositol 3-kinase as a regulator of endocytosis via activation of Rab5. Proc Natl Acad Sci USA 92:10207–10211
Li S, Couet J, Lisanti MP (1996) Src tyrosine kinases, Galpha subunits, and H-Ras share a common membrane-anchored scaffolding protein, caveolin. Caveolin binding negatively regulates the auto-activation of Src tyrosine kinases. J Biol Chem 271:29182–29190
Liu L, Pilch PF (2008) A critical role of cavin (polymerase I and transcript release factor) in caveolae formation and organization. J Biol Chem 283:4314–4322
Liu L, Brown D, McKee M, Lebrasseur NK, Yang D, Albrecht KH, Ravid K, Pilch PF (2008) Deletion of Cavin/PTRF causes global loss of caveolae, dyslipidemia, and glucose intolerance. Cell Metab 8:310–317
Liu L, Pilch PF (2016) PTRF/Cavin-1 promotes efficient ribosomal RNA transcription in response to metabolic challenges. eLife 5
Lu Z, Ghosh S, Wang Z, Hunter T (2003) Downregulation of caveolin-1 function by EGF leads to the loss of E-cadherin, increased transcriptional activity of beta-catenin, and enhanced tumor cell invasion. Cancer Cell 4:499–515
Ludwig A, Howard G, Mendoza-Topaz C, Deerinck T, Mackey M, Sandin S, Ellisman MH, Nichols BJ (2013) Molecular composition and ultrastructure of the caveolar coat complex. PLoS Biol 11:e1001640
Ludwig A, Nichols BJ, Sandin S (2016) Architecture of the caveolar coat complex. J Cell Sci 129:3077–3083
Mari M, Morales A, Colell A, Garcia-Ruiz C, Fernandez-Checa JC (2009) Mitochondrial glutathione, a key survival antioxidant. Antioxid Redox Signal 11:2685–2700
Maxwell KN, Breslow JL (2005) Proprotein convertase subtilisin kexin 9: the third locus implicated in autosomal dominant hypercholesterolemia. Curr Opin Lipidol 16:167–172
Maxwell KN, Fisher EA, Breslow JL (2005) Overexpression of PCSK9 accelerates the degradation of the LDLR in a post-endoplasmic reticulum compartment. Proc Natl Acad Sci USA 102:2069–2074
McLelland GL, Soubannier V, Chen CX, McBride HM, Fon EA (2014) Parkin and PINK1 function in a vesicular trafficking pathway regulating mitochondrial quality control. EMBO J 33:282–295

McMahon KA, Zajicek H, Li WP, Peyton MJ, Minna JD, Hernandez VJ, Luby-Phelps K, Anderson RG (2009) SRBC/cavin-3 is a caveolin adapter protein that regulates caveolae function. EMBO J 28:1001–1015

Mercier I, Jasmin JF, Pavlides S, Minetti C, Flomenberg N, Pestell RG, Frank PG, Sotgia F, Lisanti MP (2009) Clinical and translational implications of the caveolin gene family: lessons from mouse models and human genetic disorders. Lab Invest 89:614–623

Mercier I, Camacho J, Titchen K, Gonzales DM, Quann K, Bryant KG, Molchansky A, Milliman JN, Whitaker-Menezes D, Sotgia F, Jasmin JF, Schwarting R, Pestell RG, Blagosklonny MV, Lisanti MP (2012) Caveolin-1 and accelerated host aging in the breast tumor microenvironment: chemoprevention with rapamycin, an mTOR inhibitor and anti-aging drug. Am J Pathol 181:278–293

Michel JB, Feron O, Sase K, Prabhakar P, Michel T (1997) Caveolin versus calmodulin. Counterbalancing allosteric modulators of endothelial nitric oxide synthase. J Biol Chem 272:25907–25912

Mineo C, Ying YS, Chapline C, Jaken S, Anderson RG (1998) Targeting of protein kinase Calpha to caveolae. J Cell Biol 141:601–610

Minshall RD, Tiruppathi C, Vogel SM, Niles WD, Gilchrist A, Hamm HE, Malik AB (2000) Endothelial cell-surface gp60 activates vesicle formation and trafficking via G(i)-coupled Src kinase signaling pathway. J Cell Biol 150:1057–1070

Mohan J, Moren B, Larsson E, Holst MR, Lundmark R (2015) Cavin3 interacts with cavin1 and caveolin1 to increase surface dynamics of caveolae. J Cell Sci 128:979–991

Monier S, Dietzen DJ, Hastings WR, Lublin DM, Kurzchalia TV (1996) Oligomerization of VIP21-caveolin in vitro is stabilized by long chain fatty acylation or cholesterol. FEBS Lett 388:143–149

Monier S, Parton RG, Vogel F, Behlke J, Henske A, Kurzchalia TV (1995) VIP21-caveolin, a membrane protein constituent of the caveolar coat, oligomerizes in vivo and in vitro. Mol Biol Cell 6:911–927

Moren B, Shah C, Howes MT, Schieber NL, McMahon HT, Parton RG, Daumke O, Lundmark R (2012) EHD2 regulates caveolar dynamics via ATP-driven targeting and oligomerization. Mol Biol Cell 23:1316–1329

Moreno-Caceres J, Caja L, Mainez J, Mayoral R, Martin-Sanz P, Moreno-Vicente R, Del Pozo MA, Dooley S, Egea G, Fabregat I (2014) Caveolin-1 is required for TGF-beta-induced transactivation of the EGF receptor pathway in hepatocytes through the activation of the metalloprotease TACE/ADAM17. Cell Death Dis 5:e1326

Mundy DI, Machleidt T, Ying YS, Anderson RG, Bloom GS (2002) Dual control of caveolar membrane traffic by microtubules and the actin cytoskeleton. J Cell Sci 115:4327–4339

Murata T, Lin MI, Huang Y, Yu J, Bauer PM, Giordano FJ, Sessa WC (2007) Reexpression of caveolin-1 in endothelium rescues the vascular, cardiac, and pulmonary defects in global caveolin-1 knockout mice. J Exp Med 204:2373–2382

Muriel O, Echarri A, Hellriegel C, Pavon DM, Beccari L, Del Pozo MA (2011) Phosphorylated filamin A regulates actin-linked caveolae dynamics. J Cell Sci 124:2763–2776

Musiol A, Gran S, Ehrhardt C, Ludwig S, Grewal T, Gerke V, Rescher U (2013) Annexin A6-balanced late endosomal cholesterol controls influenza A replication and propagation. MBio 4:e00608–e00613

Nabi IR (2009) Cavin fever: regulating caveolae. Nat Cell Biol 11:789–791

Nakanishi N, Ogata T, Naito D, Miyagawa K, Taniguchi T, Hamaoka T, Maruyama N, Kasahara T, Nishi M, Matoba S, Ueyama T (2016) MURC deficiency in smooth muscle attenuates pulmonary hypertension. Nat Commun 7:12417

Nassoy P, Lamaze C (2012) Stressing caveolae new role in cell mechanics. Trends Cell Biol 22:381–389

Nichols B (2003) Caveosomes and endocytosis of lipid rafts. J Cell Sci 116:4707–4714

Nixon SJ, Carter A, Wegner J, Ferguson C, Floetenmeyer M, Riches J, Key B, Westerfield M, Parton RG (2007) Caveolin-1 is required for lateral line neuromast and notochord development. J Cell Sci 120:2151–2161

Norambuena A, Schwartz MA (2011) Effects of integrin-mediated cell adhesion on plasma membrane lipid raft components and signaling. Mol Biol Cell 22:3456–3464
Nystrom FH, Chen H, Cong LN, Li Y, Quon MJ (1999) Caveolin-1 interacts with the insulin receptor and can differentially modulate insulin signaling in transfected Cos-7 cells and rat adipose cells. Mol Endocrinol 13:2013–2024
Ogata T, Ueyama T, Isodono K, Tagawa M, Takehara N, Kawashima T, Harada K, Takahashi T, Shioi T, Matsubara H, Oh H (2008) MURC, a muscle-restricted coiled-coil protein that modulates the Rho/ROCK pathway, induces cardiac dysfunction and conduction disturbance. Mol Cell Biol 28:3424–3436
Ogata T, Naito D, Nakanishi N, Hayashi YK, Taniguchi T, Miyagawa K, Hamaoka T, Maruyama N, Matoba S, Ikeda K, Yamada H, Oh H, Ueyama T (2014) MURC/Cavin-4 facilitates recruitment of ERK to caveolae and concentric cardiac hypertrophy induced by α1-adrenergic receptors. Proc Natl Acad Sci USA 111:3811–3816
Okamoto T, Schlegel A, Scherer PE, Lisanti MP (1998) Caveolins, a family of scaffolding proteins for organizing "preassembled signaling complexes" at the plasma membrane. J Biol Chem 273:5419–5422
Orlichenko L, Huang B, Krueger E, McNiven MA (2006) Epithelial growth factor-induced phosphorylation of caveolin 1 at tyrosine 14 stimulates caveolae formation in epithelial cells. J Biol Chem 281:4570–4579
Osmani N, Pontabry J, Comelles J, Fekonja N, Goetz JG, Riveline D, Georges-Labouesse E, Labouesse M (2017) An Arf6- and caveolae-dependent pathway links hemidesmosome remodeling and mechanoresponse. Mol Biol Cell
Palade GE (1953) An electron microscope study of the mitochondrial structure. J Histochem Cytochem 1:188–211
Parat MO, Anand-Apte B, Fox PL (2003) Differential caveolin-1 polarization in endothelial cells during migration in two and three dimensions. Mol Biol Cell 14:3156–3168
Parton RG, del Pozo MA (2013) Caveolae as plasma membrane sensors, protectors and organizers. Nat Rev Mol Cell Biol 14:98–112
Parton RG, Howes MT (2010) Revisiting caveolin trafficking: the end of the caveosome. J Cell Biol 191:439–441
Parton RG, Simons K (2007) The multiple faces of caveolae. Nat Rev Mol Cell Biol 8:185–194
Parton RG, Joggerst B, Simons K (1994) Regulated internalization of caveolae. J Cell Biol 127:1199–1215
Parton RG, Molero JC, Floetenmeyer M, Green KM, James DE (2002) Characterization of a distinct plasma membrane macrodomain in differentiated adipocytes. J Biol Chem 277:46769–46778
Patel HH, Tsutsumi YM, Head BP, Niesman IR, Jennings M, Horikawa Y, Huang D, Moreno AL, Patel PM, Insel PA, Roth DM (2007) Mechanisms of cardiac protection from ischemia/reperfusion injury: a role for caveolae and caveolin-1. FASEB J Official Publ Fed Am Soc Exp Biol 21:1565–1574
Pelkmans L, Helenius A (2002) Endocytosis via caveolae. Traffic 3:311–320
Pelkmans L, Kartenbeck J, Helenius A (2001) Caveolar endocytosis of simian virus 40 reveals a new two-step vesicular-transport pathway to the ER. Nat Cell Biol 3:473–483
Pelkmans L, Zerial M (2005) Kinase-regulated quantal assemblies and kiss-and-run recycling of caveolae. Nature 436:128–133
Pilch PF, Liu L (2011) Fat caves: caveolae, lipid trafficking and lipid metabolism in adipocytes. Trends Endocrinol Metab 22:318–324
Pol A, Martin S, Fernandez MA, Ferguson C, Carozzi A, Luetterforst R, Enrich C, Parton RG (2004) Dynamic and regulated association of caveolin with lipid bodies: modulation of lipid body motility and function by a dominant negative mutant. Mol Biol Cell 15:99–110
Predescu SA, Predescu DN, Timblin BK, Stan RV, Malik AB (2003) Intersectin regulates fission and internalization of caveolae in endothelial cells. Mol Biol Cell 14:4997–5010
Prieto-Sanchez RM, Berenjeno IM, Bustelo XR (2006) Involvement of the Rho/Rac family member RhoG in caveolar endocytosis. Oncogene 25:2961–2973

Rahman A, Sward K (2009) The role of caveolin-1 in cardiovascular regulation. Acta Physiol (Oxf) 195:231–245

Rauch MC, Ocampo ME, Bohle J, Amthauer R, Yanez AJ, Rodriguez-Gil JE, Slebe JC, Reyes JG, Concha II (2006) Hexose transporters GLUT1 and GLUT3 are colocalized with hexokinase I in caveolae microdomains of rat spermatogenic cells. J Cell Physiol 207:397–406

Razani B, Lisanti MP (2001) Two distinct caveolin-1 domains mediate the functional interaction of caveolin-1 with protein kinase A. Am J Physiol Cell Physiol 281:C1241–C1250

Razani B, Zhang XL, Bitzer M, von Gersdorff G, Bottinger EP, Lisanti MP (2001) Caveolin-1 regulates transforming growth factor (TGF)-beta/SMAD signaling through an interaction with the TGF-beta type I receptor. J Biol Chem 276:6727–6738

Razani B, Woodman SE, Lisanti MP (2002) Caveolae: from cell biology to animal physiology. Pharmacol Rev 54:431–467

Richter T, Floetenmeyer M, Ferguson C, Galea J, Goh J, Lindsay MR, Morgan GP, Marsh BJ, Parton RG (2008) High-resolution 3D quantitative analysis of caveolar ultrastructure and caveola-cytoskeleton interactions. Traffic 9:893–909

Ritz D, Vuk M, Kirchner P, Bug M, Schutz S, Hayer A, Bremer S, Lusk C, Baloh RH, Lee H, Glatter T, Gstaiger M, Aebersold R, Weihl CC, Meyer H (2011) Endolysosomal sorting of ubiquitylated caveolin-1 is regulated by VCP and UBXD1 and impaired by VCP disease mutations. Nat Cell Biol 13:1116–1123

Rizzo V, Morton C, DePaola N, Schnitzer JE, Davies PF (2003) Recruitment of endothelial caveolae into mechanotransduction pathways by flow conditioning in vitro. Am J Physiol Heart Circ Physiol 285:H1720–H1729

Rizzo V, Sung A, Oh P, Schnitzer JE (1998) Rapid mechanotransduction in situ at the luminal cell surface of vascular endothelium and its caveolae. J Biol Chem 273:26323–26329

Robenek H, Weissen-Plenz G, Severs NJ (2008) Freeze-fracture replica immunolabelling reveals caveolin-1 in the human cardiomyocyte plasma membrane. J Cell Mol Med 12:2519–2521

Rothberg KG, Heuser JE, Donzell WC, Ying YS, Glenney JR, Anderson RG (1992) Caveolin, a protein component of caveolae membrane coats. Cell 68:673–682

Sala-Vila A, Navarro-Lerida I, Sanchez-Alvarez M, Bosch M, Calvo C, Lopez JA, Calvo E, Ferguson C, Giacomello M, Serafini A, Scorrano L, Enriquez JA, Balsinde J, Parton RG, Vazquez J, Pol A, Del Pozo MA (2016) Interplay between hepatic mitochondria-associated membranes, lipid metabolism and caveolin-1 in mice. Sci Rep 6:27351

Sanguinetti AR, Mastick CC (2003) c-Abl is required for oxidative stress-induced phosphorylation of caveolin-1 on tyrosine 14. Cell Signal 15:289–298

Sanguinetti AR, Cao H, Corley Mastick C (2003) Fyn is required for oxidative- and hyperosmotic-stress-induced tyrosine phosphorylation of caveolin-1. Biochem J 376:159–168

Santibanez JF, Blanco FJ, Garrido-Martin EM, Sanz-Rodriguez F, del Pozo MA, Bernabeu C (2008) Caveolin-1 interacts and cooperates with the transforming growth factor-beta type I receptor ALK1 in endothelial caveolae. Cardiovasc Res 77:791–799

Sarkar S, Carroll B, Buganim Y, Maetzel D, Ng AH, Cassady JP, Cohen MA, Chakraborty S, Wang H, Spooner E, Ploegh H, Gsponer J, Korolchuk VI, Jaenisch R (2013) Impaired autophagy in the lipid-storage disorder Niemann-Pick type C1 disease. Cell reports 5:1302–1315

Scott CC, Vossio S, Vacca F, Snijder B, Larios J, Schaad O, Guex N, Kuznetsov D, Martin O, Chambon M, Turcatti G, Pelkmans L, Gruenberg J (2015) Wnt directs the endosomal flux of LDL-derived cholesterol and lipid droplet homeostasis. EMBO Rep 16:741–752

Scheiffele P, Verkade P, Fra AM, Virta H, Simons K, Ikonen E (1998) Caveolin-1 and -2 in the exocytic pathway of MDCK cells. J Cell Biol 140:795–806

Scherer PE, Lisanti MP, Baldini G, Sargiacomo M, Mastick CC, Lodish HF (1994) Induction of caveolin during adipogenesis and association of GLUT4 with caveolin-rich vesicles. J Cell Biol 127:1233–1243

Scherer PE, Okamoto T, Chun M, Nishimoto I, Lodish HF, Lisanti MP (1996) Identification, sequence, and expression of caveolin-2 defines a caveolin gene family. Proc Natl Acad Sci USA 93:131–135

Scherer PE, Lewis RY, Volonte D, Engelman JA, Galbiati F, Couet J, Kohtz DS, van Donselaar E, Peters P, Lisanti MP (1997) Cell-type and tissue-specific expression of caveolin-2. Caveolins 1 and 2 co-localize and form a stable hetero-oligomeric complex in vivo. J Biol Chem 272:29337–29346
Schlegel A, Schwab RB, Scherer PE, Lisanti MP (1999) A role for the caveolin scaffolding domain in mediating the membrane attachment of caveolin-1. The caveolin scaffolding domain is both necessary and sufficient for membrane binding in vitro. J Biol Chem 274:22660–22667
Schlegel A, Arvan P, Lisanti MP (2001) Caveolin-1 binding to endoplasmic reticulum membranes and entry into the regulated secretory pathway are regulated by serine phosphorylation. Protein sorting at the level of the endoplasmic reticulum. J Biol Chem 276:4398–4408
Schlormann W, Steiniger F, Richter W, Kaufmann R, Hause G, Lemke C, Westermann M (2010) The shape of caveolae is omega-like after glutaraldehyde fixation and cup-like after cryofixation. Histochem Cell Biol 133:223–228
Schwartz EA, Reaven E, Topper JN, Tsao PS (2005) Transforming growth factor-beta receptors localize to caveolae and regulate endothelial nitric oxide synthase in normal human endothelial cells. Biochem J 390:199–206
Sedding DG, Hermsen J, Seay U, Eickelberg O, Kummer W, Schwencke C, Strasser RH, Tillmanns H, Braun-Dullaeus RC (2005) Caveolin-1 facilitates mechanosensitive protein kinase B (Akt) signaling in vitro and in vivo. Circ Res 96:635–642
Seemann E, Sun M, Krueger S, Troger J, Hou W, Haag N, Schuler S, Westermann M, Huebner CA, Romeike B, Kessels MM, Qualmann B (2017) Deciphering caveolar functions by syndapin III KO-mediated impairment of caveolar invagination. eLife 6
Senju Y, Itoh Y, Takano K, Hamada S, Suetsugu S (2011) Essential role of PACSIN2/syndapin-II in caveolae membrane sculpting. J Cell Sci 124:2032–2040
Shack S, Wang XT, Kokkonen GC, Gorospe M, Longo DL, Holbrook NJ (2003) Caveolin-induced activation of the phosphatidylinositol 3-kinase/Akt pathway increases arsenite cytotoxicity. Mol Cell Biol 23:2407–2414
Shajahan AN, Timblin BK, Sandoval R, Tiruppathi C, Malik AB, Minshall RD (2004) Role of Src-induced dynamin-2 phosphorylation in caveolae-mediated endocytosis in endothelial cells. J Biol Chem 279:20392–20400
Sharma P, Ghavami S, Stelmack GL, McNeill KD, Mutawe MM, Klonisch T, Unruh H, Halayko AJ (2010) β-Dystroglycan binds caveolin-1 in smooth muscle: a functional role in caveolae distribution and Ca^{2+} release. J Cell Sci 123:3061–3070
Shaul PW, Anderson RG (1998) Role of plasmalemmal caveolae in signal transduction. Am J Physiol 275:L843–L851
Shi F, Sottile J (2008) Caveolin-1-dependent β1 integrin endocytosis is a critical regulator of fibronectin turnover. J Cell Sci 121:2360–2371
Shin JS, Gao Z, Abraham SN (2000) Involvement of cellular caveolae in bacterial entry into mast cells. Science 289:785–788
Shvets E, Bitsikas V, Howard G, Hansen CG, Nichols BJ (2015) Dynamic caveolae exclude bulk membrane proteins and are required for sorting of excess glycosphingolipids. Nat Commun 6:6867
Siddiqi S, Sheth A, Patel F, Barnes M, Mansbach CM 2nd (2013) Intestinal caveolin-1 is important for dietary fatty acid absorption. Biochim Biophys Acta 1831:1311–1321
Simons K, Toomre D (2000) Lipid rafts and signal transduction. Nat Rev Mol Cell Biol 1:31–39
Sinha B, Koster D, Ruez R, Gonnord P, Bastiani M, Abankwa D, Stan RV, Butler-Browne G, Vedie B, Johannes L, Morone N, Parton RG, Raposo G, Sens P, Lamaze C, Nassoy P (2011) Cells respond to mechanical stress by rapid disassembly of caveolae. Cell 144:402–413
Smart EJ, Ying Y, Donzell WC, Anderson RG (1996) A role for caveolin in transport of cholesterol from endoplasmic reticulum to plasma membrane. J Biol Chem 271:29427–29435
Smart EJ, Ying YS, Anderson RG (1995) Hormonal regulation of caveolae internalization. J Cell Biol 131:929–938

Song KS, Sargiacomo M, Galbiati F, Parenti M, Lisanti MP (1997) Targeting of a G alpha subunit (Gi1 alpha) and c-Src tyrosine kinase to caveolae membranes: clarifying the role of N-myristoylation. Cell Mol Biol (Noisy-le-grand) 43:293–303

Song KS, Scherer PE, Tang Z, Okamoto T, Li S, Chafel M, Chu C, Kohtz DS, Lisanti MP (1996) Expression of caveolin-3 in skeletal, cardiac, and smooth muscle cells. Caveolin-3 is a component of the sarcolemma and co-fractionates with dystrophin and dystrophin-associated glycoproteins. J Biol Chem 271:15160–15165

Stoeber M, Stoeck IK, Hanni C, Bleck CK, Balistreri G, Helenius A (2012) Oligomers of the ATPase EHD2 confine caveolae to the plasma membrane through association with actin. EMBO J 31:2350–2364

Stoeber M, Schellenberger P, Siebert CA, Leyrat C, Helenius A, Grunewald K (2016) Model for the architecture of caveolae based on a flexible, net-like assembly of Cavin1 and Caveolin discs. Proc Natl Acad Sci USA 113:E8069–E8078

Strippoli R, Loureiro J, Moreno V, Benedicto I, Perez Lozano ML, Barreiro O, Pellinen T, Minguet S, Foronda M, Osteso MT, Calvo E, Vazquez J, Lopez Cabrera M, del Pozo MA (2014) Caveolin-1 deficiency induces a MEK-ERK1/2-Snail-1-dependent epithelial-mesenchymal transition and fibrosis during peritoneal dialysis. EMBO Mol Med 7:102–123

Strippoli R, Loureiro J, Moreno V, Benedicto I, Perez Lozano ML, Barreiro O, Pellinen T, Minguet S, Foronda M, Osteso MT, Calvo E, Vazquez J, Lopez Cabrera M, del Pozo MA (2015) Caveolin-1 deficiency induces a MEK-ERK1/2-Snail-1-dependent epithelial-mesenchymal transition and fibrosis during peritoneal dialysis. EMBO Mol Med 7:102–123

Sverdlov M, Shinin V, Place AT, Castellon M, Minshall RD (2009) Filamin A regulates caveolae internalization and trafficking in endothelial cells. Mol Biol Cell 20:4531–4540

Tagawa M, Ueyama T, Ogata T, Takehara N, Nakajima N, Isodono K, Asada S, Takahashi T, Matsubara H, Oh H (2008) MURC, a muscle-restricted coiled-coil protein, is involved in the regulation of skeletal myogenesis. Am J Physiol Cell Physiol 295:C490–C498

Tang Z, Scherer PE, Okamoto T, Song K, Chu C, Kohtz DS, Nishimoto I, Lodish HF, Lisanti MP (1996) Molecular cloning of caveolin-3, a novel member of the caveolin gene family expressed predominantly in muscle. J Biol Chem 271:2255–2261

Tang W, Feng X, Zhang S, Ren Z, Liu Y, Yang B, lv B, Cai Y, Xia J, Ge N (2015) Caveolin-1 confers resistance of hepatoma cells to anoikis by activating IGF-1 pathway. Cell Physiol Biochem 36:1223–1236

Thorn H, Stenkula KG, Karlsson M, Ortegren U, Nystrom FH, Gustavsson J, Stralfors P (2003) Cell surface orifices of caveolae and localization of caveolin to the necks of caveolae in adipocytes. Mol Biol Cell 14:3967–3976

Tourkina E, Richard M, Gooz P, Bonner M, Pannu J, Harley R, Bernatchez PN, Sessa WC, Silver RM, Hoffman S (2008) Antifibrotic properties of caveolin-1 scaffolding domain in vitro and in vivo. Am J Physiol Lung Cell Mol Physiol 294:L843–L861

Uittenbogaard A, Smart EJ (2000) Palmitoylation of caveolin-1 is required for cholesterol binding, chaperone complex formation, and rapid transport of cholesterol to caveolae. J Biol Chem 275:25595–25599

Villasenor R, Kalaidzidis Y, Zerial M (2016) Signal processing by the endosomal system. Curr Opin Cell Biol 39:53–60

Vinten J, Johnsen AH, Roepstorff P, Harpoth J, Tranum-Jensen J (2005) Identification of a major protein on the cytosolic face of caveolae. Biochim Biophys Acta 1717:34–40

Walser PJ, Ariotti N, Howes M, Ferguson C, Webb R, Schwudke D, Leneva N, Cho KJ, Cooper L, Rae J, Floetenmeyer M, Oorschot VM, Skoglund U, Simons K, Hancock JF, Parton RG (2012) Constitutive formation of caveolae in a bacterium. Cell 150:752–763

Wang XM, Zhang Y, Kim HP, Zhou Z, Feghali-Bostwick CA, Liu F, Ifedigbo E, Xu X, Oury TD, Kaminski N, Choi AM (2006) Caveolin-1: a critical regulator of lung fibrosis in idiopathic pulmonary fibrosis. J Exp Med 203:2895–2906

Wickstrom SA, Lange A, Hess MW, Polleux J, Spatz JP, Kruger M, Pfaller K, Lambacher A, Bloch W, Mann M, Huber LA, Fassler R (2010) Integrin-linked kinase controls microtubule dynamics required for plasma membrane targeting of caveolae. Dev Cell 19:574–858

Williams TM, Lisanti MP (2005) Caveolin-1 in oncogenic transformation, cancer, and metastasis. Am J Physiol Cell Physiol 288:C494–C506

Williams TM, Medina F, Badano I, Hazan RB, Hutchinson J, Muller WJ, Chopra NG, Scherer PE, Pestell RG, Lisanti MP (2004) Caveolin-1 gene disruption promotes mammary tumorigenesis and dramatically enhances lung metastasis in vivo. Role of Cav-1 in cell invasiveness and matrix metalloproteinase (MMP-2/9) secretion. J Biol Chem 279:51630–51646

Xu XL, Wu LC, Du F, Davis A, Peyton M, Tomizawa Y, Maitra A, Tomlinson G, Gazdar AF, Weissman BE, Bowcock AM, Baer R, Minna JD (2001) Inactivation of human SRBC, located within the 11p15.5-p15.4 tumor suppressor region, in breast and lung cancers. Cancer Res 61:7943–7949

Yamada E (1955) The fine structure of the gall bladder epithelium of the mouse. J Biophys Biochem Cytol 1:445–458

Yamaguchi T, Lu C, Ida L, Yanagisawa K, Usukura J, Cheng J, Hotta N, Shimada Y, Isomura H, Suzuki M, Fujimoto T, Takahashi T (2016) ROR1 sustains caveolae and survival signalling as a scaffold of cavin-1 and caveolin-1. Nat Commun 7:10060

Yamamoto M, Toya Y, Jensen RA, Ishikawa Y (1999) Caveolin is an inhibitor of platelet-derived growth factor receptor signaling. Exp Cell Res 247:380–388

Yamamoto M, Toya Y, Schwencke C, Lisanti MP, Myers MG Jr, Ishikawa Y (1998) Caveolin is an activator of insulin receptor signaling. J Biol Chem 273:26962–26968

Yamamoto H, Komekado H, Kikuchi A (2006) Caveolin is necessary for Wnt-3a-dependent internalization of LRP6 and accumulation of beta-catenin. Dev Cell 11:213–223

Yeh YC, Parekh AB (2015) Distinct structural domains of caveolin-1 independently regulate Ca^{2+} release-activated Ca^{2+} channels and Ca^{2+} microdomain-dependent gene expression. Mol Cell Biol 35:1341–1349

Yeow I, Howard G, Chadwick J, Mendoza-Topaz C, Hansen CG, Nichols BJ, Shvets E (2017) EHD proteins cooperate to generate caveolar clusters and to maintain caveolae during repeated mechanical stress. Curr Biol 27(19):2951–2962

Yu J, Bergaya S, Murata T, Alp IF, Bauer MP, Lin MI, Drab M, Kurzchalia TV, Stan RV, Sessa WC (2006) Direct evidence for the role of caveolin-1 and caveolae in mechanotransduction and remodeling of blood vessels. J Clin Invest 116:1284–1291

Yuan T, Hong S, Yao Y, Liao K (2007) Glut-4 is translocated to both caveolae and non-caveolar lipid rafts, but is partially internalized through caveolae in insulin-stimulated adipocytes. Cell Res 17:772–782

Zhang B, Peng F, Wu D, Ingram AJ, Gao B, Krepinsky JC (2007) Caveolin-1 phosphorylation is required for stretch-induced EGFR and Akt activation in mesangial cells. Cell Signal 19:1690–1700

Zhang M, Dwyer NK, Love DC, Cooney A, Comly M, Neufeld E, Pentchev PG, Blanchette-Mackie EJ, Hanover JA (2001) Cessation of rapid late endosomal tubulovesicular trafficking in Niemann-Pick type C1 disease. Proc Natl Acad Sci USA 98:4466–4471

Zilberberg A, Yaniv A, Gazit A (2004) The low density lipoprotein receptor-1, LRP1, interacts with the human frizzled-1 (HFz1) and down-regulates the canonical Wnt signaling pathway. J Biol Chem 279:17535–17542

Zimnicka AM, Husain YS, Shajahan AN, Sverdlov M, Chaga O, Chen Z, Toth PT, Klomp J, Karginov AV, Tiruppathi C, Malik AB, Minshall RD (2016) Src-dependent phosphorylation of caveolin-1 Tyr-14 promotes swelling and release of caveolae. Mol Biol Cell 27:2090–2106

Chapter 9
EGFR Trafficking in Physiology and Cancer

Giusi Caldieri, Maria Grazia Malabarba, Pier Paolo Di Fiore and Sara Sigismund

Contents

Abstract Signaling from the epidermal growth factor receptor (EGFR) elicits multiple biological responses, including cell proliferation, migration, and survival. Receptor endocytosis and trafficking are critical physiological processes that control the strength, duration, diversification, and spatial restriction of EGFR signaling through multiple mechanisms, which we review in this chapter. These mechanisms include: (i) regulation of receptor density and activation at the cell surface; (ii) concentration of receptors into distinct nascent endocytic structures; (iii) commitment of the receptor to different endocytic routes; (iv) endosomal sorting and postendocytic trafficking of the receptor through distinct pathways, and (v) recycling to restricted regions of the cell surface. We also highlight how communication between organelles controls

G. Caldieri · M. G. Malabarba · P. P. Di Fiore · S. Sigismund (✉)
Dipartimento di Oncologia ed Emato-oncologia, Università degli Studi di Milano,
Via Santa Sofia 9/1, 20122 Milan, Italy
e-mail: sara.sigismund@unimi.it

G. Caldieri · M. G. Malabarba · P. P. Di Fiore · S. Sigismund
Istituto Europeo di Oncologia, Via Ripamonti 435, 20141 Milan, Italy

C. Lamaze and I. Prior (eds.), *Endocytosis and Signaling*, Progress in Molecular and Subcellular Biology 57, https://doi.org/10.1007/978-3-319-96704-2_9

EGFR activity along the endocytic route. Finally, we illustrate how abnormal trafficking of EGFR oncogenic mutants, as well as alterations of the endocytic machinery, contributes to aberrant EGFR signaling in cancer.

9.1 Introduction: EGFR and the ErbB Family

The epidermal growth factor receptor (EGFR) belongs to the ErbB family of receptor tyrosine kinases (RTKs), which includes three other members, ErbB2, ErbB3, and ErbB4 (Lemmon et al. 2014). At the systems level, EGFR signaling is critical for developmental processes and adult tissue regeneration, while at the cellular level it elicits a number of responses, including cell proliferation, migration, and survival (Schlessinger 2014). Gain-of-function genetic lesions in the *EGFR* gene, as well as alterations in the EGFR signaling cascade, are involved in several human solid tumors, such as glioblastoma, lung, head and neck, and colon cancer. Thus, the EGFR is a target of several anti-cancer therapies [Sect. 9.4.3 and Yarden and Pines (2012)].

The readout of EGFR signaling is complicated by the fact that there are seven known EGFR ligands that are active in different physiological contexts and capable of inducing specific signaling and biological outputs (Singh and Harris 2005; Wilson et al. 2009): epidermal growth factor (EGF), transforming growth factor alpha (TGFα), amphiregulin (AREG), epiregulin (EREG), heparin-binding EGF-like (HB-EGF), betacellulin (BTC), and epigen (EPG). EGF is the best-studied EGFR ligand and—together with TGFα, AREG and EPG—is specific for the EGFR, while the other ligands also bind to ErbB4. The different EGFR ligands have distinct binding kinetics (Macdonald-Obermann and Pike 2014) and differentially influence EGFR trafficking and fate (see Sect. 9.2.3.3). Importantly, EGFR overexpression in solid tumors is often associated with increased secretion of cognate ligand(s) resulting in chronic EGFR activation (see Sect. 9.4.1).

EGFR signaling is finely tuned in cells by multiple coordinated mechanisms, including regulation by phosphatases, feedback inhibitors of the kinase, and endocytosis (Lemmon and Schlessinger 2010). Besides being the major mechanism of long-term signal attenuation—via removal of receptors from the plasma membrane (PM) and their targeting to degradation—endocytosis controls the timing, type, and strength of EGFR signaling, thanks to the spatial constraints provided by intracellular compartments through which the receptor is trafficked (Sigismund et al. 2012).

In this chapter, we first describe the different mechanisms governing EGFR endocytosis and postendocytic trafficking (Sect. 9.2). We then highlight the importance of endocytosis in controlling EGFR signaling and function in physiological processes (Sect. 9.3). Finally, we discuss how cancer cells evade endocytic control of EGFR signaling, thereby, acquiring a proliferative/migratory advantage (Sect. 9.4).

9.2 Mechanisms of EGFR Endocytosis

9.2.1 EGFR Activation at the Cell Surface

The human *EGFR* gene is located on the short arm of chromosome 7 and encodes for a 1210-residue precursor protein, which, after cleavage of the N-terminal signal peptide, yields a mature protein of 1186 residues (Ullrich et al. 1984). Herein, we adopt the amino acid numbering of the mature EGFR form.

The EGFR consists of an extracellular region responsible for ligand recognition, a transmembrane (TM) domain, and an intracellular region that includes the juxtamembrane regulatory region, the kinase domain, and the intracellular C-terminal regulatory tail containing the tyrosine residues phosphorylated upon ligand binding (Lemmon and Schlessinger 2010). The intracellular region also contains lysine acceptor residues, located primarily in the kinase domain, which are critical for receptor ubiquitination (Huang et al. 2006).

Molecular details of EGFR activation at the PM have been obtained over the last decades through the combination of biological investigations and structural studies (Garrett et al. 2002; Ogiso et al. 2002) [reviewed in Lemmon et al. (2014), Kovacs et al. (2015)]. In resting cells, the EGFR continuously shifts from an open to a closed, autoinhibited, conformation. This closed state, in which intramolecular interactions prevent receptor dimerization and spurious kinase activation, is energetically favored in the absence of ligand. Thus, in resting cells, EGFR is primarily found as an autoinhibited monomer.

Ligand binding stabilizes the open EGFR conformation, which is capable of receptor dimerization, shifting the monomer–dimer equilibrium to the dimeric state (Lemmon 2009). Dimerization, in turn, determines a series of structural rearrangements that are transmitted to the cytoplasmic domain, and allow the formation of asymmetric dimers between the juxtaposed catalytic domains, finally leading to the allosteric activation of the EGFR (Zhang et al. 2006). In the active dimer, each monomer trans-autophosphorylates specific tyrosine residues in the intracytoplasmic region of the other monomer, thereby, triggering the signaling cascade (Lemmon et al. 2014).

Notably, in the absence of ligand, EGFR moieties can spontaneously form finite-lifetime dimers, whose abundance depends on cell type and EGFR expression levels (Chung et al. 2010). These preformed dimers, although primed for ligand binding and signaling, are inactive; ligand binding is still required for receptor activation and signaling (Chung et al. 2010). Importantly, EGFR overexpression, as occurs in tumors, can increase the amount of unbound homodimers (or ErbB family heterodimers) and has been proposed as a mechanism at the basis of spurious kinase activation in the absence of ligand (Chung et al. 2010). Ligand-independent kinase activation in the presence of high numbers of surface EGFRs can however be limited by phosphatases. Indeed, constitutive trafficking of unbound/inactive EGFRs to endosomes allows receptor dephosphorylation by the phosphatase PTP1B, which is resident in the endoplasmic reticulum (ER) and interacts with endosomal EGFR via so-called ER contact sites (see Sect. 9.3.1) (Baumdick et al. 2015).

Following EGFR activation, phosphorylated tyrosine residues in the intracellular tail act as docking sites for signaling molecules and endocytic adaptors, which trigger signaling and receptor endocytosis, respectively (Lemmon and Schlessinger 2010). One protein recruited to the EGFR at the PM is the E3 ligase, Cbl, which ubiquitinates lysine residues in the kinase domain (Levkowitz et al. 1998, 1999; Huang et al. 2006). EGFR ubiquitination is a critical signal in the endocytic pathway (Umebayashi et al. 2008); at the PM, it determines the endocytic route (see Sect. 9.2.2.3), while at the endosomal sorting station it targets receptors to a degradative fate (see Sect. 9.2.3.1).

9.2.2 EGFR Internalization Routes

In the absence of ligand, EGFR is internalized at a very slow rate and is mainly recycled back to the PM at a rate that is ~5–10 times higher than its constitutive endocytic rate. This results in a predominant PM location of the receptor (Dunn et al. 1986; Carpenter and Cohen 1976; Stoscheck and Carpenter 1984). The ratio of PM versus intracellular EGFR in basal conditions, however, is highly dependent on the level of EGFR expression. As expected, ligand binding and kinase activation increase the endocytic rate constant and are indeed essential for rapid EGFR endocytosis (Sorkin and Goh 2008).

Endocytosis can occur through different pathways, broadly classified as clathrin-mediated endocytosis (CME) and non-clathrin endocytosis (NCE, Fig. 9.1). The choice of these different pathways depends on the cell context, the nature of homo-/heterodimerization of the receptor, ligand concentration, and the presence of specific endocytic signals in the intracytoplasmic tail, as discussed in the following sections.

9.2.2.1 Internalization Signals

The EGFR contains several internalization motifs and signals in its intracytoplasmic region that are unmasked/activated upon EGF binding. These include two recognition motifs for the major endocytic adaptor, adaptor protein 2 (AP2), which links cargoes to the clathrin machinery: (i) the YRAL motif, responsible for recruitment of the AP2 μ subunit; (ii) the LL motif, critical for tyrosine phosphorylation of the AP2 $\beta 2$ subunit, which is predicted to facilitate the interaction between this motif and AP2 (Goh and Sorkin 2013). Interestingly, mutation of these two AP2 binding motifs does not affect CME of the EGFR (Goh et al. 2010). Similarly, functional ablation of AP2 in cells only partially inhibits EGFR internalization (Hinrichsen et al. 2003; Motley et al. 2003), suggesting the existence of AP2-independent mechanisms responsible for EGFR-CME (see Sect. 9.2.2.2).

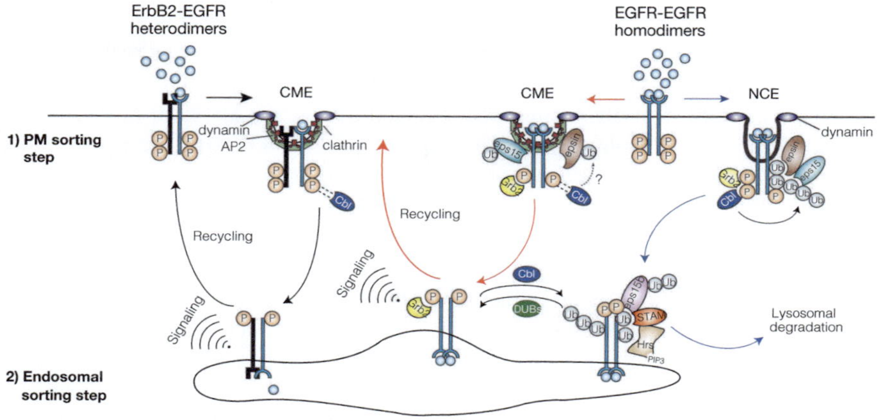

Fig. 9.1 Endocytic routes and sorting of EGFR heterodimers and homodimers. At the plasma membrane (PM), EGFR dimers can be internalized by different routes. Upon ligand binding, EGFR–ErbB2 heterodimers (left) are phosphorylated and internalized via clathrin-mediated endocytosis (CME, black lines), where the internalizing pit is coated by clathrin, AP2, and endocytic adaptor proteins. EGFR–ErbB2 heterodimers are poorly ubiquitinated due to the inefficient recruitment of Cbl. Once they reach the endosomal station, ligand dissociates from the receptor due to the more acidic pH of the endosomes, and the heterodimers are almost exclusively recycled back to the PM, while being inefficiently degraded, thus sustaining signaling. At high dose of ligand, EGFR–EGFR homodimers (right) can be internalized via both CME and non-clathrin endocytosis (NCE). EGFRs entering via CME (red lines) recruit endocytic adaptors (e.g., eps15 and epsin), AP2, and signaling proteins (e.g., Grb2) and are mainly recycled back to the PM. CME is required to sustain signaling from endosomes and/or through cycles of receptor recycling. Receptor ubiquitination by Cbl is not required for CME. In parallel, a fraction of EGFR, which is extensively ubiquitinated by Cbl, in complex with Grb2, at the PM, enter the cell via NCE and is primarily targeted to the lysosome for degradation. Receptors coming from both CME and NCE reach the endosomal station, where they are subjected to further regulation by ubiquitination/deubiquitination reactions. In the endosomes, ubiquitinated EGFRs are recognized by the ESCRT-0 complex (Hrs, STAM, EPS15b), which drives the receptor to degradation

Several tyrosine residues in the EGFR cytoplasmic tail, in addition to triggering signaling events once phosphorylated, can also recruit endocytic factors (Roskoski 2014). For instance, residues pY1068/pY1086 act as a docking site for the adaptor protein, Grb2, which bridges the phosphorylated receptor to Cbl and the endocytic machinery, as well as to the RAS/MAPK signaling cascade (Goh and Sorkin 2013; Sorkin and Goh 2008). Cbl itself, besides ubiquitinating the EGFR, also acts as an adaptor molecule for several endocytic proteins involved in receptor internalization (Schmidt and Dikic 2005; Lemmon and Schlessinger 2010). As mentioned, Cbl can be recruited to the activated EGFR indirectly via Grb2 (Waterman et al. 2002; Jiang et al. 2003), In addition, Cbl can also bind directly to pY1045 (Waterman et al. 1999b). This two-pronged interaction between Cbl and the EGFR is needed for stable Cbl recruitment and efficient receptor ubiquitination (Capuani et al. 2015; Sigismund et al. 2013).

Cooperativity between the direct and indirect binding modes results in an "off–on" threshold response in receptor ubiquitination as EGF concentration increases (Sigismund et al. 2013). Indeed, while the levels of phosphorylated EGFR (EGFR-pY) increase gradually with increasing EGF concentrations, the levels of ubiquitinated receptor (EGFR-Ub) display a sigmoidal dose-response, increasing sharply between 1 and 10 ng/ml EGF before reaching a plateau (Sigismund et al. 2013). Thus, Cbl, in complex with Grb2, effectively acts as an analogical-to-digital converter that translates a linear EGF input into an "off–on" threshold response for receptor ubiquitination (Capuani et al. 2015; Sigismund et al. 2013). This ubiquitination threshold response acts as a critical signal influencing EGFR internalization in specific cellular contexts (see Sect. 9.2.2.3) and receptor degradation (see Sect. 9.2.3.1).

The impact of ubiquitination on EGFR internalization is made more complex by the fact that the EGFR is subjected to different types of ubiquitin (Ub) modifications. Mass spectrometry studies revealed that the predominant modifications are Lys63 polyUb chains and multi-monoUb, while Lys48 and Lys11 polyUb chains are less abundant (Huang et al. 2006). Lys63 and monoUb are both critical signals in trafficking (Acconcia et al. 2009), but whether they serve different functions in EGFR endocytosis remains to be determined. The relevance of Lys48 and Lys11 polyUb chains to EGFR biology is also currently unclear.

The EGFR is also modified by the Ub-like molecule, Nedd8 (Oved et al. 2006). Neddylation is catalyzed by Cbl in complex with the Nedd8-specific E2 enzyme (Ubc12), and it is thought to occur on multiple lysine residues in the kinase domain, possibly overlapping with ubiquitination sites. Nedd8 was proposed to "prime" the EGFR for further ubiquitination reactions, and to cooperate with Ub to target EGFR to degradation (Oved et al. 2006). However, the exact involvement of neddylation in EGFR biology still needs to be clarified.

9.2.2.2 Clathrin-Mediated Endocytosis

CME has been extensively studied over the last decades, the result being a high-resolution molecular picture of the process. The EGFR is internalized via CME in all cell types and at all physiological EGF concentrations (Sigismund et al. 2008, 2013; Carpentier et al. 1982; Gorden et al. 1978; Hanover et al. 1984; Sorkin and Carpenter 1993; Jiang et al. 2003). In CME, the active receptor is recognized by adaptor molecules—primarily AP2—that bridge the cargo to clathrin, driving its internalization via clathrin-coated pits [CCPs, reviewed in Kirchhausen et al. (2014), McMahon and Boucrot (2011), Fig. 9.1]. The last step of vesicle pinching from the PM is performed by the large GTPase dynamin [reviewed in Antonny et al. (2016)]. Dynamin is also part of the scission machinery in some clathrin-independent pathways (see also Sect. 9.2.2.3).

Many accessory proteins cooperate in cargo recognition, CCP formation and vesicle release, including eps15, epsin, Grb2, Cbl, and intersectins (ITSNs) (McMahon and Boucrot 2011). The involvement of so many endocytic factors in CME and the existence of distinct internalization signals in the EGFR C-terminal tail (see

Sect. 9.2.2.1) have led to the notion that CME is controlled by several redundant mechanisms that together confer robustness to the system (Goh et al. 2010). Moreover, it has been hypothesized that the different endocytic proteins might be involved in the formation of distinct types of CCPs, specialized in cargo selection and targeting to specific intracellular fates (Lakadamyali et al. 2006).

In addition to internalization signals centered on the receptor, monoubiquitination of endocytic adaptors (e.g., eps15) has also been shown to be critical to EGFR-CME (Savio et al. 2016). Indeed, it has been proposed that cycles of ubiquitination (by the E3 ligase, NEDD4) and deubiquitination [by the deubiquitinating enzyme (DUB) enzyme, Usp9X] are necessary for EGFR-CME (Savio et al. 2016).

9.2.2.3 Non-clathrin Endocytosis

EGFR-NCE pathways were first observed over 30 years ago (Lund et al. 1990), but their study was hampered by their morphological heterogeneity, cell context dependency and peculiar growth condition requirements (Johannes et al. 2015). Despite their heterogeneity, the different EGFR-NCE mechanisms all share the common feature of being activated at high, nearly saturating, EGF doses (>10 ng/ml) (Boucrot et al. 2015; Lund et al. 1990; Orth et al. 2006; Sigismund et al. 2005).

For one EGFR-NCE pathway, the dependency on high EGF concentrations has been explained at the molecular level and directly linked to the EGFR-Ub threshold response [see Sect. 9.2.2.1 and Sigismund et al. (2005, 2013)]. It was shown that activation of EGFR-NCE occurs over the same EGF concentration range (~1–10 ng/ml) as EGFR-Ub [Sigismund et al. (2013) and Sect. 9.2.2.1]. Importantly, mutations that inhibit EGFR ubiquitination also inhibit EGFR-NCE to a similar extent, showing that EGFR-Ub and NCE are mechanistically linked [Sigismund et al. (2013) and Fig. 9.1, right]. Furthermore, proteins containing ubiquitin-binding domains (UBDs), such as eps15 and epsins, are needed to recognize EGFR-Ub and to target it to internalization via NCE [Sigismund et al. (2005) and Fig. 9.1, right], further supporting the link between receptor ubiquitination and NCE.

The above-described EGFR-NCE pathway is active in different cell lines and has a relatively slow internalization rate (≤CME) and requires cholesterol-enriched PM domains, while it is caveolin-independent (Sigismund et al. 2005, 2013). Importantly, internalization through NCE versus CME has important consequences on EGFR fate and signaling (Fig. 9.1, right), as discussed in Sect. 9.3.2.

At the molecular level, EGFR-NCE requires dynamin fission activity, and the ubiquitin-binding endocytic adaptors, eps15 and epsin. However, a molecular definition of the pathway was obtained only recently. Through a proteomic approach coupled with RNAi screening, proteins previously not suspected to participate in endocytosis were identified as specific players of the pathway, among which the ER-resident protein Reticulon3, RTN3 (Caldieri et al. 2017). The pathway relies on the formation of ER–PM contact sites that depend on RTN3 function and are required for the formation/maturation of NCE tubular invaginations. Local Ca^{2+} release at these sites, triggered by IP3-dependent activation of ER Ca^{2+} channels, is needed for the

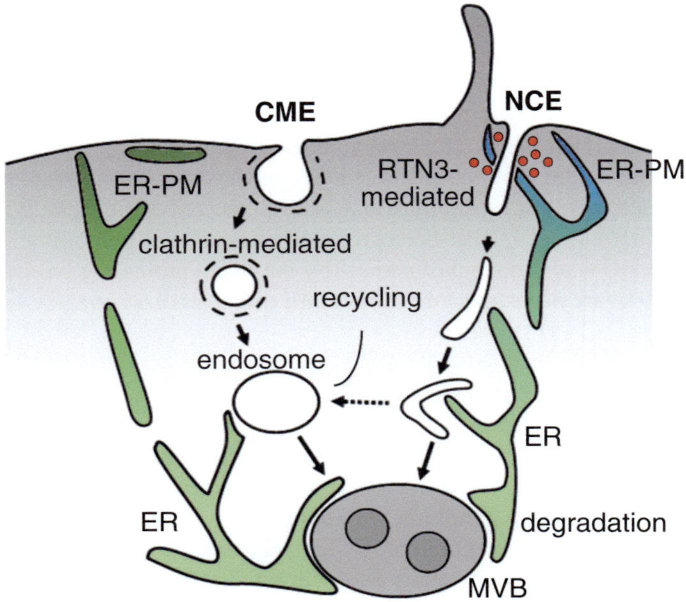

Fig. 9.2 Model for EGFR endocytosis. At high dose of EGF, EGFR is internalized through clathrin-mediated endocytosis (CME) and non-clathrin endocytosis (NCE) in some cell lines. NCE is mediated by tubular invaginations that, differently from clathrin-coated pits, need the establishment of RTN3-dependent contact sites with the endoplasmic reticulum (ER) in order to progress. In an initial phase, RTN3-dependent ER–PM contact sites are required for the formation/maturation of tubular invaginations; then, they act as sites of local calcium release (red circles), which is required for the fission of the tubular invagination and the completion of the internalization step. By inhibiting EGFR entry via NCE, RTN3 KD affects the subsequent EGFR targeting to the lysosomal compartment and delays receptor degradation. It is unclear whether RTN3 is also involved in ER contact sites with early/late endocytic stations, e.g., endosomes, multivesicular bodies (MVBs), lysosomes

completion of EGFR internalization in a positive feedback loop (Caldieri et al. 2017) (Fig. 9.2). Mechanistically, how ER–PM contacts are established, if they require a direct EGFR–RTN3 interaction and if/how EGFR ubiquitination is needed for their formation remain open issues.

Other EGFR-NCE pathways have been described in fibroblasts and migrating cells. For instance, a macropinocytic-like pathway involving large tubular structures originating from circular dorsal ruffles or "waves" has been observed in mouse and human fibroblasts that is thought to be critical for 3D cell migration and extracellular matrix degradation [see Sect. 9.3.2 and Orth et al. (2006)]. In addition, a fast-kinetic NCE pathway that mediates ligand-triggered uptake of different G protein-coupled receptors (GPCRs) and RTKs, including the EGFR, has been identified (Boucrot et al. 2015). This endophilin-dependent, clathrin-independent pathway, called fast endophilin-mediated endocytosis (FEME) appears to be active in a very specialized region of the migrating cell, i.e., the leading edge, and to be required for spatially restricted EGF-dependent signaling [see Sect. 9.3.2 and Boucrot et al. (2015)] (see also Sect. 9.3.2). At the molecular level, FEME requires the BAR domain-containing protein, endophilinA2, as well as dynamin for scission (Boucrot et al. 2015). This NCE pathway shows many similarities with the recently described Shiga toxin uptake pathway, which is also clathrin-independent, endophilinA2-, and dynamin-dependent (Renard et al. 2015). The EGFR modifications and/or the signaling cascade required to trigger these forms of EGFR-NCE are currently unknown.

9.2.3 EGFR Trafficking and Fate

Independent of the entry route, EGFRs internalized from the PM invariably reach the early endosomes (EEs), where they are sorted toward different fates [reviewed in Wandinger-Ness and Zerial (2014)]. Characteristic features of EEs include the presence of the small GTPase, Rab5, the Rab5 effector, EEA1, and an enrichment in phosphatidylinositol 3-phosphate [PI(3)P] (Wandinger-Ness and Zerial 2014). Receptors in EEs are either directed to recycling endosomes for recycling to the PM or targeted to the late endosomes (LEs) for degradation through the progressive conversion of Rab5-enriched EEs into Rab7-enriched LEs (Rink et al. 2005; Poteryaev et al. 2010).

EGFR can also be trafficked through endosomes positive for the Rab5 effector, APPL1 (Miaczynska et al. 2004). It is currently debated whether these APPL1-positive endosomes represent a distinct class of endosomes or an early compartment in the maturation of EEA1-positive endosomes (Kalaidzidis et al. 2015; Zoncu et al. 2009). The existence of different endosomal populations, characterized by distinct molecular markers and cargoes, raises the possibility that cargo-driven regulation of the endosomal compartment might be a mechanism for achieving signal diversification. Indeed, endosomes are dynamic structures that are tightly regulated by signaling. For instance, the EGF–EGFR complex regulates the location, number, and size of EEs (Collinet et al. 2010) and drives the formation of multivesicular bodies (MVBs) from LEs (White et al. 2006). These observations highlight the instructive role of EGFR signaling on endocytic progression and suggest that the endocytic pathway can be rearranged depending on the signaling input.

In line with this concept, EGF signaling induces the synthesis of the EGFR itself (Earp et al. 1986; Scharaw et al. 2016). Notably, continuous stimulation of cells with high EGF concentrations (but not low concentrations or pulse stimulation) increases the transport efficiency of newly synthesized EGFRs from the ER to the PM, via a mechanism involving the transcription factor, RNF11, normally localized in EEs (Scharaw et al. 2016). Upon continuous, high dose, EGF stimulation, a pool of RNF11 is found in the nucleus, where it activates transcription of the inner coat protein complex II (COPII) components, SEC23B, SEC24B, and SEC24D, which are specifically required for EGFR transport to the PM (Scharaw et al. 2016). Although the mechanism is still under investigation, it has been proposed that RNF11 might act as a "sensor" in the EEs, receiving signals from internalized EGFR to translocate to the nucleus (Scharaw et al. 2016). This scenario, if confirmed, would represent a new regulatory mechanism coupling EGFR degradation (that is significant at high EGF) with its biosynthesis and transport, to preserve EGFR levels at the PM.

Together, these findings suggest that the EGFR is not a passive passenger along the endocytic pathway, but, instead, it directly influences the nature of the pathway along its journey.

9.2.3.1 Ubiquitin-Dependent Sorting of EGFR to MVBs

The decision to target cargoes to recycling or degradation is critical for cell physiology, and the discriminating factor is cargo ubiquitination (Piper et al. 2014; Conte and Sigismund 2016). Following ubiquitination, EGFRs are actively trafficked along the degradative pathway by the ESCRT (endosomal sorting complexes required for transport) complexes [reviewed in Wollert et al. (2009), Raiborg and Stenmark (2009)]. Recycling, instead, appears to be the default pathway of internalized EGFRs, and escape from this fate is achieved through efficient receptor ubiquitination.

Once EGFR-Ub reaches the limiting membrane of the MVBs, it is recognized by the ESCRT-0 complex that is comprised of the UBD-containing proteins, Hrs (hepatocyte growth factor-regulated tyrosine kinase substrate) and STAM1/2 (signal transducing adaptor molecule 1 and 2). This complex retains EGFR-Ub in the limiting membrane, thus precluding its recycling (Wollert et al. 2009; Raiborg and Stenmark 2009). Retention of EGFR-Ub triggers a series of events leading to the sequential recruitment of ESCRT-I, ESCRT-II, and ESCRT-III complexes to the MVB membrane, which transfer the cargo to one another (Wollert et al. 2009; Raiborg and Stenmark 2009). Sorting along this pathway appears to rely on Lys63-polyubiquitination of the EGFR intracytoplasmic domain (Huang et al. 2013), which provides multiple binding sites for tandem UBDs present in ESCRT components.

Finally, ESCRT-III drives inward MVB membrane invagination leading to the formation of intraluminal vesicles (ILVs) into which EGFR-Ub is packed (Henne et al. 2013). Defective EGFR ubiquitination, or downregulation of the ESCRT components, results in inefficient incorporation of EGFR-Ub into ILVs, delayed receptor degradation and sustained signaling (Bache et al. 2003; Belleudi et al. 2009; Jekely and Rorth 2003). ILVs are then released from MVBs into the lumen of the lysosome, the main hydrolytic compartment of the cell. In addition to their hydrolytic role [reviewed in Scott et al. (2014)], lysosomes are also emerging as a signaling platform, where growth factor signaling, energy metabolism, and autophagic pathways are integrated (Settembre et al. 2013).

EGFR ubiquitination is finely regulated along the endocytic pathway by the coordinated action of E3 ligases and DUBs (Clague et al. 2012). The E3 ligase, Cbl, is recruited at the PM and remains associated with the EGFR all along the endocytic route (Umebayashi et al. 2008). This ensures maintenance of EGFR ubiquitination at later stages of trafficking when it is needed for receptor targeting to the ESCRT machinery. Besides Cbl, the E3 ligase Cullin3 (CUL3) is also implicated in postendocytic trafficking of the EGFR (Huotari et al. 2012). In particular, it was shown that CUL3, in complex with the substrate-specific adaptor, SPOPL, ubiquitinates eps15 (in complex with Hrs) in endosomes, thus, regulating its turnover (Gschweitl et al. 2016). Degradation of eps15 by CUL3 appears to be critical for MVB formation, EGFR sorting to MVBs, and receptor degradation (Gschweitl et al. 2016).

Interestingly, two eps15 isoforms appear to have different roles in EGFR recycling versus degradation. Eps15s, which lacks the ubiquitin-interacting motifs (UIMs), has been implicated in receptor recycling (Chi et al. 2011), while eps15b, which lacks the EH domains, interacts with Hrs and is involved in sorting of the EGFR to MVBs (Roxrud et al. 2008).

Several DUBs are also involved in EGFR trafficking and sorting. Some DUBs appear to act directly on the EGFR, such as AMSH (associated molecule with the SH3 domain of STAM) that removes Ub from the receptor at the endosomal level, protecting EGFR from degradation and favoring its recycling (McCullough et al. 2004; Ma et al. 2007). Similarly, OTUD7/Cezanne (Pareja et al. 2012) and USP2 (Liu et al. 2013) directly counteract Cbl-mediated EGFR ubiquitination and, consequently, receptor degradation. Other DUBs, instead, act directly on the endocytic machinery, such as Usp9x, which controls EGFR fate by deubiquitination of eps15 (Savio et al. 2016), and UBPY (also called USP8, Ub-specific Protease 8), which regulates the stability of Hrs and STAM, thereby impinging on EGFR degradation (Row et al. 2006).

9.2.3.2 Inducible Feedback Inhibitors Controlling EGFR Trafficking

Sustained treatment of cells with EGF induces a transcriptional response leading to entry into the G1 phase of the cell cycle (Avraham and Yarden 2011). In this phase, positive or negative feedback regulators of EGFR signaling are transcribed (Avraham and Yarden 2011). These include the feedback inhibitors SOCS4 and SOCS5 [members of the suppressor of cytokine signaling family (Kario et al. 2005; Nicholson et al. 2005)], and LRIG1 [leucine-rich and immunoglobulin-like domain 1 (Gur et al. 2004)], which increase ubiquitination and degradation of both active and ligand-free EGFR, restricting receptor activation. In contrast, MIG6 (mitogen-induced gene 6, also known as RALT) acts through a Ub-independent mechanism to inhibit EGFR signaling: it binds to the ligand-bound EGFR kinase domain and inhibits its allosteric activation (Anastasi et al. 2007; Zhang et al. 2007). MIG6 also drives endocytosis and degradation of inactivated EGFRs in a Ub-independent manner (Frosi et al. 2010), via an unknown mechanism.

Importantly, loss of LRIG1 and MIG6 in mice causes increased EGFR expression and aberrant cell proliferation, leading to tissue hyperplasia (Segatto et al. 2011) and, in the case of MIG6, to epithelial tumor formation (Ferby et al. 2006), highlighting the critical role of these feedback inhibitors in restricting EGFR activation.

9.2.3.3 Impact of Different EGFR Ligands and Heterodimers on Receptor Trafficking and Fate

Different EGFR ligands (see Sect. 9.3.2) can induce different signaling outputs by mechanisms that are still not fully defined. One mechanism controlling ligand-dependent signaling specificity appears to be the strength of the ligand–receptor interaction. This has been demonstrated for TGFα versus EGF, which display similar affinities for EGFR at the neutral pH of the PM, while in the mildly acidic endosomal environment (pH ~6–6.5) the affinity of TGFα drops causing ligand–receptor dissociation (Ebner and Derynck 1991; French et al. 1995). This results in EGFR inactivation, receptor dephosphorylation, Cbl detachment, and receptor deubiquitination (Longva et al. 2002). The TGFα-free EGFRs are then recycled to the cell surface. This propensity for receptor recycling following TGFα stimulation is consistent with the higher capacity of TGFα to induce mitogenic signaling compared with EGF (Waterman et al. 1998; Lenferink et al. 1998).

In contrast, the EGF–EGFR complex remains stable along the endocytic route and continues to be ubiquitinated by Cbl (Umebayashi et al. 2008) and to proceed toward the degradative compartments (Ebner and Derynck 1991; French et al. 1995). Of note, not all EGF–EGFRs are ubiquitinated and targeted for degradation; some EGF–EGFRs are recycled to the PM. The ratio between EGFR degradation versus recycling is finely regulated by EGF concentration and activation of different endocytic pathways (see Sect. 9.3.2).

Similar to TGFα, it was shown that EGFRs bound to EPI, EREG, and AREG are preferentially recycled back to the PM with little, if any, degradation, while BTC and HB-EGF efficiently induce EGFR ubiquitination and lysosomal degradation (Roepstorff et al. 2009; Stern et al. 2008; Wilson et al. 2012).

The nature of the EGFR homo-/heterodimers formed upon ligand binding can also influence receptor trafficking (Lenferink et al. 1998). For instance, compared with EGFR homodimers, heterodimers recruit inefficiently Cbl and the endocytic machinery (Baulida et al. 1996; Levkowitz et al. 1996; Waterman et al. 1999a). Moreover, ligand-binding affinity is reduced in the context of heterodimers, causing ligand dissociation in endosomes (Lenferink et al. 1998). Together, these properties cause the efficient recycling of heterodimers coupled with inefficient degradation (Fig. 9.1, left). Signaling from heterodimers is therefore more sustained and potentially more oncogenic than signaling from homodimers (see Sect. 9.4.1). Indeed, EGFR kinase active mutants in non-small-cell lung cancer (NSCLC) have been proposed to form heterodimers with ErbB2 to escape downregulation (see Sect. 9.4.1).

9.3 Control of EGFR Signaling by Endocytosis

Endocytosis is a mechanism to downregulate signaling by removing active receptors from the PM and targeting them to lysosomal degradation [reviewed in Sigismund et al. (2012)]. However, the impact of endocytosis on signaling extends beyond signal extinction. Endocytic recycling pathways are crucial for sustaining signaling and redirecting receptors to specific regions of the PM, while the distinct endocytic compartments provide temporal and spatial dimensions to the signaling cascade (Sigismund et al. 2012). These compartments serve at least two functions: (i) they sustain signaling originating at the PM by continuously recruiting the same PM signaling effectors; (ii) they facilitate the assembly of endomembrane-specific signaling platforms leading to diversification of the signaling response (see Sect. 9.3.3).

Endocytosis is not required for all signaling outputs. Impairment of EGFR endocytosis using a dominant-negative dynamin mutant increased PLCγ and Shc activation, and, concomitantly, decreased PI3K/Akt and Erk signaling, leading to inhibition of EGF-dependent mitogenesis (Vieira et al. 1996). Similarly, inhibition of CME in HeLa cervical cancer cells by clathrin- or AP2-knockdown curtailed Erk and Akt phosphorylation, without affecting Shc phosphorylation (Sigismund et al. 2008). In contrast, in dynamin-knockout mouse fibroblasts, inhibition of EGFR internalization did not alter Erk and Akt signaling elicited by EGF stimulation (Sousa et al. 2012), suggesting that the endocytic requirement of specific signaling outputs might be cell type specific. To further complicate the picture, the EGFR can be internalized through different internalization routes with specific fates and signaling outcomes (Fig. 9.1 and Sect. 9.3.2).

9.3.1 Regulation of EGFR Activity by Phosphatases Along the Endocytic Pathway

Along the endocytic pathway, EGFR is subjected to fine-tuned regulation of its signaling by different enzymes. For instance, DUBs, by regulating EGFR ubiquitination, influence sorting to the lysosome and receptor downmodulation (see Sect. 9.2.3.1). Protein tyrosine phosphatases (PTPs) also affect signaling at different steps of the endocytic pathway [Fig. 9.3 and Lemmon et al. (2016)].

PTPs are active in the early phases of EGFR activation at the PM [Fig. 9.3 (1) and Kleiman et al. (2011)], although at this stage the EGFR kinase activity overwhelms their action and the receptor is rapidly phosphorylated (Capuani et al. 2015). Nonetheless, the fast phosphorylation turnover at the PM is thought to increase responsiveness, providing dynamic plasticity to the system in response to different cues (Lemmon et al. 2016). PTPs have an even more prominent role in the endosomes (Kleiman et al. 2011). Here, their action is critical for maintaining a specific amount of active receptors *per* endosome (Villasenor et al. 2015), which, in turn, determines the final signaling output (Fig. 9.3 (2) and Sect. 9.3.3).

How the spatial distribution of PTPs along the endocytic pathway regulates the number of active EGFRs is exemplified by the ER-localized phosphatase, PTP1B (see Sect. 9.3.4 for EGFR regulation by PTP1B at ER-endosome contact sites). This phosphatase is unevenly distributed in the cell, with lowest concentrations found at the cell periphery and highest at perinuclear area (Eden et al. 2010), where termination

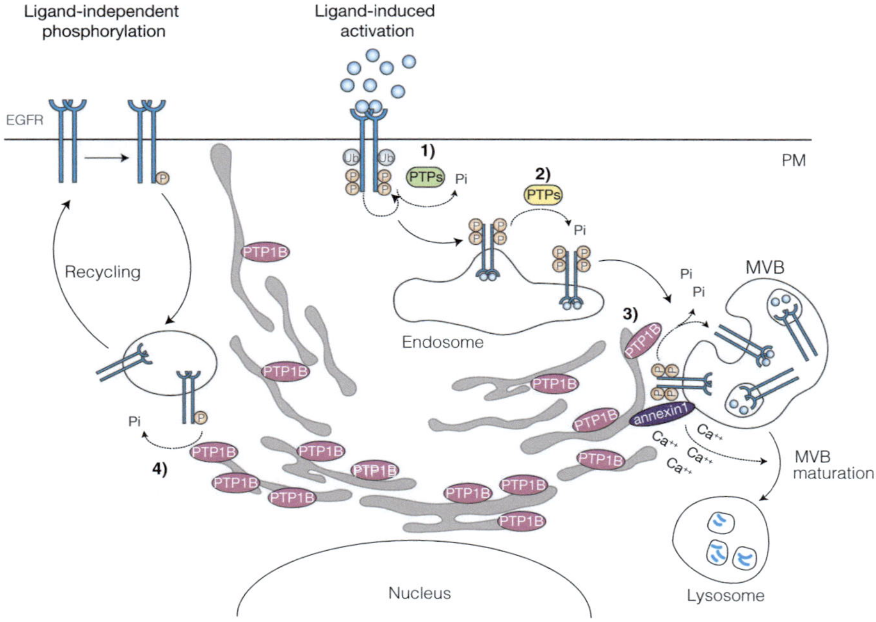

Fig. 9.3 Modulation of EGFR signaling by phosphatases. At the PM, dynamic interchange of EGFR phosphorylation by the activated EGFR kinase and dephosphorylation by phosphatases allows for rapid receptor activation, while ensuring responsiveness of the system (1). At the endosomal station, phosphatases, which are activated by the EGFR in a feedback loop, serve to maintain a constant number of active EGFRs/endosome (2). Once the EGFR has been internalized and reaches the multiple vesicular bodies (MVBs), the phosphatase, PTP1B, located at the cytosolic face of the endoplasmic reticulum (ER), dephosphorylates the receptor at the ER–MVB contact sites, prior to its targeting to the intraluminal vesicles of MVBs (3). ER–MVB contact sites are tethered by annexin1—localized in the ER—through its binding to the EGFR in the MVBs. Annexin1 is regulated by calcium (Ca^{++}) release at contact sites and is involved in intraluminal vesicle formation and MVB maturation. PTP1B is also involved in the dephosphorylation of unliganded receptors, which have been internalized via the constitutive pathway to the endosomal station where they are dephosphorylated by PTP1B at ER-endosome contact sites and are then recycled back to the PM (4). This mechanism has been proposed to limit spurious kinase activation

of signaling in LEs takes place. PTP1B dephosphorylates ligand-activated EGFRs trafficking en route toward the LEs prior to degradation in the lysosome [Fig. 9.3 (3) and Baumdick et al. (2015)], as well as EGFRs activated independently of ligand (phosphorylated at Y845) that have reached the perinuclear compartment [Fig. 9.3 (4)], prompting their recycling back to the PM. This latter mechanism is thought to suppress spurious kinase activation, while maintaining sensitivity to EGF at the PM.

These studies imply that dephosphorylation by PTPs is a way to restrict EGFR signaling and to maintain physiological levels of active receptors. This regulatory function of PTPs is in agreement with their role as tumor suppressors (Zhao et al. 2015). However, PTPs can also function as positive regulators of RTKs, as in the case of PTPD1, a FERM (four-point-one, ezrin, radixin, moesin) domain-containing PTP that has been shown to promote EGFR signaling (Cardone et al. 2004; Carlucci et al. 2010). In cell monolayers, PTPD1 is excluded from E-cadherin rich cell–cell contacts, while in isolated cells it relocalizes from the cytosol to the PM regions by binding to phosphoinositides through its FERM domain (Roda-Navarro and Bastiaens 2014). Specifically, PTPD1 is transiently recruited to EGF-induced membrane ruffles and is released just before the formation of active EGFR-containing micropinosomes. Although the exact mechanism is unclear, functional data suggest that PTPD1 has a positive role in the propagation of EGFR signaling at early stages of the pathway (Roda-Navarro and Bastiaens 2014).

9.3.2 Regulation of EGFR Signaling by the Internalization Route

The internalization route taken by the EGFR at the PM is critical in determining receptor fate. Depending on the concentration of the ligand, different endocytic pathways (CME and NCE) can be activated (see Sect. 9.2.2.3). In HeLa cells, CME and NCE counteract each other by determining opposing (recycling vs. degradation) receptor fates [Fig. 9.1, right, and Sigismund et al. (2008)]. CME, which is active at all ligand concentrations, preferentially targets the EGFR for recycling to the PM (around 70%), with a minor portion directed to lysosomes for degradation (around 30%). In contrast, NCE is sharply activated at sub-saturating EGF doses following receptor ubiquitination (Sect. 9.2.2.3) and targets the majority of EGFRs for degradation (>90%), resulting in signal attenuation in conditions of excessive stimulus (Sigismund et al. 2008). The integrated function of CME and NCE determines the final EGFR signaling response: a mechanism that also applies to other receptors, such as TGFβR (Di Guglielmo et al. 2003), Notch (Shimizu et al. 2014) and Wnt (Yamamoto et al. 2006, 2008).

The mechanisms by which CME influences EGFR signaling are multiple. By promoting recycling, CME prolongs the EGFR signaling response and protects the receptor from degradation in conditions of limited ligand availability. Additionally, receptors can be recycled to specific regions of the PM where signaling is needed. These two properties highlight CME as a mechanism providing spatial and temporal control to EGFR signaling. Consistently, CME is required for sustaining the later decay-phase of EGFR signaling and for EGFR-mediated DNA synthesis [see Sect. 9.3 (Vieira et al. 1996; Sigismund et al. 2008)].

CME also contributes to the "early phase" of EGFR signaling at the PM. A single particle tracking study investigating the correlation between EGFR mobility/aggregation at the PM and receptor signaling activity showed that immobile EGFR is clustered in CCPs that act as platforms for enhanced receptor phosphorylation and consequently signal amplification (Ibach et al. 2015). This allows the formation of local gradients of active receptors that spatially constrain EGFR signaling in response to local stimuli.

In contrast, NCE appears to be responsible for EGFR degradation and long-term signaling attenuation in conditions of high EGF in specific cellular contexts (Sigismund et al. 2008). One hypothesis is that NCE represents a mechanism to protect cells from overstimulation. Thus, loss of this route could lead to aberrant EGFR signaling and contribute to tumorigenesis.

The upstream signal triggering NCE is the sharp increase in EGFR ubiquitination at high EGF concentrations [see Sect. 9.2.2.1 and Sigismund et al. (2013)], which seals receptor fate already at the PM. A study integrating mathematical modeling and wet-laboratory experiments revealed that EGFR ubiquitination—and consequently its recruitment to NCE—is controlled by EGFR levels (Capuani et al. 2015). In physiological conditions, EGFR phosphorylation is counterbalanced by its ubiquitination, limiting receptor activation. However, at supraphysiological EGFR/EGF levels, EGFR phosphorylation and ubiquitination become uncoupled, leading to increased receptor signaling that is no longer counteracted by degradation (Capuani et al. 2015). Under these conditions, EGFR would evade NCE-mediated downmodulation, providing cancer cells with a proliferative advantage (see Sect. 9.4.1).

As in the case of CME, where CCPs were shown to function as platforms for local amplification of EGFR signaling, NCE routes have also been shown to be confined to specific PM regions where they execute polarized functions. For example, the FEME pathway (see Sect. 9.2.2.3) was shown to act locally at the leading edge of migrating cells (Boucrot et al. 2015) to ensure the rapid internalization of receptors through tubular-vesicular structures and, possibly, to promote EGF-dependent directed cell migration. Additionally, in mouse and human fibroblasts, EGFR is internalized through clathrin-independent macropinocytic-like pathways mediated by circular dorsal ruffles or "waves" in specific regions of the PM, which generate tubular-vesicular structures (Orth et al. 2006). The ability of cells to internalize large numbers of EGFRs might be relevant for signaling and polarized processes. It has been hypothesized that "waves" might contribute to three-dimensional cell migration and to extracellular matrix degradation, two critical processes in tumor cell invasion (Suetsugu et al. 2003).

9.3.3 Regulation of EGFR Signaling at the Level of the Endosomes

Endosomes, in addition to being critical sorting stations, are thought to be important platforms for signaling events, where signals elicited at the PM can be sustained and/or diversified (Villasenor et al. 2016). This notion was first proposed in the 1990s, when RTKs and connected signaling molecules were detected in endosomes (Di Guglielmo et al. 1994; Grimes et al. 1996), and was later reinforced and extended to other receptors [see, for instance, Schenck et al. (2008), Coumailleau et al. (2009),

Fortian and Sorkin (2014), Calebiro (2009, #8), Ferrandon et al. (2009), Nakamura (2014, #37), Irannejad et al. (2013), Lampugnani et al. (2006)]. The concept of a "signaling endosome" originated from neurons, in which NGF binding to its receptor, TrkA, in axon terminals initiates a signaling response that is then transmitted to the neuronal cell body through a long distance, retrograde, transport of endosomes carrying activated TrkA (Grimes et al. 1996; Beattie et al. 1996; Howe and Mobley 2005; Cosker et al. 2008).

More recently, the concept of "signaling endosomes" was corroborated by studies on both RTKs and GPCRs [reviewed in Irannejad et al. (2015)]. Three mechanisms have been proposed by which endosomes control signaling: "scaffolding," "sequestration," and "catalysis" (Irannejad et al. 2015). In the "scaffolding" mechanism, growth factor receptors confined in endosomes engage signaling adaptors that act as a scaffold for downstream effectors, promoting their activation. For example, EGFR engages the adaptor Grb2 which recruits and activates Erk (Di Guglielmo et al. 1994; Fortian and Sorkin 2014). Similarly, GPCRs use beta-arrestin as an endosomal scaffold to continue signaling after internalization (Shenoy and Lefkowitz 2005), and phosphorylated C-Met engages the nucleotide exchange factor, Vav2, which leads to sustained Rac signaling (Menard et al. 2014).

In the "sequestration" mechanism, signal amplification is achieved by entrapping cytoplasmic, negative, signaling regulators in endosomes. For instance, in Wnt/wingless signaling, the inhibitory enzyme, GSK3, is physically sequestered into the endosomal lumen, leading to reduced cytosolic GSK3 activity and, consequently, to enhanced beta-catenin signaling (Taelman et al. 2010).

Finally, the "catalysis" mechanism involves activation of enzymes in endosomes to augment signaling. For example, heterotrimeric G proteins from GPCRs are activated not only at the PM, but also in the limiting endosomal membranes where they promote downstream signaling through production of second messengers, such as cyclic AMP (cAMP) (Irannejad et al. 2013; Tsvetanova and von Zastrow 2014). In this case, endosomal signaling acts to sustain the cAMP response observed at the PM, and to determine the final cAMP-dependent transcriptional response (Tsvetanova and von Zastrow 2014).

In the case of the EGFR, a quantitative high-resolution microscopy approach revealed that the endosomal system works as an analog–digital converter (Villasenor et al. 2015). Active phosphorylated EGFRs form clusters of ~80 molecules *per* EE. The endosomal fusion machinery works to keep the number of active EGFRs/endosome constant: Increasing the EGF concentration does not produce larger EGFR clusters, rather, a higher number of EGFR-positive endosomes. Notably, inhibition of endosome fusion enhances the number of EGFR clusters and determines a different signaling outcome, i.e., prolonged EGFR activation and Erk signaling response. These clusters represent the *quanta* of signaling that provide robustness to the cellular response in case of fluctuations in ligand or receptor levels. This mechanism applies also to other RTKs, such as the hepatocyte growth factor receptor (HGFR) and the nerve growth factor receptor (NGFR) (Villasenor et al. 2015).

Also in this case, phosphatases have a pivotal role in fine-tuning EGFR signaling: Phosphorylated EGFR in endosomes recruits and activates, through phosphorylation, the phosphatase SHP2, forming a negative feedback loop to maintain a constant number of phosphorylated EGFRs/endosome (Villasenor et al. 2015).

9.3.4 How the Endoplasmic Reticulum Modulates EGFR Signaling

Multiple cell compartments act in concert to control intracellular signals. This integrated function is mainly achieved by membrane–membrane contact sites: regions of close apposition (<30 nm) between the membranes of organelles (Phillips and Voeltz 2016). In recent years, contact sites have emerged as platforms of signaling regulation and places where materials, such as lipids and Ca^{2+}, can be rapidly exchanged [reviewed in Levine and Patel (2016)]. ER-endosomal contact sites have been detected in mammalian cells, while ER-vacuole contact sites have been identified in yeast (Eden et al. 2010; Rocha et al. 2009; West et al. 2011). A study based on high-resolution, three-dimensional, electron microscopy showed that endosomes trafficking along microtubules are wrapped by ER tubules. These contacts are maintained, and actually increase, as endosomes traffic and mature (Friedman et al. 2013). Importantly, the ER-endosomal contact sites determine the timing and position of endosome fission events during cargo sorting (Rowland et al. 2014).

Crosstalk between the endosomal compartment and the ER also has a role in the modulation of RTK signaling. For instance, upon internalization, the EGFR—and other RTKs (e.g., insulin receptor and Met)—interacts with the phosphatase PTP1B localized at the cytosolic face of the ER (Haj et al. 2002; Sangwan et al. 2008; Romsicki et al. 2004). PTP1B regulates both constitutively internalized and ligand-activated EGFR (see Sect. 9.3.2). EGFR–PTP1B proximity was shown to occur at sites of physical contact between the ER and the limiting membrane of MVBs (Eden et al. 2010). At these sites, ER-resident PTP1B dephosphorylates MVB-localized EGFR "in trans" [Fig. 9.3 (3)]. The formation of ER–MVB contacts is mediated by annexin-1 and its Ca^{2+}-dependent binding partner S100A11, in a Ca^{2+}-dependent fashion (Eden et al. 2016; Kilpatrick et al. 2017). Ca^{2+} is released from the endolysosomal compartment by the two-pore channel (TPC), which localizes at ER-endosome contact sites and is regulated by nicotinic acid adenine dinucleotide phosphate (NAADP). Affecting these contacts delayed EGFR dephosphorylation by PTP1B and its subsequent degradation, enhancing signaling (Kilpatrick et al. 2017).

In addition to dephosphorylating internalized RTKs after endocytosis, PTP1B was also shown to act on EGFR localized at the PM, through the formation of ER–PM contact sites (Haj et al. 2012). This interaction appears to be restricted to regions of cell–cell contacts, identified as sites of PTP1B-mediated signaling regulation.

As we discussed (Sect. 9.2.2.3), ER–PM contact sites are also critical at early step of EGFR endocytosis. Indeed, the ER-resident protein RTN3 mediates the formation of contacts between the ER and sites of EGFR internalization at the PM at high ligand concentration (Fig. 9.2), a mechanism that leads to EGFR-NCE receptor degradation and signal extinction (Caldieri et al. 2017). Thus, ER contact sites control EGFR fate at multiple levels; e.g., at the PM and the endosomal stations. Ca^{2+} signaling appears to be involved in both cases, although through different mechanisms (Kilpatrick et al. 2017; Caldieri et al. 2017). Whether RTN3 is acting only at PM–ER interface or it has a role also at later step is unclear. Furthermore, a possible interplay between PTP1B and RTN3-dependent ER–PM contact sites in EGFR regulation remains to be established.

Finally, ER-based ubiquitination has been proposed to regulate levels of newly synthesized ErbB3 receptors by promoting their ER-associated degradation (ERAD) (Fry et al. 2011). Indeed, the ER-localized E3 ubiquitin ligase, Nrdp1, interacts with and ubiquitinates the nascent form of ErbB3, thereby regulating the steady-state levels of the receptor (Fry et al. 2011). Whether this mechanism also applies to the EGFR or other RTKs remains to be established.

9.4 EGFR Trafficking and Cancer

In addition to being a critical regulator of physiological cellular processes, EGFR signaling has a crucial role in the development and progression of many types of cancer; a condition where normal cellular homeostasis is subverted (Zwick et al. 2001). The first evidence linking altered EGFR signaling to cancer came in the early eighties when the viral-erbB (v-erbB) oncogene product was found to be homologous to the amplified *EGFR* gene in the human A431 epidermal carcinoma cell line (Ullrich et al. 1984). Since then, numerous studies characterizing the role of the EGFR in cancer have been conducted. The emerging concept is that there is a tight relationship between the oncogenic forms of the receptor and the trafficking routes the protein takes inside the cell.

9.4.1 How Different Oncogenic Forms of EGFR Are Influenced by Trafficking

Neoplastic transformation induced by the EGFR can be triggered by gene amplification and/or protein overexpression, mutations, or in-frame deletions (Roskoski 2014). These genetic lesions frequently occur concomitantly with increased EGFR ligand production triggered by autocrine or paracrine loops (Wilson et al. 2009). Autocrine secretion is often the result of positive feedback loops downstream of excessive EGFR activation that ultimately lead to the induction of the promoter of EGF family ligands (Avraham and Yarden 2011). Additionally, some solid tumors upregulate metalloproteases leading to enhanced cleavage of EGF ligand precursors (Wilson et al. 2009).

EGFR genetic alterations have been reported to cause altered trafficking of the receptor, which contributes to aberrant signaling and oncogenesis. For instance, gene amplification or receptor overexpression leads to increased EGFR density on the PM, which favors receptor dimerization and spurious kinase activation (Wiley 1988; Sawano et al. 2002; Chung et al. 2010). It has been suggested that the excessive number of activated EGFRs causes saturation of the endocytic machinery, increasing the residence time of surface EGFRs, delaying downregulation and, ultimately, leading to sustained signal (French et al. 1994; Wiley 1988).

Additionally, saturation of the endocytic/ubiquitination machinery has been proposed as a mechanism underlying sustained signaling in EGFR-overexpressing cancer cells (Capuani et al. 2015). As receptor levels increase, there is a progressive uncoupling between EGFR phosphorylation and ubiquitination (see Sect. 9.3.2). This uncoupling is due to the limiting amount of Cbl, which becomes saturated in conditions of high numbers of activated EGFR (i.e., EGFR overexpression coupled with ligand overproduction). This situation is predicted to cause sustained EGFR signaling and impaired receptor downregulation, which, however, can be partially restored by overexpressing Cbl in the cell (Capuani et al. 2015). Thus, Cbl is the weak and critical element in the system and, consistently, escape from Cbl-dependent degradation is one of the most common mechanisms enacted by oncogenic EGFR mutants (see below).

Finally, EGFR overexpression favors heterodimerization with the other ErbB family members, which influences trafficking (Arteaga and Engelman 2014). In particular, heterodimerization of EGFR with the ligand–orphan receptor ErbB2 enhances recycling [Fig. 9.1, left, and Ebner and Derynck (1991), French et al. (1995)]. ErbB2, besides being constitutively active when engaged in a heterodimer, evades ubiquitination, thereby, favoring recycling and sustained signaling over degradation and signal attenuation (see Sect. 9.2.3.3). Therefore, the formation of EGFR–ErbB2 heterodimers shifts the signaling output toward proliferation (Lenferink et al. 1998; Worthylake et al. 1999).

One of the best-described mechanisms of oncogenic activation of the EGFR is mutation. Large genetic rearrangements, as well as single base mutations, have been described, which produce oncogenic forms of the EGFR, whose expression often correlates with poor prognosis (Yarden and Pines 2012). In some cases, aberrant endocytosis and trafficking of these mutated receptors have been shown to contribute to their deregulated signaling (Yarden and Pines 2012).

A well-characterized truncated form of the EGFR is EGFRvIII, which has been detected in brain, most glioblastomas, and lung, breast and ovarian cancers (Ekstrand et al. 1992; Moscatello et al. 1995; Wong et al. 1992). EGFRvIII is a deletion mutant that lacks exons 2–7, resulting in a receptor with a truncated extracellular domain. This truncation mutant dimerizes and undergoes autophosphorylation in absence of ligands, while being poorly internalized and efficiently recycled back to the PM rather than being degraded. The end result is the excessive and sustained activation of the EGFR signaling cascade (Grandal et al. 2007). Although the truncation affects the extracellular portion of the receptor, the sustained EGFR activation is thought to be caused by impairment of receptor ubiquitination due to hypo-phosphorylation of pY1045, the direct Cbl binding site. This leads to slow receptor turnover and increased signaling (Schmidt et al. 2003; Han et al. 2006; Grandal et al. 2007). Despite possessing the same intracytoplasmic tail as the wild-type receptor, EGFRvIII activates different signaling networks compared to the wild-type receptor possibly attributable to altered internalization and trafficking (Johnson et al. 2012).

Reduced downmodulation of EGFR has also been described for other mutant forms of the receptor, EGFRvIV and EGFRvV, which lack the portion of the cytoplasmic tail encompassing the Cbl binding site, Y1045 (Roskoski 2014). These mutants, whose activation is still ligand-dependent, retain the potential to modulate oncogenic signaling pathways, e.g., Ras/MAPK signaling, commonly elicited by the wild-type receptor (Grovdal et al. 2004).

Other somatic activating mutations in the EGFR have been identified in NSCLC and patients carrying these mutations are treated with EGFR kinase inhibitors (e.g., Gefinitib and Erlotinib) as the first-line therapy (Lynch et al. 2004; Paez et al. 2004; Pao et al. 2004; Roskoski 2014). These activating mutations appear to lock the receptor in an active conformational state, causing ligand-independent firing and signaling up to 50-fold above the basal unliganded receptor activity (Yun et al. 2007). The EGFR-L834R mutant exemplifies the connection between EGFR ubiquitination-dependent trafficking and human cancers. EGFR–L834R possesses an intact Cbl binding site that is more highly phosphorylated compared with the wild-type receptor. Nevertheless, Cbl recruitment and receptor ubiquitination are impaired, causing reduced degradation and sustained activation of downstream signaling molecules, including Ras, MEK, and Erk (Kon et al. 2014; Shtiegman et al. 2007). One hypothesis to explain these observations is that EGFR–L834R forms heterodimers with ErbB2, even in the absence of ligand (Kon et al. 2014). NCSLC EGFR mutants in exons 18–21 also show a higher propensity to heterodimerize with ErbB3 (Rothenberg et al. 2008), which, as in the case of ErbB2 heterodimers, might divert EGFR mutants from a degradative toward a recycling fate, thereby enhancing signaling.

9.4.2 Mutations in Trafficking Genes Influencing EGFR Oncogenic Potential

Besides EGFR mutations that affect Cbl recruitment and activity toward the receptor, Cbl itself is mutated in human cancers [reviewed in Sigismund et al. (2012)]. Missense homozygous mutations of Cbl targeting its E3 ligase activity have been described in ~5% of myeloid neoplasms (Caligiuri et al. 2007; Dunbar et al. 2008; Sargin et al. 2007). In these cases, however, Cbl activity is primarily directed toward the RTK, FLT3 (Grand et al. 2009; Sargin et al. 2007; Sanada et al. 2009), with no connection to EGFR ubiquitination and trafficking. Similarly, heterozygous germline mutations of Cbl are found in patients affected by Noonan Syndrome (NS), a clinically variable disease [reviewed in Allanson (2007), Tartaglia et al. (2011)]. As in myeloid malignancies, Cbl mutations are missense mutations that alter the region responsible for ligase activity, but in this case they are heterozygous and thus predicted to act in a dominant-negative fashion (Martinelli et al. 2010). When overexpressed in COS-1 cells, these mutants affect EGFR ubiquitination and cause prolonged Ras–MAPK signaling (Martinelli et al. 2010). However, the relevance of these mutations to EGFR ubiquitination and trafficking in vivo has not yet been established.

In addition to Cbl, several oncogenes have been proposed to influence EGFR signaling by altering its trafficking, thereby contributing to EGFR oncogenic potential. For instance, in NSCLC cell lines, Src, a non-receptor tyrosine kinase (nRTK) that is directly activated by the EGFR (and other RTKs), cooperates with mutated EGFR to generate aberrant signaling and to induce cell transformation (Chung et al. 2009; Leung et al. 2009). Furthermore, aberrant Src activation, as observed in many cancer cells or cells transformed by the viral oncogene, v-Src, interferes with Cbl-mediated EGFR ubiquitination and receptor downmodulation (Bao et al. 2003; Wu et al. 2003; Feng et al. 2006).

Another oncogene that influences EGFR endocytosis is ACK1 (activated Cdc42-associated Kinase), a nRTK that interacts with EGFR-Ub through its UBD (Shen et al. 2007), facilitating receptor degradation (Kelley and Weed 2012). When mutated, ACK1 retains EGFR at the PM, sustaining its signaling (Chua et al. 2010; Kelley and Weed 2012). Similarly, the oncogenic form of Vav, a Rho GTPase guanine nucleotide exchange factor (GEF), causes increased Erk and Akt phosphorylation upon EGFR activation by delaying receptor endocytosis (Thalappilly et al. 2010).

In addition to PM signaling, aberrant EGFR signaling from intracellular compartments can also be oncogenic. This was shown in cancer cells with loss of function mutations of the Von Hippel-Lindau (VHL) protein or in cells subjected to hypoxic conditions (Wang et al. 2009). In both cases, reduced expression of the Rab5 effector, rabaptin-5, was observed, which determines inefficient Rab5-mediated endosome fusion and persistent retention of active EGFR in EEs, leading to prolonged pro-survival signaling from intracellular compartments (Wang et al. 2009).

Finally, alterations of proteins not directly involved in EGFR regulation, but active in other cellular pathways subverted in cancer, can interfere with EGFR (and MET) signaling (Muller et al. 2009, 2013). This is the case of p53 gain-of-function mutants that have lost tumor-suppressor activity, but have acquired endocytosis-related phenotypes, which interfere with EGFR trafficking and signaling. Expression of these mutants enhances co-trafficking and recycling of the β1-integrin/EGFR complex, via a mechanism dependent on the Rab11-effector, Rab-coupling protein (RCP), resulting in constitutive activation of EGFR/integrin signaling. Consequently, mutant p53 expression promotes tumor cell invasion, random cell migration and metastatic dissemination (Muller et al. 2009).

In conclusion, there are several, although scattered, evidences linking the oncogenic potential of known endocytic/signaling molecules to an altered EGFR trafficking. Yet, a direct conclusive link is missing. Given the potential relevance of this issue to cancer, future investigations are warranted.

9.4.3 Pharmacological Targeting of EGFR: Harnessing EGFR Endocytosis

Given the crucial role of the EGFR in different cancers, much effort has been placed on the discovery of target-specific drugs that modulate its activity (Arteaga and Engelman 2014). These include monoclonal humanized antibodies (mAbs) directed against the extracellular domain of the EGFR and selective small molecule inhibitors that target the intracellular tyrosine kinase domain.

By targeting the ATP-binding domain of the EGFR, small molecule inhibitors impair phosphorylation of the receptor C-terminal tail causing repression of ligand-induced signals (Arteaga and Engelman 2014). Interestingly, these inhibitors show a higher affinity for mutated forms of EGFR, meaning inhibition is achieved at lower drug concentrations compared to those needed for inhibition of the wild-type receptor (Carey et al. 2006). Examples of small molecule EGFR inhibitors include Gefinitib, Erlotinib, and Afatinib, which are approved for lung cancer treatment (Hirsch et al. 2013; Cohen et al. 2005; Thatcher et al. 2005). Interestingly, EGFR kinase inhibitors, like Gefitinib, increase the formation of inactive dimers through an inside-out signaling transmitted from the kinase domain to the extracellular dimerization domain (Arteaga et al. 1997; Gan et al. 2007). Gefitinib-induced EGFR dimers display increased ligand-binding affinity and peculiar binding kinetics (Bjorkelund et al. 2011). Whether the increase in dimer formation might stimulate EGFR endocytosis and, thus, contribute to signal extinction in parallel to kinase inhibition, remains to be clarified.

Cetuximab and Panitumumab are the most widely used EGFR-neutralizing mAbs. Their effectiveness has been proven in the treatment of head and neck cancer, and metastatic colon cancer (Peeters et al. 2015; Licitra et al. 2013; Pierotti et al. 2010). Mechanistically, these mAbs act by preventing ligand binding, thereby, inhibiting receptor activation and downstream signaling (Bou-Assaly and Mukherji 2010; Dubois and Cohen 2009; Vincenzi et al. 2008). The mAbs also induce EGFR dimerization and, thus, it has been proposed that they stimulate EGFR endocytosis and downmodulation (Fan et al. 1993). However, experiments with radiolabeled Cetuximab showed that antibody-bound EGFRs are internalized at a lower rate compared with ligand-induced endocytosis and are more efficiently recycled compared with EGF-bound dimers (Jaramillo et al. 2006). Interestingly, the combination of anti-EGFR antibodies directed against non-overlapping antigens was more efficient in interfering with ligand binding, and in accelerating EGFR endocytosis and degradation (Friedman et al. 2005; Pedersen et al. 2010) or increasing receptor recycling (Spangler et al. 2010). Although the mechanism is still unclear, combinatorial EGFR antibody treatment might improve anti-tumor efficacy through the regulation of EGFR trafficking.

9.5 Concluding Remarks

A wealth of evidence points to the relevance of endocytosis and trafficking in determining EGFR signaling outcome and in governing cell behavior, as also supported by the frequent alterations of EGFR endocytic and trafficking routes in human cancers. Yet, there are many aspects of the EGFR pathway that still need to be decoded, both in physiological contexts as well as in cancer. A major challenge is to clarify how EGFR signaling is interpreted in space and time, and how it is integrated with other cellular processes and signaling pathways to determine a specific cellular outcome. This should be clarified not only at the population level, but also at single-cell level. Indeed, single-cell heterogeneity in a population context was shown to be critical for the final cellular response (Elowitz et al. 2002; Frechin et al. 2015; Snijder et al. 2009). Notably, EGFR endocytosis and its downstream signaling are strongly population context dependent (Cohen-Saidon et al. 2009; Snijder et al. 2009; Liberali et al. 2014).

There is also pressing need to follow the trafficking and fate of individual receptors in unperturbed conditions (i.e., without ablation of critical factors or treatment with chemical inhibitors), in order to illuminate the contribution of the different factors to EGFR endocytosis in physiological settings. This area of investigation, which is being greatly advanced by technologies for single-molecule tracking, is particularly relevant for endocytosis. Endocytosis is a highly modular process with many alternative (and redundant) signals, adaptors, and fission machineries. As a consequence, it is highly plastic and can be efficiently and rapidly rewired through adaptive modifications of the availability of endocytic factors, PM lipid/cholesterol level, and changes in membrane tension. Consequently, compensation among different endocytic pathways is likely and it has actually has been reported (Kalia et al. 2006; Nevins and Thurmond 2006; Damke et al. 1995; Guha et al. 2003; Chaudhary et al. 2014), rendering the analysis in unperturbed conditions highly needed.

Finally, increasingly advanced models of EGFR signaling and endocytosis are needed to achieve system-level understanding. Mathematical models of both the EGFR signaling cascade (Kholodenko et al. 1999) and EGFR trafficking (Sorkin et al. 1991; Wiley et al. 1991) have been generated in the past; however, they were treated initially as separated processes. Attempts to integrate EGFR activation, ubiquitination, and trafficking were undertaken only more recently and have unveiled peculiar, unexpected, characteristics of the system [see for instance Wiley et al. (2003), Resat et al. (2003), Capuani et al. (2015), Kleiman et al. (2011)]. Such an approach is critical, as it will also help to identify the weak elements of the network that are hijacked by cancer cells and that could represent critical points of therapeutic intervention.

Acknowledgements We thank Rosalind Gunby for critically reading the manuscript. This work was supported by grants from the Associazione Italiana per la Ricerca sul Cancro to PPDF (IG 14404 and MCO 10.000); MIUR (the Italian Ministry of University and Scientific Research), the Italian Ministry of Health, and The Monzino Foundation to PPDF; and the WWCR (Worldwide Cancer Research) to SS (16-1245).

References

Acconcia F, Sigismund S, Polo S (2009) Ubiquitin in trafficking: the network at work. Exp Cell Res 315(9):1610–1618. https://doi.org/10.1016/j.yexcr.2008.10.014

Allanson JE (2007) Noonan syndrome. Am J Med Genet C Semin Med Genet 145C(3):274–279. https://doi.org/10.1002/ajmg.c.30138

Anastasi S, Baietti MF, Frosi Y, Alema S, Segatto O (2007) The evolutionarily conserved EBR module of RALT/MIG6 mediates suppression of the EGFR catalytic activity. Oncogene 26(57):7833–7846. https://doi.org/10.1038/sj.onc.1210590

Antonny B, Burd C, De Camilli P, Chen E, Daumke O, Faelber K, Ford M, Frolov VA, Frost A, Hinshaw JE, Kirchhausen T, Kozlov MM, Lenz M, Low HH, McMahon H, Merrifield C, Pollard TD, Robinson PJ, Roux A, Schmid S (2016) Membrane fission by dynamin: what we know and what we need to know. EMBO J 35(21):2270–2284. https://doi.org/10.15252/embj.201694613

Arteaga CL, Engelman JA (2014) ERBB receptors: from oncogene discovery to basic science to mechanism-based cancer therapeutics. Cancer Cell 25(3):282–303. https://doi.org/10.1016/j.ccr.2014.02.025

Arteaga CL, Ramsey TT, Shawver LK, Guyer CA (1997) Unliganded epidermal growth factor receptor dimerization induced by direct interaction of quinazolines with the ATP binding site. J Biol Chem 272(37):23247–23254

Avraham R, Yarden Y (2011) Feedback regulation of EGFR signalling: decision making by early and delayed loops. Nat Rev Mol Cell Biol 12(2):104–117. https://doi.org/10.1038/nrm3048

Bache KG, Brech A, Mehlum A, Stenmark H (2003) Hrs regulates multivesicular body formation via ESCRT recruitment to endosomes. J Cell Biol 162(3):435–442. https://doi.org/10.1083/jcb.200302131

Bao J, Gur G, Yarden Y (2003) Src promotes destruction of c-Cbl: implications for oncogenic synergy between Src and growth factor receptors. Proc Natl Acad Sci U S A 100(5):2438–2443. https://doi.org/10.1073/pnas.0437945100

Baulida J, Kraus MH, Alimandi M, Di Fiore PP, Carpenter G (1996) All ErbB receptors other than the epidermal growth factor receptor are endocytosis impaired. J Biol Chem 271(9):5251–5257

Baumdick M, Bruggemann Y, Schmick M, Xouri G, Sabet O, Davis L, Chin JW, Bastiaens PI (2015) EGF-dependent re-routing of vesicular recycling switches spontaneous phosphorylation suppression to EGFR signaling. Elife 4. https://doi.org/10.7554/elife.12223

Beattie EC, Zhou J, Grimes ML, Bunnett NW, Howe CL, Mobley WC (1996) A signaling endosome hypothesis to explain NGF actions: potential implications for neurodegeneration. Cold Spring Harb Symp Quant Biol 61:389–406

Belleudi F, Leone L, Maggio M, Torrisi MR (2009) Hrs regulates the endocytic sorting of the fibroblast growth factor receptor 2b. Exp Cell Res 315(13):2181–2191. https://doi.org/10.1016/j.yexcr.2009.03.022

Bjorkelund H, Gedda L, Barta P, Malmqvist M, Andersson K (2011) Gefitinib induces epidermal growth factor receptor dimers which alters the interaction characteristics with (1)(2)(5)I-EGF. PLoS ONE 6(9):e24739. https://doi.org/10.1371/journal.pone.0024739

Bou-Assaly W, Mukherji S (2010) Cetuximab (erbitux). AJNR Am J Neuroradiol 31(4):626–627. https://doi.org/10.3174/ajnr.A2054

Boucrot E, Ferreira AP, Almeida-Souza L, Debard S, Vallis Y, Howard G, Bertot L, Sauvonnet N, McMahon HT (2015) Endophilin marks and controls a clathrin-independent endocytic pathway. Nature 517(7535):460–465. https://doi.org/10.1038/nature14067

Caldieri G, Barbieri E, Nappo G, Raimondi A, Bonora M, Conte A, Verhoef LGGC, Confalonieri S, Malabarba MG, Bianchi F, Cuomo A, Bonaldi T, Martini E, Mazza D, Pinton P, Tacchetti C, Polo S, Di Fiore PP, Sigismund S (2017) Reticulon 3-dependent ER-PM contact sites control EGFR nonclathrin endocytosis. Science 356(6338):617–624. https://doi.org/10.1126/science.aah6152

Caligiuri MA, Briesewitz R, Yu J, Wang L, Wei M, Arnoczky KJ, Marburger TB, Wen J, Perrotti D, Bloomfield CD, Whitman SP (2007) Novel c-CBL and CBL-b ubiquitin ligase mutations in human acute myeloid leukemia. Blood 110(3):1022–1024. https://doi.org/10.1182/blood-2006-12-061176

Calebiro D, Nikolaev VO, Gagliani MC, de Filippis T, Dees C, Tacchetti C, Persani L, Lohse MJ (2009) Persistent cAMP-signals triggered by internalized G-protein-coupled receptors. PLoS Biol 7(8):e1000172. https://doi.org/10.1371/journal.pbio.1000172

Capuani F, Conte A, Argenzio E, Marchetti L, Priami C, Polo S, Di Fiore PP, Sigismund S, Ciliberto A (2015) Quantitative analysis reveals how EGFR activation and downregulation are coupled in normal but not in cancer cells. Nat Commun 6:7999. https://doi.org/10.1038/ncomms8999

Cardone L, Carlucci A, Affaitati A, Livigni A, DeCristofaro T, Garbi C, Varrone S, Ullrich A, Gottesman ME, Avvedimento EV, Feliciello A (2004) Mitochondrial AKAP121 binds and targets protein tyrosine phosphatase D1, a novel positive regulator of src signaling. Mol Cell Biol 24(11):4613–4626. https://doi.org/10.1128/MCB.24.11.4613-4626.2004

Carey KD, Garton AJ, Romero MS, Kahler J, Thomson S, Ross S, Park F, Haley JD, Gibson N, Sliwkowski MX (2006) Kinetic analysis of epidermal growth factor receptor somatic mutant proteins shows increased sensitivity to the epidermal growth factor receptor tyrosine kinase inhibitor, erlotinib. Cancer Res 66(16):8163–8171. https://doi.org/10.1158/0008-5472.CAN-06-0453

Carlucci A, Porpora M, Garbi C, Galgani M, Santoriello M, Mascolo M, di Lorenzo D, Altieri V, Quarto M, Terracciano L, Gottesman ME, Insabato L, Feliciello A (2010) PTPD1 supports receptor stability and mitogenic signaling in bladder cancer cells. J Biol Chem 285(50):39260–39270. https://doi.org/10.1074/jbc.M110.174706

Carpenter G, Cohen S (1976) 125I-labeled human epidermal growth factor. Binding, internalization, and degradation in human fibroblasts. J Cell Biol 71(1):159–171

Carpentier JL, Gorden P, Anderson RG, Goldstein JL, Brown MS, Cohen S, Orci L (1982) Co-localization of 125I-epidermal growth factor and ferritin-low density lipoprotein in coated pits: a quantitative electron microscopic study in normal and mutant human fibroblasts. J Cell Biol 95(1):73–77

Chaudhary N, Gomez GA, Howes MT, Lo HP, McMahon KA, Rae JA, Schieber NL, Hill MM, Gaus K, Yap AS, Parton RG (2014) Endocytic crosstalk: cavins, caveolins, and caveolae regulate clathrin-independent endocytosis. PLoS Biol 12(4):e1001832. https://doi.org/10.1371/journal.pbio.1001832

Chi S, Cao H, Wang Y, McNiven MA (2011) Recycling of the epidermal growth factor receptor is mediated by a novel form of the clathrin adaptor protein Eps15. J Biol Chem 286(40):35196–35208. https://doi.org/10.1074/jbc.M111.247577

Chua BT, Lim SJ, Tham SC, Poh WJ, Ullrich A (2010) Somatic mutation in the ACK1 ubiquitin association domain enhances oncogenic signaling through EGFR regulation in renal cancer derived cells. Mol Oncol 4(4):323–334. https://doi.org/10.1016/j.molonc.2010.03.001

Chung BM, Dimri M, George M, Reddi AL, Chen G, Band V, Band H (2009) The role of cooperativity with Src in oncogenic transformation mediated by non-small cell lung cancer-associated EGF receptor mutants. Oncogene 28(16):1821–1832. https://doi.org/10.1038/onc.2009.31

Chung I, Akita R, Vandlen R, Toomre D, Schlessinger J, Mellman I (2010) Spatial control of EGF receptor activation by reversible dimerization on living cells. Nature 464(7289):783–787. https://doi.org/10.1038/nature08827

Clague MJ, Liu H, Urbe S (2012) Governance of endocytic trafficking and signaling by reversible ubiquitylation. Dev Cell 23(3):457–467. https://doi.org/10.1016/j.devcel.2012.08.011

Cohen MH, Johnson JR, Chen YF, Sridhara R, Pazdur R (2005) FDA drug approval summary: erlotinib (Tarceva) tablets. Oncologist 10(7):461–466. https://doi.org/10.1634/theoncologist.10-7-461

Cohen-Saidon C, Cohen AA, Sigal A, Liron Y, Alon U (2009) Dynamics and variability of ERK2 response to EGF in individual living cells. Mol Cell 36(5):885–893. https://doi.org/10.1016/j.molcel.2009.11.025

Collinet C, Stoter M, Bradshaw CR, Samusik N, Rink JC, Kenski D, Habermann B, Buchholz F, Henschel R, Mueller MS, Nagel WE, Fava E, Kalaidzidis Y, Zerial M (2010) Systems survey of endocytosis by multiparametric image analysis. Nature 464(7286):243–249. https://doi.org/10.1038/nature08779

Conte A, Sigismund S (2016) Chapter Six—The ubiquitin network in the control of EGFR Endocytosis and Signaling. Prog Mol Biol Transl Sci 141:225–276. https://doi.org/10.1016/bs.pmbts.2016.03.002

Cosker KE, Courchesne SL, Segal RA (2008) Action in the axon: generation and transport of signaling endosomes. Curr Opin Neurobiol 18(3):270–275. https://doi.org/10.1016/j.conb.2008.08.005

Coumailleau F, Furthauer M, Knoblich JA, Gonzalez-Gaitan M (2009) Directional Delta and Notch trafficking in Sara endosomes during asymmetric cell division. Nature 458(7241):1051–1055. https://doi.org/10.1038/nature07854

Damke H, Baba T, van der Bliek AM, Schmid SL (1995) Clathrin-independent pinocytosis is induced in cells overexpressing a temperature-sensitive mutant of dynamin. J Cell Biol 131(1):69–80

Di Guglielmo GM, Baass PC, Ou WJ, Posner BI, Bergeron JJ (1994) Compartmentalization of SHC, GRB2 and mSOS, and hyperphosphorylation of Raf-1 by EGF but not insulin in liver parenchyma. EMBO J 13(18):4269–4277

Di Guglielmo GM, Le Roy C, Goodfellow AF, Wrana JL (2003) Distinct endocytic pathways regulate TGF-beta receptor signalling and turnover. Nat Cell Biol 5(5):410–421. https://doi.org/10.1038/ncb975

Dubois EA, Cohen AF (2009) Panitumumab. Br J Clin Pharmacol 68(4):482–483. https://doi.org/10.1111/j.1365-2125.2009.03492.x

Dunbar AJ, Gondek LP, O'Keefe CL, Makishima H, Rataul MS, Szpurka H, Sekeres MA, Wang XF, McDevitt MA, Maciejewski JP (2008) 250K single nucleotide polymorphism array karyotyping identifies acquired uniparental disomy and homozygous mutations, including novel missense substitutions of c-Cbl, in myeloid malignancies. Cancer Res 68(24):10349–10357. https://doi.org/10.1158/0008-5472.CAN-08-2754

Dunn WA, Connolly TP, Hubbard AL (1986) Receptor-mediated endocytosis of epidermal growth factor by rat hepatocytes: receptor pathway. J Cell Biol 102(1):24–36

Earp HS, Austin KS, Blaisdell J, Rubin RA, Nelson KG, Lee LW, Grisham JW (1986) Epidermal growth factor (EGF) stimulates EGF receptor synthesis. J Biol Chem 261(11):4777–4780

Ebner R, Derynck R (1991) Epidermal growth factor and transforming growth factor-alpha: differential intracellular routing and processing of ligand-receptor complexes. Cell Regul 2(8):599–612

Eden ER, White IJ, Tsapara A, Futter CE (2010) Membrane contacts between endosomes and ER provide sites for PTP1B-epidermal growth factor receptor interaction. Nat Cell Biol 12(3):267–272. https://doi.org/10.1038/ncb2026

Eden ER, Sanchez-Heras E, Tsapara A, Sobota A, Levine TP, Futter CE (2016) Annexin A1 tethers membrane contact sites that mediate ER to endosome cholesterol transport. Dev Cell 37(5):473–483. https://doi.org/10.1016/j.devcel.2016.05.005

Ekstrand AJ, Sugawa N, James CD, Collins VP (1992) Amplified and rearranged epidermal growth factor receptor genes in human glioblastomas reveal deletions of sequences encoding portions of the N- and/or C-terminal tails. Proc Natl Acad Sci U S A 89(10):4309–4313

Elowitz MB, Levine AJ, Siggia ED, Swain PS (2002) Stochastic gene expression in a single cell. Science 297(5584):1183–1186. https://doi.org/10.1126/science.1070919

Fan Z, Masui H, Altas I, Mendelsohn J (1993) Blockade of epidermal growth factor receptor function by bivalent and monovalent fragments of 225 anti-epidermal growth factor receptor monoclonal antibodies. Cancer Res 53(18):4322–4328

Feng Q, Baird D, Peng X, Wang J, Ly T, Guan JL, Cerione RA (2006) Cool-1 functions as an essential regulatory node for EGF receptor- and Src-mediated cell growth. Nat Cell Biol 8(9):945–956. https://doi.org/10.1038/ncb1453

Ferby I, Reschke M, Kudlacek O, Knyazev P, Pante G, Amann K, Sommergruber W, Kraut N, Ullrich A, Fassler R, Klein R (2006) Mig6 is a negative regulator of EGF receptor-mediated skin morphogenesis and tumor formation. Nat Med 12(5):568–573. https://doi.org/10.1038/nm1401

Ferrandon S, Feinstein TN, Castro M, Wang B, Bouley R, Potts JT, Gardella TJ, Vilardaga JP (2009) Sustained cyclic AMP production by parathyroid hormone receptor endocytosis. Nat Chem Biol 5(10):734–742. https://doi.org/10.1038/nchembio.206

Fortian A, Sorkin A (2014) Live-cell fluorescence imaging reveals high stoichiometry of Grb2 binding to the EGF receptor sustained during endocytosis. J Cell Sci 127(Pt 2):432–444. https://doi.org/10.1242/jcs.137786

Frechin M, Stoeger T, Daetwyler S, Gehin C, Battich N, Damm EM, Stergiou L, Riezman H, Pelkmans L (2015) Cell-intrinsic adaptation of lipid composition to local crowding drives social behaviour. Nature 523(7558):88–91. https://doi.org/10.1038/nature14429

French AR, Sudlow GP, Wiley HS, Lauffenburger DA (1994) Postendocytic trafficking of epidermal growth factor-receptor complexes is mediated through saturable and specific endosomal interactions. J Biol Chem 269(22):15749–15755

French AR, Tadaki DK, Niyogi SK, Lauffenburger DA (1995) Intracellular trafficking of epidermal growth factor family ligands is directly influenced by the pH sensitivity of the receptor/ligand interaction. J Biol Chem 270(9):4334–4340

Friedman LM, Rinon A, Schechter B, Lyass L, Lavi S, Bacus SS, Sela M, Yarden Y (2005) Synergistic down-regulation of receptor tyrosine kinases by combinations of mAbs: implications for cancer immunotherapy. Proc Natl Acad Sci U S A 102(6):1915–1920. https://doi.org/10.1073/pnas.0409610102

Friedman JR, Dibenedetto JR, West M, Rowland AA, Voeltz GK (2013) Endoplasmic reticulum-endosome contact increases as endosomes traffic and mature. Mol Biol Cell 24(7):1030–1040. https://doi.org/10.1091/mbc.E12-10-0733

Frosi Y, Anastasi S, Ballaro C, Varsano G, Castellani L, Maspero E, Polo S, Alema S, Segatto O (2010) A two-tiered mechanism of EGFR inhibition by RALT/MIG6 via kinase suppression and receptor degradation. J Cell Biol 189(3):557–571. https://doi.org/10.1083/jcb.201002032

Fry WH, Simion C, Sweeney C, Carraway KL 3rd (2011) Quantity control of the ErbB3 receptor tyrosine kinase at the endoplasmic reticulum. Mol Cell Biol 31(14):3009–3018. https://doi.org/10.1128/MCB.05105-11

Gan HK, Walker F, Burgess AW, Rigopoulos A, Scott AM, Johns TG (2007) The epidermal growth factor receptor (EGFR) tyrosine kinase inhibitor AG1478 increases the formation of inactive untethered EGFR dimers. Implications for combination therapy with monoclonal antibody 806. J Biol Chem 282(5):2840–2850. https://doi.org/10.1074/jbc.M605136200

Garrett TP, McKern NM, Lou M, Elleman TC, Adams TE, Lovrecz GO, Zhu HJ, Walker F, Frenkel MJ, Hoyne PA, Jorissen RN, Nice EC, Burgess AW, Ward CW (2002) Crystal structure of a truncated epidermal growth factor receptor extracellular domain bound to transforming growth factor alpha. Cell 110(6):763–773

Goh LK, Sorkin A (2013) Endocytosis of receptor tyrosine kinases. Cold Spring Harb Perspect Biol 5(5):a017459. https://doi.org/10.1101/cshperspect.a017459

Goh LK, Huang F, Kim W, Gygi S, Sorkin A (2010) Multiple mechanisms collectively regulate clathrin-mediated endocytosis of the epidermal growth factor receptor. J Cell Biol 189(5):871–883. https://doi.org/10.1083/jcb.201001008

Gorden P, Carpentier JL, Cohen S, Orci L (1978) Epidermal growth factor: morphological demonstration of binding, internalization, and lysosomal association in human fibroblasts. Proc Natl Acad Sci U S A 75(10):5025–5029

Grand FH, Hidalgo-Curtis CE, Ernst T, Zoi K, Zoi C, McGuire C, Kreil S, Jones A, Score J, Metzgeroth G, Oscier D, Hall A, Brandts C, Serve H, Reiter A, Chase AJ, Cross NC (2009) Frequent CBL mutations associated with 11q acquired uniparental disomy in myeloproliferative neoplasms. Blood 113(24):6182–6192. https://doi.org/10.1182/blood-2008-12-194548
Grandal MV, Zandi R, Pedersen MW, Willumsen BM, van Deurs B, Poulsen HS (2007) EGFRvIII escapes down-regulation due to impaired internalization and sorting to lysosomes. Carcinogenesis 28(7):1408–1417. https://doi.org/10.1093/carcin/bgm058
Grimes ML, Zhou J, Beattie EC, Yuen EC, Hall DE, Valletta JS, Topp KS, LaVail JH, Bunnett NW, Mobley WC (1996) Endocytosis of activated TrkA: evidence that nerve growth factor induces formation of signaling endosomes. J Neurosci 16(24):7950–7964
Grovdal LM, Stang E, Sorkin A, Madshus IH (2004) Direct interaction of Cbl with pTyr 1045 of the EGF receptor (EGFR) is required to sort the EGFR to lysosomes for degradation. Exp Cell Res 300(2):388–395. https://doi.org/10.1016/j.yexcr.2004.07.003
Gschweitl M, Ulbricht A, Barnes CA, Enchev RI, Stoffel-Studer I, Meyer-Schaller N, Huotari J, Yamauchi Y, Greber UF, Helenius A, Peter M (2016) A SPOPL/Cullin-3 ubiquitin ligase complex regulates endocytic trafficking by targeting EPS15 at endosomes. Elife 5:e13841. https://doi.org/10.7554/eLife.13841
Guha A, Sriram V, Krishnan KS, Mayor S (2003) Shibire mutations reveal distinct dynamin-independent and -dependent endocytic pathways in primary cultures of Drosophila hemocytes. J Cell Sci 116(Pt 16):3373–3386. https://doi.org/10.1242/jcs.00637
Gur G, Rubin C, Katz M, Amit I, Citri A, Nilsson J, Amariglio N, Henriksson R, Rechavi G, Hedman H, Wides R, Yarden Y (2004) LRIG1 restricts growth factor signaling by enhancing receptor ubiquitylation and degradation. EMBO J 23(16):3270–3281. https://doi.org/10.1038/sj.emboj.7600342
Haj FG, Verveer PJ, Squire A, Neel BG, Bastiaens PI (2002) Imaging sites of receptor dephosphorylation by PTP1B on the surface of the endoplasmic reticulum. Science 295(5560):1708–1711. https://doi.org/10.1126/science.1067566
Haj FG, Sabet O, Kinkhabwala A, Wimmer-Kleikamp S, Roukos V, Han HM, Grabenbauer M, Bierbaum M, Antony C, Neel BG, Bastiaens PI (2012) Regulation of signaling at regions of cell-cell contact by endoplasmic reticulum-bound protein-tyrosine phosphatase 1B. PLoS ONE 7(5):e36633. https://doi.org/10.1371/journal.pone.0036633
Han W, Zhang T, Yu H, Foulke JG, Tang CK (2006) Hypophosphorylation of residue Y1045 leads to defective downregulation of EGFRvIII. Cancer Biol Ther 5(10):1361–1368
Hanover JA, Willingham MC, Pastan I (1984) Kinetics of transit of transferrin and epidermal growth factor through clathrin-coated membranes. Cell 39(2 Pt 1):283–293
Henne WM, Stenmark H, Emr SD (2013) Molecular mechanisms of the membrane sculpting ESCRT pathway. Cold Spring Harb Perspect Biol 5(9). https://doi.org/10.1101/cshperspect.a016766
Hinrichsen L, Harborth J, Andrees L, Weber K, Ungewickell EJ (2003) Effect of clathrin heavy chain- and alpha-adaptin-specific small inhibitory RNAs on endocytic accessory proteins and receptor trafficking in HeLa cells. J Biol Chem 278(46):45160–45170. https://doi.org/10.1074/jbc.M307290200
Hirsch FR, Janne PA, Eberhardt WE, Cappuzzo F, Thatcher N, Pirker R, Choy H, Kim ES, Paz-Ares L, Gandara DR, Wu YL, Ahn MJ, Mitsudomi T, Shepherd FA, Mok TS (2013) Epidermal growth factor receptor inhibition in lung cancer: status 2012. J Thorac Oncol 8(3):373–384. https://doi.org/10.1097/JTO.0b013e31827ed0ff
Howe CL, Mobley WC (2005) Long-distance retrograde neurotrophic signaling. Curr Opin Neurobiol 15(1):40–48. https://doi.org/10.1016/j.conb.2005.01.010
Huang F, Kirkpatrick D, Jiang X, Gygi S, Sorkin A (2006) Differential regulation of EGF receptor internalization and degradation by multiubiquitination within the kinase domain. Mol Cell 21(6):737–748. https://doi.org/10.1016/j.molcel.2006.02.018
Huang F, Zeng X, Kim W, Balasubramani M, Fortian A, Gygi SP, Yates NA, Sorkin A (2013) Lysine 63-linked polyubiquitination is required for EGF receptor degradation. Proc Natl Acad Sci U S A 110(39):15722–15727. https://doi.org/10.1073/pnas.1308014110

Huotari J, Meyer-Schaller N, Hubner M, Stauffer S, Katheder N, Horvath P, Mancini R, Helenius A, Peter M (2012) Cullin-3 regulates late endosome maturation. Proc Natl Acad Sci U S A 109(3):823–828. https://doi.org/10.1073/pnas.1118744109

Ibach J, Radon Y, Gelleri M, Sonntag MH, Brunsveld L, Bastiaens PI, Verveer PJ (2015) Single particle tracking reveals that EGFR signaling activity is amplified in clathrin-coated pits. PLoS ONE 10(11):e0143162. https://doi.org/10.1371/journal.pone.0143162

Irannejad R, Tomshine JC, Tomshine JR, Chevalier M, Mahoney JP, Steyaert J, Rasmussen SG, Sunahara RK, El-Samad H, Huang B, von Zastrow M (2013) Conformational biosensors reveal GPCR signalling from endosomes. Nature 495(7442):534–538. https://doi.org/10.1038/nature12000

Irannejad R, Tsvetanova NG, Lobingier BT, von Zastrow M (2015) Effects of endocytosis on receptor-mediated signaling. Curr Opin Cell Biol 35:137–143. https://doi.org/10.1016/j.ceb.2015.05.005

Jaramillo ML, Leon Z, Grothe S, Paul-Roc B, Abulrob A, O'Connor McCourt M (2006) Effect of the anti-receptor ligand-blocking 225 monoclonal antibody on EGF receptor endocytosis and sorting. Exp Cell Res 312(15):2778–2790. https://doi.org/10.1016/j.yexcr.2006.05.008

Jekely G, Rorth P (2003) Hrs mediates downregulation of multiple signalling receptors in Drosophila. EMBO Rep 4(12):1163–1168. https://doi.org/10.1038/sj.embor.7400019

Jiang X, Huang F, Marusyk A, Sorkin A (2003) Grb2 regulates internalization of EGF receptors through clathrin-coated pits. Mol Biol Cell 14(3):858–870. https://doi.org/10.1091/mbc.E02-08-0532

Johannes L, Parton RG, Bassereau P, Mayor S (2015) Building endocytic pits without clathrin. Nat Rev Mol Cell Biol 16(5):311–321. https://doi.org/10.1038/nrm3968

Johnson H, Del Rosario AM, Bryson BD, Schroeder MA, Sarkaria JN, White FM (2012) Molecular characterization of EGFR and EGFRvIII signaling networks in human glioblastoma tumor xenografts. Mol Cell Proteomics 11(12):1724–1740. https://doi.org/10.1074/mcp.M112.019984

Kalaidzidis I, Miaczynska M, Brewinska-Olchowik M, Hupalowska A, Ferguson C, Parton RG, Kalaidzidis Y, Zerial M (2015) APPL endosomes are not obligatory endocytic intermediates but act as stable cargo-sorting compartments. J Cell Biol 211(1):123–144. https://doi.org/10.1083/jcb.201311117

Kalia M, Kumari S, Chadda R, Hill MM, Parton RG, Mayor S (2006) Arf6-independent GPI-anchored protein-enriched early endosomal compartments fuse with sorting endosomes via a Rab5/phosphatidylinositol-3'-kinase-dependent machinery. Mol Biol Cell 17(8):3689–3704. https://doi.org/10.1091/mbc.E05-10-0980

Kario E, Marmor MD, Adamsky K, Citri A, Amit I, Amariglio N, Rechavi G, Yarden Y (2005) Suppressors of cytokine signaling 4 and 5 regulate epidermal growth factor receptor signaling. J Biol Chem 280(8):7038–7048. https://doi.org/10.1074/jbc.M408575200

Kelley LC, Weed SA (2012) Cortactin is a substrate of activated Cdc42-associated kinase 1 (ACK1) during ligand-induced epidermal growth factor receptor downregulation. PLoS ONE 7(8):e44363. https://doi.org/10.1371/journal.pone.0044363

Kholodenko BN, Demin OV, Moehren G, Hoek JB (1999) Quantification of short term signaling by the epidermal growth factor receptor. J Biol Chem 274(42):30169–30181

Kilpatrick BS, Eden ER, Hockey LN, Yates E, Futter CE, Patel S (2017) An endosomal NAADP-sensitive two-pore Ca^{2+} channel regulates ER-endosome membrane contact sites to control growth factor signaling. Cell Rep 18(7):1636–1645. https://doi.org/10.1016/j.celrep.2017.01.052

Kirchhausen T, Owen D, Harrison SC (2014) Molecular structure, function, and dynamics of clathrin-mediated membrane traffic. Cold Spring Harb Perspect Biol 6(5):a016725. https://doi.org/10.1101/cshperspect.a016725

Kleiman LB, Maiwald T, Conzelmann H, Lauffenburger DA, Sorger PK (2011) Rapid phospho-turnover by receptor tyrosine kinases impacts downstream signaling and drug binding. Mol Cell 43(5):723–737. https://doi.org/10.1016/j.molcel.2011.07.014

Kon S, Kobayashi N, Satake M (2014) Altered trafficking of mutated growth factor receptors and their associated molecules: implication for human cancers. Cell Logist 4:e28461. https://doi.org/10.4161/cl.28461

Kovacs E, Zorn JA, Huang Y, Barros T, Kuriyan J (2015) A structural perspective on the regulation of the epidermal growth factor receptor. Annu Rev Biochem 84:739–764. https://doi.org/10.1146/annurev-biochem-060614-034402

Lakadamyali M, Rust MJ, Zhuang X (2006) Ligands for clathrin-mediated endocytosis are differentially sorted into distinct populations of early endosomes. Cell 124(5):997–1009. https://doi.org/10.1016/j.cell.2005.12.038

Lampugnani MG, Orsenigo F, Gagliani MC, Tacchetti C, Dejana E (2006) Vascular endothelial cadherin controls VEGFR-2 internalization and signaling from intracellular compartments. J Cell Biol 174(4):593–604. https://doi.org/10.1083/jcb.200602080

Lemmon MA (2009) Ligand-induced ErbB receptor dimerization. Exp Cell Res 315(4):638–648. https://doi.org/10.1016/j.yexcr.2008.10.024

Lemmon MA, Schlessinger J (2010) Cell signaling by receptor tyrosine kinases. Cell 141(7):1117–1134. https://doi.org/10.1016/j.cell.2010.06.011

Lemmon MA, Schlessinger J, Ferguson KM (2014) The EGFR family: not so prototypical receptor tyrosine kinases. Cold Spring Harb Perspect Biol 6(4):a020768. https://doi.org/10.1101/cshperspect.a020768

Lemmon MA, Freed DM, Schlessinger J, Kiyatkin A (2016) The dark side of cell signaling: positive roles for negative regulators. Cell 164(6):1172–1184. https://doi.org/10.1016/j.cell.2016.02.047

Lenferink AE, Pinkas-Kramarski R, van de Poll ML, van Vugt MJ, Klapper LN, Tzahar E, Waterman H, Sela M, van Zoelen EJ, Yarden Y (1998) Differential endocytic routing of homo- and hetero-dimeric ErbB tyrosine kinases confers signaling superiority to receptor heterodimers. EMBO J 17(12):3385–3397. https://doi.org/10.1093/emboj/17.12.3385

Leung EL, Tam IY, Tin VP, Chua DT, Sihoe AD, Cheng LC, Ho JC, Chung LP, Wong MP (2009) SRC promotes survival and invasion of lung cancers with epidermal growth factor receptor abnormalities and is a potential candidate for molecular-targeted therapy. Mol Cancer Res 7(6):923–932. https://doi.org/10.1158/1541-7786.MCR-09-0003

Levine TP, Patel S (2016) Signalling at membrane contact sites: two membranes come together to handle second messengers. Curr Opin Cell Biol 39:77–83. https://doi.org/10.1016/j.ceb.2016.02.011

Levkowitz G, Klapper LN, Tzahar E, Freywald A, Sela M, Yarden Y (1996) Coupling of the c-Cbl protooncogene product to ErbB-1/EGF-receptor but not to other ErbB proteins. Oncogene 12(5):1117–1125

Levkowitz G, Waterman H, Zamir E, Kam Z, Oved S, Langdon WY, Beguinot L, Geiger B, Yarden Y (1998) c-Cbl/Sli-1 regulates endocytic sorting and ubiquitination of the epidermal growth factor receptor. Genes Dev 12(23):3663–3674

Levkowitz G, Waterman H, Ettenberg SA, Katz M, Tsygankov AY, Alroy I, Lavi S, Iwai K, Reiss Y, Ciechanover A, Lipkowitz S, Yarden Y (1999) Ubiquitin ligase activity and tyrosine phosphorylation underlie suppression of growth factor signaling by c-Cbl/Sli-1. Mol Cell 4(6):1029–1040

Liberali P, Snijder B, Pelkmans L (2014) A hierarchical map of regulatory genetic interactions in membrane trafficking. Cell 157(6):1473–1487. https://doi.org/10.1016/j.cell.2014.04.029

Licitra L, Storkel S, Kerr KM, Van Cutsem E, Pirker R, Hirsch FR, Vermorken JB, von Heydebreck A, Esser R, Celik I, Ciardiello F (2013) Predictive value of epidermal growth factor receptor expression for first-line chemotherapy plus cetuximab in patients with head and neck and colorectal cancer: analysis of data from the EXTREME and CRYSTAL studies. Eur J Cancer 49(6):1161–1168. https://doi.org/10.1016/j.ejca.2012.11.018

Liu Z, Zanata SM, Kim J, Peterson MA, Di Vizio D, Chirieac LR, Pyne S, Agostini M, Freeman MR, Loda M (2013) The ubiquitin-specific protease USP2a prevents endocytosis-mediated EGFR degradation. Oncogene 32(13):1660–1669. https://doi.org/10.1038/onc.2012.188

Longva KE, Blystad FD, Stang E, Larsen AM, Johannessen LE, Madshus IH (2002) Ubiquitination and proteasomal activity is required for transport of the EGF receptor to inner membranes of multivesicular bodies. J Cell Biol 156(5):843–854. https://doi.org/10.1083/jcb.200106056
Lund KA, Opresko LK, Starbuck C, Walsh BJ, Wiley HS (1990) Quantitative analysis of the endocytic system involved in hormone-induced receptor internalization. J Biol Chem 265(26):15713–15723
Lynch TJ, Bell DW, Sordella R, Gurubhagavatula S, Okimoto RA, Brannigan BW, Harris PL, Haserlat SM, Supko JG, Haluska FG, Louis DN, Christiani DC, Settleman J, Haber DA (2004) Activating mutations in the epidermal growth factor receptor underlying responsiveness of non-small-cell lung cancer to gefitinib. N Engl J Med 350(21):2129–2139. https://doi.org/10.1056/NEJMoa040938
Ma YM, Boucrot E, Villen J, el Affar B, Gygi SP, Gottlinger HG, Kirchhausen T (2007) Targeting of AMSH to endosomes is required for epidermal growth factor receptor degradation. J Biol Chem 282(13):9805–9812. https://doi.org/10.1074/jbc.M611635200
Macdonald-Obermann JL, Pike LJ (2014) Different epidermal growth factor (EGF) receptor ligands show distinct kinetics and biased or partial agonism for homodimer and heterodimer formation. J Biol Chem 289(38):26178–26188. https://doi.org/10.1074/jbc.M114.586826
Martinelli S, De Luca A, Stellacci E, Rossi C, Checquolo S, Lepri F, Caputo V, Silvano M, Buscherini F, Consoli F, Ferrara G, Digilio MC, Cavaliere ML, van Hagen JM, Zampino G, van der Burgt I, Ferrero GB, Mazzanti L, Screpanti I, Yntema HG, Nillesen WM, Savarirayan R, Zenker M, Dallapiccola B, Gelb BD, Tartaglia M (2010) Heterozygous germline mutations in the CBL tumor-suppressor gene cause a Noonan syndrome-like phenotype. Am J Hum Genet 87(2):250–257. https://doi.org/10.1016/j.ajhg.2010.06.015
McCullough J, Clague MJ, Urbe S (2004) AMSH is an endosome-associated ubiquitin isopeptidase. J Cell Biol 166(4):487–492. https://doi.org/10.1083/jcb.200401141
McMahon HT, Boucrot E (2011) Molecular mechanism and physiological functions of clathrin-mediated endocytosis. Nat Rev Mol Cell Biol 12(8):517–533. https://doi.org/10.1038/nrm3151
Menard L, Parker PJ, Kermorgant S (2014) Receptor tyrosine kinase c-Met controls the cytoskeleton from different endosomes via different pathways. Nat Commun 5:3907. https://doi.org/10.1038/ncomms4907
Miaczynska M, Christoforidis S, Giner A, Shevchenko A, Uttenweiler-Joseph S, Habermann B, Wilm M, Parton RG, Zerial M (2004) APPL proteins link Rab5 to nuclear signal transduction via an endosomal compartment. Cell 116(3):445–456
Moscatello DK, Holgado-Madruga M, Godwin AK, Ramirez G, Gunn G, Zoltick PW, Biegel JA, Hayes RL, Wong AJ (1995) Frequent expression of a mutant epidermal growth factor receptor in multiple human tumors. Cancer Res 55(23):5536–5539
Motley A, Bright NA, Seaman MN, Robinson MS (2003) Clathrin-mediated endocytosis in AP-2-depleted cells. J Cell Biol 162(5):909–918. https://doi.org/10.1083/jcb.200305145
Muller PA, Caswell PT, Doyle B, Iwanicki MP, Tan EH, Karim S, Lukashchuk N, Gillespie DA, Ludwig RL, Gosselin P, Cromer A, Brugge JS, Sansom OJ, Norman JC, Vousden KH (2009) Mutant p53 drives invasion by promoting integrin recycling. Cell 139(7):1327–1341. https://doi.org/10.1016/j.cell.2009.11.026
Muller PA, Trinidad AG, Timpson P, Morton JP, Zanivan S, van den Berghe PV, Nixon C, Karim SA, Caswell PT, Noll JE, Coffill CR, Lane DP, Sansom OJ, Neilsen PM, Norman JC, Vousden KH (2013) Mutant p53 enhances MET trafficking and signalling to drive cell scattering and invasion. Oncogene 32(10):1252–1265. https://doi.org/10.1038/onc.2012.148
Nakamura N, Lill JR, Phung Q, Jiang Z, Bakalarski C, de Mazière A, Klumperman J, Schlatter M, Delamarre L, Mellman I (2014) Endosomes are specialized platforms for bacterial sensing and NOD2 signalling. Nature 509(7499):240–244. https://doi.org/10.1038/nature13133
Nevins AK, Thurmond DC (2006) Caveolin-1 functions as a novel Cdc42 guanine nucleotide dissociation inhibitor in pancreatic beta-cells. J Biol Chem 281(28):18961–18972. https://doi.org/10.1074/jbc.M603604200

Nicholson SE, Metcalf D, Sprigg NS, Columbus R, Walker F, Silva A, Cary D, Willson TA, Zhang JG, Hilton DJ, Alexander WS, Nicola NA (2005) Suppressor of cytokine signaling (SOCS)-5 is a potential negative regulator of epidermal growth factor signaling. Proc Natl Acad Sci U S A 102(7):2328–2333. https://doi.org/10.1073/pnas.0409675102

Ogiso H, Ishitani R, Nureki O, Fukai S, Yamanaka M, Kim JH, Saito K, Sakamoto A, Inoue M, Shirouzu M, Yokoyama S (2002) Crystal structure of the complex of human epidermal growth factor and receptor extracellular domains. Cell 110(6):775–787

Orth JD, Krueger EW, Weller SG, McNiven MA (2006) A novel endocytic mechanism of epidermal growth factor receptor sequestration and internalization. Cancer Res 66(7):3603–3610. https://doi.org/10.1158/0008-5472.CAN-05-2916

Oved S, Mosesson Y, Zwang Y, Santonico E, Shtiegman K, Marmor MD, Kochupurakkal BS, Katz M, Lavi S, Cesareni G, Yarden Y (2006) Conjugation to Nedd8 instigates ubiquitylation and down-regulation of activated receptor tyrosine kinases. J Biol Chem 281(31):21640–21651. https://doi.org/10.1074/jbc.M513034200

Paez JG, Janne PA, Lee JC, Tracy S, Greulich H, Gabriel S, Herman P, Kaye FJ, Lindeman N, Boggon TJ, Naoki K, Sasaki H, Fujii Y, Eck MJ, Sellers WR, Johnson BE, Meyerson M (2004) EGFR mutations in lung cancer: correlation with clinical response to gefitinib therapy. Science 304(5676):1497–1500. https://doi.org/10.1126/science.1099314

Pao W, Miller V, Zakowski M, Doherty J, Politi K, Sarkaria I, Singh B, Heelan R, Rusch V, Fulton L, Mardis E, Kupfer D, Wilson R, Kris M, Varmus H (2004) EGF receptor gene mutations are common in lung cancers from "never smokers" and are associated with sensitivity of tumors to gefitinib and erlotinib. Proc Natl Acad Sci U S A 101(36):13306–13311. https://doi.org/10.1073/pnas.0405220101

Pareja F, Ferraro DA, Rubin C, Cohen-Dvashi H, Zhang F, Aulmann S, Ben-Chetrit N, Pines G, Navon R, Crosetto N, Kostler W, Carvalho S, Lavi S, Schmitt F, Dikic I, Yakhini Z, Sinn P, Mills GB, Yarden Y (2012) Deubiquitination of EGFR by Cezanne-1 contributes to cancer progression. Oncogene 31(43):4599–4608. https://doi.org/10.1038/onc.2011.587

Pedersen MW, Jacobsen HJ, Koefoed K, Hey A, Pyke C, Haurum JS, Kragh M (2010) Sym004: a novel synergistic anti-epidermal growth factor receptor antibody mixture with superior anticancer efficacy. Cancer Res 70(2):588–597. https://doi.org/10.1158/0008-5472.CAN-09-1417

Peeters M, Karthaus M, Rivera F, Terwey JH, Douillard JY (2015) Panitumumab in metastatic colorectal cancer: the importance of tumour RAS status. Drugs 75(7):731–748. https://doi.org/10.1007/s40265-015-0386-x

Phillips MJ, Voeltz GK (2016) Structure and function of ER membrane contact sites with other organelles. Nat Rev Mol Cell Biol 17(2):69–82. https://doi.org/10.1038/nrm.2015.8

Pierotti MA, Negri T, Tamborini E, Perrone F, Pricl S, Pilotti S (2010) Targeted therapies: the rare cancer paradigm. Mol Oncol 4(1):19–37. https://doi.org/10.1016/j.molonc.2009.10.003

Piper RC, Dikic I, Lukacs GL (2014) Ubiquitin-dependent sorting in endocytosis. Cold Spring Harb Perspect Biol 6(1). https://doi.org/10.1101/cshperspect.a016808

Poteryaev D, Datta S, Ackema K, Zerial M, Spang A (2010) Identification of the switch in early-to-late endosome transition. Cell 141(3):497–508. https://doi.org/10.1016/j.cell.2010.03.011

Raiborg C, Stenmark H (2009) The ESCRT machinery in endosomal sorting of ubiquitylated membrane proteins. Nature 458(7237):445–452. https://doi.org/10.1038/nature07961

Renard HF, Simunovic M, Lemiere J, Boucrot E, Garcia-Castillo MD, Arumugam S, Chambon V, Lamaze C, Wunder C, Kenworthy AK, Schmidt AA, McMahon HT, Sykes C, Bassereau P, Johannes L (2015) Endophilin-A2 functions in membrane scission in clathrin-independent endocytosis. Nature 517(7535):493–496. https://doi.org/10.1038/nature14064

Resat H, Ewald JA, Dixon DA, Wiley HS (2003) An integrated model of epidermal growth factor receptor trafficking and signal transduction. Biophys J 85(2):730–743

Rink J, Ghigo E, Kalaidzidis Y, Zerial M (2005) Rab conversion as a mechanism of progression from early to late endosomes. Cell 122(5):735–749. https://doi.org/10.1016/j.cell.2005.06.043

Rocha N, Kuijl C, van der Kant R, Janssen L, Houben D, Janssen H, Zwart W, Neefjes J (2009) Cholesterol sensor ORP1L contacts the ER protein VAP to control Rab7-RILP-p150 Glued and late endosome positioning. J Cell Biol 185(7):1209–1225. https://doi.org/10.1083/jcb.200811005

Roda-Navarro P, Bastiaens PI (2014) Dynamic recruitment of protein tyrosine phosphatase PTPD1 to EGF stimulation sites potentiates EGFR activation. PLoS ONE 9(7):e103203. https://doi.org/10.1371/journal.pone.0103203

Roepstorff K, Grandal MV, Henriksen L, Knudsen SL, Lerdrup M, Grovdal L, Willumsen BM, van Deurs B (2009) Differential effects of EGFR ligands on endocytic sorting of the receptor. Traffic 10(8):1115–1127. https://doi.org/10.1111/j.1600-0854.2009.00943.x

Romsicki Y, Reece M, Gauthier JY, Asante-Appiah E, Kennedy BP (2004) Protein tyrosine phosphatase-1B dephosphorylation of the insulin receptor occurs in a perinuclear endosome compartment in human embryonic kidney 293 cells. J Biol Chem 279(13):12868–12875. https://doi.org/10.1074/jbc.M309600200

Roskoski R Jr (2014) The ErbB/HER family of protein-tyrosine kinases and cancer. Pharmacol Res 79:34–74. https://doi.org/10.1016/j.phrs.2013.11.002

Rothenberg SM, Engelman JA, Le S, Riese DJ 2nd, Haber DA, Settleman J (2008) Modeling oncogene addiction using RNA interference. Proc Natl Acad Sci U S A 105(34):12480–12484. https://doi.org/10.1073/pnas.0803217105

Row PE, Prior IA, McCullough J, Clague MJ, Urbe S (2006) The ubiquitin isopeptidase UBPY regulates endosomal ubiquitin dynamics and is essential for receptor down-regulation. J Biol Chem 281(18):12618–12624. https://doi.org/10.1074/jbc.M512615200

Rowland AA, Chitwood PJ, Phillips MJ, Voeltz GK (2014) ER contact sites define the position and timing of endosome fission. Cell 159(5):1027–1041. https://doi.org/10.1016/j.cell.2014.10.023

Roxrud I, Raiborg C, Pedersen NM, Stang E, Stenmark H (2008) An endosomally localized isoform of Eps15 interacts with Hrs to mediate degradation of epidermal growth factor receptor. J Cell Biol 180(6):1205–1218. https://doi.org/10.1083/jcb.200708115

Sanada M, Suzuki T, Shih LY, Otsu M, Kato M, Yamazaki S, Tamura A, Honda H, Sakata-Yanagimoto M, Kumano K, Oda H, Yamagata T, Takita J, Gotoh N, Nakazaki K, Kawamata N, Onodera M, Nobuyoshi M, Hayashi Y, Harada H, Kurokawa M, Chiba S, Mori H, Ozawa K, Omine M, Hirai H, Nakauchi H, Koeffler HP, Ogawa S (2009) Gain-of-function of mutated C-CBL tumour suppressor in myeloid neoplasms. Nature 460(7257):904–908. https://doi.org/10.1038/nature08240

Sangwan V, Paliouras GN, Abella JV, Dube N, Monast A, Tremblay ML, Park M (2008) Regulation of the Met receptor-tyrosine kinase by the protein-tyrosine phosphatase 1B and T-cell phosphatase. J Biol Chem 283(49):34374–34383. https://doi.org/10.1074/jbc.M805916200

Sargin B, Choudhary C, Crosetto N, Schmidt MH, Grundler R, Rensinghoff M, Thiessen C, Tickenbrock L, Schwable J, Brandts C, August B, Koschmieder S, Bandi SR, Duyster J, Berdel WE, Muller-Tidow C, Dikic I, Serve H (2007) Flt3-dependent transformation by inactivating c-Cbl mutations in AML. Blood 110(3):1004–1012. https://doi.org/10.1182/blood-2007-01-066076

Savio MG, Wollscheid N, Cavallaro E, Algisi V, Di Fiore PP, Sigismund S, Maspero E, Polo S (2016) USP9X controls EGFR fate by deubiquitinating the endocytic adaptor Eps15. Curr Biol 26(2):173–183. https://doi.org/10.1016/j.cub.2015.11.050

Sawano A, Takayama S, Matsuda M, Miyawaki A (2002) Lateral propagation of EGF signaling after local stimulation is dependent on receptor density. Dev Cell 3(2):245–257

Scharaw S, Iskar M, Ori A, Boncompain G, Laketa V, Poser I, Lundberg E, Perez F, Beck M, Bork P, Pepperkok R (2016) The endosomal transcriptional regulator RNF11 integrates degradation and transport of EGFR. J Cell Biol 215(4):543–558. https://doi.org/10.1083/jcb.201601090

Schenck A, Goto-Silva L, Collinet C, Rhinn M, Giner A, Habermann B, Brand M, Zerial M (2008) The endosomal protein Appl1 mediates Akt substrate specificity and cell survival in vertebrate development. Cell 133(3):486–497. https://doi.org/10.1016/j.cell.2008.02.044

Schlessinger J (2014) Receptor tyrosine kinases: legacy of the first two decades. Cold Spring Harb Perspect Biol 6(3). https://doi.org/10.1101/cshperspect.a008912

Schmidt MH, Dikic I (2005) The Cbl interactome and its functions. Nat Rev Mol Cell Biol 6(12):907–918. https://doi.org/10.1038/nrm1762

Schmidt MH, Furnari FB, Cavenee WK, Bogler O (2003) Epidermal growth factor receptor signaling intensity determines intracellular protein interactions, ubiquitination, and internalization. Proc Natl Acad Sci U S A 100(11):6505–6510. https://doi.org/10.1073/pnas.1031790100
Scott CC, Vacca F, Gruenberg J (2014) Endosome maturation, transport and functions. Semin Cell Dev Biol 31:2–10. https://doi.org/10.1016/j.semcdb.2014.03.034
Segatto O, Anastasi S, Alema S (2011) Regulation of epidermal growth factor receptor signalling by inducible feedback inhibitors. J Cell Sci 124(Pt 11):1785–1793. https://doi.org/10.1242/jcs.083303
Settembre C, Fraldi A, Medina DL, Ballabio A (2013) Signals from the lysosome: a control centre for cellular clearance and energy metabolism. Nat Rev Mol Cell Biol 14(5):283–296. https://doi.org/10.1038/nrm3565
Shen F, Lin Q, Gu Y, Childress C, Yang W (2007) Activated Cdc42-associated kinase 1 is a component of EGF receptor signaling complex and regulates EGF receptor degradation. Mol Biol Cell 18(3):732–742. https://doi.org/10.1091/mbc.E06-02-0142
Shenoy SK, Lefkowitz RJ (2005) Seven-transmembrane receptor signaling through beta-arrestin. Sci STKE 2005(308):cm10. https://doi.org/10.1126/stke.2005/308/cm10
Shimizu H, Woodcock SA, Wilkin MB, Trubenova B, Monk NA, Baron M (2014) Compensatory flux changes within an endocytic trafficking network maintain thermal robustness of Notch signaling. Cell 157(5):1160–1174. https://doi.org/10.1016/j.cell.2014.03.050
Shtiegman K, Kochupurakkal BS, Zwang Y, Pines G, Starr A, Vexler A, Citri A, Katz M, Lavi S, Ben-Basat Y, Benjamin S, Corso S, Gan J, Yosef RB, Giordano S, Yarden Y (2007) Defective ubiquitinylation of EGFR mutants of lung cancer confers prolonged signaling. Oncogene 26(49):6968–6978. https://doi.org/10.1038/sj.onc.1210503
Sigismund S, Woelk T, Puri C, Maspero E, Tacchetti C, Transidico P, Di Fiore PP, Polo S (2005) Clathrin-independent endocytosis of ubiquitinated cargos. Proc Natl Acad Sci U S A 102(8):2760–2765. https://doi.org/10.1073/pnas.0409817102
Sigismund S, Argenzio E, Tosoni D, Cavallaro E, Polo S, Di Fiore PP (2008) Clathrin-mediated internalization is essential for sustained EGFR signaling but dispensable for degradation. Dev Cell 15(2):209–219. https://doi.org/10.1016/j.devcel.2008.06.012
Sigismund S, Confalonieri S, Ciliberto A, Polo S, Scita G, Di Fiore PP (2012) Endocytosis and signaling: cell logistics shape the eukaryotic cell plan. Physiol Rev 92(1):273–366. https://doi.org/10.1152/physrev.00005.2011
Sigismund S, Algisi V, Nappo G, Conte A, Pascolutti R, Cuomo A, Bonaldi T, Argenzio E, Verhoef LG, Maspero E, Bianchi F, Capuani F, Ciliberto A, Polo S, Di Fiore PP (2013) Threshold-controlled ubiquitination of the EGFR directs receptor fate. EMBO J 32(15):2140–2157. https://doi.org/10.1038/emboj.2013.149
Singh AB, Harris RC (2005) Autocrine, paracrine and juxtacrine signaling by EGFR ligands. Cell Signal 17(10):1183–1193. https://doi.org/10.1016/j.cellsig.2005.03.026
Snijder B, Sacher R, Ramo P, Damm EM, Liberali P, Pelkmans L (2009) Population context determines cell-to-cell variability in endocytosis and virus infection. Nature 461(7263):520–523. https://doi.org/10.1038/nature08282
Sorkin A, Carpenter G (1993) Interaction of activated EGF receptors with coated pit adaptins. Science 261(5121):612–615
Sorkin A, Goh LK (2008) Endocytosis and intracellular trafficking of ErbBs. Exp Cell Res 314(17):3093–3106. https://doi.org/10.1016/j.yexcr.2008.08.013
Sorkin A, Krolenko S, Kudrjavtceva N, Lazebnik J, Teslenko L, Soderquist AM, Nikolsky N (1991) Recycling of epidermal growth factor-receptor complexes in A431 cells: identification of dual pathways. J Cell Biol 112(1):55–63
Sousa LP, Lax I, Shen H, Ferguson SM, De Camilli P, Schlessinger J (2012) Suppression of EGFR endocytosis by dynamin depletion reveals that EGFR signaling occurs primarily at the plasma membrane. Proc Natl Acad Sci U S A 109(12):4419–4424. https://doi.org/10.1073/pnas.1200164109

Spangler JB, Neil JR, Abramovitch S, Yarden Y, White FM, Lauffenburger DA, Wittrup KD (2010) Combination antibody treatment down-regulates epidermal growth factor receptor by inhibiting endosomal recycling. Proc Natl Acad Sci U S A 107(30):13252–13257. https://doi.org/10.1073/pnas.0913476107

Stern KA, Place TL, Lill NL (2008) EGF and amphiregulin differentially regulate Cbl recruitment to endosomes and EGF receptor fate. Biochem J 410(3):585–594. https://doi.org/10.1042/BJ20071505

Stoscheck CM, Carpenter G (1984) Characterization of the metabolic turnover of epidermal growth factor receptor protein in A-431 cells. J Cell Physiol 120(3):296–302. https://doi.org/10.1002/jcp.1041200306

Suetsugu S, Yamazaki D, Kurisu S, Takenawa T (2003) Differential roles of WAVE1 and WAVE2 in dorsal and peripheral ruffle formation for fibroblast cell migration. Dev Cell 5(4):595–609

Taelman VF, Dobrowolski R, Plouhinec JL, Fuentealba LC, Vorwald PP, Gumper I, Sabatini DD, De Robertis EM (2010) Wnt signaling requires sequestration of glycogen synthase kinase 3 inside multivesicular endosomes. Cell 143(7):1136–1148. https://doi.org/10.1016/j.cell.2010.11.034

Tartaglia M, Gelb BD, Zenker M (2011) Noonan syndrome and clinically related disorders. Best Pract Res Clin Endocrinol Metab 25(1):161–179. https://doi.org/10.1016/j.beem.2010.09.002

Thalappilly S, Soubeyran P, Iovanna JL, Dusetti NJ (2010) VAV2 regulates epidermal growth factor receptor endocytosis and degradation. Oncogene 29(17):2528–2539. https://doi.org/10.1038/onc.2010.1

Thatcher N, Chang A, Parikh P, Rodrigues Pereira J, Ciuleanu T, von Pawel J, Thongprasert S, Tan EH, Pemberton K, Archer V, Carroll K (2005) Gefitinib plus best supportive care in previously treated patients with refractory advanced non-small-cell lung cancer: results from a randomised, placebo-controlled, multicentre study (Iressa Survival Evaluation in Lung Cancer). Lancet 366(9496):1527–1537. https://doi.org/10.1016/S0140-6736(05)67625-8

Tsvetanova NG, von Zastrow M (2014) Spatial encoding of cyclic AMP signaling specificity by GPCR endocytosis. Nat Chem Biol 10(12):1061–1065. https://doi.org/10.1038/nchembio.1665

Ullrich A, Coussens L, Hayflick JS, Dull TJ, Gray A, Tam AW, Lee J, Yarden Y, Libermann TA, Schlessinger J et al (1984) Human epidermal growth factor receptor cDNA sequence and aberrant expression of the amplified gene in A431 epidermoid carcinoma cells. Nature 309(5967):418–425

Umebayashi K, Stenmark H, Yoshimori T (2008) Ubc4/5 and c-Cbl continue to ubiquitinate EGF receptor after internalization to facilitate polyubiquitination and degradation. Mol Biol Cell 19(8):3454–3462. https://doi.org/10.1091/mbc.E07-10-0988

Vieira AV, Lamaze C, Schmid SL (1996) Control of EGF receptor signaling by clathrin-mediated endocytosis. Science 274(5295):2086–2089

Villasenor R, Nonaka H, Del Conte-Zerial P, Kalaidzidis Y, Zerial M (2015) Regulation of EGFR signal transduction by analogue-to-digital conversion in endosomes. Elife 4. https://doi.org/10.7554/elife.06156

Villasenor R, Kalaidzidis Y, Zerial M (2016) Signal processing by the endosomal system. Curr Opin Cell Biol 39:53–60. https://doi.org/10.1016/j.ceb.2016.02.002

Vincenzi B, Schiavon G, Silletta M, Santini D, Tonini G (2008) The biological properties of cetuximab. Crit Rev Oncol Hematol 68(2):93–106. https://doi.org/10.1016/j.critrevonc.2008.07.006

Wandinger-Ness A, Zerial M (2014) Rab proteins and the compartmentalization of the endosomal system. Cold Spring Harb Perspect Biol 6(11):a022616. https://doi.org/10.1101/cshperspect.a022616

Wang Y, Roche O, Yan MS, Finak G, Evans AJ, Metcalf JL, Hast BE, Hanna SC, Wondergem B, Furge KA, Irwin MS, Kim WY, Teh BT, Grinstein S, Park M, Marsden PA, Ohh M (2009) Regulation of endocytosis via the oxygen-sensing pathway. Nat Med 15(3):319–324. https://doi.org/10.1038/nm.1922

Waterman H, Sabanai I, Geiger B, Yarden Y (1998) Alternative intracellular routing of ErbB receptors may determine signaling potency. J Biol Chem 273(22):13819–13827

Waterman H, Alroy I, Strano S, Seger R, Yarden Y (1999a) The C-terminus of the kinase-defective neuregulin receptor ErbB-3 confers mitogenic superiority and dictates endocytic routing. EMBO J 18(12):3348–3358. https://doi.org/10.1093/emboj/18.12.3348

Waterman H, Levkowitz G, Alroy I, Yarden Y (1999b) The RING finger of c-Cbl mediates desensitization of the epidermal growth factor receptor. J Biol Chem 274(32):22151–22154

Waterman H, Katz M, Rubin C, Shtiegman K, Lavi S, Elson A, Jovin T, Yarden Y (2002) A mutant EGF-receptor defective in ubiquitylation and endocytosis unveils a role for Grb2 in negative signaling. EMBO J 21(3):303–313. https://doi.org/10.1093/emboj/21.3.303

West M, Zurek N, Hoenger A, Voeltz GK (2011) A 3D analysis of yeast ER structure reveals how ER domains are organized by membrane curvature. J Cell Biol 193(2):333–346. https://doi.org/10.1083/jcb.201011039

White IJ, Bailey LM, Aghakhani MR, Moss SE, Futter CE (2006) EGF stimulates annexin 1-dependent inward vesiculation in a multivesicular endosome subpopulation. EMBO J 25(1):1–12. https://doi.org/10.1038/sj.emboj.7600759

Wiley HS (1988) Anomalous binding of epidermal growth factor to A431 cells is due to the effect of high receptor densities and a saturable endocytic system. J Cell Biol 107(2):801–810

Wiley HS, Herbst JJ, Walsh BJ, Lauffenburger DA, Rosenfeld MG, Gill GN (1991) The role of tyrosine kinase activity in endocytosis, compartmentation, and down-regulation of the epidermal growth factor receptor. J Biol Chem 266(17):11083–11094

Wiley HS, Shvartsman SY, Lauffenburger DA (2003) Computational modeling of the EGF-receptor system: a paradigm for systems biology. Trends Cell Biol 13(1):43–50

Wilson KJ, Gilmore JL, Foley J, Lemmon MA, Riese DJ 2nd (2009) Functional selectivity of EGF family peptide growth factors: implications for cancer. Pharmacol Ther 122(1):1–8. https://doi.org/10.1016/j.pharmthera.2008.11.008

Wilson KJ, Mill C, Lambert S, Buchman J, Wilson TR, Hernandez-Gordillo V, Gallo RM, Ades LM, Settleman J, Riese DJ 2nd (2012) EGFR ligands exhibit functional differences in models of paracrine and autocrine signaling. Growth Factors 30(2):107–116. https://doi.org/10.3109/08977194.2011.649918

Wollert T, Yang D, Ren X, Lee HH, Im YJ, Hurley JH (2009) The ESCRT machinery at a glance. J Cell Sci 122(Pt 13):2163–2166. https://doi.org/10.1242/jcs.029884

Wong AJ, Ruppert JM, Bigner SH, Grzeschik CH, Humphrey PA, Bigner DS, Vogelstein B (1992) Structural alterations of the epidermal growth factor receptor gene in human gliomas. Proc Natl Acad Sci U S A 89(7):2965–2969

Worthylake R, Opresko LK, Wiley HS (1999) ErbB-2 amplification inhibits down-regulation and induces constitutive activation of both ErbB-2 and epidermal growth factor receptors. J Biol Chem 274(13):8865–8874

Wu WJ, Tu S, Cerione RA (2003) Activated Cdc42 sequesters c-Cbl and prevents EGF receptor degradation. Cell 114(6):715–725

Yamamoto H, Komekado H, Kikuchi A (2006) Caveolin is necessary for Wnt-3a-dependent internalization of LRP6 and accumulation of beta-catenin. Dev Cell 11(2):213–223. https://doi.org/10.1016/j.devcel.2006.07.003

Yamamoto H, Sakane H, Michiue T, Kikuchi A (2008) Wnt3a and Dkk1 regulate distinct internalization pathways of LRP6 to tune the activation of beta-catenin signaling. Dev Cell 15(1):37–48. https://doi.org/10.1016/j.devcel.2008.04.015

Yarden Y, Pines G (2012) The ERBB network: at last, cancer therapy meets systems biology. Nat Rev Cancer 12(8):553–563. https://doi.org/10.1038/nrc3309

Yun CH, Boggon TJ, Li Y, Woo MS, Greulich H, Meyerson M, Eck MJ (2007) Structures of lung cancer-derived EGFR mutants and inhibitor complexes: mechanism of activation and insights into differential inhibitor sensitivity. Cancer Cell 11(3):217–227. https://doi.org/10.1016/j.ccr.2006.12.017

Zhang X, Gureasko J, Shen K, Cole PA, Kuriyan J (2006) An allosteric mechanism for activation of the kinase domain of epidermal growth factor receptor. Cell 125(6):1137–1149. https://doi.org/10.1016/j.cell.2006.05.013

Zhang X, Pickin KA, Bose R, Jura N, Cole PA, Kuriyan J (2007) Inhibition of the EGF receptor by binding of MIG6 to an activating kinase domain interface. Nature 450(7170):741–744. https://doi.org/10.1038/nature05998
Zhao S, Sedwick D, Wang Z (2015) Genetic alterations of protein tyrosine phosphatases in human cancers. Oncogene 34(30):3885–3894. https://doi.org/10.1038/onc.2014.326
Zoncu R, Perera RM, Balkin DM, Pirruccello M, Toomre D, De Camilli P (2009) A phosphoinositide switch controls the maturation and signaling properties of APPL endosomes. Cell 136(6):1110–1121. https://doi.org/10.1016/j.cell.2009.01.032
Zwick E, Bange J, Ullrich A (2001) Receptor tyrosine kinase signalling as a target for cancer intervention strategies. Endocr Relat Cancer 8(3):161–173

Chapter 10
Evolving View of Membrane Trafficking and Signaling Systems for G Protein-Coupled Receptors

Silvia Sposini and Aylin C. Hanyaloglu

Contents

Abstract The G protein-coupled receptor (GPCR) superfamily activates complex signal pathways, yet untangling these signaling systems to understand how specificity in receptor signaling pathways is achieved, has been a challenging question. The roles of membrane trafficking in GPCR signal regulation has undergone a recent paradigm shift, from a mechanism that programs the plasma membrane G protein signaling profile to providing distinct signaling platforms critical for specifying receptor function in vivo. In this chapter, we discuss this evolution of our understanding in the endocytic trafficking systems employed by GPCRs, and how such systems play a deeply integrated role with signaling. We describe recent studies that suggest that the endomembrane compartment can provide a mechanism to both specify, and yet also diversify, GPCR signal transduction. These new evolving models could aid mechanistic understanding of complex disease and provide novel therapeutic avenues.

S. Sposini · A. C. Hanyaloglu (✉)
Institute of Reproductive and Developmental Biology, Department of Surgery and Cancer, Hammersmith Campus, Imperial College London, Du Cane Road, London W12 0NN, UK
e-mail: a.hanyaloglu@imperial.ac.uk

C. Lamaze and I. Prior (eds.), *Endocytosis and Signaling*, Progress in Molecular and Subcellular Biology 57, https://doi.org/10.1007/978-3-319-96704-2_10

Abbreviations

AIP1	Actin-interacting protein 1
AKT (PKB)	Protein kinase B
ALIX	ALG-interacting protein X
AMSH	Associated molecule with the SH3 domain of STAM
AP2	Adaptor protein 2
APPL1	Adaptor protein containing PH domain, PTB domain, and leucine zipper motif
ARRDC	Arresting domain containing
B1AR, B2AR	Beta adrenergic receptor 1 or 2
CB1	Cannabinoid receptor 1
CCP	Clathrin-coated pit
CXCR4	Chemokine receptor 4
DOR	δ-opioid receptor
EE	Early endosome
EGF	Epidermal growth factor
EGFR	EGF receptor
EPB 50	ERM-binding phosphoprotein 50
ERM	Ezrin–radixin–moesin
ESCRT	Endosomal sorting complex required for transport
GAP	GTPase-activating protein
GASP-1	GPCR-associated sorting protein-1
GDP	Guanosine diphosphate
GEF	GDP exchange factor
GIPC	Gαi-interacting protein C-terminus
GPCR	G protein-coupled receptors
GRK	GPCR kinase
GSK3	Glycogen synthase kinase 3
GTP	Guanosine triphosphate
HGF	Hepatocyte growth factor
Hrs	HGF-regulated tyrosine kinase substrate
LHR	Luteinizing hormone receptor
MOR	μ-opioid receptor
MVB	Multivesicular body
PAR	Protease-activated receptor
PDZ	Post-synaptic density 95/disk large/zonula occludens-1
PI3P	Phosphatidylinositol-3 phosphate
PKA	Protein kinase A
PKC	Protein kinase C
PSD	Post-synaptic density protein 95
PTHR	Parathyroid hormone receptor
RGS	Regulator of G protein signaling
SNX27	Sorting nexin 27

STAM	Signal-transducing adaptor protein
STAT	Signal transducer and activator of transcription
TAB1	TGF-beta-activated kinase 1-binding protein
TGN	Trans-Golgi network
TM	Transmembrane
TSHR	Thyroid-stimulating hormone receptor
UBD	Ubiquitin-binding domain
UBPY	Ubiquitin-specific processing protease Y
V2R	V2 vasopressin receptor
VASP	Vasodilator-stimulated phosphoprotein
VEE	Very early endosome
Vps	Vacuolar protein sorting
WASH	Wiskott–Aldrich syndrome protein and SCAR homolog
Wnt	Wingless-related integration site

10.1 Introduction

G protein-coupled receptors (GPCRs) represent the largest family of signaling receptors, encoded by more than 2% of the human genome (Fredriksson et al. 2003). Classified by their common architecture of a single polypeptide that traverses the membrane seven times, with an extracellular N-terminus and intracellular C-terminus of highly varying size and structure, they are also referred to as 7 transmembrane (7TM) receptors. GPCRs are categorized into six separate sub-families based on their sequence and functional similarities. The majority of receptors are in Class A/rhodopsin-family, and the remainder classified into a further four families of secretin (Class B), glutamate (Class C), adhesion, and frizzled (Fredriksson et al. 2003). GPCR ligands are chemically diverse, ranging from photons, single amino acids, ions, odorants to hormones, enzymes, neurotransmitters, and lipids. With >800 GPCRs in humans, coupled with their widespread expression, they represent key signaling molecules in all physiological systems. Their reputation as a tractable and successful drug target (Stevens et al. 2013) means that fundamental mechanisms controlling these receptors are of high interest to identify avenues that increase target specificity in vivo.

The classic view of GPCR signaling pathways is inconsistent with the diverse functions they play in vivo. Cellular 'atlases' of GPCR expression in diverse tissues suggest >100 different GPCRs are expressed in any one cell type (Insel et al. 2015); however, there are only four main heterotrimeric G protein pathways these receptors activate (see below). The current models depict GPCR signaling to be highly complex, and so now the challenge is to understand how cells decode this complexity into downstream specific responses (Stallaert et al. 2011; Ferre et al. 2014; Rankovic et al. 2016). One mechanism that has emerged not only to diversify signaling, but also as a means for cells to 'decode' or specify complex GPCR signaling pathways is endocytic membrane trafficking. In this chapter, we will give an overview of how membrane trafficking has evolved our understanding of the fundamental mechanisms driving GPCR signal activity in vivo.

10.2 From Classic to Complex—The Pluridimensionality of GPCR Signaling

10.2.1 Heterotrimeric G Protein Signaling Activated by GPCRs

The primary mechanism for GPCRs to signal is via coupling to heterotrimeric G proteins composed of Gα, Gβ, and Gγ with Gα binding the guanine nucleotides GTP and GDP. The prototypic model of GPCR signal activation depicts individual cell surface receptors binding its extracellular ligand and subsequently activating a specific, but linear, heterotrimeric G protein pathway that converge on to common downstream signal pathways. Both Gα and Gγ are lipidated and so are tethered to the plasma membrane, with Gβ tightly associated with Gγ. Although G proteins activate distinct effectors they share a core mechanism of activation. In the inactive state, the Gα subunit is bound to GDP and is associated with Gβγ subunits, with the latter heterodimer increasing the affinity of Gα to GDP. Ligand-induced activation of a GPCR causes conformational changes enabling the receptor to act as a guanine nucleotide exchange factor (GEF) leading to the Gα-bound GDP to be exchanged for GTP (Fig. 10.1a). The GTP-bound Gα dissociates from the Gβγ dimer where each can then go on to modulate the activity of various downstream effectors. The hydrolysis of GTP to GDP is induced by GTPase-activating proteins (GAPs). Gα has weak intrinsic GTPase activity, thus additional molecules called Regulator of G protein signaling (RGS) facilitate this process, triggering heterotrimeric complex re-association and the return of the system to the basal state (Fig. 10.1a). There are four main types of Gα family proteins, $G\alpha_s$, $G\alpha_{i/o}$, $G\alpha_{q/11}$, and $G\alpha_{12}/G\alpha_{13}$, that initiate a specific signaling cascade through their respective effectors (Fig. 10.1b) (Birnbaumer 2007). Although primarily ubiquitous in expression, certain members of the Gα family are cell type specific. However, given the physiological significance and large number of receptors, drove the field to try and understand how a limited number of G proteins mediate such diverse responses in distinct tissues. It is now appreciated that GPCR signaling is highly complex (see below). The question now is how does an individual cell translate such complexity in signal pathways into specific downstream responses. Over the past decade, the spatial control of GPCR signaling, via endocytic trafficking, has emerged as a mechanism to contribute to GPCR signal complexity yet, also provide signal specificity, both of which will be discussed further in this chapter.

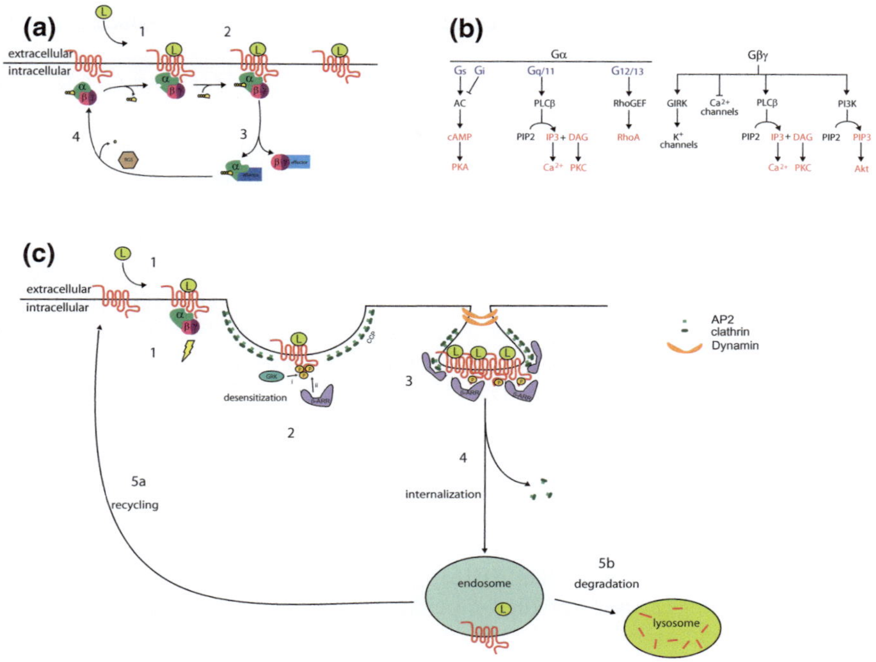

Fig. 10.1 **a** Model of heterotrimeric G protein activation. Following ligand binding (1), the GPCR activates its cognate heterotrimeric (α,β,γ) G protein by promoting the exchange of GDP for GTP on the α subunit (2). α and β/γ subunits disassociate to activate signal transduction via distinct cellular effectors (3). Regulators of G protein signaling (RGS) stimulate the GTPase activity of Gα, thus the re-association with β/γ and the return to the basal state (4). **b** Schematic representation of different G protein subunits and the signaling pathways they activate. There are four main Gα classes (blue): stimulatory Gα (Gαs) activates adenylate cyclase (AC) which converts ATP into cAMP; inhibitory Gα (Gαi), inhibits AC; Gαq/11 activates phospholipase C beta (PLCβ) that converts phosphatidylinositol phosphate 2 (PIP2) into diacylglycerol (DAG) and inositol triphosphate (IP3) which activate PKC and trigger calcium (Ca^{2+}) release from intracellular stores, respectively; Gα11/12 activates members of the Rho family of GTPases. Second messengers are in red. **c** Archetypal model of GPCR activation, internalization, and post-endocytic sorting. The ligand-activated receptor (1) is phosphorylated on multiple intracellular sites by GPCR kinases (GRKs) (2i) resulting in recruitment and binding of β-arrestin (2ii). β-arrestin uncouples receptors from their G proteins, leading to desensitization and enables receptor clustering to clathrin-coated pits (CCPs) (3). Dynamin-mediated CCP scission from the plasma membrane and subsequent clathrin uncoating results in internalized receptor sorted to endosomes (4). From here GPCRs can be divergently sorted back to the plasma membrane (5a) or to lysosomes (5b). Figure modified from (Hanyaloglu 2018)

10.2.2 Mechanisms Mediating GPCR Signal Diversity

As stated above, GPCR signal pathways are complex, and in order to discuss the key role of membrane trafficking in this process, one must first briefly describe the potential mechanisms that result in pleiotropic signaling, referring the reader to recent reviews that cover specific mechanisms of GPCR signal diversity, as GPCR endocytic trafficking is the primary focus.

One well-studied, though still controversial, mechanism for how GPCRs can diversify their signaling is via receptor dimerization or oligomerization, to form either homomeric or heteromeric complexes. The controversy in this field of study stems in part to the technical approaches and cellular systems employed; however, it has been demonstrated that homomerization of certain GPCRs is a physiologically relevant form of signaling in vivo. The ability of GPCRs to form homomers has been proposed as a means to amplify signal responses, enabling a platform for multiple, or even distinct, G protein couplings (Jonas et al. 2015; Guo et al. 2008). The ability of two, or more, distinct GPCRs to associate as heteromers can diversify signaling as essentially the heteromer is a new functional receptor unit, with properties distinct from their homomeric counterparts (Ji et al. 2017; Coke et al. 2016; Feng et al. 2013). These complexes have been reported to exhibit altered ability to traffic newly synthesized receptors to the cell surface, distinct pharmacological profiles, and altered G protein-coupling and endocytosis. Targeting GPCR homomers versus heteromers has attracted interest as a potential pharmacological target as certain off-target/side effects of therapeutic drugs, or distinct diseases, involve the organization of receptors into these complexes (Gomes et al. 2016).

GPCR signaling can be diversified via a number of additional mechanisms, including coupling to multiple G proteins, receptor splice variants that exhibit distinct signal and regulatory properties, or by GPCRs that have >1 endogenous ligand that may activate distinct signaling profiles (pharmacological bias). However, a key mechanism of current interest is the ability of receptors to signal in a G protein-independent manner via a family of well-studied GPCR adaptor proteins, the arrestins. Arrestins are key molecules in GPCR endocytic trafficking and will be discussed below in Sect. 10.3; however, their function as signaling scaffolds is an area of strategic focus in drug discovery. There are four arrestin isoforms, with arrestins 1 and 4 specifically restricted to the retina. Arrestins 2 and 3 (also termed β-arrestin1 and β-arrestin2 historically due to initial studies first identifying roles in regulating β2-adrenergic receptor signaling) are ubiquitously found outside of the retina. These non-visual arrestins interact with a number of different GPCRs, specifically by binding to the activated and phosphorylated receptor to mediate receptor rapid desensitization of G protein signaling (Luttrell and Lefkowitz 2002; Pierce and Lefkowitz 2001). However, their ability to act as scaffolds for a variety of signaling proteins (e.g., ERK1/2, AKT, JNKs) to mediate distinct cellular functions such as survival, apoptosis, and cell migration (Xiao et al. 2007, 2010) has attracted attention for the potential to identify ligands that could discriminate between G protein and arrestin-dependent signaling (Rankovic et al. 2016; Reiter et al. 2012). While it was thought that GPCRs can couple to this G protein-independent form of signaling without detectable G protein signaling, recent studies have challenged this model, demonstrating that active G proteins are still essential for arrestin-dependent signaling for certain receptors (Grundmann et al. 2018; Wang et al. 2017). These recent studies will assist in assessing how biased signaling is exploited by the drug discovery industry. For GPCRs such as the μ-opioid receptor (MOR), biased signaling is a valid therapeutic strategy as it has been demonstrated in vivo that ligands biased to G protein, but not arrestin-dependent signaling (or at least recruit arrestin to the active receptor), have potent analgesic properties without the unwanted side effects of respiratory depression and dysfunctional gastrointestinal actions (De Wire et al. 2013).

10.3 Endocytic and Post-endocytic Sorting of GPCRs

In this section, we will provide an overview of the endocytic pathways GPCRs traverse following ligand-dependent activation, and our current understanding of the mechanisms mediating divergent sorting between the plasma membrane recycling pathway and lysosomal degradation. One core principle of these mechanisms is that GPCR post-endocytic sorting is a tightly regulated process that occurs via multisteps, enabling receptor sorting and thus signal activity of a GPCR, to be reprogrammed (Hanyaloglu and von Zastrow 2008). The impact of post-endocytic sorting on GPCR signaling classically involves regulation of cell signaling described in Sect. 10.2.1. In latter sections, we will discuss how studies over the past decade have unveiled additional roles of the endomembrane system in generating GPCR signaling, and its physiological/pathophysiological role.

10.3.1 GPCR-Mediated Internalization to Desensitize or not Desensitize

Following ligand-dependent activation of cell surface GPCR/G protein signaling, many GPCRs undergo rapid desensitization via a very well-studied mechanism involving receptor phosphorylation on intracellular serine/threonines, via a member of the GPCR kinase (GRK) family and recruitment of arrestins (Fig. 10.1c). Arrestins not only mediate rapid desensitization, via uncoupling the receptor from its cognate G protein, but also mediate rapid ligand-induced internalization via clathrin-coated pits (CCPs). This contributes to acute signal termination and occurs via arrestin's ability to bind GPCR, the β-subunit of the adaptor protein 2 (AP2) and clathrin heavy chain (Moore et al. 2007; Barki-Harrington and Rockman 2008). Following the recruitment of GPCRs into CCPs, the scission of CCPs to clathrin-coated vesicles from the plasma membrane requires the large GTPase dynamin, resulting in GPCR internalization to the endosomal compartment (Fig. 10.1c).

While the above described model is utilized by many receptors, GPCRs greatly differ in how they engage this mechanism, from the type of GRK and arrestin employed, to whether arrestin co-internalizes or transiently associates with its GPCR at the plasma membrane and to internalization via CCPs independent of GRKs and arrestins, e.g., the gonadotrophin-releasing hormone receptor and protease-activated receptor-1 (McArdle et al. 2002; Paing et al. 2002). Receptor-dependent differences can contribute to signal diversity in GPCR/G protein signaling by regulating the spatial-temporal patterns of these pathways. Another example is the reported heterogeneity in CCPs, whereby GPCRs can be differentially organized within distinct CCPs to mediate the subsequent post-endocytic fate of the receptor between recycling or degradative pathways. CCPs with GPCRs targeted to the recycling pathway contained GRKs/arrestins, while CCPs where GPCRs were phosphorylated by second messenger kinases mediated sorting to the lysosome for degradation (Mundell et al. 2006; Lakadamyali et al. 2006). CCP specialization may also function as signaling microdomains and supported by the observations that GPCRs temporally regulate their time in CCPs prior to dynamin-dependent endocytosis, termed residency time. By interacting with scaffold proteins of the post-synaptic density 95 (PSD95)/disks large (Dlg)/zonula occludens-1 (Zo-1) (PDZ) family, the β2-adrenergic receptor (β2AR) extends its residency time via the GPCR/PDZ interaction with cortical actin and thus delaying the dynamin recruitment (Puthenveedu and von Zastrow 2006). Other GPCRs such as the MOR, which do not have a PDZ-interacting sequence within its intracellular C-terminal tail (C-tail), delay its residency time by regulating the scission of the CCP by dynamin. Interestingly, this mechanism is mediated by receptor ubiquitination to dictate MOR's CCP residency time, with non-ubiquitinated receptors displaying a pronged residency time, while ubiquitination on lysine residues on the first intracellular loop of MOR by the E3 ligase Smurf (recruited with arrestin to the receptor) triggers scission of the CCP (Henry et al. 2012). The significance of altering CCP residency time of a GPCR to downstream cellular function has been demonstrated for the cannabinoid receptor type 1 (CB1). For CB1, its CCP residency time enables formation of arrestin signalosomes to spatial-temporally regulate mitogenic signals that impacts the regulation of genes mediating contrasting pro-survival and pro-apoptotic functions (Delgado-Peraza et al. 2016; Flores-Otero et al. 2014). Temporally sustained arrestin signalosomes at the CCP may not even require maintaining receptor within the CCP. Recent studies have demonstrated that the β1-adrenergic receptor (β1AR), which internalizes more slowly and clusters poorly compared to GPCRs such as the β2AR, recruits arrestin-3 to the activated receptor yet rapidly dissociates leaving arrestin in its active form to CCPs (Eichel et al. 2016). A recent follow-up study has identified the structural mechanism mediating this transient association between GPCR and arrestin, yet sustained between arrestin and the CCP. The transient association of arrestin with β1AR occurs via the receptor core, without involving the C-tail. This interaction results in arrestin maintaining an active conformation and retention of arrestin in CCPs through interactions with membrane phospholipids and clathrin proteins, even after receptor has disassociated (Eichel et al. 2018). This is consistent with prior reports of visual arrestin exhibiting abilities to insert into the membrane (Lally et al. 2017). Although not assessed in these recent studies, the β1AR C-tail has known interactions with specific PDZ proteins (PSD95, MAGI-2, GIPC (He et al. 2006)), which could regulate clustering prior to receptor/arrestin disassociation or facilitate a plasma membrane scaffold between β1AR and activated arrestin in CCPs.

Although arrestin and CCP adaptors AP2 represent core machinery that drive receptor clustering into CCPs and regulate internalization for many different receptors, whether there are additional factors that can positively, or negatively, direct receptor clustering and internalization remains an outstanding question. Post-translational modifications such as ubiquitination and palmitoylation are known to regulate endocytosis, suggesting additional adaptor proteins are likely involved. Although arrestin is also a key factor in accessibility of enzymes, such as E3 ubiquitin ligases, to induce these post-translational modifications of the receptor (Girnita et al. 2005; Shenoy et al. 2009; Shenoy 2014), the identify of additional binding partners is poorly understood, particularly for those GPCRs that do not employ the ubiquitous GPCR adaptor protein arrestin. Studies on clathrin and arrestin-independent endocytic pathways have provided understanding on how additional pathways mediate GPCR internalization. Endophilin-A, previously thought to be involved in clathrin-mediated endocytosis, in fact drives a distinct dynamin-dependent, but clathrin/arrestin-independent internalization pathway termed fast endophilin-mediated endocytosis (FEME) of a subset of GPCRs. This rapid (seconds) endocytosis is characterized by their ability to bind the SH3 domain of endophilin-A to the third intracellular loop of different receptors, including the β1AR, D3, and D4 dopamine receptors and muscarinic acetylcholine receptor 4. As FEME mediates endocytosis of distinct cargoes, in addition to GPCRs, suggests this form of endocytosis could potentially mediate/regulate diverse cellular functions (Boucrot et al. 2015).

10.3.2 Post-endocytic Sorting of GPCRs to the Recycling Pathway

Following endocytosis, many kinds of cargo, including GPCRs, are trafficked to early endosomes (EEs) from which they are then sorted to distinct opposing fates; plasma membrane recycling or lysosomal-mediated degradation. The decision to divergently sort GPCRs has significant consequences in shaping the signaling response from that activated receptor, and thus are important pathways physiologically and in disease where GPCR sorting is perturbed (Dunn and Ferguson 2015; Hanyaloglu and von Zastrow 2008). As will be discussed in Sect. 10.4, their role in signaling is not only important in temporally defining plasma membrane G protein signaling (Hanyaloglu and von Zastrow 2008) but also by providing additional intracellular platforms to mediate signaling. For more than 15 years, researchers have been deciphering the machinery that directs receptor fate, and what is apparent from this body of work is that post-endocytic sorting is tightly regulated at multiple levels. GPCRs sorted to a plasma membrane recycling pathway are important in resensitization, or recovery, of cell surface signaling. This impacts receptor function both physiologically and pathophysiologically, as β2AR recycling is important in regulating cardiac myocyte contractility (Xiang and Kobilka 2003), while for the MOR, recycling is implicated in opiate tolerance (Roman-Vendrell et al. 2012). However, our current understanding of how GPCRs undergo this sorting event depicts a mechanism requiring a highly complex series of events. A common requirement for GPCR recycling, compared to other membrane cargo that traverse the same recycling pathways in a 'default' manner (such as the transferrin receptor), is that is regulated via specific cis-acting sorting sequences in the GPCR distal C-tail (also termed sequence-directed recycling) (Hanyaloglu and von Zastrow 2008; Trejo 2005). This regulated form of recycling was first identified for the β2AR, where the C-tail recycling sequence conferring to a type 1 PDZ ligand was essential for recycling, without which the receptor was rerouted to the degradative pathway (Hirakawa et al. 2003b; Cao et al. 1999). Another feature of these recycling sequences was that they are alone sufficient for recycling, as transplanting them on to GPCRs that are targeted for degradation, such as the δ-opioid receptor (DOR), reroutes their post-endocytic trafficking to the recycling pathway (Cao et al. 1999; Gage et al. 2001). The initial studies with β2AR suggested the PDZ protein that mediated this regulated recycling was ezrin–radixin–moesin (ERM)-binding phosphoprotein-50 (EPB50), also called Na^+/H^+ exchange regulatory factor 1 (NHERF-1). However, GPCRs with PDZ ligands have been shown to interact with more than one PDZ protein (He et al. 2006), and EPB50/NHERF-1 is a PDZ protein primarily localized to the plasma membrane, or juxta-membrane, regions of the cell. This then directed subsequent studies to focus on the PDZ protein sorting nexin-27 (SNX27), due to its ability to bind PI3P, an EE membrane-enriched lipid. SNX27 was found to be essential for β2AR recycling (Lauffer et al. 2010). Many GPCRs targeted to the recycling pathway have specific sequences in their C-tails that are essential for trafficking to this pathway. Intriguingly, they are heterogeneous with many not corresponding to PDZ type 1 ligands, and thus many have unknown interacting partners. We have previously proposed that this high diversity in recycling sequences suggests receptor specificity in regulating GPCR recycling, in other words, a cell has the potential to modulate sorting of specific receptors, or indeed more globally, by altering expression of receptor-specific or core endosomal sorting machinery respectively (Marchese et al. 2008; Hanyaloglu and von Zastrow 2008).

That recycling of membrane cargo could occur via two distinct modes, regulated and default, combined with the knowledge that multiple distinct GPCR recycling sequences exist, led to the pursuit of whether there are distinct mechanisms of recycling for these different cargos and/or common machinery. The studies to date suggest involvement of both core- and receptor-specific mechanisms and unexpectedly identified that regulated GPCR recycling did not only require a protein interaction with the distal C-tail recycling sequence, but a more complex mechanism, involving additional upstream sequences in the receptor C-tail and post-translational modifications driven through the receptors own signal activity (Hanyaloglu and von Zastrow 2008). One protein that represents a component of the core mechanism mediating GPCR post-endocytic sorting is the early endosome-localized HGF-regulated tyrosine kinase substrate (Hrs), also termed Vps27 (a component of ESCRT 0—see Sect. 10.3.3 below). Hrs was established as a protein mediating the sorting of different membrane cargo, including GPCRs to the lysosome for degradation, but is also essential for sequence-directed, or regulated, recycling of the β2AR, MOR, and calcitonin-receptor-like receptor (Hanyaloglu et al. 2005; Hasdemir et al. 2007). The N-terminal VHS domain of Hrs, so termed by its conserved presence in yeast Hrs, Vps27, Hrs, and the Hrs-interacting partner STAM, had prior to these studies an unknown function, yet this region of Hrs mediated the recycling of these distinct GPCRs, although whether this was through distinct protein interactions via the VHS domain remains to be determined (Hanyaloglu et al. 2005; Hasdemir et al. 2007; Huang et al. 2009). Additional core sorting proteins could also be the arrestins, although opposing actions on receptor recycling for different GPCRs have been reported. While recycling of the N-formyl peptide receptor requires β-arrestin (Vines et al. 2003), for the glucose-dependent insulinotropic polypeptide receptor (GIPR), arrestin negatively regulates rapid recycling and instead promotes its slow sorting via the trans-Golgi network (TGN), leading to downregulation of this GPCR (Abdullah et al. 2016). The retromer is a core endosomal complex well-characterized in endosomal-TGN transport of membrane cargo, but has recently been shown to be a key component of sorting GPCRs to the plasma membrane recycling pathway and directing receptors into specific endosomal tubules that mediate this regulated recycling (Puthenveedu et al. 2010). For the β2AR, its interaction with the PDZ protein SNX27 enables this receptor to be targeted to the retromer complex, via the ability of SNX27 to interact with Wiskott–Aldrich syndrome protein and SCAR Homolog (WASH) protein complex (Temkin et al. 2011; McGarvey et al. 2016). These studies also identified the endosomal mechanisms for GPCR sorting between two modes of recycling, regulated and default, via the physical organization into tubules differing in both their protein complement, with endosomal actin, WASH and retromer present in tubules mediating regulated GPCR recycling, and kinetics, with default recycling tubules forming in a more transient manner compared to regulated recycling (Puthenveedu et al. 2010). At the level of the receptor, there are also specific sequences that dictate sorting between these two modes of recycling. For the β2AR, these represent acidic dileucine-like sequences located upstream in the C-tail of the distal β2AR PDZ ligand, whereby mutation of this dileucine-like sequence did not affect overall ability of this receptor to recycle, but rerouted the receptor to a PDZ-independent, default pathway (Hanyaloglu and von Zastrow 2007). Likewise, the phosphorylation on the proximal region of the β2AR C-tail by PKA switches the receptor between default and sequence-dependent recycling tubules, highlighting the role of β2AR-Gαs-cAMP pathway which negatively regulates entry into the sequence-dependent recycling tubule (Vistein and Puthenveedu 2013). In contrast, PKA phosphorylation of the β1AR C-tail drives regulated recycling (Nooh and Bahouth 2017). Collectively, these studies highlight the complexity and intricate regulation mediating the sorting of these receptors into the plasma membrane recycling pathway. It also indicates that there are perhaps additional cellular functions of this complex pathway beyond returning receptors back to the plasma membrane for signal resensitization, especially when there is a simpler, default, recycling pathway employed by other membrane cargo. While we have previously proposed a model for this complexity in the recycling pathway in enabling reprogramming of GPCRs to multiple distinct fates, particularly in cells exposed to dynamic extracellular environment (Hanyaloglu and von Zastrow 2008; West and Hanyaloglu 2015), it is also possible that there are additional, perhaps receptor-specific signal functions in this multistep regulated recycling pathway. Certainly, this ability to reprogram the trafficking fate of a GPCR is being assessed at a pharmacological level with a recent study proposing these sorting pathways could represent a screening platform to identify novel compounds that discriminate between default and regulated recycling (Nooh et al. 2016).

The above studies have identified intricate molecular mechanisms directing regulated GPCR recycling, where the EE is the primary sorting platform. As stated above, the sequences that mediate GPCR recycling are heterogenous, possibly suggesting these distinct GPCR recycling sequences, and potentially distinct interacting partners, may have additional functions. One function that has been identified is to organize GPCRs into endosome compartments distinct from the classic EE for its post-endocytic sorting. The human luteinizing hormone receptor (LHR), follicle-stimulating hormone receptor and the β1AR, primarily traffic to physically smaller endosomes (~third of the size of EEs) that do not contain the EE markers such as the GTPase Ras-related protein Rab5, PI3P, and EEA1. We have termed these endosomes, very early endosomes (VEEs) (Jean-Alphonse et al. 2014), not only to distinguish from the EE, but also as these GPCRs must traffic to VEEs to be sorted to the recycling pathway. Here, the interaction between GPCR recycling sequence and the PDZ protein, GIPC (Gαi-interacting protein C-terminus, (Hirakawa et al. 2003b)), mediates recycling because it directs the receptor to the VEE away from the EE (Jean-Alphonse et al. 2014). This may suggest that the term 'recycling sequences' may be a too limited description for their role in receptor trafficking, as certainly these sequences may interact with more than one binding partner, and for the VEE, it is directing receptors to this compartment through associations early on in endocytosis, at the level of receptor clustering in CCPs (Jean-Alphonse et al. 2014). While the VEE does not contain class EE markers, a subpopulation is positive for the adaptor protein containing PH domain, PTB domain, and leucine zipper motif (APPL1) (Jean-Alphonse et al. 2014). Although APPL1 was known as a component of the EE intermediate with Rab5 (Zoncu et al. 2009), the lack of role for Rab5 in VEE-targeted receptors suggested this is a distinct APPL1 compartment (Jean-Alphonse et al. 2014). Recently, we have demonstrated that APPL1 is essential for the recycling of GPCRs from the VEE, a new role for this adaptor protein, and mechanistically this rapid recycling involved the cAMP/PKA signaling activated by the receptor to phosphorylate APPL1 on serine 410 (Sposini et al. 2017), again demonstrating the integrated nature of membrane trafficking and signaling.

10.3.3 Post-endocytic Sorting of GPCRs to the Degradative Pathway

Post-endocytic sorting of GPCRs to the lysosomes for degradation is a mechanism for cells to terminate or downregulate signaling from that receptor, particularly when that signal if remained active, would be detrimental to the cell. A dramatic example of this is for the chemokine receptor CXCR4, where overactive signaling due to altered post-endocytic sorting is associated with invasive breast cancer (Marchese et al. 2008). In addition, GPCRs that undergo recycling if chronically stimulated are rerouted to a degradative pathway as part of the process of receptor downregulation. This is a protective mechanism for cells, but when pharmacologically targeting a GPCR, these innate mechanisms in regulating signaling through downregulation may play a role in drug tachyphylaxis (loss of drug response over time) or tolerance (Tappe-Theodor et al. 2007; Hanyaloglu and von Zastrow 2008). Therefore, the clinical significance of this sorting pathway for GPCRs has driven studies to identify the underlying molecular mechanisms. The mechanisms that sort GPCRs to the lysosomal pathway, like recycling, are intricately regulated, involve multiple steps, and begin early in the endocytic pathway. GPCRs internalized to EEs traffic to Rab7 late endosomes, which then result in involution of receptors to form multivesicular bodies (MVBs). The fusion of MVBs with lysosomes results in cargo degradation. The most well-characterized mechanism for sorting of membrane cargo to a degradative pathway is via ubiquitination at lysine residues and endosomal sorting complex required for transport (ESCRT)-dependent degradation. ESCRT represents four distinct protein complexes (ESCRT-0, I, II, III) with ESCRT-0, I, and II containing proteins with ubiquitin-binding domains (UBDs) that act sequentially to bind and retain ubiquitinated cargo, with the final step leading to involution of cargo in MVBs. The ESCRT complexes are then disassociated from the maturing MVB via the AAA-ATPase Vps4 and the cycle is repeated (Rusten et al. 2011; Henne et al. 2013). For GPCRs that employ these core mechanisms of lysosomal sorting, e.g., CXCR4, protease-activated receptor 2 (PAR2) and β2AR, their ubiquitination occurs early during their endocytosis, whereby ubiquitin E3 ligases associate directly with the GPCR or via arrestin (Kennedy and Marchese 2015). Interestingly, there are also GPCRs that employ distinct mechanisms, whereby even if they are ubiquitinated, this modification is not required for degradation, or for dependence on certain components of the ESCRT machinery (Henry et al. 2011; Hislop et al. 2011; Dores and Trejo 2015). For these GPCRs, their degradation involves interacting with additional machinery via the receptor C-tails with proteins such as GPCR-associated sorting protein-1 (GASP-1) and the autophagy protein Beclin-2 (He et al. 2013). Furthermore, receptor ubiquitination may have functions at distinct steps of the endocytic pathway leading to degradation. For the DOR, ubiquitination via the E3 ligase AIP4 is required for efficient involution of receptor into MVBs to mediate its rapid degradation (Henry et al. 2011; Hislop et al. 2009). Similarly, following chronic agonist stimulation of MOR, receptor ubiquitination also promotes receptor involution into MVBs (Hislop et al. 2011). Interestingly, a recent study has suggested that sorting of GPCRs such as DOR at the level of MVB involution may not necessarily dictate a fate of receptor degradation, and there may be mechanisms to redirect involuted receptors back to the limiting membrane of the MVBs and subsequent traffic to the TGN for plasma membrane recycling (Charfi et al. 2018). Targeting of GPCRs to the degradative pathway may involve ubiquitination of associated adaptor proteins rather than the GPCR, e.g., arrestin, (Mosser et al. 2008), or the ubiquitous adaptor protein ALG-interacting protein X (ALIX) employed by PAR1. PAR1 requires neither receptor ubiquitination, ESCRT 0, I nor GASP-1 for its lysosomal sorting, instead ubiquitination of ALIX enables the formation of an endosomal complex with the arrestin-domain-containing proteins (ARRDCs) ARRDC1 and ARRDC3 (Dores et al. 2012a, b, 2015). This pathway is critical in regulating signaling from PAR1, as loss of ARRDC3 in breast cancer results in inhibition of lysosomal sorting and aberrant persistent signaling from this receptor (Arakaki et al. 2018).

As for sorting of GPCRs to the regulated recycling pathway (Sect. 10.3.2), lysosomal sorting is also regulated by GPCR signaling and even the GPCR signal machinery itself can have direct roles in sorting. Phosphorylation of the dopamine D2 and D3 receptors by the second messenger kinase PKC is involved in mediating its degradation (Zhang et al. 2016; Cho et al. 2013), while the Gαs subunit of heterotrimeric G proteins regulates post-endocytic targeting of GPCRs to a degradative pathway, independent of its GTPase activity (Rosciglione et al. 2014). Gαs was essential for the degradation of CXCR4, DOR, and the angiotensin 1A receptor, which interestingly are GPCRs that exhibit differential dependence on ubiquitin on their sorting. Gαs was shown to be required for receptor involution into MVBs. Although this suggests Gαs mediates a late step in the lysosomal post-endocytic sorting pathway, this G protein formed complexes at the EE with GASP-1, dysbindin, and Hrs, possibly suggesting additional roles upstream of receptor involution into MVBs (Rosciglione et al. 2014). It is clear that as for sorting to the recycling pathway, the mechanisms of lysosomal sorting for GPCRs are also complex and involve multiple steps. Therefore, this also provides a platform for reprogramming the fate of GPCRs targeted for degradation, as evidenced recently with the DOR where receptor can be rerouted at a late stage of sorting (Charfi et al. 2018) and in breast cancer, where lysosomal sorting of PAR1 and CXCR4 is inhibited at an upstream step of sorting (receptor tyrosine kinase Her-2/ErB2 inhibits receptor ubiquitination), resulting in increased surface expression and promotion of more aggressive tumors (Marchese et al. 2008). Overall, the plasticity in the GPCR post-endocytic sorting system represents a key mechanism for regulating cellular sensitivity via altering receptor targeting fate (Mukai et al. 2010; Cadigan 2010). This will be discussed further in Sect. 10.4 in the context of GPCR endosomal signaling.

10.4 Endosomes as GPCR Signaling Platforms

As described in Sect. 10.2, the models of GPCR signaling have significantly evolved. Likewise, so has our understanding in the complexity of post-endocytic trafficking pathways, yet its impact on regulating GPCR signaling was based on a model where GPCRs activate signaling from the plasma membrane. These models have also dramatically changed over the past 9 years whereby post-endocytic pathways not only shape the cell surface signaling response but also directly function as distinct signaling platforms with unique downstream cellular functions. This section will review our current understanding of GPCR endosomal signaling across distinct intracellular compartments, and even endosomal microdomains, and its potential applications in health and disease.

10.4.1 Early Endosomes (EE) as a Key Heterotrimeric G Protein Signaling Hub

The endomembrane provides a distinct platform for receptor signaling from the plasma membrane, as it can provide access to specialized compartments with distinct substrates that in turn can diversify downstream cellular responses, by activating either unique pathways or even how the cell translates signals from the same signaling pathway (e.g., Gαs-cAMP pathway). The first examples of GPCR/G protein signaling from endosomes was for the thyrotropin-stimulating hormone receptor (TSHR) and the parathyroid hormone receptor (PTHR), both Gαs-coupled GPCRs that exhibited a sustained, or persistent levels of cAMP, even when extracellular ligand had been removed and the receptor had been internalized. For these receptors, endosomal G protein signaling contributed to a second phase of cAMP signaling following the more transient plasma membrane signaling response (Calebiro et al. 2009; Ferrandon et al. 2009). For PTHR, subsequent studies have shown that endosomal cAMP signaling is enhanced by arrestin. Despite the well-established role of arrestin in G protein uncoupling at the plasma membrane, recent molecular studies have identified that arrestin can bind receptors at distinct sites, whereby interaction with the receptor core mediates G protein uncoupling, while sustained association with the phosphorylated GPCR C-tail enables formation of what was termed a 'megaplex,' enabling receptor to simultaneously associate with both G protein and arrestin (Thomsen et al. 2016). As for plasma membrane signaling, endosomal G protein signaling must also be regulated and for the PTHR, this signaling is attenuated by endosomal acidification and the retromer complex, which like B2AR is also required to direct this receptor to the recycling pathway (Ferrandon et al. 2009; Feinstein et al. 2011; Gidon et al. 2014; McGarvey et al. 2016) (Fig. 10.2).

Although these studies directly demonstrate that second messenger signaling was persistent and required receptor internalization, the first direct evidence that endosomes exhibit both active receptor and active heterotrimeric G proteins following ligand-induced activation of a GPCR, was for the β2AR. The development of nanobody-based biosensors was instrumental for this, by their ability to bind the active, nucleotide-free form of Gαs (termed nanobody-37) and the active conformation of β2AR (nanobody-80) (Irannejad et al. 2013; Pardon et al. 2014). Recent studies on β2AR endosomal Gαs/cAMP reveal this signaling may be highly organized into endosomal signaling microdomains. Association of the α-arrestin ARRDC3 with EE-localized β2AR enhanced endosomal signaling by negatively regulated receptor sorting into regulated recycling tubules (characterized by SNX27 and retromer) (Fig. 10.2) (Tian et al. 2016). It is unclear whether this function of ARRDC3 is by physical constraint of the receptor in signal microdomains, and/or a direct role of ARRDC3 in facilitating G protein endosomal signaling, akin to the arrestin megaplexes. A distinct study on β2AR endosomal signaling demonstrated that the active β2AR in EEs (visualized by nanobody-80) was detected throughout the limiting membrane of the endosome, while active G protein, via nanobody-37, was specifically in regulated SNX27/WASH recycling tubules (Bowman et al. 2016).

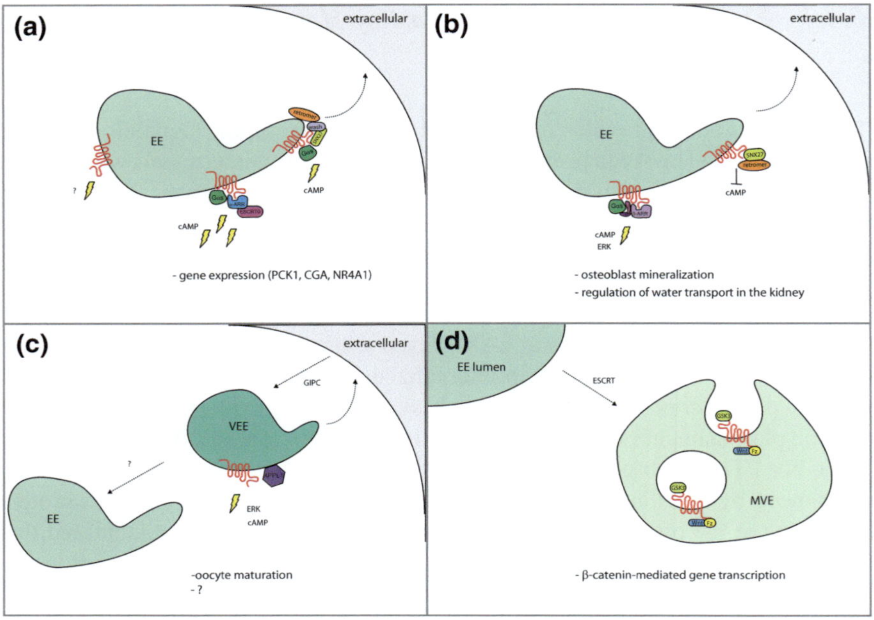

Fig. 10.2 GPCR intracellular signaling at distinct endosomal compartments. **a** GPCRs that transiently associate with arrestins (e.g., β2AR) are internalized to early endosomes (EEs) where active receptors are present throughout the endomembrane; Gαs-mediated cAMP production is elicited by receptors localized in recycling tubules marked by sorting nexin 27 (SNX27), retromer, and wash complexes and increased by retention of receptors outside these tubules by interaction with α-arrestin and the endosomal sorting complex required for transport 0 (ESCRT0). Endosomal signaling of β2AR has a unique transcriptional profile. **b** GPCRs that exhibit a sustained association with arrestin during internalization (e.g., V2R, PTHR) are also targeted to the EE, where they continue to activate cAMP and ERK responses via the interaction with both the G protein heterotrimer and β-arrestin. Interaction of GPCR with retromer and SNX27 dictates its localization into recycling tubules and terminates endosomal signaling. There is evidence that endosomal signaling from these receptors has key physiological roles. **c** Certain GPCRs (e.g., LHR) are internalized into very early endosomes (VEEs) marked by APPL1 through the interaction with GAIP-interacting protein, C-terminus (GIPC) where they sustain ERK and cAMP signaling. It is unknown what the downstream consequences of such endosomal signals are. **d** GPCRs, such as frizzled, are sorted to the lysosomal pathway and can induce a sustained Wnt-mediated β-catenin response via ESCRT-dependent involution into multivesicular endosomes (MVE) with glycogen synthase kinase 3 (GSK3) (see text). Figured modified from (Hanyaloglu 2018)

Even though the role of ARRDC3 in promoting endosomal signaling by inhibiting the sorting of β2AR into these recycling tubules was demonstrated, collectively these studies suggest the functional site of ARRDC3, in terms of regulating GPCR/Gαs signaling, is at microdomains within an individual regulated recycling tubule. This is consistent with the finding that ARRDC3 can regulate interactions between β2AR with SNX27 (Tian et al. 2016 and Fig. 10.2). A key outstanding question from these studies is whether there are additional signaling pathways activated at the endomembrane by the β2AR given the homogenous endosomal organization of nanobody-80, including both default and regulated recycling tubules.

Other heterotrimeric G protein pathways also require receptor internalization for persistent signaling, including Gαq/11-calcium signaling from the kisspeptin receptor, a pathway with a key role in puberty onset given the activating receptor mutation, R386P, exhibits enhanced endosomal signaling (Min et al. 2014; Bianco et al. 2011). Endosomal Gαq/11 signaling is also important for the calcium-sensing receptor, yet intriguingly, this GPCR is able to couple to multiple G protein pathways, including Gαi/o and Gα12/13, yet endosomal signaling from this receptor was predominantly mediated by Gαq/11 signaling, providing the first demonstration that the endocytic pathway could selectively activate distinct G protein pathways from a GPCR (Gorvin et al. 2018). Both the G protein α and βγ subunits are effectors for distinct enzymes (Fig. 10.1b). For the PTHR, the Gβγ, associated with Gαi/o at the plasma membrane, dissociates from its α subunit following activation and stimulates endosomal adenylate cyclase 2 activity to prolong nuclear cAMP and PKA activity (Jean-Alphonse et al. 2016). The pool of Gβγ recruited was found to be from β2AR/Gαi/o, demonstrating crosstalk between plasma membrane and endomembrane signaling and that endosomal signaling is part of a network of signaling events. Physiologically these two GPCRs are co-expressed in osteoblasts and provide mechanistic insight into prior studies reporting promotion of bone mineralization of PTH via β2AR activation (Hanyu et al. 2012; Jean-Alphonse et al. 2016).

10.4.2 Additional Endosomal Compartments as GPCR Signalosomes

Are EEs the only compartment capable of signaling of internalized GPCRs given the complexity of both GPCR signaling and the post-endocytic pathways GPCRs traverse? To date, there are limited studies that directly address this for GPCRs; however, the distinct endosomal compartment of the VEEs (see Sect. 10.3.2) (Jean-Alphonse et al. 2014; Sposini et al. 2017) also represents GPCR signal platforms. Initial studies with human LHR demonstrated that persistent ERK signaling required (1) receptor internalization, as inhibition of endocytosis completely blocked ligand-induced ERK signaling, and (2) receptor targeting to the VEE, as rerouting receptor from the VEE to the EE resulted in a transient ERK signal profile (Jean-Alphonse et al. 2014). Recently, we have demonstrated that GPCRs can acutely activate Gαs-cAMP signaling from a subset of VEEs, differentiating this endosomal signaling from the temporally sustained, or persistent, endosomal G protein signaling observed with the PTHR, TSHR, and β2AR. VEE-derived endosomal cAMP signaling of LHR, FSHR, and the β1AR, was negatively regulated by APPL1, but via an opposing mechanism to the role of APPL1 in driving receptor recycling from the VEE, as the unphosphorylated form of APPL1 was required for this negative regulation (Sposini et al. 2017). Thus, the concept of 'location bias' in heterotrimeric G protein signaling can be extended at multiple levels, i.e., between distinct endosomal compartments as well within endosomal populations. The downstream roles of VEE signaling remain to be determined for these VEE-targeted receptors. There is evidence that the mouse LHR exhibits a persistent cAMP profile in ovarian follicles and that inhibition of internalization impairs LH-mediated resumption of meiosis (Lyga et al. 2016), although the role of the VEE is unknown as rodent LHRs do not interact with GIPC and are primarily targeted to a degradative pathway (Galet et al. 2003, 2004; Hirakawa et al. 2003a; Nakamura et al. 2000).

Late endosomes/MVBs could also directly control GPCR signaling from this compartment. Indeed, a signaling role for late endosomes, defined by Rab7 localization, has been demonstrated for certain growth factors and nutrient signals such as EGF, insulin and amino acids, via endosomal signaling through Ras GTPases and mTORC-1 (Flinn and Backer 2010; Lu et al. 2009). The strongest evidence to date for GPCRs is the requirement of receptor organization within MVBs in Wnt signal regulation. Wnts activate members of the GPCR superfamily, frizzled and the involution of glycogen synthase kinase 3 (GSK3) with Wnt/frizzled resulted in sustained WNT signaling, a pathway leading to accumulation and nuclear translocation of the β-catenin transcription factor, and subsequent activation of gene transcription (Fig. 10.2) (Dobrowolski et al. 2012; Taelman et al. 2010).

10.4.3 Untangling the Cellular and Physiological Significance of GPCR Endosomal Signaling

While it is clear that the endocytic system can shape plasma membrane signaling and facilitate endomembrane signaling to enable both tight control and diversification in GPCR activity, there are important outstanding questions that need to be addressed. What are the downstream functions of not only GPCR plasma membrane versus endosomal signaling, but also across endosomal compartments and microdomains? Furthermore, what are the mechanisms that enable cells to decode these common signals to translate into specific downstream responses? The downstream cellular and physiological functions for endosomal signaling have been demonstrated for the TSHR and PTHR. TSHR, in response to its ligand TSH, mediates secretion of thyroxine from the thyroid gland, a hormone that is converted to triiodothyronine and has important roles in metabolism. Sustained endosomal TSHR/cAMP signaling in thyroid follicles was required for reorganization of the actin cytoskeleton via the cAMP/PKA effector vasodilator-stimulated phosphoprotein (VASP) (Calebiro et al. 2009). This reorganization of the actin cytoskeleton in turn impacts reuptake of thyroglobulin and the subsequent activation of thyroid-specific genes (Calebiro et al. 2009). For PTHR, the in vivo impact of sustained endosomal signaling is on trabecular bone volume and enhancing cortical bone turnover through direct actions at the level of the bone and kidney. Thus, there are clinical implications in the use of PTH analogs in osteoporosis that display differential ability in inducing sustained/endosomal signaling (Ferrandon et al. 2009; Okazaki et al. 2008). Given the broad roles of the GPCR superfamily, endosomal cAMP signaling is emerging as an important component of the physiological actions of these receptors (Kuna et al. 2013; Feinstein et al. 2013; Merriam et al. 2013). Recently, the endosomal signaling properties of two GPCRs the neurokinin-1 receptor and the calcitonin-gene related peptide receptor were implicated in transmission of pain from spinal cord neurons. For both receptors, the sustained endosomal signal pathway activated was PKC/ERK. These studies not only elegantly demonstrated the physiological significance of endosomal signaling in vivo but also as a tractable therapeutic target, as cell permeable antagonists to these GPCRs accumulated in endosomes and specifically inhibited sustained neuronal excitation and nociceptive responses in animal models (Jensen et al. 2017; Yarwood et al. 2017).

Evidence that cells can differentiate the same signal type (e.g., cAMP) from distinct subcellular activation points have been provided from studies of β2AR endosomal signaling. By employing optogenetic adenylate cyclase probes targeted to the plasma membrane, endosome, or cytoplasm, activation of downstream genes was profiled following compartment-specific adenylate cyclase activation and demonstrated a specific subset of genes were uniquely transcribed by endosomal Gαs-cAMP signaling (Tsvetanova and von Zastrow 2014). Furthermore, these endosomally driven transcriptional responses must emanate from endosomal microdomains by receptor organization to regulated, and not default, recycling tubules consistent with prior findings that active Gαs signaling is visible from these structures (Bowman et al. 2016 and Fig. 10.2) and providing a signaling function for targeting GPCRs between regulated and default recycling (Bowman et al. 2016). The β2AR is rapidly trafficked through the regulated recycling pathway, so it is noteworthy to highlight that this endosomal-driven gene expression is highly sensitive, thus β2AR ligands with pharmacological distinct profiles will activate endosomal signals in a similar manner (Tsvetanova et al. 2017). It remains to be determined how this would apply to other G protein signal pathways activated at endosomes, or for GPCRs, such as LHR, where its acute signaling is driven at the endosomal level. Overall, this body of work supports an evolving model that fundamental signaling molecules such as cAMP may not operate as a long-range second messenger, but rather signal specificity is critically dependent on compartmentalization (Musheshe et al. 2018).

10.5 Summary and Perspective

The role of membrane trafficking in GPCR signaling has evolved from a mechanism to regulate ligand sensitivity of a tissue, to an increasingly complex model involving an intricate integration of receptor signaling throughout the endocytic network. These advances have been made possible through the development of novel tools and imaging modalities that will likely unlock further insight into the role of endosomal signaling across the superfamily of signaling receptors, and ultimately aid in the pursuit of these as a valid therapeutic target. Indeed, nanobody tools that recognize the active MOR have recently identified that spatial control of signaling of this GPCR could be programmed by specific ligands or drugs (Stoeber et al. 2018).

Identifying the physiological significance of endosomal signaling to GPCR function more broadly across the GPCR superfamily is likely to be further uncovered in the near future, including examples where spatial-temporal control of signaling, via membrane trafficking, is perturbed in disease. The studies to date clearly illustrate that even minor alterations in receptor location such as altering the endosomal compartment a GPCR is sorted to, even at the level of an individual endosomal microdomain, could have profound effects on overall signaling response. The translational road from cell biology to drug development is often slow, for example, the time from identification of arrestins as GPCR signaling molecules (Luttrell et al. 1999) to creation of 'biased' ligands given to patients (Soergel et al. 2013) is nearly 14 years. However, the promise that targeting GPCRs at the level of endosomal signaling has recently been demonstrated to be a viable approach (Jensen et al. 2017; Yarwood et al. 2017), paving the way for novel avenues to identify drugs with increased specificity in activity and thus, perhaps the future therapeutic targeting of novel GPCR endocytic signaling systems is likely be expedited.

Acknowledgements This work was supported by grants from the Wellcome Trust (WT085099MA), Genesis Research Trust (P15844) to A.C.H, and an Imperial College London President's Scholarship to S.S.

References

Abdullah N, Beg M, Soares D, Dittman JS, McGraw TE (2016) Downregulation of a GPCR by β-arrestin2-mediated switch from an endosomal to a TGN recycling pathway. Cell Rep 17(11):2966–2978. https://doi.org/10.1016/j.celrep.2016.11.050

Arakaki AKS, Pan WA, Lin H, Trejo J (2018) The α-arrestin ARRDC3 suppresses breast carcinoma invasion by regulating G protein-coupled receptor lysosomal sorting and signaling. J Biol Chem 293(9):3350–3362. https://doi.org/10.1074/jbc.RA117.001516

Barki-Harrington L, Rockman HA (2008) β-Arrestins: multifunctional cellular mediators. Physiology 23:17–22. https://doi.org/10.1152/physiol.00042.2007

Bianco SD, Vandepas L, Correa-Medina M, Gereben B, Mukherjee A, Kuohung W, Carroll R, Teles MG, Latronico AC, Kaiser UB (2011) KISS1R intracellular trafficking and degradation: effect of the Arg386Pro disease-associated mutation. Endocrinology 152(4):1616–1626. https://doi.org/10.1210/en.2010-0903

Birnbaumer L (2007) The discovery of signal transduction by G proteins: a personal account and an overview of the initial findings and contributions that led to our present understanding. Biochim Biophys Acta 1768(4):756–771. https://doi.org/10.1016/j.bbamem.2006.09.027

Boucrot E, Ferreira AP, Almeida-Souza L, Debard S, Vallis Y, Howard G, Bertot L, Sauvonnet N, McMahon HT (2015) Endophilin marks and controls a clathrin-independent endocytic pathway. Nature 517(7535):460–465. https://doi.org/10.1038/nature14067

Bowman SL, Shiwarski DJ, Puthenveedu MA (2016) Distinct G protein-coupled receptor recycling pathways allow spatial control of downstream G protein signaling. J Cell Biol 214(7):797–806. https://doi.org/10.1083/jcb.201512068

Cadigan KM (2010) Receptor endocytosis: frizzled joins the ubiquitin club. EMBO J 29(13):2099–2100. https://doi.org/10.1038/emboj.2010.132

Calebiro D, Nikolaev VO, Gagliani MC, de Filippis T, Dees C, Tacchetti C, Persani L, Lohse MJ (2009) Persistent cAMP-signals triggered by internalized G-protein-coupled receptors. PLoS Biol 7(8):e1000172. https://doi.org/10.1371/journal.pbio.1000172

Cao TT, Deacon HW, Reczek D, Bretscher A, von Zastrow M (1999) A kinase-regulated PDZ-domain interaction controls endocytic sorting of the β2-adrenergic receptor. Nature 401(6750):286–290. https://doi.org/10.1038/45816

Charfi I, Abdallah K, Gendron L, Pineyro G (2018) Delta opioid receptors recycle to the membrane after sorting to the degradation path. Cell Mol Life Sci: CMLS 75(12):2257–2271. https://doi.org/10.1007/s00018-017-2732-5

Cho DI, Zheng M, Min C, Kwon KJ, Shin CY, Choi HK, Kim KM (2013) ARF6 and GASP-1 are post-endocytic sorting proteins selectively involved in the intracellular trafficking of dopamine D(2) receptors mediated by GRK and PKC in transfected cells. Br J Pharmacol 168(6):1355–1374. https://doi.org/10.1111/bph.12025

Coke CJ, Scarlett KA, Chetram MA, Jones KJ, Sandifer BJ, Davis AS, Marcus AI, Hinton CV (2016) Simultaneous activation of induced heterodimerization between CXCR4 chemokine receptor and cannabinoid receptor 2 (CB2) reveals a mechanism for regulation of tumor progression. J Biol Chem 291(19):9991–10005. https://doi.org/10.1074/jbc.M115.712661

Delgado-Peraza F, Ahn KH, Nogueras-Ortiz C, Mungrue IN, Mackie K, Kendall DA, Yudowski GA (2016) Mechanisms of biased β-arrestin-mediated signaling downstream from the cannabinoid 1 receptor. Mol Pharmacol 89(6):618–629. https://doi.org/10.1124/mol.115.103176

DeWire SM, Yamashita DS, Rominger DH, Liu G, Cowan CL, Graczyk TM, Chen XT, Pitis PM, Gotchev D, Yuan C, Koblish M, Lark MW, Violin JD (2013) A G protein-biased ligand at the μ-opioid receptor is potently analgesic with reduced gastrointestinal and respiratory dysfunction compared with morphine. J Pharmacol Exp Ther 344(3):708–717. https://doi.org/10.1124/jpet.112.201616

Dobrowolski R, Vick P, Ploper D, Gumper I, Snitkin H, Sabatini DD, De Robertis EM (2012) Presenilin deficiency or lysosomal inhibition enhances Wnt signaling through relocalization of GSK3 to the late-endosomal compartment. Cell Rep 2(5):1316–1328. https://doi.org/10.1016/j.celrep.2012.09.026

Dores MR, Trejo J (2015) GPCR sorting at multivesicular endosomes. Methods Cell Biol 130:319–332. https://doi.org/10.1016/bs.mcb.2015.05.006

Dores MR, Chen B, Lin H, Soh UJ, Paing MM, Montagne WA, Meerloo T, Trejo J (2012a) ALIX binds a YPX(3)L motif of the GPCR PAR1 and mediates ubiquitin-independent ESCRT-III/MVB sorting. J Cell Biol 197(3):407–419. https://doi.org/10.1083/jcb.201110031

Dores MR, Paing MM, Lin H, Montagne WA, Marchese A, Trejo J (2012b) AP-3 regulates PAR1 ubiquitin-independent MVB/lysosomal sorting via an ALIX-mediated pathway. Mol Biol Cell 23(18):3612–3623. https://doi.org/10.1091/mbc.E12-03-0251

Dores MR, Lin H, Grimsey NJ, Mendez F, Trejo J (2015) The α-arrestin ARRDC3 mediates ALIX ubiquitination and G protein-coupled receptor lysosomal sorting. Mol Biol Cell 26(25):4660–4673. https://doi.org/10.1091/mbc.E15-05-0284

Dunn HA, Ferguson SS (2015) PDZ protein regulation of G protein-coupled receptor trafficking and signaling pathways. Mol Pharmacol 88(4):624–639. https://doi.org/10.1124/mol.115.098509

Eichel K, Jullie D, von Zastrow M (2016) β-Arrestin drives MAP kinase signaling from clathrin-coated structures after GPCR dissociation. Nat Cell Biol 18(3):303–310. https://doi.org/10.1038/ncb3307

Eichel K, Jullie D, Barsi-Rhyne B, Latorraca NR, Masureel M, Sibarita JB, Dror RO, von Zastrow M (2018) Catalytic activation of β-arrestin by GPCRs. Nature. https://doi.org/10.1038/s41586-018-0079-1

Feinstein TN, Wehbi VL, Ardura JA, Wheeler DS, Ferrandon S, Gardella TJ, Vilardaga JP (2011) Retromer terminates the generation of cAMP by internalized PTH receptors. Nat Chem Biol 7(5):278–284. https://doi.org/10.1038/nchembio.545

Feinstein TN, Yui N, Webber MJ, Wehbi VL, Stevenson HP, King JD Jr, Hallows KR, Brown D, Bouley R, Vilardaga JP (2013) Noncanonical control of vasopressin receptor type 2 signaling by retromer and arrestin. J Biol Chem 288(39):27849–27860. https://doi.org/10.1074/jbc.M112.445098

Feng X, Zhang M, Guan R, Segaloff DL (2013) Heterodimerization between the lutropin and follitropin receptors is associated with an attenuation of hormone-dependent signaling. Endocrinology 154(10):3925–3930. https://doi.org/10.1210/en.2013-1407

Ferrandon S, Feinstein TN, Castro M, Wang B, Bouley R, Potts JT, Gardella TJ, Vilardaga JP (2009) Sustained cyclic AMP production by parathyroid hormone receptor endocytosis. Nat Chem Biol 5(10):734–742. https://doi.org/10.1038/nchembio.206

Ferre S, Casado V, Devi LA, Filizola M, Jockers R, Lohse MJ, Milligan G, Pin JP, Guitart X (2014) G protein-coupled receptor oligomerization revisited: functional and pharmacological perspectives. Pharmacol Rev 66(2):413–434. https://doi.org/10.1124/pr.113.008052

Flinn RJ, Backer JM (2010) mTORC1 signals from late endosomes: taking a TOR of the endocytic system. Cell Cycle 9(10):1869–1870. https://doi.org/10.4161/cc.9.10.11679

Flores-Otero J, Ahn KH, Delgado-Peraza F, Mackie K, Kendall DA, Yudowski GA (2014) Ligand-specific endocytic dwell times control functional selectivity of the cannabinoid receptor 1. Nature Commun 5:4589. https://doi.org/10.1038/ncomms5589

Fredriksson R, Lagerstrom MC, Lundin LG, Schioth HB (2003) The G-protein-coupled receptors in the human genome form five main families. Phylogenetic analysis, paralogon groups, and fingerprints. Mol Pharmacol 63(6):1256–1272. https://doi.org/10.1124/mol.63.6.1256

Gage RM, Kim KA, Cao TT, von Zastrow M (2001) A transplantable sorting signal that is sufficient to mediate rapid recycling of G protein-coupled receptors. J Biol Chem 276(48):44712–44720. https://doi.org/10.1074/jbc.M107417200

Galet C, Min L, Narayanan R, Kishi M, Weigel NL, Ascoli M (2003) Identification of a transferable two-amino-acid motif (GT) present in the C-terminal tail of the human lutropin receptor that redirects internalized G protein-coupled receptors from a degradation to a recycling pathway. Mol Endocrinol 17(3):411–422. https://doi.org/10.1210/me.2002-0161

Galet C, Hirakawa T, Ascoli M (2004) The postendocytotic trafficking of the human lutropin receptor is mediated by a transferable motif consisting of the C-terminal cysteine and an upstream leucine. Mol Endocrinol 18(2):434–446. https://doi.org/10.1210/me.2003-0293

Gidon A, Al-Bataineh MM, Jean-Alphonse FG, Stevenson HP, Watanabe T, Louet C, Khatri A, Calero G, Pastor-Soler NM, Gardella TJ, Vilardaga JP (2014) Endosomal GPCR signaling turned off by negative feedback actions of PKA and v-ATPase. Nat Chem Biol 10(9):707–709. https://doi.org/10.1038/nchembio.1589

Girnita L, Shenoy SK, Sehat B, Vasilcanu R, Girnita A, Lefkowitz RJ, Larsson O (2005) β-Arrestin is crucial for ubiquitination and down-regulation of the insulin-like growth factor-1 receptor by acting as adaptor for the MDM2 E3 ligase. J Biol Chem 280(26):24412–24419. https://doi.org/10.1074/jbc.M501129200

Gomes I, Ayoub MA, Fujita W, Jaeger WC, Pfleger KD, Devi LA (2016) G protein-coupled receptor heteromers. Annu Rev Pharmacol Toxicol 56:403–425. https://doi.org/10.1146/annurev-pharmtox-011613-135952

Gorvin CM, Rogers A, Hastoy B, Tarasov AI, Frost M, Sposini S, Inoue A, Whyte MP, Rorsman P, Hanyaloglu AC, Breitwieser GE, Thakker RV (2018) AP2? Mutations impair calcium-sensing receptor trafficking and signaling, and show an endosomal pathway to spatially direct G-protein selectivity. Cell Rep 22(4):1054–1066. https://doi.org/10.1016/j.celrep.2017.12.089

Grundmann M, Merten N, Malfacini D, Inoue A, Preis P, Simon K, Ruttiger N, Ziegler N, Benkel T, Schmitt NK, Ishida S, Muller I, Reher R, Kawakami K, Inoue A, Rick U, Kuhl T, Imhof D, Aoki J, Konig GM, Hoffmann C, Gomeza J, Wess J, Kostenis E (2018) Lack of β-arrestin signaling in the absence of active G proteins. Nature Commun 9(1):341. https://doi.org/10.1038/s41467-017-02661-3

Guo W, Urizar E, Kralikova M, Mobarec JC, Shi L, Filizola M, Javitch JA (2008) Dopamine D2 receptors form higher order oligomers at physiological expression levels. EMBO J 27(17):2293–2304. https://doi.org/10.1038/emboj.2008.153

Hanyaloglu AC (2018) Advances in Membrane Trafficking and Endosomal Signaling of G Protein-Coupled Receptors. Int Rev Cell Mol Biol 339:93–131. https://doi.org/10.1016/bs.ircmb.2018.03.001

Hanyaloglu AC, von Zastrow M (2007) A novel sorting sequence in the β2-adrenergic receptor switches recycling from default to the Hrs-dependent mechanism. J Biol Chem 282(5):3095–3104. https://doi.org/10.1074/jbc.M605398200

Hanyaloglu AC, von Zastrow M (2008) Regulation of GPCRs by endocytic membrane trafficking and its potential implications. Annu Rev Pharmacol Toxicol 48:537–568. https://doi.org/10.1146/annurev.pharmtox.48.113006.094830

Hanyaloglu AC, McCullagh E, von Zastrow M (2005) Essential role of Hrs in a recycling mechanism mediating functional resensitization of cell signaling. EMBO J 24(13):2265–2283. https://doi.org/10.1038/sj.emboj.7600688

Hanyu R, Wehbi VL, Hayata T, Moriya S, Feinstein TN, Ezura Y, Nagao M, Saita Y, Hemmi H, Notomi T, Nakamoto T, Schipani E, Takeda S, Kaneko K, Kurosawa H, Karsenty G, Kronenberg HM, Vilardaga JP, Noda M (2012) Anabolic action of parathyroid hormone regulated by the β2-adrenergic receptor. Proc Natl Acad Sci USA 109(19):7433–7438. https://doi.org/10.1073/pnas.1109036109

Hasdemir B, Bunnett NW, Cottrell GS (2007) Hepatocyte growth factor-regulated tyrosine kinase substrate (HRS) mediates post-endocytic trafficking of protease-activated receptor 2 and calcitonin receptor-like receptor. J Biol Chem 282(40):29646–29657. https://doi.org/10.1074/jbc.M702974200

He J, Bellini M, Inuzuka H, Xu J, Xiong Y, Yang X, Castleberry AM, Hall RA (2006) Proteomic analysis of β1-adrenergic receptor interactions with PDZ scaffold proteins. J Biol Chem 281(5):2820–2827. https://doi.org/10.1074/jbc.M509503200

He C, Wei Y, Sun K, Li B, Dong X, Zou Z, Liu Y, Kinch LN, Khan S, Sinha S, Xavier RJ, Grishin NV, Xiao G, Eskelinen EL, Scherer PE, Whistler JL, Levine B (2013) Beclin 2 functions in autophagy, degradation of G protein-coupled receptors, and metabolism. Cell 154(5):1085–1099. https://doi.org/10.1016/j.cell.2013.07.035

Henne WM, Stenmark H, Emr SD (2013) Molecular mechanisms of the membrane sculpting ESCRT pathway. Cold Spring Harb Perspect Biol 5(9). https://doi.org/10.1101/cshperspect.a016766

Henry AG, White IJ, Marsh M, von Zastrow M, Hislop JN (2011) The role of ubiquitination in lysosomal trafficking of delta-opioid receptors. Traffic 12(2):170–184. https://doi.org/10.1111/j.1600-0854.2010.01145.x

Henry AG, Hislop JN, Grove J, Thorn K, Marsh M, von Zastrow M (2012) Regulation of endocytic clathrin dynamics by cargo ubiquitination. Dev Cell 23(3):519–532. https://doi.org/10.1016/j.devcel.2012.08.003

Hirakawa T, Galet C, Kishi M, Ascoli M (2003a) GIPC binds to the human lutropin receptor (hLHR) through an unusual PDZ domain binding motif, and it regulates the sorting of the internalized human choriogonadotropin and the density of cell surface hLHR. J Biol Chem 278(49):49348–49357. https://doi.org/10.1074/jbc.M306557200

Hirakawa T, Galet C, Kishi M, Ascoli M (2003b) GIPC binds to the human lutropin receptor (hLHR) through an unusual PDZ domain binding motif, and it regulates the sorting of the internalized human choriogonadotropin and the density of cell surface hLHR. J Biol Chem 278(49):49348–49357. https://doi.org/10.1074/jbc.M306557200

Hislop JN, Henry AG, Marchese A, von Zastrow M (2009) Ubiquitination regulates proteolytic processing of G protein-coupled receptors after their sorting to lysosomes. J Biol Chem 284(29):19361–19370. https://doi.org/10.1074/jbc.M109.001644

Hislop JN, Henry AG, von Zastrow M (2011) Ubiquitination in the first cytoplasmic loop of μ-opioid receptors reveals a hierarchical mechanism of lysosomal down-regulation. J Biol Chem 286(46):40193–40204. https://doi.org/10.1074/jbc.M111.288555

Huang SH, Zhao L, Sun ZP, Li XZ, Geng Z, Zhang KD, Chao MV, Chen ZY (2009) Essential role of Hrs in endocytic recycling of full-length TrkB receptor but not its isoform TrkB.T1. J Biol Chem 284(22):15126–15136. https://doi.org/10.1074/jbc.M809763200

Insel PA, Wilderman A, Zambon AC, Snead AN, Murray F, Aroonsakool N, McDonald DS, Zhou S, McCann T, Zhang L, Sriram K, Chinn AM, Michkov AV, Lynch RM, Overland AC, Corriden R (2015) G protein-coupled receptor (GPCR) expression in native cells: "novel" endoGPCRs as physiologic regulators and therapeutic targets. Mol Pharmacol 88(1):181–187. https://doi.org/10.1124/mol.115.098129

Irannejad R, Tomshine JC, Tomshine JR, Chevalier M, Mahoney JP, Steyaert J, Rasmussen SG, Sunahara RK, El-Samad H, Huang B, von Zastrow M (2013) Conformational biosensors reveal GPCR signaling from endosomes. Nature 495(7442):534–538. https://doi.org/10.1038/nature12000

Jean-Alphonse FG, Bowersox S, Chen S, Beard G, Puthenveedu MA, Hanyaloglu AC (2014) Spatially restricted G protein-coupled receptor activity via divergent endocytic compartments. J Biol Chem 289(7):3960–3977. https://doi.org/10.1074/jbc.M113.526350

Jean-Alphonse FG, Wehbi VL, Chen J, Noda M, Taboas JM, Xiao K, Vilardaga J-P (2016) β2-adrenergic receptor control of endosomal PTH receptor signaling via Gβγ. Nat Chem Biol. https://doi.org/10.1038/nchembio.2267

Jensen DD, Lieu T, Halls ML, Veldhuis NA, Imlach WL, Mai QN, Poole DP, Quach T, Aurelio L, Conner J, Herenbrink CK, Barlow N, Simpson JS, Scanlon MJ, Graham B, McCluskey A, Robinson PJ, Escriou V, Nassini R, Materazzi S, Geppetti P, Hicks GA, Christie MJ, Porter CJH, Canals M, Bunnett NW (2017) Neurokinin 1 receptor signaling in endosomes mediates sustained nociception and is a viable therapeutic target for prolonged pain relief. Sci Transl Med 9(392). https://doi.org/10.1126/scitranslmed.aal3447

Ji B, Liu H, Zhang R, Jiang Y, Wang C, Li S, Chen J, Bai B (2017) Novel signalling of dynorphin at κ-opioid receptor/bradykinin B2 receptor heterodimers. Cell Signal 31:66–78. https://doi.org/10.1016/j.cellsig.2017.01.005

Jonas KC, Fanelli F, Huhtaniemi IT, Hanyaloglu AC (2015) Single molecule analysis of functionally asymmetric G protein-coupled receptor (GPCR) oligomers reveals diverse spatial and structural assemblies. J Biol Chem 290(7):3875–3892. https://doi.org/10.1074/jbc.M114.622498

Kennedy JE, Marchese A (2015) Regulation of GPCR trafficking by ubiquitin. Prog Mol Biol Transl Sci 132:15–38. https://doi.org/10.1016/bs.pmbts.2015.02.005

Kuna RS, Girada SB, Asalla S, Vallentyne J, Maddika S, Patterson JT, Smiley DL, DiMarchi RD, Mitra P (2013) Glucagon-like peptide-1 receptor-mediated endosomal cAMP generation promotes glucose-stimulated insulin secretion in pancreatic β-cells. Am J Physiol Endocrinol Metab 305(2):161–170. https://doi.org/10.1152/ajpendo.00551.2012

Lakadamyali M, Rust MJ, Zhuang X (2006) Ligands for clathrin-mediated endocytosis are differentially sorted into distinct populations of early endosomes. Cell 124(5):997–1009. https://doi.org/10.1016/j.cell.2005.12.038

Lally CC, Bauer B, Selent J, Sommer ME (2017) C-edge loops of arrestin function as a membrane anchor. Nature Commun 8:14258. https://doi.org/10.1038/ncomms14258

Lauffer BE, Melero C, Temkin P, Lei C, Hong W, Kortemme T, von Zastrow M (2010) SNX27 mediates PDZ-directed sorting from endosomes to the plasma membrane. J Cell Biol 190(4):565–574. https://doi.org/10.1083/jcb.201004060

Lu A, Tebar F, Alvarez-Moya B, Lopez-Alcala C, Calvo M, Enrich C, Agell N, Nakamura T, Matsuda M, Bachs O (2009) A clathrin-dependent pathway leads to KRas signaling on late endosomes en route to lysosomes. J Cell Biol 184(6):863–879. https://doi.org/10.1083/jcb.200807186

Luttrell LM, Lefkowitz RJ (2002) The role of β-arrestins in the termination and transduction of G-protein-coupled receptor signals. J Cell Sci 115(Pt 3):455–465

Luttrell LM, Ferguson SS, Daaka Y, Miller WE, Maudsley S, Della Rocca GJ, Lin F, Kawakatsu H, Owada K, Luttrell DK, Caron MG, Lefkowitz RJ (1999) β-Arrestin-dependent formation of β2 adrenergic receptor-Src protein kinase complexes. Science 283(5402):655–661

Lyga S, Volpe S, Werthmann RC, Gotz K, Sungkaworn T, Lohse MJ, Calebiro D (2016) Persistent cAMP signaling by internalized LH receptors in ovarian follicles. Endocrinology 157(4):1613–1621. https://doi.org/10.1210/en.2015-1945

Marchese A, Paing MM, Temple BR, Trejo J (2008) G protein-coupled receptor sorting to endosomes and lysosomes. Annu Rev Pharmacol Toxicol 48:601–629. https://doi.org/10.1146/annurev.pharmtox.48.113006.094646

McArdle CA, Franklin J, Green L, Hislop JN (2002) The gonadotrophin-releasing hormone receptor: signalling, cycling and desensitisation. Arch Physiol Biochem 110(1–2):113–122. https://doi.org/10.1076/apab.110.5.5.113.19855

McGarvey JC, Xiao K, Bowman SL, Mamonova T, Zhang Q, Bisello A, Sneddon WB, Ardura JA, Jean-Alphonse F, Vilardaga JP, Puthenveedu MA, Friedman PA (2016) Actin-sorting nexin 27 (SNX27)-retromer complex mediates rapid parathyroid hormone receptor recycling. J Biol Chem 291(21):10986–11002. https://doi.org/10.1074/jbc.M115.697045

Merriam LA, Baran CN, Girard BM, Hardwick JC, May V, Parsons RL (2013) Pituitary adenylate cyclase 1 receptor internalization and endosomal signaling mediate the pituitary adenylate cyclase activating polypeptide-induced increase in guinea pig cardiac neuron excitability. J Neurosci: Official J Soc Neurosci 33(10):4614–4622. https://doi.org/10.1523/JNEUROSCI.4999-12.2013

Min L, Soltis K, Reis AC, Xu S, Kuohung W, Jain M, Carroll RS, Kaiser UB (2014) Dynamic kisspeptin receptor trafficking modulates kisspeptin-mediated calcium signaling. Mol Endocrinol 28(1):16–27. https://doi.org/10.1210/me.2013-1165

Moore CA, Milano SK, Benovic JL (2007) Regulation of receptor trafficking by GRKs and arrestins. Annu Rev Physiol 69:451–482. https://doi.org/10.1146/annurev.physiol.69.022405.154712

Mosser VA, Jones KT, Hoffman KM, McCarty NA, Jackson DA (2008) Differential role of β-arrestin ubiquitination in agonist-promoted down-regulation of M1 vs M2 muscarinic acetylcholine receptors. J Mol Signaling 3:20. https://doi.org/10.1186/1750-2187-3-20

Mukai A, Yamamoto-Hino M, Awano W, Watanabe W, Komada M, Goto S (2010) Balanced ubiquitylation and deubiquitylation of Frizzled regulate cellular responsiveness to Wg/Wnt. EMBO J 29(13):2114–2125. https://doi.org/10.1038/emboj.2010.100

Mundell SJ, Luo J, Benovic JL, Conley PB, Poole AW (2006) Distinct clathrin-coated pits sort different G protein-coupled receptor cargo. Traffic 7(10):1420–1431. https://doi.org/10.1111/j.1600-0854.2006.00469.x

Musheshe N, Schmidt M, Zaccolo M (2018) cAMP: From long-range second messenger to nanodomain signalling. Trends Pharmacol Sci 39(2):209–222. https://doi.org/10.1016/j.tips.2017.11.006

Nakamura K, Liu X, Ascoli M (2000) Seven non-contiguous intracellular residues of the lutropin/choriogonadotropin receptor dictate the rate of agonist-induced internalization and its sensitivity to non-visual arrestins. J Biol Chem 275(1):241–247

Nooh MM, Bahouth SW (2017) Two barcodes encoded by the type-1 PDZ and by phospho-Ser312 regulate retromer/WASH-mediated sorting of the ss1-adrenergic receptor from endosomes to the plasma membrane. Cell Signal 29:192–208. https://doi.org/10.1016/j.cellsig.2016.10.014

Nooh MM, Mancarella S, Bahouth SW (2016) Identification of novel transplantable GPCR recycling motif for drug discovery. Biochem Pharmacol 120:22–32. https://doi.org/10.1016/j.bcp.2016.09.011

Okazaki M, Ferrandon S, Vilardaga JP, Bouxsein ML, Potts JT Jr, Gardella TJ (2008) Prolonged signaling at the parathyroid hormone receptor by peptide ligands targeted to a specific receptor conformation. Proc Natl Acad Sci USA 105(43):16525–16530. https://doi.org/10.1073/pnas.0808750105

Paing MM, Stutts AB, Kohout TA, Lefkowitz RJ, Trejo J (2002) β-Arrestins regulate protease-activated receptor-1 desensitization but not internalization or Down-regulation. J Biol Chem 277(2):1292–1300. https://doi.org/10.1074/jbc.M109160200

Pardon E, Laeremans T, Triest S, Rasmussen SG, Wohlkonig A, Ruf A, Muyldermans S, Hol WG, Kobilka BK, Steyaert J (2014) A general protocol for the generation of nanobodies for structural biology. Nat Protoc 9(3):674–693. https://doi.org/10.1038/nprot.2014.039

Pierce KL, Lefkowitz RJ (2001) Classical and new roles of β-arrestins in the regulation of G-protein-coupled receptors. Nat Rev Neurosci 2(10):727–733. https://doi.org/10.1038/35094577

Puthenveedu MA, von Zastrow M (2006) Cargo regulates clathrin-coated pit dynamics. Cell 127(1):113–124. https://doi.org/10.1016/j.cell.2006.08.035

Puthenveedu MA, Lauffer B, Temkin P, Vistein R, Carlton P, Thorn K, Taunton J, Weiner OD, Parton RG, Von Mark Z (2010) Sequence-dependent sorting of recycling proteins by actin-stabilized endosomal microdomains. Cell 143(5):761–773. https://doi.org/10.1016/j.cell.2010.10.003

Rankovic Z, Brust TF, Bohn LM (2016) Biased agonism: an emerging paradigm in GPCR drug discovery. Bioorg Med Chem Lett 26(2):241–250. https://doi.org/10.1016/j.bmcl.2015.12.024

Reiter E, Ahn S, Shukla A, Lefkowitz R (2012) Molecular mechanism of β-arrestin-biased agonism at seven-transmembrane receptors. Annual Rev Pharmacol Toxicol 52:179–276. https://doi.org/10.1146/annurev.pharmtox.010909.105800

Roman-Vendrell C, Yu YJ, Yudowski GA (2012) Fast modulation of μ-opioid receptor (MOR) recycling is mediated by receptor agonists. J Biol Chem 287(18):14782–14791. https://doi.org/10.1074/jbc.M111.319616

Rosciglione S, Theriault C, Boily MO, Paquette M, Lavoie C (2014) Gαs regulates the post-endocytic sorting of G protein-coupled receptors. Nature Commun 5:4556. https://doi.org/10.1038/ncomms5556

Rusten TE, Vaccari T, Stenmark H (2011) Shaping development with ESCRTs. Nat Cell Biol 14(1):38–45. https://doi.org/10.1038/ncb2381

Shenoy SK (2014) Arrestin interaction with E3 ubiquitin ligases and deubiquitinases: functional and therapeutic implications. Handb Exp Pharmacol 219:187–203. https://doi.org/10.1007/978-3-642-41199-1_10

Shenoy SK, Modi AS, Shukla AK, Xiao K, Berthouze M, Ahn S, Wilkinson KD, Miller WE, Lefkowitz RJ (2009) B-arrestin-dependent signaling and trafficking of 7-transmembrane receptors is reciprocally regulated by the deubiquitinase USP33 and the E3 ligase Mdm2. Proc Natl Acad Sci USA 106(16):6650–6655. https://doi.org/10.1073/pnas.0901083106

Soergel DG, Subach RA, Cowan CL, Violin JD, Lark MW (2013) First clinical experience with TRV027: pharmacokinetics and pharmacodynamics in healthy volunteers. J Clin Pharmacol 53(9):892–899. https://doi.org/10.1002/jcph.111

Sposini S, Jean-Alphonse FG, Ayoub MA, Oqua A, West C, Lavery S, Brosens JJ, Reiter E, Hanyaloglu AC (2017) Integration of GPCR signaling and sorting from very early endosomes via opposing APPL1 mechanisms. Cell Rep 21(10):2855–2867. https://doi.org/10.1016/j.celrep.2017.11.023

Stallaert W, Christopoulos A, Bouvier M (2011) Ligand functional selectivity and quantitative pharmacology at G protein-coupled receptors. Expert Opin Drug Discov 6(8):811–825. https://doi.org/10.1517/17460441.2011.586691

Stevens RC, Cherezov V, Katritch V, Abagyan R, Kuhn P, Rosen H, Wuthrich K (2013) The GPCR network: a large-scale collaboration to determine human GPCR structure and function. Nat Rev Drug Discov 12(1):25–34. https://doi.org/10.1038/nrd3859

Stoeber M, Jullie D, Lobingier BT, Laeremans T, Steyaert J, Schiller PW, Manglik A, von Zastrow M (2018) A genetically encoded biosensor reveals location bias of opioid drug action. Neuron. https://doi.org/10.1016/j.neuron.2018.04.021

Taelman VF, Dobrowolski R, Plouhinec JL, Fuentealba LC, Vorwald PP, Gumper I, Sabatini DD, De Robertis EM (2010) Wnt signaling requires sequestration of glycogen synthase kinase 3 inside multivesicular endosomes. Cell 143(7):1136–1148. https://doi.org/10.1016/j.cell.2010.11.034

Tappe-Theodor A, Agarwal N, Katona I, Rubino T, Martini L, Swiercz J, Mackie K, Monyer H, Parolaro D, Whistler J, Kuner T, Kuner R (2007) A molecular basis of analgesic tolerance to cannabinoids. J Neurosci: Official J Soc Neurosci 27(15):4165–4177. https://doi.org/10.1523/JNEUROSCI.5648-06.2007

Temkin P, Lauffer B, Jager S, Cimermancic P, Krogan NJ, von Zastrow M (2011) SNX27 mediates retromer tubule entry and endosome-to-plasma membrane trafficking of signalling receptors. Nat Cell Biol 13(6):715–721. https://doi.org/10.1038/ncb2252

Thomsen ARB, Plouffe B, Cahill TJ, Shukla AK, Tarrasch JT, Dosey AM, Kahsai AW, Strachan RT, Pani B, Mahoney JP, Huang L, Breton B, Heydenreich FM, Sunahara RK, Skiniotis G, Bouvier M, Lefkowitz RJ (2016) GPCR-G protein-β-arrestin super-complex mediates sustained G protein signaling. Cell 166(4):907–919. https://doi.org/10.1016/j.cell.2016.07.004

Tian X, Irannejad R, Bowman SL, Du Y, Puthenveedu MA, von Zastrow M, Benovic JL (2016) The α-arrestin ARRDC3 regulates the endosomal residence time and intracellular signaling of the β2-adrenergic receptor. J Biol Chem 291(28):14510–14525. https://doi.org/10.1074/jbc.M116.716589

Trejo J (2005) Internal PDZ ligands: novel endocytic recycling motifs for G protein-coupled receptors. Mol Pharmacol 67(5):1388–1390. https://doi.org/10.1124/mol.105.011288

Tsvetanova NG, von Zastrow M (2014) Spatial encoding of cyclic AMP signaling specificity by GPCR endocytosis. Nat Chem Biol 10(12):1061–1065. https://doi.org/10.1038/nchembio.1665

Tsvetanova NG, Trester-Zedlitz M, Newton BW, Riordan DP, Sundaram AB, Johnson JR, Krogan NJ, von Zastrow M (2017) G protein-coupled receptor endocytosis confers uniformity in responses to chemically distinct ligands. Mol Pharmacol 91(2):145–156. https://doi.org/10.1124/mol.116.106369

Vines CM, Revankar CM, Maestas DC, LaRusch LL, Cimino DF, Kohout TA, Lefkowitz RJ, Prossnitz ER (2003) N-formyl peptide receptors internalize but do not recycle in the absence of arrestins. J Biol Chem 278(43):41581–41584. https://doi.org/10.1074/jbc.C300291200

Vistein R, Puthenveedu MA (2013) Reprogramming of G protein-coupled receptor recycling and signaling by a kinase switch. Proc Natl Acad Sci USA 110(38):15289–15294. https://doi.org/10.1073/pnas.1306340110

Wang J, Hanada K, Staus DP, Makara MA, Dahal GR, Chen Q, Ahles A, Engelhardt S, Rockman HA (2017) Gαi is required for carvedilol-induced β1 adrenergic receptor β-arrestin biased signaling. Nature Commun 8(1):1706. https://doi.org/10.1038/s41467-017-01855-z

West C, Hanyaloglu AC (2015) Minireview: spatial programming of G Protein-coupled receptor activity: decoding signaling in health and disease. Mol Endocrinol 29(8):1095–1106. https://doi.org/10.1210/ME.2015-1065

Xiang Y, Kobilka B (2003) The PDZ-binding motif of the β2-adrenoceptor is essential for physiologic signaling and trafficking in cardiac myocytes. Proc Natl Acad Sci USA 100(19):10776–10781. https://doi.org/10.1073/pnas.1831718100

Xiao K, McClatchy DB, Shukla AK, Zhao Y, Chen M, Shenoy SK, Yates JR 3rd, Lefkowitz RJ (2007) Functional specialization of β-arrestin interactions revealed by proteomic analysis. Proc Natl Acad Sci USA 104(29):12011–12016. https://doi.org/10.1073/pnas.0704849104

Xiao K, Sun J, Kim J, Rajagopal S, Zhai B, Villen J, Haas W, Kovacs JJ, Shukla AK, Hara MR, Hernandez M, Lachmann A, Zhao S, Lin Y, Cheng Y, Mizuno K, Maayan A, Gygi SP, Lefkowitz RJ (2010) Global phosphorylation analysis of β-arrestin-mediated signaling downstream of a seven transmembrane receptor (7TMR). Proc Natl Acad Sci USA 107(34):15299–15304. https://doi.org/10.1073/pnas.1008461107

Yarwood RE, Imlach WL, Lieu T, Veldhuis NA, Jensen DD, Klein Herenbrink C, Aurelio L, Cai Z, Christie MJ, Poole DP, Porter CJH, McLean P, Hicks GA, Geppetti P, Halls ML, Canals M, Bunnett NW (2017) Endosomal signaling of the receptor for calcitonin gene-related peptide mediates pain transmission. Proc Natl Acad Sci USA 114(46):12309–12314. https://doi.org/10.1073/pnas.1706656114

Zhang X, Sun N, Zheng M, Kim KM (2016) Clathrin-mediated endocytosis is responsible for the lysosomal degradation of dopamine D3 receptor. Biochem Biophys Res Commun 476(4):245–251. https://doi.org/10.1016/j.bbrc.2016.05.104

Zoncu R, Perera RM, Balkin DM, Pirruccello M, Toomre D, De Camilli P (2009) A phosphoinositide switch controls the maturation and signaling properties of APPL endosomes. Cell 136(6):1110–1121. https://doi.org/10.1016/j.cell.2009.01.032

Chapter 11
Endosomal Trafficking During Mitosis and Notch-Dependent Asymmetric Division

Alicia Daeden and Marcos Gonzalez-Gaitan

Contents

Abstract Endocytosis is key in a number of cell events. In particular, its role during cell division has been a challenging question: while early studies examined whether endocytosis occurs during cell division, recent works show that, during division, cells do perform endocytosis actively. More importantly, during asymmetric cell division, endocytic pathways also control Notch signaling: endocytic vesicles regulate the presence, at the plasma membrane, of receptors and ligands at different levels between the two-daughter cells. Both early and late endocytic compartments have been shown to exert key regulatory controls by up-regulating or down-regulating Notch signaling in those cells. This biased Notch signaling enable finally cell fate assignation and specification which play a central role in development and physiology. In this chapter, we cover a number of significant works on endosomal trafficking evincing the importance of endocytosis in Notch-mediated cell fate specification during development.

A. Daeden · M. Gonzalez-Gaitan (✉)
Biochemistry Department, University of Geneva, 30 Quai E. Ansermet, 1211 Geneva, Switzerland
e-mail: Marcos.Gonzalez@unige.ch

C. Lamaze and I. Prior (eds.), *Endocytosis and Signaling*, Progress in Molecular and Subcellular Biology 57, https://doi.org/10.1007/978-3-319-96704-2_11

Endocytosis and mitosis are major events in the life of a cell. The former empowers cells to internalize nutrients, recycle, and degrade cargo proteins, while the latter enables cells to proliferate by splitting their two sets of chromosomes, cytosolic contents, and intracellular organelles into the two daughter cells. Historically, the discovery of cell division was made in 1832 by Dumortier (Fenner 1999; Guerrini 1995) and it is more than 50 years later, in 1883, that endocytosis via phagocytosis, an endocytic pathway, was first observed by Metchnikoff (1883). This chapter discusses how these two processes are tightly linked, and how they regulate each other.

During cytokinesis, the mother cell is partitioned into two daughters without a major change in total volume (Boucrot and Kirchhausen 2008), implying that the cell surface must significantly increase to accommodate two cells within the same volume. A common concept in the field is that endocytosis can regulate these critical plasma membrane dynamics required for mitosis. Indeed, a cell surface increase could be accomplished by a reduction in the endocytic rates. Here, we will consider current developments about the role of endocytosis to allow mitosis to happen, in particular how endocytosis does take place during mitosis and how endosomal membrane dynamics plays a key role during cytokinesis.

It is believed that endocytosis also regulates the concentrations of ligands and receptors at the plasma membrane required to regulate signaling. A common concept is that signaling happens *before* mitosis, in interphase, and that the result of such signaling cascades is, among other things, to decide whether a cell divides or not. However, we will consider here the case of Notch signaling, where the signaling event is set up *during* mitosis so that newly formed daughter cells perform asymmetric signaling. We will review here the current knowledge of how asymmetric signaling is mediated by a number of asymmetric endocytic events, both at the plasma membrane and in signaling endosomes.

11.1 Endocytosis During Mitosis and Cytokinesis

11.1.1 Endocytosis or No Endocytosis During Mitosis?

Several early studies suggested that endocytosis was either downregulated, completely arrested or transiently inhibited during mitosis. However, recent studies tackled these ideas by showing that endocytosis was mainly unperturbed during all stages of mitosis and that the former experimental conditions had a considerably impact on the endocytosis rate measured.

The first evidence of endocytic trafficking inhibition during mitosis was reported in the late 70s in macrophages and mouse embryonic fibroblast cells: Internalization by phagocytosis and internalization of the fluid phase in general were severely downregulated (Berlin et al. 1978). This was supported by the observation of an absence of coated pits in these cells (Fielding and Royle 2013; Fielding et al. 2012; Pypaert et al. 1987), lower levels of receptor-mediated endocytosis (Warren et al. 1983), a decrease of endosomal recycling (Boucrot and Kirchhausen 2007), a reduction of fluid-phase pinocytosis (Berlin and Oliver 1980; Oliver and Seagrave 1987) as well as an inhibition of autophagy (Eskelinen et al. 2002).

To rationalize why endocytosis was reduced during mitosis, it has been argued that cell rounding during division was associated with an increase in membrane tension that builds up until metaphase (Fischer-Friedrich et al. 2014). Increased membrane tension would then oppose membrane invagination during endocytosis. Consistent with this idea, it has been shown in HeLa cells that endocytosis is blocked at the beginning of mitosis, but resumes after metaphase, when the furrow separating the two cells starts to ingress during cytokinesis (Schweitzer et al. 2005). Furthermore, a strong correlation has been detected between the plasma membrane tension, measured by optical tweezers, and the endocytic rate (Raucher and Sheetz 1999b). Indeed, in interphase cells, increasing the plasma membrane tension by hypo-osmotic shocks blocks endocytosis, while this interruption can be relieved by adding amphiphilic compounds that decrease membrane tension (Boulant et al. 2011; Heuser and Anderson 1989; Nieland et al. 2005; Raucher and Sheetz 1999a). Similar assays performed in giant unilamellar vesicles (GUVs) showed that high tension fully inhibits clathrin polymerization at the surface membrane (Saleem et al. 2015). It has been, however, now cleared that while an increase of membrane tension can downregulate endocytosis, clathrin-mediated endocytosis can still occur even if membrane tension is high: Mechanical forces generated by the actin cytoskeleton at the endocytic pit can indeed counteract membrane tension (Boulant et al. 2011).

Another explanation of the transient inhibition of endocytosis at mitosis has been rationalized by the limiting pool of actin monomers available for endocytosis due to the massive actin polymerization required for mitotic events (such as formation of the actin ring at the furrow and thickening of the cell actin cortex) (Kaur et al. 2014). Indeed, multiple steps of the clathrin-mediated, caveolae-mediated, CLIC/GEEC or Flotillin endocytosis (membrane invagination and formation, scission and the final cut of the endocytic vesicle) require actin filaments (Durrbach et al. 1996; Robertson et al. 2009; Sirotkin 2011; Smythe and Ayscough 2006; Yarar et al. 2005). During mitosis, a thickening of the actin cortex is observed and a polymerization wave of actin filaments emerges at the equatorial region to form the equatorial contractile furrow (Bray and White 1988; Champion et al. 2017; Maddox and Burridge 2003; Salbreux et al. 2012; Stewart et al. 2011). Therefore, the cell cortex competes with the endocytic pathways for a common pool of actin monomers within the cell, which could, in principle, explain why endocytosis stops transiently at the onset of mitosis (Kaur et al. 2014).

While a number of these early reports claimed that the rate of endocytosis was reduced during mitosis, two recent studies challenged this idea (Boucrot and Kirchhausen 2007; Devenport et al. 2011; Tacheva-Grigorova et al. 2013). They showed that earlier studies were based on mitotic synchronization, which was experimentally achieved by means of chemical treatments, temperature shocks, and starvations. These treatments actually have a strong direct impact on the formation of clathrin-mediated vesicles. Under such conditions, endocytosis is indeed stopped or delayed (Boucrot and Kirchhausen 2007; Pypaert et al. 1991; Tacheva-Grigorova et al. 2013). Consistently, it has been observed by electron microscopy that endocytosis (through caveolae) was not perturbed during normal mitosis (Boucrot et al. 2011). Furthermore, the internalization assays in previous works that suggest a mitotic arrest of clathrin-mediated endocytosis are mostly based on internalization of transferrin receptor (Fielding et al. 2012). However, transferrin receptors are transiently internalized during metaphase leading to its disappearance from the cell surface (Warren et al. 1984), ultimately explaining the apparent observed reduction of transferrin receptor endocytosis. This decrease of transferrin receptor at the surface membrane of mitotic cells has been further shown by mass spectrometry study (Cocucci et al. 2012; Özlü et al. 2015). Thus, it has been confirmed that endocytosis does take place during normal mitosis by following fluid-phase endocytosis through fluorescent dextrans and endocytic clathrin-coated pits via fluorescently tagged AP2 adaptor proteins (Tacheva-Grigorova et al. 2013). Further analysis showed that, in metaphase, a moderate 25% decrease of the overall coated-pit abundance can be observed under these conditions, while the dynamics of clathrin-coated-pit formation remain normal (Aguet et al. 2016). Interestingly, under physiological conditions, recycling endosome dynamics slow down during an early mitotic stage (Boucrot and Kirchhausen 2007). This reduced recycling rate in the presence of active endocytosis would allow mitotic cells to decrease its surface area and round up in metaphase (Boucrot and Kirchhausen 2007).

It is therefore now well established that endocytosis and mitosis are linked processes: Mitosis can influence the endocytic pathways (e.g., recycling route). This prompted the question whether, conversely, endocytosis has a direct impact on cell division.

11.1.2 Importance of the Endocytic Pathways During Mitosis

Knock out of the endocytic coat protein, clathrin causes cytokinesis defects in Dictyostelium cells (Gerald et al. 2001). Similarly, treatments of Zebrafish embryos with drugs inhibiting endocytosis prevent normal completion of cytokinesis (Feng et al. 2002). From Dictyostelium, through Zebrafish up to mammalian cells, clathrin-mediated endocytosis is required for successful cytokinesis (Boucrot and Kirchhausen 2007; Gerald et al. 2001; Motley et al. 2003; Schweitzer et al. 2005) (Table 11.1). Thus, an emerging concept is that endocytosis is essential in membrane remodeling and in the control of the cell surface area (Boucrot and Kirchhausen 2007). As considered above, at the entry of mitosis, plasma membrane reduces its surface area by transiently blocking recycling routes back to the cell surface (Boucrot and Kirchhausen 2007; Tacheva-Grigorova et al. 2013). This enables cell rounding during metaphase while creating internal membrane reservoirs. Later, in anaphase, to support polar relaxation, cells recover their plasma membrane by reactivating recycling and exocytosis which lead to the rapid fusion of these membrane reservoirs to the surface (Boucrot 2008; Dyer et al. 2007). Accordingly, abolishing endocytosis and/or activating exocytosis at the entry of mitosis using mutants or drugs impairs cell rounding and can generate multinuclear cells reflecting a failure in cytokinesis (Boucrot and Kirchhausen 2007) (Table 11.1). Ultimately, the correct balance between endocytosis and exocytosis maintains a low membrane tension required for an accurate cytokinesis (Finger and White 2002).

11.1.3 Endosomal Dynamics During Mitosis

In addition to endocytosis, endosomal trafficking is also involved in the regulation of mitosis (Table 11.1). Originally, it was thought that endosomal dynamics were deeply decreased during division because endosomal fusion had been reported to be interrupted during mitosis (Tuomikoski et al. 1989). Fast image acquisition as well as new tools to follow individual endosomal compartments has, however, uncovered a key role for endosomal trafficking during cytokinesis. In a study of eight different Rab proteins, it was found that many Rabs are associated with distinct compartments that move and are located differently when comparing interphase and mitotic cells (Yu et al. 2007). Further works (discussed below) argue where endocytosis takes place during mitosis: Some studies suggest that endocytosis occurs at specific locations of the plasma membrane while a recent study showed that endocytosis occurs all over the plasma membrane. Downstream, endocytic vesicles, and endosomes move and are targeted to specific regions differentially in metaphase, anaphase, and during abscission.

During both prophase and metaphase, endocytosis occurs all around the cell contour, causing the plasma membrane to reduce its surface area leading to cell rounding (Boucrot and Kirchhausen 2007). At this time, Rab11 recycling endosomes cluster around microtubule organizing centers in a dynein-dependent manner (Takatsu et al. 2013).

In anaphase, based on internalized transferrin assay, endocytosis has been previously observed to occur at the cell poles, near the mitotic spindle poles (Schweitzer et al. 2005). However, recent study using lattice light sheet fluorescence microscope showed a uniform distribution of coated pits all over the entire cell surface during division (Aguet et al. 2016). In contrast, there is some evidence that endocytosis also takes place specifically at the cleavage furrow, as both clathrin- and caveolae-mediated endocytosis have been observed in the furrow region (Feng et al. 2002). In particular, in Zebrafish, it has been observed that endocytosis takes place during the whole process of cytokinesis at the cleavage furrow (Feng et al. 2002; Gerald et al. 2001; Thompson et al. 2002). Moreover, many endocytic regulating proteins including ARF6, caveolin, and dynamin have also been observed at the cleavage furrow midzone (Dyer et al. 2007; Feng et al. 2002; Kogo and Fujimoto 2000; Schweitzer and D'Souza-Schorey 2002a). This furrow localization of the endocytic machinery seems to be functionally relevant for cytokinesis: for instance, in Dictyostelium discoideum, cells lacking clathrin undergo cleavage furrow instability and cytokinesis failure (Gerald et al. 2001). These observations can be related to the phenomenon of polar relaxation characterized by the disappearance of cortical actin at the poles

Table 11.1 Trafficking factors required for cytokinesis

GTPases	Role in cytokinesis	References
Rab5	Downregulation of Rab5 causes binucleated cell phenotypes	Kouranti et al. (2006)
	Rab5 dominant-negative mutant embryos showed cleavage furrow progression failure	Pelissier et al. (2003)
	Downregulation of Rab5 caused defects in chromosome segregation and exhibited abnormal delays in prometaphase	Serio et al. (2011)
	Rab5 endosomes display a role in the disassembly of the nuclear envelope and in the cortical recruitment of the Mud protein (the *Drosophila* counterpart of mammalian NuMA), a key protein known to organize spindle positioning during mitosis	Capalbo et al. (2011), Lanzetti (2012)
Rab6	Rab6 misregulation causes cytokinesis failure in mice and in HeLa cells. Similarly, overexpression of Rab6-KIFL, the Rab6 binding kinesin, caused cell division defects resulting in cell death	Bardin et al. (2015), Hill et al. (2000), Miserey-Lenkei and Colombo (2016)
Rab8A	Rab8A localizes to the midbody during division: mistargeting Rab8A vesicles localization by knocking down proteins involved in their transport led to an increase of multinucleated cells	Kaplan and Reiner (2011)
Rab8	Rab8 organizes the furrow ingression by controlling and targeting membrane addition to the cleavage furrow: Disruption of Rab8 led to furrow formation and ingression failures	Mavor et al. (2016)
Rab11	Depletion of Rab11 causes binucleated cell phenotypes	Kouranti et al. (2006)
	Rab11 endosomes contribute to mitotic spindle organization by modulating important microtubule-nucleating and -anchoring roles: Depletion of Rab11 led to spindle defect as well as misalignment of the spindle with respect to the cell division plane	Hehnly and Doxsey (2014), Zhang et al. (2008)
	Rab11 is required for furrow formation: downregulation of Rab11 caused furrow defects	Riggs et al. (2003), Skop et al. (2001)

(continued)

Table 11.1 (continued)

GTPases	Role in cytokinesis	References
	Rab11 is essential for a successful abscission: Rab11 inactivation led to an abnormally prolonged midbody stage and furrow regression	Fielding et al. (2005), Skop et al. (2001), Wilson et al. (2005), Yu et al. (2007)
Rab21	Inhibition of Rab21 led to cytokinesis failure and multinucleated cells	Högnäs et al. (2011), Pellinen et al. (2008)
Rab24	Downregulation of Rab24 caused an increase in the number of multinucleated cells, abnormal spindle, and cytokinesis failures	Militello et al. (2013)
Rab35	Depletion of Rab35 caused delays in late anaphase and inhibit abscission	Dambournet et al. (2011), Prekeris (2011)
	Rab35 is required for cleavage furrow formation and ingression: Rab35 controls the recruitment of actin filaments and the localization of PtdIns(4,5)P2, a regulator of the actin cytoskeleton, to the cleavage furrow to enable furrow ingression	Chesneau et al. (2012), Kouranti et al. (2006)
	Depletion of Rab35 and dominant-negative mutant caused binucleated phenotypes	Kouranti et al. (2006)
	Inactivation of Rab35 caused cytokinesis defects through mislocalization of Septin2, a key protein in the stability of the positioning of the cleavage furrow and the intercellular bridge	Chesneau et al. (2012)
Clathrin, dynamin, α-adaptin, syntaxin 1, syntaxin 2 and VAMP8	Cytokinesis failures when knocked down	Cayrol et al. (2002), Chanez et al. (2006), Jantsch-Plunger and Glotzer (1999), Low et al. (2003)
Arf6	Arf6, a trafficking factor regulating endocytosis and the recycling route, is required for cleavage furrow. *arf6* mutants caused cleavage furrow regression and cytokinesis failure in *Drosophila*	Chesneau et al. (2012), Dyer et al. (2007)
	In HeLa cells, downregulation of Arf6 caused late cytokinesis arrest	Schweitzer and D'Souza-Schorey (2002b, 2005), Schweitzer et al. (2011)

(Gudejko et al. 2012; Kunda et al. 2012). This decrease in polar actin would then limit the formation of actin-dependent endocytic vesicles at those polar regions in strong contrast with the actin-rich contractile ring at the furrow. What is the relative importance of endocytosis and recycling at the furrow versus the poles is still unclear.

Nonetheless, at the poles of the dividing cell, new membrane addition occurs from an endosomal recycling pool to the plasma membrane involving exocytic events (Boucrot and Kirchhausen 2007). Recycling endosomes labeled by Rab11 are targeted to the plasma membrane poles within a short period of time (Schweitzer et al. 2005); however, how these vesicles are specifically targeted into these polar sites remains unknown. As anaphase progresses, Rab11-containing (as well as Rab35) endosomes are also seen at the ingressing furrow, where it provides the membranes for exocytosis at the cleavage region (Goss and Toomre 2008; Kouranti et al.; Wilson et al. 2005). These exocytosis events have been observed from plant to mammalian dividing cells (Bluemink and de Laat 1973; Gromley et al. 2005; Shuster and Burgess 2002; Skop et al. 2001; Wilson et al. 2005). Recycling endosomes have been proposed to be specifically transported to the cleavage furrow to provide new membrane (Albertson et al. 2008; Albertson et al. 2005; Fielding et al. 2005; Montagnac and Chavrier 2008; Neto et al. 2011; Wilson et al. 2005). In particular, trafficking from the centrosomes to the cleavage furrow has been shown to be a key for a successful cytokinesis (Gromley et al. 2005). Consistently, Drosophila embryos lacking Rab11 fail to recruit membrane and actin to the cleavage region during furrow formation leading to cytokinesis failure (Riggs et al. 2003).

Targeting and accumulation of endocytic vesicles at the cleavage furrow or near the cytoplasmic bridge have also been observed during later stage of cytokinesis until the abscission cut (Bluemink and de Laat 1973). These recruitments of vesicles to the midbody region are required to complete abscission at the end of cytokinesis (Conner and Wessel 1999; Danilchik et al. 2003; Jantsch-Plunger and Glotzer 1999). Indeed, it has been suggested that Rab11, which is targeted to the midbody region (Wilson et al. 2005), mediates the mitotic progression from furrowing to abscission by transporting key factors required for the last abscission step (Schiel and Prekeris 2010).

In summary, it is now clear that endocytosis and endosomal trafficking occur during and are important for mitosis. During cell-to-cell communication, endosomal trafficking is also known to play a key role in the control of the signaling levels, at least in interphase cells (Dobrowolski and De Robertis 2011). Is signaling also fine-tuned by endocytosis in mitotic cells? In the case of the signaling Notch pathway, endocytosis is key to regulate the presence of the receptors at the plasma membrane and, surprisingly, to activate the ligand too. On the other hand, asymmetric Notch signaling is also known to take place during asymmetric cell division in order to dictate the cell fate of the newly formed daughter cells. What is then the role of endocytic trafficking during asymmetric mitosis?

11.2 Endocytic Regulation of Notch Signaling

In recent years, the existence of specialized signaling endosomes as distinct platforms for transduction during cell-to-cell communication became a recurrent topic in signaling and vesicular trafficking studies (Sorkin and von Zastrow 2009; Barbieri et al. 2016). In the case of Notch signaling, the concentrations of both the ligand Delta and the receptor Notch at the plasma membrane are precisely regulated by means of endocytic trafficking and their behavior in endosomal compartments plays important roles in signaling events. The Notch signaling field started in 1914 when a notch was first observed in the wings of a mutant fly (Dexter 1914). It was only decades later that Notch was established as a transmembrane protein with a key role in signaling (F. 1939) and tumorigenesis (Hu et al. 2012). The evolutionarily conserved Notch pathway has also been shown to maintain tissue homeostasis and cell diversity by controlling cell differentiation (Chapouton et al. 2010). Notably, several different types of cancers such as leukemia (Aster et al. 2008), lung (Xu 2010), breast (Imatani and Callahan 2000), and cervical cancers (Gray et al. 1999; Zagouras et al. 1995) can be caused by a misregulation of Notch signaling.

The activation of Notch signaling starts by the interaction between Notch receptors in signal receiving cells and its Notch ligands from signal sending cells. This binding results in the gamma secretase proteolytic cleavage of Notch and the subsequent release of the Notch intracellular domain (NICD) into the cytoplasm of the signal receiving cell. The NICD then translocates into the nucleus, interacts with Notch target genes, and activates their transcription.

Throughout the animal kingdom, Notch has been shown to mediate cell fate assignation events (Lewis 1998; Muskavitch 1994), including the specification of feathers in chicken (Crowe et al. 1998), neural stem cell determination (Lehmann et al. 1983) or retinal development (Henrique et al. 1997), among others. In particular, cell specification within the sensory organ precursor (SOP) lineage in *Drosophila* became a model of choice. The SOP cell divides sequentially to give rise to four cells (a neuron, a sheath, a socket, and a hair) forming the mature mechanosensory organ in adult animals. In this system, the first asymmetric division of the SOP cell gives rise to an anterior pIIb and a posterior pIIa cell, which acquire different cell fates. Indeed, the pIIa cell will divide again to give rise to the two external cells (the socket and the hair), while pIIb give rise to two internal cells (the neuron and the sheath). The identity of the pIIa and the pIIb defines therefore the different downstream lineages of the four final cells. It is then easy to follow any cell fate specification defect generated by a failure in Notch signaling as two of the four cells in the sensory organ are external and can be observed directly in the back of the fly with a simple stereoscopic microscope. Upon cell division, Delta is activated in pIIb and thereby triggers Notch signaling in pIIa. This signaling event is the key cell communication episode that endows these two cells with different fates. In *Notch* mutants, cell fate specification fails to generate the four different cells in the sensory organ lineage and generate instead four neurons, leaving the cuticle bald of bristles (Hartenstein and Posakony 1990).

Notch signaling is heavily controlled by endocytic trafficking. The first evidence of a role of endocytosis during Notch signaling was reported in the late 90s, when a *Drosophila* mutant for *dynamin* (a GTPase responsible for membrane fission during endocytosis, named *shibire* in flies) was found to show a neurogenic phenotype similar to Notch mutants (Poodry 1990; Ramaswami et al. 1993; Seugnet et al. 1997). This is explained by the fact that internalization of both Notch and Delta are a key for their activation (Chapman et al. 2016). Their endocytosis affects most obviously their concentration at the plasma membrane, which ultimately determines the signaling levels at the receiving cell. Trafficking thus plays a key role by regulating the number of active ligands in the signal sending cell as well as the number of receptor in the receiving cell at their plasma membrane. Indeed, the ratio between ligand and receptor concentration regulates the signaling level: Double mutants for ligand and receptor showed enhanced Notch loss of function phenotypes, whereas gene duplication of the ligand or receptor showed enhanced Notch gain of function phenotypes (de-la-Concha et al. 1988; de Celis and Bray 2000; Ramos et al. 1989). Consistently, impairing endocytic pathways led to defective Notch activation (Valapala et al. 2013). In particular, in *dynamin* mutant mosaic animals, a mutant phenotype is seen regardless whether the mutant territory affects the signal sending or the signal receiving cell. Endocytosis is thus required both to activate the ligand in the sending cells and to allow signaling to happen in the receiving cell (Seugnet et al. 1997). Further studies of inhibition or dysregulation of endocytosis have been corroborating this idea by revealing cell proliferation phenotypes, cell fate determination defects, and even tumorigenesis which might have an origin in defective Notch signaling (Mosesson et al. 2008).

Because of its central role in development and physiology, Notch is regulated by a number of redundant and independent control mechanisms, which confer strong robustness to the system. Four core control mechanisms have been discovered to date (Summarized in Fig. 11.1). Two of them involve cortical factors, Numb (Fig. 11.1a) (Guo et al. 1996; Spana and Doe 1996) and Neuralized (Fig. 11.1b) (Pavlopoulos et al. 2001); two others involve endosomal compartments, Rab11 (Fig. 11.1c) (Emery et al. 2005) and Sara endosomes (Fig. 11.1d) (Coumailleau et al. 2009).

11.2.1 Notch: Numb-Mediated Endocytosis and Recycling Versus Degradation

The transmembrane protein Numb is accumulated in the anterior cortex of the SOP and segregates into the pIIb cell (Fig. 11.1b). There, Numb binds and antagonizes Notch activity by triggering its endocytosis, taking it away from the plasma membrane. Numb binds also alpha-adaptin, a protein part of the adaptor protein-2 (AP2) complex involved in clathrin-mediated endocytosis (Berdnik et al. 2002; Gonzalez-Gaitan and Jackle 1997). It also binds Eps15, another component of the clathrin adaptor complex AP2 machinery (Benmerah et al. 1996; Benmerah et al.

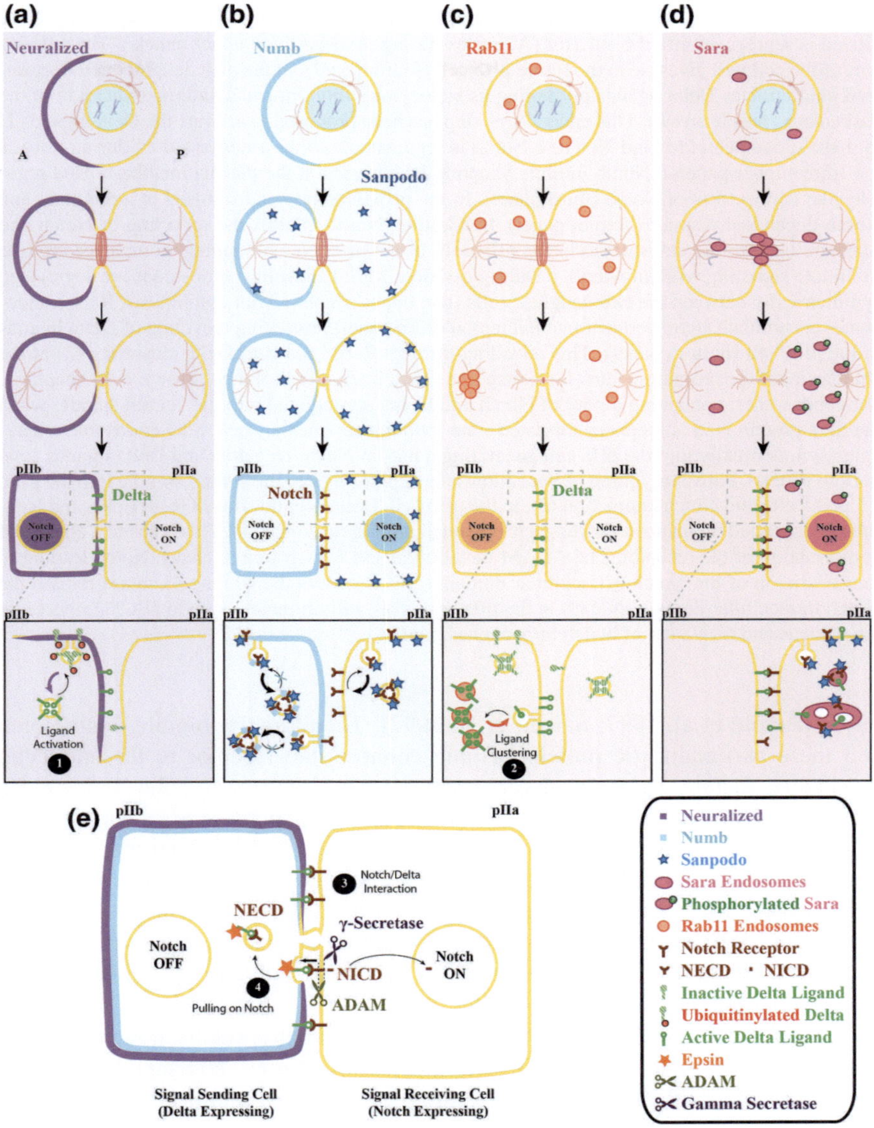
(a)
Neuralized
A
P
pIIb
pIIa
Delta
Notch OFF
Notch ON
Ligand Activation
1
(b)
Numb
Sanpodo
Notch
(c)
Rab11
Ligand Clustering
2
(d)
Sara
(e)
Notch/Delta Interaction
3
NECD
γ-Secretase
NICD
ADAM
4
Pulling on Notch
Signal Sending Cell (Delta Expressing)
Signal Receiving Cell (Notch Expressing)
Neuralized
Numb
Sanpodo
Sara Endosomes
Phosphorylated Sara
Rab11 Endosomes
Notch Receptor
NECD
NICD
Inactive Delta Ligand
Ubiquitinylated Delta
Active Delta Ligand
Epsin
ADAM
Gamma Secretase

◄**Fig. 11.1 Four endocytic pathways regulating Notch signaling: the SOP model system. a** Neuralized is segregated into the anterior (A) cortex during mitosis (three upper panels). The anterior part of the cell will give rise to the future pIIb cell (fourth panel). In the pIIb daughter cell, Neuralized ubiquitinates Delta ligands promoting its endocytosis (bottom panel and magnification of the dashed box; purple arrow). This endocytosis step has been proposed to activate the Delta ligand (**1**) to make it competent to bind Notch. **b** Numb is segregated into the anterior cortex during mitosis too (three upper panels). Numb inhibits Sanpodo localization at the plasma membrane and regulates the endocytosis of Notch (third panel). In the pIIb, Notch is endocytosed in a Sanpodo and Numb-dependent manner (bottom panel). In addition, Numb inhibits the recycling of Notch and Sanpodo back to the cell surface in the pIIb cell. In the pIIa, Notch/Sanpodo complexes are free to initiate signaling without Numb. **c** During division, Rab11 recycling endosomes are segregated symmetrically between the two daughter cells (two upper panels). After cytokinesis, Rab11 endosomes accumulate at the pericentrosomal region (third panel) promoting recycling of Delta ligands in the pIIb cell (bottom panel). This recycling through Rab11 leads to Delta clustering (**2**) at the cellular interface promoting sites for Notch and Delta interaction (orange arrow). **d** In anaphase, Sara endosomes, containing a pool of Notch and Delta, are targeted into the central spindle when the Sara protein is in an unphosphorylated state (two upper panels). Then, Sara endosomes segregate asymmetrically into the pIIa cell generating a bias in Notch receptors and Delta ligands (two bottom panels). Phosphorylation of Sara allows the departure from the central spindle into the pIIa. **e** Upon Delta and Notch interaction (**3**), in the pIIb cell, Delta is endocytosed in an Epsin and Neuralized depended on manner, exerting a mechanical pulling on Notch (**4**). This leads to a change of conformation of the receptor. The ADAM metalloprotease is then free to cleave Notch leading to the separation of the Notch extracellular domain (NECD) while the gamma secretase release the Notch intracellular domain (NICD) in the pIIa cell. This will ultimately lead to NICD import into the nucleus and activation of Notch target genes in pIIa

1995; Iannolo et al. 1997; Salcini et al. 1997). Therefore, by binding both Notch and these two endocytic proteins, Numb connects the receptor to the endocytic machinery, thereby triggering its internalization and downregulation. Indeed, the pIIb cell, where Numb accumulates, becomes then unresponsive to Notch signaling and the pIIa cell, lacking Numb, activates Notch and adopt the corresponding pIIa fate (Couturier et al. 2012). Numb then has a direct impact on cell fate assignation (Bhat et al. 2011; Le Borgne and Schweisguth 2003; Ming Guo 1996; Rhyu et al. 1994; Uemura et al. 1989). Indeed, *Numb* mutants show bristle phenotypes consistent with gain of function Notch defects, corresponding to a pIIa/pIIa cell fate lineage (Berdnik et al.; Guo et al. 1996).

In the pIIb cell, Numb also inhibits Sanpodo, a membrane protein known to interact with the Notch receptor (Fig. 11.1b). The interaction of Numb and Sanpodo promotes the endocytosis of the Notch/Sanpodo/Numb complex and inhibits its recycling by redirecting them into Rab7 endosomes (Fig. 11.1b) (Cotton et al. 2013; Couturier et al. 2012; O'Connor-Giles and Skeath 2003). Thus, only in the pIIa cell, Sanpodo is free to activate Notch at the plasma membrane and finally promote once more Notch signaling (Fig. 11.1b).

Interestingly, once localized in endosomes, Numb is also responsible for their endosomal trafficking and sorting (Couturier et al. 2013; Jafar-Nejad et al. 2002; McGill et al. 2009; Santolini et al. 2000). Indeed, Numb has been recently shown to control the sorting of Notch receptors between Rab11 recycling endosomes and Rab7 late endosomes (Johnson et al. 2016). In the pIIb cell, Numb downregulates the recycling of Notch, by deflecting this receptor preferentially into Rab7-positive late endosomes, thereby preventing their presence at the plasma membrane (Johnson et al. 2016). Other observations have revealed the importance of the localization of Notch in late endosomes. Indeed, Deltex, another E3-ubiquitin ligase, regulates Notch by ubiquitinating its intracellular domain. Notch ubiquitination targets the receptor into late endosomes. This late-endosomal targeting is essential to control the right levels of Notch signaling (Hori 2004). Recently, new regulators of Notch among the Bardet–Biedl syndrome (BBS) protein family have also been shown to regulate Notch trafficking to the late endosomes (Leitch et al. 2014).

Moreover, some regulators of Notch, such as Rme-8, have been uncovered to traffic through recycling Rab11 endosomes. Rme-8 is a retromer-associated DNAJ protein, which plays a role in regulating the recycling of Notch. Indeed, in *Rme-8* mutant, Notch accumulates in enlarged tubulated Rab4-positive endosomes, leading to an aberrant trafficking of Notch (Gomez-Lamarca et al. 2015). Another factor, Dmon1, is also involved in Notch targeting to late endosomes (Yousefian et al. 2013). Interestingly, Dmon1 is implicated in the maturation of early endosomes into late endosomes, a process in which the vesicular compartment changes from accumulating Rab5 to Rab7 (Shao et al. 2016; Yousefian et al. 2013). In *Dmon1* mutants, Notch accumulates in enlarge endosomes void of intraluminal vesicles which lost the Rab7 label. Surprisingly, accumulation of Notch in those aberrant endosomes did not cause any Notch signaling defect. This suggests that the targeting of activated Notch receptors in intraluminal vesicles is not sufficient to downregulate Notch signaling. Whether other mechanisms compensate Notch signaling during this accumulation of trapped receptors away from the plasma membrane remains an open question.

11.2.2 Delta: Neuralized-Mediated Endocytosis and Rab11 Mediated Recycling

While Numb internalizes Notch, Neuralized internalizes Delta (Fig. 11.1b). Neuralized is one of the E3 ubiquitin ligases which ubiquitinates Delta (Deblandre et al. 2001; Lai et al. 2001; Le Borgne and Schweisguth 2003; Pavlopoulos et al. 2001). Ubiquitination of Delta is essential for its endocytosis, recycling, and activation (Chitnis 2006; Wang and Struhl 2004). Like Numb, Neuralized accumulates in the anterior cortical region of the SOP, which ultimately will be segregated into the pIIb cell (Fig. 11.1a). Thus, in the pIIb, the ubiquitination of Delta by Neuralized stimulates its endocytosis (Fig. 11.1a) (Deblandre et al. 2001; Le Borgne and Schweisguth 2003; Pavlopoulos et al. 2001). Surprisingly, this endocytosis activates Delta in the pIIb cell, which can then interact with Notch receptors in pIIa, where signaling is thereby promoted (Le Borgne and Schweisguth 2003). Indeed, *Delta* mutants that cannot be efficiently endocytosed showed impaired Notch signaling (Itoh et al. 2003; Parks et al. 2000). Likewise, Neuralized has also been shown to have a direct impact on cell fate assignation (Bhat et al. 2011; Le Borgne and Schweisguth 2003; Ming

Guo 1996): *Neur* mutant, with defect in Delta endocytosis in the pIIb daughter cell, was leading to neurogenic phenotype and a pIIb/pIIb cell fate lineage. This suggests that endocytosis is regulated in the pIIb signal sending cell in order to assign the fate of pIIa (Le Borgne and Schweisguth 2003). Both Neuralized (Neur) and Mind-bomb1 (Mib1), another Delta E3-ubiquitin ligase, regulate Delta endocytosis and have been shown to mediate Notch activation (Pavlopoulos et al. 2001). However, more recently, a new study proposed that Delta ligands could also activate Notch signaling through two ubiquitination independent pathways (Berndt et al. 2017). Further studies will clarify the role of Delta ubiquitination in its activation.

Endocytosis of Delta should decrease its concentration at the plasma membrane. How could then Delta endocytosis cause activation of the ligand? The reason is that, for Delta to be competent to interact with Notch, it must first transit through a specific endocytic recycling route (Fig. 11.1) (Wang and Struhl 2004). Indeed, studies of ubiquitination-defective mutant of Delta ligand have shown that while ubiquitination is not absolutely required for its endocytosis, mono-ubiquitination is mandatory for targeting of Delta into Rab11 endosomes from where it can be recycled (Heuss et al. 2008; Stenmark 2009; Wang and Struhl 2004). This specific targeting of Delta into the recycling route is indeed essential for Notch signaling activation (Heuss et al. 2008). Consistently, the recycling of Delta seems to be required only in the signal sending cell for proper signaling (Emery et al. 2005; Jafar-Nejad et al. 2005; Rajan et al. 2009).

Upon division, Rab11 recycling endosomes accumulate in the pericentrosomal region of the pIIb cell just after cytokinesis (Fig. 11.1c). This asymmetry of Rab11 endosomes in the pIIb cell generates a bias in the recycling and activation of Delta in one of the daughters, thereby leading to the asymmetric activation of Notch in pIIa (Emery et al. 2005; Wang and Struhl 2004). This asymmetry of Rab11 endosomes has been proposed to generate a bias in active Delta between the two daughter cells leading to directional Notch signaling. Indeed, inducing an accumulation of Rab11 endosomes in the wrong daughter cell (pIIa) has been shown to affect cell fate specification (Emery et al. 2005). Further studies confirmed that impairing Delta trafficking in recycling endosomes caused indeed aberrant cell fate assignation (Jafar-Nejad et al. 2005). Conversely, overexpression of Rab11 has been reported to generate lineage defects in the scutellar and dorsocentral macrochaeta in the back of the fly (Abdelilah-Seyfried et al. 2000). These data showed the importance of the Rab11 endosome asymmetry in regulating Notch activity and assigning the correct cell fate (Abdelilah-Seyfried et al. 2000; Wang and Struhl 2004).

Once the interaction between Delta and Notch has happened, endocytosis of Delta bound to the Notch receptor can exert mechanical pulling force that will unfold the receptor to expose the a disintegrin and metalloprotease (ADAM) proteolytic cleavage site (Fig. 11.1e) (Meloty-Kapella et al. 2012). Indeed, the endocytosis of Delta has then been suggested to generate enough force to physically unfold Notch receptor for this cleavage (Nichols et al. 2007; Parks et al. 2000). NECD is then trans-endocytosed together with Delta into the signal sending cell, leading to the activation of Notch in the signal receiving cell (Nichols et al. 2007; Parks et al. 2000). This model has been confirmed by structural studies suggesting that, only upon ligand interaction, the protease site on the Notch extracellular domain become exposed (Gordon et al. 2007). The ADAM cleavage in Notch allows subsequently a second γ–secretase proteolytic cleavage releasing the NICD fragment (Fig. 11.1e). NICD migrates then to the nucleus activating the transcription of Notch target genes (Fig. 11.1e) (Wilkin and Baron 2005). Importantly, it has been shown that Delta endocytosis depends on Epsin, also named Liquid Facets (Overstreet et al. 2004; Wang and Struhl 2004), in order to exert sufficient force on Notch (Langridge and Struhl 2017). Indeed, ligands that do not enter via an Epsin pathway can still bind Notch, but fail to activate the receptor. Consistently, mutant *Epsin* animals (whether *Drosophila*, C.Elegans or mice) show phenotypes akin to those in Notch loss of function mutants (Chen et al. 2009; Overstreet et al. 2004; Tian et al. 2004). All these observations led to the proposal that ligand endocytosis activates Delta through the recycling route and is able to activate Notch in the adjacent cell through pulling (Fig. 11.1a, c, e).

In epithelial cells, Delta recycling through transcytosis has been observed in the signal sending cell. Transcytosis involves the endocytosis of basal Delta and its recycling to the apical membrane in epithelial cells (Benhra et al. 2010). Delta transcytosis has been proposed to activate the ligand by promoting clustering (Hicks et al. 2002) as well as helping relocate it to the apical plasma membrane where it can meet Notch from the adjacent cell (Benhra et al. 2010; Jafar-Nejad et al. 2005; Rajan et al. 2009).

11.3 Notch and Delta Trafficking Through Sara Endosomes

Both Notch and Delta traffic through recycling and late endosomes. To reach these compartments, they are previously sorted in early endosomes. Indeed, 25% of transmembrane Notch is associated with intracellular vesicles, most of which are endosomal compartments (Loubery et al. 2014). A key subpopulation of this endosomal Notch, about 75% of endosomal Notch, appears in Sara endosomes, a subpopulation of early endosomes characterized by the presence of Sara (Fig. 11.1d). Smad Anchor for Receptor Activation (Sara) contains a FYVE domain which binds PI(3)P, a phosphoinositide which is only present in early endosomes. Sara was first discovered as an adaptor protein that mediates transforming growth factor β (TGFβ) signal transduction in mammalian cells (Itoh et al. 2002; Tsukazaki et al. 1998). It was then also implicated in bone morphogenetic protein (BMP) signaling: Sara endosomes contain BMP signaling molecules and are dispatched symmetrically during normal symmetric divisions, thereby ensuring that the two daughter cells receive the same levels of BMP signaling (Bokel et al. 2006).

During the asymmetric division of the SOP cells, internalized Delta ligands and Notch receptors also traffic through Sara endosomes (Coumailleau et al. 2009). In the early steps of SOP mitosis, Sara endosomes contain both uncleaved Notch (NECD and NICD are both present) as well as Delta. Then, during asymmetric cytokinesis, Sara endosomes (and their Notch/Delta cargo) segregate asymmetrically in the cytoplasm to be targeted only to the pIIa daughter cell (Coumailleau et al. 2009; Derivery et al. 2015; Loubery et al. 2017; Loubéry et al. 2015). Ultimately, about 90% of Delta and Notch in these vesicles have been found in the pIIa cell (Coumailleau et al. 2009).

After SOP mitosis, in pIIa, only NECD is present in Sara endosomes; in contrast, Sara endosomes which were dispatched in pIIb contain both NECD and NICD (Coumailleau et al. 2009). This implies that Notch cleavage, reflecting signaling activation, is seen in Sara endosomes, but only in those of pIIa, not pIIb. This directional targeting of Sara endosomes generates asymmetric Notch/Delta signaling, which contributes to the cell fate decision of pIIa versus pIIb (Coumailleau et al. 2009; Derivery et al. 2015; Loubéry et al. 2015).

11.3.1 Asymmetric Motility of Sara Endosomes

How is endosomal dynamics regulated during mitosis? What motility events underlie the targeting of Sara endosomes to the pIIa daughter cell? It is now well established that endosomes move actively along microtubules or even actin filament powered by dynein, kinesin, or myosin molecular motors (Loubery et al. 2008; Matteoni and Kreis 1987). Indeed, endosomal movements are severely impaired by microtubule depolymerization (Rapp et al. 1996), by inhibiting or mutating motor proteins (Burkhardt et al. 1997; Firestone et al. 2012), by depleting ATP (De Brabander et al. 1988; Swanson et al. 1992) or changing proton homeostasis (Murray et al. 2017). Furthermore, in conditions where the number of actin filaments is reduced, a decrease in endosomal motility and a lack of endosomal fusion have been observed (Murphy et al. 1996). In particular, the vesicular transport on actin filaments has been shown to be governed by actin regulator proteins such as Rho protein and Diaphanous (Murphy et al. 1996; Randazzo 2003).

Some vesicle populations have been reported to be more dynamic than others (Fernando Aniento 1993). For instance, Rab11 endosome velocity has been found to be around 0.8 μm/s when moving toward the spindle pole (Hehnly and Doxsey 2014), while Rab5 early endosomes have been reported to move at a slower speed of 0.17 μm/s on average (Flores-Rodriguez et al. 2011). However, Rab5 endosomes have also been shown to be capable of very high instantaneous speeds up to 4 μm/s (Zajac et al. 2013). These velocities depend on the number, directionality, and processivity of the molecular motors attached to vesicles. Interestingly, because of the competition between kinesin and dynein motors, bidirectional movements happen much more frequently in late endosomes than in early endosomes (Zajac et al. 2013).

Bidirectional movements have also been observed for Sara early endosomes during asymmetric cell division (Derivery et al. 2015). Notably, Sara endosome motility is mediated solely by a plus-end directed kinesin motor, Klp98a (Fig. 11.2). The central spindle is composed of antiparallel microtubules, with their plus-end in the center of the dividing cell and their minus-ends toward the poles. Therefore, the kinesin targets Sara endosomes toward the center of the central spindle. In *klp98a* mutants, endosomes fail to move toward the central spindle and cannot be asymmetrically dispatched. This causes a phenotype in the process of asymmetric cell fate assignation.

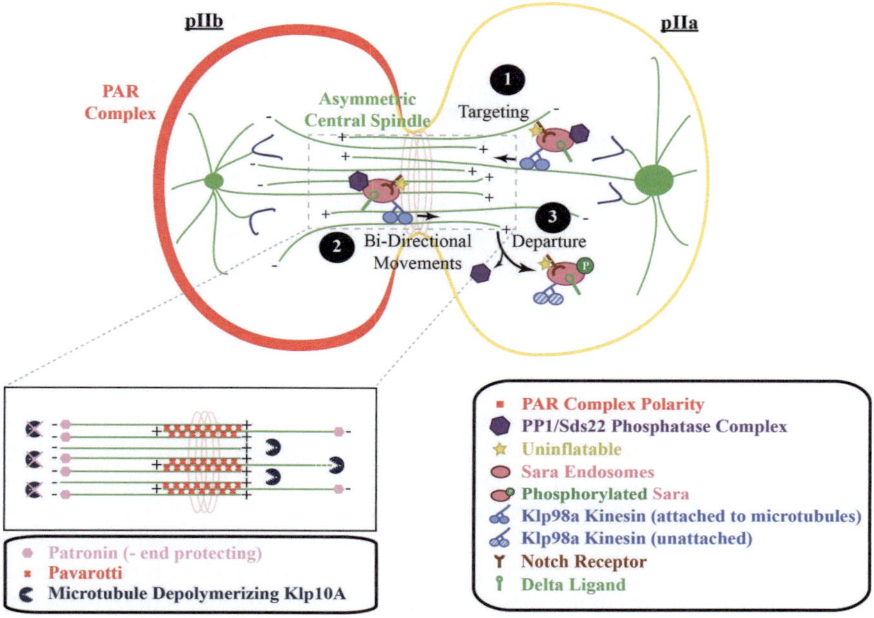

Fig. 11.2 Targeting, motility, and departure of Sara endosomes. Three steps (**1**, **2**, **3**) for the asymmetric dispatch of Sara endosomes. **1** Sara endosomes, containing a pool of Notch and Delta, are associated to Klp98a, a plus-end-directed kinesin which binds early endosomes as cargo. This kinesin is the sole motor responsible for the motility of the Sara endosomes. Because of the configuration of the central spindle microtubules (with their plus-end toward the center of the dividing cell and minus-end toward the poles—See magnified Box), Sara endosomes are targeted to the center of the dividing cell. This targeting to the central spindle is also mediated by the phosphorylated state of the Sara protein. PP1/Sds22 phosphatase complex dephosphorylates Sara allowing the targeting of endosomes to microtubules. **2** The central spindle is organized in an antiparallel array of microtubules labeled by Pavarotti (a kinesin-like protein part of the Centralspindlin complex). Due to this antiparallel organization, Sara endosomes undergo bidirectional movements within the central spindle overlap. **3** Patronin sits on the minus-end of the microtubules protecting them from Klp10A, a depolymerizing microtubule factor (magnified box). This enrichment of Patronin in the pIIb side of the central spindle is downstream to the PAR complex polarity. This ultimately leads to an asymmetry of the microtubule density between the two sides of the central spindle. The targeting of the Sara endosomes to pIIa is directly linked to this asymmetry of microtubule density. Departure of Sara endosomes from the central spindle is dependent on Uninflatable, Notch, and the phosphorylation state of Sara. The phosphorylation of Sara enables the departure of the Sara endosomes from the central spindle

The center of the central spindle is composed of antiparallel microtubules where microtubules from the two poles of the cell meet with opposite polarity. In this configuration, Sara endosomes can undergo bidirectional movements by switching microtubules and thus direction in the microtubule array (Derivery et al. 2015). Furthermore, in SOP cells, the central spindle is asymmetric, with more microtubules in the pIIb side (Fig. 11.2). This asymmetry of microtubule density occurs downstream of positional cues by the asymmetric Par complex (Chen and Zhang 2013) at the cortical poles of the cell (Derivery et al. 2015) which generates the asymmetric accumulation of the microtubule-associated protein Patronin (Fig. 11.2). Patronin binds to the minus end of microtubules and stabilize them against depolymerization (Fig. 11.2). This leads to the accumulation of microtubules (pointing toward the center of the cell) in the pIIb cell. This asymmetric density of microtubules generates a bias in the residence time of endosomes in the two daughter cells, which ultimately causes the dispatch of Sara endosomes into pIIa (Derivery et al. 2015). Impairing either the kinesin-mediated motility or the central spindle asymmetry affects the asymmetric targeting of Sara endosomes and Notch signaling during SOP mitosis (Derivery et al. 2015). Therefore, the asymmetric targeting depends on (1) a plus-end kinesin and (2) the asymmetric topology of the central spindle, in the center of the dividing cell.

Sara endosome trafficking has been followed in endogenous conditions (without over-expressing a molecular marker) by means of a fluorescent Delta antibody uptake assay *ex-vivo* in a primary culture system (Loubery and Gonzalez-Gaitan 2014). In flies, asymmetric segregation of Sara endosomes has also been observed in intestinal stem cells (Montagne and Gonzalez-Gaitan 2014) and neural stem cells of the central nervous system (Coumailleau et al. 2009). In Zebrafish, it has also been seen in neural precursors of the spinal cord (Kressmann et al. 2015). In all these systems, endosomal cargo included Notch molecules and their asymmetric targeting in mitosis is essential for asymmetric cell fate assignation. Indeed, *sara* mutants phenocopy Notch loss of function in the fly intestine, causing an amplification of the stem cell compartment (Montagne and Gonzalez-Gaitan 2014). Likewise, in Zebrafish, Sara endosomes partition can control the number of neural precursor cells as well as differentiating neurons during neurogenesis, reflecting the lack of asymmetric cell fate assignation. Indeed, in asymmetric neural precursor lineages, the daughter cell that inherits Sara remains a neural progenitor that is able to divide again, whereas the other daughter cell, devoid of Sara, undergoes neuronal differentiation (Kressmann et al. 2015). Also, in mammalian cells, endocytosed Notch1 is present in Sara endosomes (Chapman et al. 2016).

It is worth noting that Sara and the motility of its endosomes seem to contribute to Notch signaling beyond the setup of asymmetric division. Null mutants in *sara* or *klp98a* do not cause an obvious phenotype, consistent with the idea that control of asymmetric Notch signaling is heavily redundant. However, if the endocytic systems described above (e.g., Neuralized or Numb) are compromised to levels in which the phenotype is mild, loss of function in Sara, or Klp98a leads to a dramatic synthetic phenotype. These phenotypes are manifested in a failure in asymmetric Notch signaling within the SOP lineage, which causes a phenotype where flies do not have external sensory cells, but only neurons underneath the epidermis: The cuticle is bald. In addition, an earlier so-called lateral inhibition phenotype, which is not associated to dividing cells or asymmetric division, is also seen in these sensitized conditions. This implies that Sara endosomes, and their motility are required for a core Notch function. It is still unclear how Sara endosomes contribute to core Notch signaling, but the fact that it does explain why the asymmetric distribution of those endosomes contribute to asymmetric Notch signaling and asymmetric cell fate assignation.

11.3.2 Departure of Sara Endosomes from the Central Spindle

Studies on Sara overexpression have shown that the presence of Sara at the surface of Rab5 endosomes is required for a proper regulation of Rab5 endosomal trafficking (Hu et al. 2002; Loubery et al. 2017). Indeed, the Sara protein itself seems to play a role in endosomal dynamics (Arias et al. 2015; Hu et al. 2002; Itoh et al. 2002; Loubery et al. 2017). The targeting of Sara endosomes to the central spindle as well as the departure of the central spindle generating this asymmetric dispatch of Sara endosomes is controlled by a phosphorylation switch (Fig. 11.2) (Loubery et al. 2017). To begin with, the Sara protein itself is essential for the movement of these endosomes on microtubules toward the central spindle. Furthermore, if Sara is dephosphorylated, Sara endosomes move on microtubules toward the central spindle; phosphorylation of Sara mediates the detachment from the endosomes at the central spindle leading to the departure into the pIIa cell. A Sara phospho-mimetic mutant fails to depart its Sara endosomes and causes a symmetric distribution of endosomes that leads to asymmetric specification phenotype: Instead of the four cells in the lineage, these mutants show a duplication of the external cells, that is two sockets. The dephosphorylation event depends on the PP2 phosphatase and phosphorylation at the center of the cell may involve the AuroraB kinase, which itself is associated to the central spindle. The two molecular mechanisms by which Sara itself and its phosphorylation state determine these two motility events (Sara-dependent engagement on microtubules and phosphorylation-dependent departure from microtubules) are still unknown.

Two other factors contribute to the final departure of Sara endosomes to the pIIa: Notch itself and its binding factor Uninflatable (Fig. 11.2). Uninflatable is a trans-membrane protein that, like Notch, contains extracellular EGF repeats. Uninflatable can modulate Notch activity by antagonizing Notch (Xie et al. 2012). It also traffics together with Notch through Sara endosomes. Notably, Uninflatable is essential for the departure of Sara endosomes from the central spindle into the pIIa (Loubery et al. 2014): In the absence of Uninflatable, endosomes go to the spindle, but fail to depart to pIIa, causing a symmetric phenotype visible in the postorbital bristles around the eye. Notch knockdown conditions cause the same lack of endosomal dispatch in the SOP (Loubery et al. 2014). Interestingly, neither Notch nor Uninflatable has been shown to be essential for the targeting of Notch and Delta to Sara endosomes, nor the targeting of those endosomes to the central spindle. While the machinery has now been established (Sara phosphorylation, Notch, Uninflatable), the molecular mechanism of Sara endosome departure from the spindle remains to be unraveled.

11.4 Concluding Remarks

While early studies questioned whether endocytosis happens during cell division, a body of recent work shows that mitotic cells perform endocytosis actively. Furthermore, endocytic trafficking plays key roles during mitosis such as regulating the membrane surface or targeting key players to the midbody for the abscission event, among others. Endocytic pathways and endosomal motility have been therefore shown to be essential for cytokinesis.

Since endocytosis happens in mitosis, this opens a world of regulatory possibilities to direct signaling during cell division, in particular during asymmetric cell division where asymmetric signaling makes two-daughter cells different from each other. In the case of Notch signaling, receptors and ligands traffic through endocytic compartments to regulate the signaling levels. Prima facie endocytosis can indeed be used to regulate the abundance of Notch and Delta in the plasma membrane. It became clear that both the early and late endocytic compartments exert key regulatory controls for upregulating or attenuating the trafficking to and from the plasma membrane. Beyond that, Notch activation has been shown to also depend on endocytic events: the activation of Delta, which requires the trafficking of Delta through the recycling Rab1 route; the clustering and concentration of ligands in the cellular interface for more efficient Notch/Delta interaction; the NECD and NICD cleavage capitalizing on a mechanical pulling force mediated by the endocytosis of Delta and finally, the dispatch of Sara endosomes containing a pool of Notch and Delta to only one of the two daughter cells. Therefore, routing the Notch complex through endocytic compartments is a key for the regulation of Notch signaling. Many of these situations also happen during asymmetric cell division, where endocytic events are themselves asymmetric and are the core on how the signaling of Notch can be biased. In particular, the asymmetrical targeting of these cell fate determinants enables a correct cell fate specification. Altogether, these studies uncovered the importance of endocytosis in Notch-mediated cell fate specification during development.

References

Abdelilah-Seyfried S et al (2000) A gain-of-function screen for genes that affect the development of the *Drosophila* adult external sensory organ. Genetics 155:733–752

Aguet F et al (2016) Membrane dynamics of dividing cells imaged by lattice light-sheet microscopy. Mol Biol Cell 27:3418–3435. https://doi.org/10.1091/mbc.E16-03-0164

Albertson R, Cao J, Hsieh TS, Sullivan W (2008) Vesicles and actin are targeted to the cleavage furrow via furrow microtubules and the central spindle. J Cell Biol 181:777–790. https://doi.org/10.1083/jcb.200803096

Albertson R, Riggs B, Sullivan W (2005) Membrane traffic: a driving force in cytokinesis. Trends Cell Biol 15:92–101. https://doi.org/10.1016/j.tcb.2004.12.008

Arias CI, Siri SO, Conde C (2015) Involvement of SARA in axon and dendrite growth. PLoS ONE 10:e0138792. https://doi.org/10.1371/journal.pone.0138792

Aster JC, Pear WS, Blacklow SC (2008) Notch signaling in leukemia. Annu Rev Pathol 3:587–613. https://doi.org/10.1146/annurev.pathmechdis.3.121806.154300

Barbieri E, Di Fiore PP, Sigismund S (2016) Endocytic control of signaling at the plasma membrane. Curr Opin Cell Biol 39:21–27. https://doi.org/10.1016/j.ceb.2016.01.012

Bardin S, Miserey-Lenkei S, Hurbain I, Garcia-Castillo D, Raposo G, Goud B (2015) Phenotypic characterization of RAB6A knockout mouse embryonic fibroblasts vol 107. https://doi.org/10.1111/boc.201400083

Benhra N, Vignaux F, Dussert A, Schweisguth F, Le Borgne R (2010) Neuralized promotes basal to apical transcytosis of delta in epithelial cells. Mol Biol Cell 21:2078–2086. https://doi.org/10.1091/mbc.E09-11-0926

Benmerah A, Begue B, Dautry-Varsat A, Cerf-Bensussan N (1996) The ear of alpha-adaptin interacts with the COOH-terminal domain of the Eps 15 protein. J Biol Chem 271:12111–12116

Benmerah A, Gagnon J, Begue B, Megarbane B, Dautry-Varsat A, Cerf-Bensussan N (1995) The tyrosine kinase substrate eps15 is constitutively associated with the plasma membrane adaptor AP-2. J Cell Biol 131:1831–1838

Berdnik D, Torok T, Gonzalez-Gaitan M, Knoblich JA (2002) The endocytic protein alpha-adaptin is required for numb-mediated asymmetric cell division in *Drosophila*. Dev Cell 3:221–231. https://doi.org/10.1016/s1534-5807(02)00215-0

Berlin RD, Oliver JM (1980) Surface functions during mitosis. II. Quantitation of pinocytosis and kinetic characterization of the mitotic cycle with a new fluorescence technique. J Cell Biol 85:660–671

Berlin RD, Oliver JM, Walter RJ (1978) Surface functions during Mitosis I: phagocytosis, pinocytosis and mobility of surface-bound Con A. Cell 15:327–341

Berndt N et al (2017) Ubiquitylation-independent activation of Notch signalling by delta. Elife 6:e27346. https://doi.org/10.7554/eLife.27346

Bhat KM, Gaziova I, Katipalla S (2011) Neuralized mediates asymmetric division of neural precursors by two distinct and sequential events: promoting asymmetric localization of numb and enhancing activation of Notch-signaling. Dev Biol 351:186–198. https://doi.org/10.1016/j.ydbio.2010.12.008

Bluemink JG, de Laat SW (1973) New membrane formation during cytokinesis in normal and cytochalasin B-treated eggs of *Xenopus laevis*. I. Electron microscope observations. J Cell Biol 59:89–108

Bokel C, Schwabedissen A, Entchev E, Renaud O, Gonzalez-Gaitan M (2006) Sara endosomes and the maintenance of Dpp signaling levels across mitosis. Science 314:1135–1139. https://doi.org/10.1126/science.1132524

Boucrot E (2008) What does the plasma membrane do during mitosis? Med Sci (Paris) 24:34–36. https://doi.org/10.1051/medsci/200824134

Boucrot E, Howes MT, Kirchhausen T, Parton RG (2011) Redistribution of caveolae during mitosis. J Cell Sci 124:1965–1972. https://doi.org/10.1242/jcs.076570

Boucrot E, Kirchhausen T (2007) Endosomal recycling controls plasma membrane area during mitosis. Proc Natl Acad Sci U S A 104:7939–7944. https://doi.org/10.1073/pnas.0702511104

Boucrot E, Kirchhausen T (2008) Mammalian cells change volume during mitosis. PLoS ONE 3:e1477. https://doi.org/10.1371/journal.pone.0001477

Boulant S, Kural C, Zeeh JC, Ubelmann F, Kirchhausen T (2011) Actin dynamics counteract membrane tension during clathrin-mediated endocytosis. Nat Cell Biol 13:1124–1131. https://doi.org/10.1038/ncb2307

Bray D, White JG (1988) Cortical flow in animal cells. Science 239:883–888

Burkhardt JK, Echeverri CJ, Nilsson T, Vallee RB (1997) Overexpression of the dynamitin (p50) subunit of the dynactin complex disrupts dynein-dependent maintenance of membrane organelle distribution. J Cell Biol 139:469–484

Capalbo L, D'Avino PP, Archambault V, Glover DM (2011) Rab5 GTPase controls chromosome alignment through Lamin disassembly and relocation of the NuMA-like protein Mud to the poles during mitosis. Proc Natl Acad Sci U S A 108:17343–17348. https://doi.org/10.1073/pnas.1103720108

Cayrol C, Cougoule C, Wright M (2002) The beta2-adaptin clathrin adaptor interacts with the mitotic checkpoint kinase BubR1. Biochem Biophys Res Commun 298:720–730

Champion L, Linder MI, Kutay U (2017) Cellular reorganization during mitotic entry. Trends cell biol 27:26–41. https://doi.org/10.1016/j.tcb.2016.07.004

Chanez AL, Hehl AB, Engstler M, Schneider A (2006) Ablation of the single dynamin of *T. brucei* blocks mitochondrial fission and endocytosis and leads to a precise cytokinesis arrest. J Cell Sci 119:2968–2974. https://doi.org/10.1242/jcs.03023

Chapman G, Major JA, Iyer K, James AC, Pursglove SE, Moreau JL, Dunwoodie SL (2016) Notch1 endocytosis is induced by ligand and is required for signal transduction. Biochim Biophys Acta 1863:166–177. https://doi.org/10.1016/j.bbamcr.2015.10.021

Chapouton P et al (2010) Notch activity levels control the balance between quiescence and recruitment of adult neural stem cells. J Neurosci 30:7961–7974. https://doi.org/10.1523/JNEUROSCI.6170-09.2010

Chen H et al (2009) Embryonic arrest at midgestation and disruption of Notch signaling produced by the absence of both epsin 1 and epsin 2 in mice. Proc Natl Acad Sci U S A 106:13838–13843. https://doi.org/10.1073/pnas.0907008106

Chen J, Zhang M (2013) The Par3/Par6/aPKC complex and epithelial cell polarity. Exp Cell Res 319:1357–1364. https://doi.org/10.1016/j.yexcr.2013.03.021

Chesneau L, Dambournet D, Machicoane M, Kouranti I, Fukuda M, Goud B, Echard A (2012) An ARF6/Rab35 GTPase cascade for endocytic recycling and successful cytokinesis. Curr Biol 22:147–153. https://doi.org/10.1016/j.cub.2011.11.058

Chitnis A (2006) Why is delta endocytosis required for effective activation of Notch? Dev Dyn 235:886–894. https://doi.org/10.1002/dvdy.20683

Cocucci E, Aguet F, Boulant S, Kirchhausen T (2012) The first five seconds in the life of a clathrin-coated pit. Cell 150:495–507. https://doi.org/10.1016/j.cell.2012.05.047

Conner SD, Wessel GM (1999) Syntaxin is required for cell division. Mol Biol Cell 10:2735–2743

Cotton M, Benhra N, Le Borgne R (2013) Numb inhibits the recycling of Sanpodo in *Drosophila* sensory organ precursor. Curr Biol 23:581–587. https://doi.org/10.1016/j.cub.2013.02.020

Coumailleau F, Furthauer M, Knoblich JA, Gonzalez-Gaitan M (2009) Directional delta and Notch trafficking in Sara endosomes during asymmetric cell division. Nature 458:1051–1055. https://doi.org/10.1038/nature07854

Couturier L, Mazouni K, Schweisguth F (2013) Numb localizes at endosomes and controls the endosomal sorting of Notch after asymmetric division in *Drosophila*. Curr Biol 23:588–593. https://doi.org/10.1016/j.cub.2013.03.002

Couturier L, Vodovar N, Schweisguth F (2012) Endocytosis by numb breaks Notch symmetry at cytokinesis. Nat Cell Biol 14:131–139. https://doi.org/10.1038/ncb2419

Crowe R, Henrique D, Ish-Horowicz D, Niswander L (1998) A new role for Notch and delta in cell fate decisions: patterning the feather array. Development 125:767–775

Dambournet D et al (2011) Rab35 GTPase and OCRL phosphatase remodel lipids and F-actin for successful cytokinesis. Nat Cell Biol 13:981–988. https://doi.org/10.1038/ncb2279

Danilchik MV, Bedrick SD, Brown EE, Ray K (2003) Furrow microtubules and localized exocytosis in cleaving *Xenopus laevis* embryos. J Cell Sci 116:273–283

de-la-Concha A, Dietrich U, Weigel D, Campos-Ortega JA (1988) Functional interactions of neurogenic genes of *Drosophila melanogaster* Genetics 118:499–508

De Brabander M, Nuydens R, Geerts H, Hopkins CR (1988) Dynamic behavior of the transferrin receptor followed in living epidermoid carcinoma (A431) cells with nanovid microscopy. Cell Motil Cytoskeleton 9:30–47. https://doi.org/10.1002/cm.970090105

de Celis JF, Bray SJ (2000) The abruptex domain of Notch regulates negative interactions between Notch, its ligands and fringe. Development 127:1291–1302

Deblandre GA, Lai EC, Kintner C (2001) Xenopus neuralized is a ubiquitin ligase that interacts with XDelta1 and regulates Notch signaling. Dev Cell 1:795–806. https://doi.org/10.1016/S1534-5807(01)00091-0

Derivery E, Seum C, Daeden A, Loubery S, Holtzer L, Julicher F, Gonzalez-Gaitan M (2015) Polarized endosome dynamics by spindle asymmetry during asymmetric cell division. Nature 528:280–285. https://doi.org/10.1038/nature16443

Devenport D, Oristian D, Heller E, Fuchs E (2011) Mitotic internalization of planar cell polarity proteins preserves tissue polarity. Nat Cell Biol 13:893–902. https://doi.org/10.1038/ncb2284

Dexter JS (1914) The analysis of a case of continuous variation in drosophila by a study of its linkage relations. Am Nat 48:712–758

Dobrowolski R, De Robertis EM (2011) Endocytic control of growth factor signalling: multivesicular bodies as signalling organelles. Nat Rev Mol Cell Biol 13:53–60. https://doi.org/10.1038/nrm3244

Durrbach A, Louvard D, Coudrier E (1996) Actin filaments facilitate two steps of endocytosis. J Cell Sci 109(Pt 2):457–465

Dyer N et al (2007) Spermatocyte cytokinesis requires rapid membrane addition mediated by ARF6 on central spindle recycling endosomes. Development 134:4437–4447. https://doi.org/10.1242/dev.010983

Emery G, Hutterer A, Berdnik D, Mayer B, Wirtz-Peitz F, Gaitan MG, Knoblich JA (2005) Asymmetric Rab 11 endosomes regulate delta recycling and specify cell fate in the *Drosophila* nervous system. Cell 122:763–773 https://doi.org/10.1016/j.cell.2005.08.017

Eskelinen EL et al (2002) Inhibition of autophagy in mitotic animal cells. Traffic 3:878–893

F. PD (1939) Effects of Notch deficiencies. Drosoph Inf Serv 12

Feng B, Schwarz H, Jesuthasan S (2002) Furrow-specific endocytosis during cytokinesis of zebrafish blastomeres. Exp Cell Res 279:14–20

Fenner F (1999) The birth of the cell. Immunol Cell Biol 77:468

Fernando Aniento NE, Grittiths G, Gruenberg J (1993) Cytoplasmic dynein-dependent vesicular trans port from early to late endosomes. J Cell Biol 123(6):1373–1387. (Part 1)

Fielding AB, Royle SJ (2013) Mitotic inhibition of clathrin-mediated endocytosis. Cell Mol Life Sci 70:3423–3433. https://doi.org/10.1007/s00018-012-1250-8

Fielding AB et al (2005) Rab11-FIP3 and FIP4 interact with Arf6 and the exocyst to control membrane traffic in cytokinesis. EMBO J 24:3389–3399. https://doi.org/10.1038/sj.emboj.7600803

Fielding AB, Willox AK, Okeke E, Royle SJ (2012) Clathrin-mediated endocytosis is inhibited during mitosis. Proc Natl Acad Sci U S A 109:6572–6577. https://doi.org/10.1073/pnas.1117401109

Finger FP, White JG (2002) Fusion and fission: membrane trafficking in animal cytokinesis. Cell 108:727–730

Firestone AJ et al (2012) Small-molecule inhibitors of the AAA+ ATPase motor cytoplasmic dynein. Nature 484:125–129. https://doi.org/10.1038/nature10936

Fischer-Friedrich E, Hyman AA, Julicher F, Muller DJ, Helenius J (2014) Quantification of surface tension and internal pressure generated by single mitotic cells. Sci Rep 4:6213. https://doi.org/10.1038/srep06213

Flores-Rodriguez N, Rogers SS, Kenwright DA, Waigh TA, Woodman PG, Allan VJ (2011) Roles of dynein and dynactin in early endosome dynamics revealed using automated tracking and global analysis. PLoS ONE 6:e24479. https://doi.org/10.1371/journal.pone.0024479

Gerald NJ, Damer CK, O'Halloran TJ, De Lozanne A (2001) Cytokinesis failure in clathrin-minus cells is caused by cleavage furrow instability. Cell Motil Cytoskeleton 48:213–223. https://doi.org/10.1002/1097-0169(200103)48:3<213::AID-CM1010>3.0.CO;2-V

Gomez-Lamarca MJ, Snowdon LA, Seib E, Klein T, Bray SJ (2015) Rme-8 depletion perturbs Notch recycling and predisposes to pathogenic signaling. J Cell Biol 210:303–318. https://doi.org/10.1083/jcb.201411001

Gonzalez-Gaitan M, Jackle H (1997) Role of *Drosophila* alpha-adaptin in presynaptic vesicle recycling. Cell 88:767–776

Gordon WR, Vardar-Ulu D, Histen G, Sanchez-Irizarry C, Aster JC, Blacklow SC (2007) Structural basis for autoinhibition of Notch. Nat Struct Mol Biol 14:295. https://doi.org/10.1038/nsmb1227

Goss JW, Toomre DK (2008) Both daughter cells traffic and exocytose membrane at the cleavage furrow during mammalian cytokinesis. J Cell Biol 181:1047–1054. https://doi.org/10.1083/jcb.200712137

Gray GE et al (1999) Human ligands of the Notch receptor. Am J Pathol 154:785–794. https://doi.org/10.1016/S0002-9440(10)65325-4

Gromley A et al (2005) Centriolin anchoring of exocyst and SNARE complexes at the midbody is required for secretory-vesicle-mediated abscission. Cell 123:75–87. https://doi.org/10.1016/j.cell.2005.07.027

Gudejko HF, Alford LM, Burgess DR (2012) Polar expansion during cytokinesis. Cytoskeleton (Hoboken) 69:1000–1009. https://doi.org/10.1002/cm.21078

Guerrini A (1995) A history of the life sciences. Lois N. Magner Isis 86:89. https://doi.org/10.1086/357086

Guo M, Jan LY, Jan YN (1996) Control of daughter cell fates during asymmetric division: interaction of numb and Notch. Neuron 17:27–41

Hartenstein V, Posakony JW (1990) A dual function of the Notch gene in *Drosophila sensillum* development. Dev Biol 142:13–30

Hehnly H, Doxsey S (2014) Rab11 endosomes contribute to mitotic spindle organization and orientation. Dev Cell 28:497–507. https://doi.org/10.1016/j.devcel.2014.01.014

Henrique D, Hirsinger E, Adam J, Le Roux I, Pourquie O, Ish-Horowicz D, Lewis J (1997) Maintenance of neuroepithelial progenitor cells by delta-Notch signalling in the embryonic chick retina. Curr Biol 7:661–670

Heuser JE, Anderson RG (1989) Hypertonic media inhibit receptor-mediated endocytosis by blocking clathrin-coated pit formation. J Cell Biol 108:389–400

Heuss SF, Ndiaye-Lobry D, Six EM, Israel A, Logeat F (2008) The intracellular region of Notch ligands Dll1 and Dll3 regulates their trafficking and signaling activity. Proc Natl Acad Sci U S A 105:11212–11217. https://doi.org/10.1073/pnas.0800695105

Hicks C, Ladi E, Lindsell C, Hsieh JJ, Hayward SD, Collazo A, Weinmaster G (2002) A secreted Delta1-Fc fusion protein functions both as an activator and inhibitor of Notch1 signaling. J Neurosci Res 68:655–667. https://doi.org/10.1002/jnr.10263

Hill E, Clarke M, Barr FA (2000) The Rab6-binding kinesin, Rab6-KIFL, is required for cytokinesis. EMBO J 19:5711–5719. https://doi.org/10.1093/emboj/19.21.5711

Högnäs G et al (2011) Cytokinesis failure due to derailed integrin traffic induces aneuploidy and oncogenic transformation in vitro and in vivo. Oncogene 31:3597. https://doi.org/10.1038/onc.2011.527

Hori K (2004) *Drosophila* deltex mediates suppressor of hairless-independent and late-endosomal activation of Notch signaling. Development 131:5527–5537. https://doi.org/10.1242/dev.01448

Hu Y-Y, Zheng M-h, Zhang R, Liang Y-M, Han H (2012) Notch signaling pathway and cancer metastasis. In: Reichrath J, Reichrath S (eds) Notch Signaling in embryology and cancer. Springer US, New York, NY, pp 186–198. https://doi.org/10.1007/978-1-4614-0899-4_14

Hu Y, Chuang J-Z, Xu K, McGraw TG, Sung C-H (2002) SARA, a FYVE domain protein, affects Rab5-mediated endocytosis. J Cell Sci 115:4755–4763

Iannolo G et al (1997) Mapping of the molecular determinants involved in the interaction between eps15 and AP-2. Cancer Res 57:240–245

Imatani A, Callahan R (2000) Identification of a novel NOTCH-4/INT-3 RNA species encoding an activated gene product in certain human tumor cell lines. Oncogene 19:223–231. https://doi.org/10.1038/sj.onc.1203295

Itoh F et al (2002) The FYVE domain in Smad anchor for receptor activation (SARA) is sufficient for localization of SARA in early endosomes and regulates TGF-β/Smad signalling. Genes Cells 7:321–331. https://doi.org/10.1046/j.1365-2443.2002.00519.x

Itoh M et al (2003) Mind bomb is a ubiquitin ligase that is essential for efficient activation of Notch signaling by delta. Dev Cell 4:67–82

Jafar-Nejad H et al (2005) Sec15, a component of the exocyst, promotes Notch signaling during the asymmetric division of *Drosophila* sensory organ precursors. Dev Cell 9:351–363. https://doi.org/10.1016/j.devcel.2005.06.010

Jafar-Nejad H, Norga K, Bellen H (2002) Numb: "adapting" Notch for endocytosis. Dev Cell 3:155–156

Jantsch-Plunger V, Glotzer M (1999) Depletion of syntaxins in the early *Caenorhabditis elegans* embryo reveals a role for membrane fusion events in cytokinesis. Curr Biol 9:738–745

Johnson SA, Zitserman D, Roegiers F (2016) Numb regulates the balance between Notch recycling and late-endosome targeting in *Drosophila* neural progenitor cells. Mol Biol Cell 27:2857–2866. https://doi.org/10.1091/mbc.E15-11-0751

Kaplan A, Reiner O (2011) Linking cytoplasmic dynein and transport of Rab8 vesicles to the midbody during cytokinesis by the doublecortin domain-containing 5 protein. J Cell Sci 124:3989–4000. https://doi.org/10.1242/jcs.085407

Kaur S, Fielding AB, Gassner G, Carter NJ, Royle SJ (2014) An unmet actin requirement explains the mitotic inhibition of clathrin-mediated endocytosis. Elife 3:e00829. https://doi.org/10.7554/eLife.00829

Kogo H, Fujimoto T (2000) Concentration of caveolin-1 in the cleavage furrow as revealed by time-lapse analysis. Biochem Biophys Res Commun 268:82–87. https://doi.org/10.1006/bbrc.1999.2058

Kouranti I, Sachse M, Arouche N, Goud B, Echard A (2006) Rab35 regulates an endocytic recycling pathway essential for the terminal steps of cytokinesis. Curr Biol 16:1719–1725. https://doi.org/10.1016/j.cub.2006.07.020

Kressmann S, Campos C, Castanon I, Furthauer M, Gonzalez-Gaitan M (2015) Directional Notch trafficking in sara endosomes during asymmetric cell division in the spinal cord. Nat Cell Biol 17:333–339. https://doi.org/10.1038/ncb3119

Kunda P, Rodrigues NT, Moeendarbary E, Liu T, Ivetic A, Charras G, Baum B (2012) PP1-mediated moesin dephosphorylation couples polar relaxation to mitotic exit. Curr Biol 22:231–236. https://doi.org/10.1016/j.cub.2011.12.016

Lai EC, Deblandre GA, Kintner C, Rubin GM (2001) *Drosophila* neuralized is a ubiquitin ligase that promotes the internalization and degradation of Delta. Dev Cell 1:783–794. https://doi.org/10.1016/S1534-5807(01)00092-2

Langridge PD, Struhl G (2017) Epsin-dependent ligand endocytosis activates Notch by force. Cell 171:1383–1396. e1312. https://doi.org/10.1016/j.cell.2017.10.048

Lanzetti L (2012) A novel function of Rab5 in mitosis. Small GTPases 3:168–172. https://doi.org/10.4161/sgtp.19987

Le Borgne R, Schweisguth F (2003) Unequal segregation of neuralized biases Notch activation during asymmetric cell division. Dev Cell 5:139–148

Lehmann R, Jimenez F, Dietrich U, Camposortega JA (1983) On the phenotype and development of mutants of early neurogenesis in *Drosophila-Melanogaster*. Wilhelm Rouxs Arch Dev Biol 192:62–74. https://doi.org/10.1007/Bf00848482

Leitch CC, Lodh S, Prieto-Echague V, Badano JL, Zaghloul NA (2014) Basal body proteins regulate Notch signaling through endosomal trafficking. J Cell Sci 127:2407–2419. https://doi.org/10.1242/jcs.130344

Lewis J (1998) Notch signalling and the control of cell fate choices in vertebrates. Semin Cell Dev Biol 9:583–589. https://doi.org/10.1006/scdb.1998.0266

Loubery S et al (2017) Sara phosphorylation state controls the dispatch of endosomes from the central spindle during asymmetric division. Nat Commun 8:15285. https://doi.org/10.1038/ncomms15285

Loubery S, Gonzalez-Gaitan M (2014) Monitoring Notch/delta endosomal trafficking and signaling in *Drosophila*. Methods Enzymol 534:301–321. https://doi.org/10.1016/B978-0-12-397926-1.00017-2

Loubery S, Seum C, Moraleda A, Daeden A, Furthauer M, Gonzalez-Gaitan M (2014) Uninflatable and Notch control the targeting of Sara endosomes during asymmetric division. Curr Biol 24:2142–2148. https://doi.org/10.1016/j.cub.2014.07.054

Loubéry S, Seum C, Moraleda A, Daeden A, Fürthauer M, Gonzalez-Gaitan M (2015) Uninflatable and Notch control the targeting of Sara Endosomes during asymmetric division. Curr Biol 25:817–818. https://doi.org/10.1016/j.cub.2015.02.053

Loubery S, Wilhelm C, Hurbain I, Neveu S, Louvard D, Coudrier E (2008) Different microtubule motors move early and late endocytic compartments. Traffic 9:492–509. https://doi.org/10.1111/j.1600-0854.2008.00704.x

Low SH, Li X, Miura M, Kudo N, Quinones B, Weimbs T (2003) Syntaxin 2 and endobrevin are required for the terminal step of cytokinesis in mammalian cells. Dev Cell 4:753–759

Maddox AS, Burridge K (2003) RhoA is required for cortical retraction and rigidity during mitotic cell rounding. J Cell Biol 160:255–265. https://doi.org/10.1083/jcb.200207130

Matteoni R, Kreis TE (1987) Translocation and clustering of endosomes and lysosomes depends on microtubules. J Cell Biol 105:1253–1265

Mavor LM, Miao H, Zuo Z, Holly RM, Xie Y, Loerke D, Blankenship JT (2016) Rab8 directs furrow ingression and membrane addition during epithelial formation in *Drosophila melanogaster*. Development 143:892–903. https://doi.org/10.1242/dev.128876

McGill MA, Dho SE, Weinmaster G, McGlade CJ (2009) Numb regulates post-endocytic trafficking and degradation of Notch1. J Biol Chem 284:26427–26438. https://doi.org/10.1074/jbc.M109.014845

Meloty-Kapella L, Shergill B, Kuon J, Botvinick E, Weinmaster G (2012) Notch ligand endocytosis generates mechanical pulling force dependent on dynamin, epsins, and actin. Dev Cell 22:1299–1312. https://doi.org/10.1016/j.devcel.2012.04.005

Metchnikoff E (1883) Untersuchungen über die mesodermalen Phagocyten einiger. Wirbeltiere Biol Zentralbl 3:560–565

Militello RD, Munafo DB, Beron W, Lopez LA, Monier S, Goud B, Colombo MI (2013) Rab24 is required for normal cell division. Traffic 14:502–518. https://doi.org/10.1111/tra.12057

Ming Guo LYJ, Yuh Nung Jan (1996) Control of daughter cell fates during asymmetric division: interaction of numb and Notch. Neuron

Miserey-Lenkei S, Colombo MI (2016) Small RAB GTPases regulate multiple steps of mitosis. Front Cell Dev Biol 4:2. https://doi.org/10.3389/fcell.2016.00002

Montagnac G, Chavrier P (2008) Endosome positioning during cytokinesis. Biochem Soc Trans 36:442–443. https://doi.org/10.1042/BST0360442

Montagne C, Gonzalez-Gaitan M (2014) Sara endosomes and the asymmetric division of intestinal stem cells. Development 141:2014–2023. https://doi.org/10.1242/dev.104240

Mosesson Y, Mills GB, Yarden Y (2008) Derailed endocytosis: an emerging feature of cancer. Nat Rev Cancer 8:835–850. http://www.nature.com/nrc/journal/v8/n11/suppinfo/nrc2521_S1.html

Motley A, Bright NA, Seaman MN, Robinson MS (2003) Clathrin-mediated endocytosis in AP-2-depleted cells. J Cell Biol 162:909–918. https://doi.org/10.1083/jcb.200305145

Murphy C et al (1996) Endosome dynamics regulated by a Rho protein. Nature 384:427–432. https://doi.org/10.1038/384427a0

Murray JW, Yin D, Wolkoff AW (2017) Reduction of organelle motility by removal of potassium and other solutes. PLOS ONE 12:e0184898. https://doi.org/10.1371/journal.pone.0184898

Muskavitch MA (1994) Delta-Notch signaling and *Drosophila* cell fate choice. Dev Biol 166:415–430. https://doi.org/10.1006/dbio.1994.1326

Neto H, Collins LL, Gould GW (2011) Vesicle trafficking and membrane remodelling in cytokinesis. Biochem J 437:13–24. https://doi.org/10.1042/BJ20110153

Nichols JT, Miyamoto A, Olsen SL, D'Souza B, Yao C, Weinmaster G (2007) DSL ligand endocytosis physically dissociates Notch1 heterodimers before activating proteolysis can occur. J Cell Biol 176:445–458. https://doi.org/10.1083/jcb.200609014

Nieland TJ, Ehrlich M, Krieger M, Kirchhausen T (2005) Endocytosis is not required for the selective lipid uptake mediated by murine SR-BI. Biochim Biophys Acta 1734:44–51. https://doi.org/10.1016/j.bbalip.2005.02.007

O'Connor-Giles KM, Skeath JB (2003) Numb inhibits membrane localization of Sanpodo, a four-pass transmembrane protein, to promote asymmetric divisions in *Drosophila*. Dev Cell 5:231–243

Oliver JM, Seagrave JC (1987) The blocked pinocytic activity of mitotic cells is restored in mitotic-interphase hybrids. Exp Cell Res 171:475–482

Overstreet E, Fitch E, Fischer JA (2004) Fat facets and liquid facets promote delta endocytosis and delta signaling in the signaling cells. Development 131:5355–5366. https://doi.org/10.1242/dev.01434

Özlü N et al (2015) Quantitative comparison of a human cancer cell surface proteome between interphase and mitosis. EMBO J 34:251–265. https://doi.org/10.15252/embj.201385162

Parks AL, Klueg KM, Stout JR, Muskavitch MA (2000) Ligand endocytosis drives receptor dissociation and activation in the Notch pathway. Development 127:1373–1385

Pavlopoulos E, Pitsouli C, Klueg KM, Muskavitch MA, Moschonas NK, Delidakis C (2001) Neuralized encodes a peripheral membrane protein involved in delta signaling and endocytosis. Dev Cell 1:807–816

Pelissier A, Chauvin J-P, Lecuit T (2003) Trafficking through Rab11 endosomes is required for cellularization during *Drosophila* embryogenesis. Curr Biol 13:1848–1857. https://doi.org/10.1016/j.cub.2003.10.023

Pellinen T et al (2008) Integrin trafficking regulated by Rab21 is necessary for cytokinesis. Dev Cell 15:371–385. https://doi.org/10.1016/j.devcel.2008.08.001

Poodry CA (1990) Shibire, a neurogenic mutant of *Drosophila*. Dev Biol 138:464–472

Prekeris R (2011) Actin regulation during abscission: unexpected roles of Rab35 and endocytic transport. Cell Res 21:1283–1285. https://doi.org/10.1038/cr.2011.131

Pypaert M, Lucocq JM, Warren G (1987) Coated pits in interphase and mitotic A431 cells. Eur J Cell Biol 45:23–29

Pypaert M, Mundy D, Souter E, Labbe JC, Warren G (1991) Mitotic cytosol inhibits invagination of coated pits in broken mitotic cells. J Cell Biol 114:1159–1166

Rajan A, Tien AC, Haueter CM, Schulze KL, Bellen HJ (2009) The Arp2/3 complex and WASp are required for apical trafficking of delta into microvilli during cell fate specification of sensory organ precursors. Nat Cell Biol 11:815–824. https://doi.org/10.1038/ncb1888

Ramaswami M, Rao S, van der Bliek A, Kelly RB, Krishnan KS (1993) Genetic studies on dynamin function in *Drosophila*. J Neurogenet 9:73–87

Ramos RG, Grimwade BG, Wharton KA, Scottgale TN, Artavanis-Tsakonas S (1989) Physical and functional definition of the *Drosophila* Notch locus by P element transformation Genetics 123:337–348

Randazzo PA (2003) RhoD, Src, and hDia2C in endosome motility. Dev Cell 4:287–288

Rapp S, Saffrich R, Anton M, Jakle U, Ansorge W, Gorgas K, Just WW (1996) Microtubule-based peroxisome movement. J Cell Sci 109(Pt 4):837–849

Raucher D, Sheetz MP (1999a) Membrane expansion increases endocytosis rate during mitosis. J Cell Biol 144:497–506

Raucher D, Sheetz MP (1999b) Membrane expansion increases endocytosis rate during mitosis. J Cell Biol 144:497–506. https://doi.org/10.1083/jcb.144.3.497

Rhyu MS, Jan LY, Jan YN (1994) Asymmetric distribution of numb protein during division of the sensory organ precursor cell confers distinct fates to daughter cells. Cell 76:477–491

Riggs B et al (2003) Actin cytoskeleton remodeling during early *Drosophila* furrow formation requires recycling endosomal components Nuclear-fallout and Rab11. J Cell Biol 163:143–154. https://doi.org/10.1083/jcb.200305115

Robertson AS, Smythe E, Ayscough KR (2009) Functions of actin in endocytosis. Cell Mol Life Sci 66:2049–2065. https://doi.org/10.1007/s00018-009-0001-y

Salbreux G, Charras G, Paluch E (2012) Actin cortex mechanics and cellular morphogenesis. Trends Cell Biol 22:536–545. https://doi.org/10.1016/j.tcb.2012.07.001

Salcini AE et al (1997) Binding specificity and in vivo targets of the EH domain, a novel protein-protein interaction module. Genes Dev 11:2239–2249
Saleem M, Morlot S, Hohendahl A, Manzi J, Lenz M, Roux A (2015) A balance between membrane elasticity and polymerization energy sets the shape of spherical clathrin coats. Nat Commun 6:6249. https://doi.org/10.1038/ncomms7249
Santolini E, Puri C, Salcini AE, Gagliani MC, Pelicci PG, Tacchetti C, Di Fiore PP (2000) Numb is an endocytic protein. J Cell Biol 151:1345–1352
Schiel JA, Prekeris R (2010) Making the final cut—mechanisms mediating the abscission step of cytokinesis. Sci World J 10:1424–1434. https://doi.org/10.1100/tsw.2010.129
Schweitzer JK, Burke EE, Goodson HV, D'Souza-Schorey C (2005) Endocytosis resumes during late mitosis and is required for cytokinesis. J Biol Chem 280:41628–41635. https://doi.org/10.1074/jbc.M504497200
Schweitzer JK, D'Souza-Schorey C (2002a) Localization and activation of the ARF6 GTPase during cleavage furrow ingression and cytokinesis. J Biol Chem 277:27210–27216. https://doi.org/10.1074/jbc.M201569200
Schweitzer JK, D'Souza-Schorey C (2002b) Localization and activation of the ARF6 GTPase during cleavage furrow ingression and cytokinesis. J Biol Chem 277:27210–27216. https://doi.org/10.1074/jbc.M201569200
Schweitzer JK, D'Souza-Schorey C (2005) A requirement for ARF6 during the completion of cytokinesis. Exp Cell Res 311:74–83. https://doi.org/10.1016/j.yexcr.2005.07.033
Schweitzer JK, Sedgwick AE, D'Souza-Schorey C (2011) ARF6-mediated endocytic recycling impacts cell movement, cell division and lipid homeostasis. Semin Cell Dev Biol 22:39–47. https://doi.org/10.1016/j.semcdb.2010.09.002
Serio G, Margaria V, Jensen S, Oldani A, Bartek J, Bussolino F, Lanzetti L (2011) Small GTPase Rab5 participates in chromosome congression and regulates localization of the centromere-associated protein CENP-F to kinetochores. Proc Natl Acad Sci U S A 108:17337–17342. https://doi.org/10.1073/pnas.1103516108
Seugnet L, Simpson P, Haenlin M (1997) Requirement for dynamin during Notch signaling in *Drosophila* neurogenesis. Dev Biol 192:585–598. https://doi.org/10.1006/dbio.1997.8723
Shao X et al (2016) Numb regulates vesicular docking for homotypic fusion of early endosomes via membrane recruitment of Mon1b. Cell Res 26:593–612. https://doi.org/10.1038/cr.2016.34
Shuster CB, Burgess DR (2002) Targeted new membrane addition in the cleavage furrow is a late, separate event in cytokinesis. Proc Natl Acad Sci U S A 99:3633–3638. https://doi.org/10.1073/pnas.052342699
Sirotkin V (2011) Cell biology: actin keeps endocytosis on a short leash. Curr Biol 21:R552–R554. https://doi.org/10.1016/j.cub.2011.06.029
Skop AR, Bergmann D, Mohler WA, White JG (2001) Completion of cytokinesis in C. elegans requires a brefeldin A-sensitive membrane accumulation at the cleavage furrow apex. Curr Biol CB 11:735–746
Smythe E, Ayscough KR (2006) Actin regulation in endocytosis. J Cell Sci 119:4589–4598. https://doi.org/10.1242/jcs.03247
Sorkin A, von Zastrow M (2009) Endocytosis and signalling: intertwining molecular networks. Nat Rev Mol Cell Biol 10:609–622. https://doi.org/10.1038/nrm2748
Spana EP, Doe CQ (1996) Numb antagonizes Notch signaling to specify sibling neuron cell fates. Neuron 17:21–26
Stenmark H (2009) Rab GTPases as coordinators of vesicle traffic. Nat Rev Mol Cell Biol 10:513–525. https://doi.org/10.1038/nrm2728
Stewart MP, Helenius J, Toyoda Y, Ramanathan SP, Muller DJ, Hyman AA (2011) Hydrostatic pressure and the actomyosin cortex drive mitotic cell rounding. Nature 469:226–230. https://doi.org/10.1038/nature09642
Swanson JA, Locke A, Ansel P, Hollenbeck PJ (1992) Radial movement of lysosomes along microtubules in permeabilized macrophages. J Cell Sci 103(Pt 1):201–209

Tacheva-Grigorova SK, Santos AJ, Boucrot E, Kirchhausen T (2013) Clathrin-mediated endocytosis persists during unperturbed mitosis. Cell Rep 4:659–668. https://doi.org/10.1016/j.celrep.2013.07.017

Takatsu H et al (2013) Mitosis-coupled, microtubule-dependent clustering of endosomal vesicles around centrosomes. Cell Struct Func 38:31–41. https://doi.org/10.1247/csf.12028

Thompson HM, Skop AR, Euteneuer U, Meyer BJ, McNiven MA (2002) The large GTPase dynamin associates with the spindle midzone and is required for cytokinesis. Curr Biol 12:2111–2117

Tian X, Hansen D, Schedl T, Skeath JB (2004) Epsin potentiates Notch pathway activity in *Drosophila* and *C. elegans*. Development 131:5807–5815. https://doi.org/10.1242/dev.01459

Tsukazaki T, Chiang TA, Davison AF, Attisano L, Wrana JL (1998) SARA, a FYVE domain protein that recruits Smad2 to the TGFbeta receptor. Cell 95:779–791

Tuomikoski T, Felix MA, Doree M, Gruenberg J (1989) Inhibition of endocytic vesicle fusion in vitro by the cell-cycle control protein kinase cdc2. Nature 342:942–945. https://doi.org/10.1038/342942a0

Uemura T, Shepherd S, Ackerman L, Jan LY, Jan YN (1989) Numb, a gene required in determination of cell fate during sensory organ formation in *Drosophila* embryos. Cell 58:349–360

Valapala M, Hose S, Gongora C, Dong L, Wawrousek EF, Samuel Zigler J, Sinha D (2013) Impaired endolysosomal function disrupts Notch signalling in optic nerve astrocytes. Nat Commun 4:1629. https://doi.org/10.1038/ncomms2624

Wang W, Struhl G (2004) *Drosophila* epsin mediates a select endocytic pathway that DSL ligands must enter to activate Notch. Development 131:5367–5380. https://doi.org/10.1242/dev.01413

Warren G, Davoust J, Cockcroft A (1984) Recycling of transferrin receptors in A431 cells is inhibited during mitosis. EMBO J 3:2217–2225

Warren G, Featherstone C, Griffiths G, Burke B (1983) Newly synthesized G protein of vesicular stomatitis virus is not transported to the cell surface during mitosis. J Cell Biol 97:1623–1628

Wilkin MB, Baron M (2005) Endocytic regulation of Notch activation and down-regulation (review). Mol Membr Biol 22:279–289. https://doi.org/10.1080/09687860500129778

Wilson GM et al (2005) The FIP3-Rab11 protein complex regulates recycling endosome targeting to the cleavage furrow during late cytokinesis. Mol Biol Cell 16:849–860. https://doi.org/10.1091/mbc.E04-10-0927

Xie G, Zhang H, Du G, Huang Q, Liang X, Ma J, Jiao R (2012) Uif, a large transmembrane protein with EGF-like repeats, can antagonize Notch signaling in *Drosophila*. PLoS ONE 7:e36362. https://doi.org/10.1371/journal.pone.0036362

Xu KMN, Egan SE (2010) Notch signaling in lung development and disease. In: Curie Madame (ed) Bioscience. Landes Bioscience, Database Austin (TX)

Yarar D, Waterman-Storer CM, Schmid SL (2005) A dynamic actin cytoskeleton functions at multiple stages of clathrin-mediated endocytosis. Mol Biol Cell 16:964–975. https://doi.org/10.1091/mbc.E04-09-0774

Yousefian J, Troost T, Grawe F, Sasamura T, Fortini M, Klein T (2013) Dmon1 controls recruitment of Rab7 to maturing endosomes in *Drosophila*. J Cell Sci 126:1583–1594. https://doi.org/10.1242/jcs.114934

Yu X, Prekeris R, Gould GW (2007) Role of endosomal Rab GTPases in cytokinesis. Eur J Cell Biol 86:25–35. https://doi.org/10.1016/j.ejcb.2006.10.002

Zagouras P, Stifani S, Blaumueller CM, Carcangiu ML, Artavanis-Tsakonas S (1995) Alterations in Notch signaling in neoplastic lesions of the human cervix. Proc Natl Acad Sci U S A 92:6414–6418

Zajac AL, Goldman YE, Holzbaur EL, Ostap EM (2013) Local cytoskeletal and organelle interactions impact molecular-motor- driven early endosomal trafficking. Curr Biol 23:1173–1180. https://doi.org/10.1016/j.cub.2013.05.015

Zhang H, Squirrell JM, White JG (2008) RAB-11 permissively regulates spindle alignment by modulating metaphase microtubule dynamics in *Caenorhabditis elegans* early embryos. Mol Biol Cell 19:2553–2565. https://doi.org/10.1091/mbc.E07-09-0862

MIX
Papier aus verantwortungsvollen Quellen
Paper from responsible sources
FSC® C105338

If you have any concerns about our products,
you can contact us on
ProductSafety@springernature.com

In case Publisher is established outside the EU,
the EU authorized representative is:
Springer Nature Customer Service Center GmbH
Europaplatz 3, 69115 Heidelberg, Germany

Printed by Libri Plureos GmbH
in Hamburg, Germany